HANDBOOK OF
CLASSICAL CONDITIONING

HANDBOOK OF CLASSICAL CONDITIONING

by

David G. Lavond
University of Southern California
Los Angeles, California

Joseph E. Steinmetz
Indiana University
Bloomington, Indiana

KLUWER ACADEMIC PUBLISHERS
Boston / New York / Dordrecht / London

Distributors for North, Central and South America:
Kluwer Academic Publishers
101 Philip Drive
Assinippi Park
Norwell, Massachusetts 02061 USA
Telephone (781) 871-6600
Fax (781) 681-9045
E-Mail: kluwer@wkap.com

Distributors for all other countries:
Kluwer Academic Publishers Group
Post Office Box 322
3300 AH Dordrecht, THE NETHERLANDS
Telephone 31 786 576 000
Fax 31 786 576 254
E-Mail: services@wkap.nl

 Electronic Services <http://www.wkap.nl>

Library of Congress Cataloging-in-Publication Data

Lavond, David G., 1952-
  Handbook of classical conditioning / by David G. Lavond, Joseph E. Steinmetz.
    p. cm.
  Includes bibliographical references and index.
  ISBN 1-40207-269-4
  1. Classical conditioning. I. Steinmetz, Joseph E., 1955- II. Title.

QP416 .L38 2002
  153.1'526—dc21

                                                                2002034093

Copyright © 2003 by Kluwer Academic Publishers

All rights reserved. No part of this work may be reproduced, stored in a retrieval system, or transmitted in any form or by any means, electronic, mechanical, photocopying, microfilming, recording, or otherwise, without the written permission from the Publisher, with the exception of any material supplied specifically for the purpose of being entered and executed on a computer system, for exclusive use by the purchaser of the work.

Permission for books published in Europe: permissions@wkap.nl
Permission for books published in the United States of America: permissions@wkap.com

Printed on acid-free paper. Printed in the United States of America.

All figures with the exception of figures 1.1 and 1.2 are reproduced with the permission of the authors.

TABLE OF CONTENTS

List of Figures vii

List of Tables xv

Preface xvii

Acknowledgments xxiii

1. The Classical Conditioning Paradigm 1

2. The Delivery of Stimuli 35

3. Measuring Behavioral Responses 71

4. Recording Neuronal Data During Classical Conditioning
   Experiments 129

5. Collecting and Analyzing Behavioral and Neural Data 161

6. Other Behavioral Paradigms 199

7. Surgical Methods and Techniques 223

8. Lesion Techniques for Behavioral Experiments 249

9. Brain Stimulation Techniques 277

10. Histological Methods 299

11. Controlling Classical Conditioning and Other Behavioral
    Neuroscience Experiments 323

12. Important Electronics for Classical Conditioning Experiments 369

Appendix A. Suppliers 389

Appendix B. Rabbit Atlas 393

References 423

Index 437

LIST OF FIGURES

Figure 1.1: Ivan Petrovich Pavlov (1849-1936). Photograph courtesy of and copyright © The Nobel Foundation 1904. For further information see *http://www.nobel.se/medicine/laureates/1904/*.

Figure 1.2: Early human eyeblink classical conditioning experiment. Graduate student Ernest Hilgard is the subject (seated) while Clark Hull (wearing the visor) looks on as Walter C. Shipley adjusts the automated wooden face-slapper (the unconditioned stimulus) with the help of postdoctoral student Helen Peak. Photograph courtesy of the collection of Richard F. Thompson.

Figure 1.3: Relationship of timing of the stimuli with the behavioral response during classical conditioning.

Figure 1.4: Relationship of timing of the stimuli during classical conditioning using the delay conditioning paradigm.

Figure 1.5: Relationship of timing of the stimuli during classical conditioning using the trace conditioning paradigm.

Figure 1.6: Relationship of timing of the stimuli during classical conditioning using the long delay paradigm.

Figure 1.7: Hypothetical development of learned responses during classical conditioning using a delay paradigm.

Figure 1.8: Definitions of the behavioral measurements during PreCS, CS and US periods for delay classical conditioning. Note that the CS and US are both present at the beginning of the US period, and that there are no stimuli present during the latter part of the US period. The latency and peak amplitudes during the CS and US periods hold the greatest interest.

Figure 1.9: Examples of behavioral responses that could be considered to be 'bad trials' that the experimenter might consider excluding from analysis.

Figure 2.1: Relationship of elements to deliver a tone stimulus.

Figure 2.2: Morse code (twin T oscillator) tone source.

Figure 2.3: Versatile tone source.

Figure 2.4a: Circuit for programmable frequency selection.

Figure 2.4b: Notes for programmable frequency selection.

Figure 2.5: White noise circuits.

Figure 2.6: Basic rise/fall switch to control the onset and offset of auditory stimuli.

Figure 2.7: Basic attenuator circuit.

Figure 2.8: Attenuator circuit based on an op amp used as a mixer.

Figure 2.9: Attenuator based on rise/fall switch.

Figure 2.10: Relay control of light stimulus.

Figure 2.11: Transistor control of light stimulus.

Figure 2.12: Flashing light circuit.

Figure 2.13: Hayes' fistula for delivering conditioned stimulus in taste aversion experiments.

Figure 2.14: Interface to solenoid for air puff circuit.

Figure 2.15: Basic shocker circuit. We recommend that the user purchase commercially available shock sources for safety.

Figure 2.16: Interface to control stimulator.

Figure 3.1: Simple eyeblink transducer using a minitorque potentiometer and current-limiting resistors for protection against excessive currents.

Figure 3.2: Eyeblink transducer using the relative registration of polarized plates to vary the amount of transmitted light. Based on a design by Solomon.

Figure 3.3: Rigid linkage of the nictitating membrane to the arm of the potentiometer.

Figure 3.4: Solomon's intermittent infrared circuit for measuring eye blinks.

Figure 3.5: Disterhoft's reflected light circuit for measuring eye blinks.

Figure 3.6: Clark's reflected light circuit for measuring eye blinks.

Figure 3.7: Circuit for converting eyelid EMG into a continuous wave, nictitating membrane-like signal.

List of Figures

Figure 3.8: "NM box" or "NM circuit" to automatically filter, adjust the baseline and amplify the eye blink into a computer-compatible signal. This circuit allows for continuous or sampled filtering for classical or operant eye blink conditioning, respectively.

Figure 3.9: Simple stabilimeter (jiggle stand) for measuring gross movements.

Figure 3.10: Differential recording heart rate amplifier.

Figure 3.11: Rorick's stylish rat-sized vest for recording heart rate.

Figure 3.12: Latched drinkometer (contact) detection with positive reset.

Figure 3.13: Timed or latched drinkometer (contact) detection with negative reset.

Figure 3.14: Circuit for detecting lever pressing as an analog rather than digital event.

Figure 3.15: Plexiglas rabbit restrainer.

Figure 3.16: Plexiglas at restrainer.

Figure 3.17: Plexiglas rat restrainer.

Figure 3.18: Wrap for rabbit restraint.

Figure 3.19: Commutator for recording from a freely-moving rat.

Figure 3.20: Amplifier to minimize movement artifact in the freely-moving rat.

Figure 3.21: Traditional elastic bands with hooks to retract eyelids during eye blink conditioning.

Figure 3.22: Ophthalmic eye-clips to retract eyelids during eye blink conditioning.

Figure 3.23: Parts and assembly of harness for attaching headstage to an unoperated rabbit.

Figure 3.24: Parts and assembly headstage that holds the airpuff delivery system (unconditioned stimulus) and minitorque potentiometer for behavioral measurement.

Figure 3.25: Two views of assembled headstage.

Figure 3.26: Rabbit carrier.

Figure 4.1: Acid etching the tips of stainless steel wire or insect pins for creating recording electrodes.

Figure 4.2: The 'zapper' for blowing off the insulation at the tip of a recording electrode.

Figure 4.3: Circuit for creating a calibration signal for testing amplifiers and filters used for recording neural unit activity.

Figure 4.4: The simplest low and high pass circuits using passive resistor and capacitor components.

Figure 4.5: Designs for creating second order low- and high-pass filters.

Figure 4.6: Designs for creating fourth order low- and high-pass filters.

Figure 4.7: Designs for creating sixth order low- and high-pass filters.

Figure 4.8: Buffering plus selective, tunable high- and low-pass filters using transconductance op amps combined with selectable 60 Hz notch filter and gain stages.

Figure 4.9: Overview of the components for recording neural unit activity and simple data analysis using a height discriminator.

Figure 4.10: Gain, polarity and comparator functions for unit discrimination.

Figure 4.11: Multiplexed display for unit discrimination.

Figure 4.12: The headstage is equipped with dual op-amps configured as voltage followers to eliminate movement artifact in the neural recordings from a freely-moving animal.

Figure 4.13: Unit activity recorded from the interpositus nucleus of a freely-moving rat without having any movement artifacts.

Figure 5.1: Schematic of experimental layout and data collection.

Figure 5.2: Common measurements for behavior can be made on individual trials or on an average of good trials. Here the square in the upper left indicates this is a 'block' of trials. CR onset is the first time a criterion amplitude is met (here 0.5 mm) in the proper time frame (here between 25 msec after CS onset to US onset). CR and UR peak amplitudes are the greatest values found in the CS and US periods, respectively. The display uses dots to indicate these measurements. Note that these measurements may not always be obvious in the display because the pixels on the monitor cannot display the full range of A/D values.

Figure 5.3: Data collected during a single trial showing eyelid position (top), timing marks (CS onset on left, US onset on right), and three levels of discriminated unit activity.

Figure 5.4: Average of a block of trials in which only 'good trials' have been included.

List of Figures

Figure 5.5: Z-score analysis of unit activity for a block of trials compared with averaged behavioral response.

Figure 5.6: Comparison of fictitious raw behavioral and unit data (left) converted for analysis by computer (right).

Figure 5.7: Comparison of the original behavioral and unit data (top) with successive shifting of unit activity for computation of cross-correlations.

Figure 5.8: Definitions of unit clusters.

Figure 5.9: Cluster analysis for a single trial.

Figure 5.10: Cluster analysis for a block of trials.

Figure 6.1: Operant chamber or Skinner box.

Figure 6.2: Runway or straight maze.

Figure 6.3: T-maze.

Figure 6.4: Olton 8-arm radial maze.

Figure 6.5: Morris water maze.

Figure 7.1: Example of surgical record.

Figure 7.2: Layout for a surgical suite.

Figure 7.3: Biela rabbit stereotaxic headholder, consisting of a single nose/mouth piece through which halothane gas anesthesia can be delivered, mounted on Kopf stereotax frame.

Figure 7.4: Kopf rat stereotaxic headholder, consisting of ear bars and nose/mouth piece, mounted on Knopf stereotax frame.

Figure 7.5: Reading the Vernier scale on the stereotax allows measurements to the tenth of a millimeter. The value here is 32.6 mm. Notice that the line for 0.6 lines up with a line on the centimeter scale, and that the lines for 0.5 and 0.7 are slightly offset and within lines on the centimeter scale.

Figure 7.6: Rat skull showing the positions of bregma and lambda landmarks. Bregma is the intersection of the coronal and sagittal (longitudinal) sutures. Alternatively, bregma can be defined as the point where the sutures *would* meet if they were straight. Lambda is the intersection of the sagittal (longitudinal) and transverse sutures. Alternatively, lambda can be defined as the highest point posterior to the apex where the transverse suture meets the sagittal (longitudinal) suture. The fact that the sutures 'wander' makes the definitions of both bregma and lambda subject to one's best guess. As short-hand notations, we usually use β (beta) for bregma and λ (lamdba) for itself. Figure based on Paxinos and Watson (1982).

Figure 7.7: Angled electrode placement using calculated values to correct for anterior-posterior (AP) and dorsal-ventral (DV) offset.

Figure 7.8: Angled electrode placement using empirically-determined values to correct for anterior-posterior (AP) offset and dorsal-ventral (DV) length. The pointer (a blunted injection needle) is placed at two different positions: First, at some arbitrary point (here 50.0 mm), and second at a vertical position representing the intended vertical depth of the electrode. The AP and DV coordinates are measured at both positions. The differences between these two positions for AP and DV values represent the corrections needed for the angled placement from the original vertical placements.

Figure 7.9: Introduction to knots. We have seen any of knots that use a double overhand knot with an overhand or double-overhand used for the eyelid suture. Knots that are asymmetric (surgeon's knot, eyelid knot) work well with slippery nylon suture.

Figure 8.1: Aspirator created from a smple vacuum using water pressure.

Figure 8.2: Fundamental method for creating a constant current lesion.

Figure 8.3: Thompson lesion maker: Active method used to create constant current lesion.

Figure 8.4: Modification of Thompson constant current lesion maker using a digital display.

Figure 8.5: Simple cooling probe with soldered Y-joint.

Figure 8.6: Makena's simple cooling probe with manifold Y-joint made by Small Parts.

Figure 8.7: Exploded view of the components for a full version of a cooling probe.

Figure 8.8: Clark's modification of circuitry to control cooling probe's internal heating coil.

Figure 8.9: Assembled full version of a cooling probe with soldered Y-joint and internal heating coil.

Figure 8.10: Multiplexed controller for multiple cold probes.

Figure 8.11: Dual multiplexed controller for multiple cold probes.

Figure 9.1: Current flow from the stimulating electrode to surrounding tissue, the recording electrode and an axon.

Figure 9.2: Monophasic and biphasic stimulation. Biphasic pulses mitigate metal electrode polarization during stimulation and glass electrodes clogging during iontophoretic injections.

List of Figures

Figure 9.3: Stimulation parameters include current (i) intensity, duration and frequency. Temporal parameters of stimulation include single, twin (double) and train patterns. In double stimulation, the first pulse is the conditioning (c) pulse and the second pulse is the test (t) pulse. Both c-t and c-c intervals can be varied. Trains are repeated patterns of stimulation.

Figure 9.4: Stimulation artifact in the recording is controlled by confining the flow of current in the respective paths. Here the stimulator is shown as one battery source and the neuron as a second battery source for two separate circuits. Isolation is typically achieved by controlling the stimulator through an optical (optoisolator) or inductive (transformer) element. These and other concepts for stimulation and recording are expanded further in Figure 9.7.

Figure 9.5: Current flow through axons oriented longitudinally or transversely to bipolar stimulating electrodes. Stimulation is more effective (i.e., less injected current evokes a neural response) with a longitudinal orientation.

Figure 9.6: Overlap of the stimulation current and the neuronal population is one of the factors that determines the number of neurons that are affected by the stimulation. Other factors are neuron density, the size and shape of the stimulating electrode tip, and stimulation frequency.

Figure 9.7: Features of stimulus isolation, recording isolation, avoidance of ground loops through a single ground point, and shielding.

Figure 11.1: Analog controller.

Figure 11.2: Decoded bin controller.

Figure 11.3: Bin controller.

Figure 11.4: ROM controller.

Figure 11.5: Forth interface card.

Figure 12.1: Regulating battery supply to ±5 volt.

Figure 12.2: Linear ±5 volt power supply.

Figure 12.3: Linear +5 volt power supply using a center-tapped transformer.

Figure 12.4: Linear +5 volt power supply using simple transformer and bridge diode.

Figure 12.5: Linear -5 volt power supply using simple transformer and bridge diode.

Figure 12.6: Variable +2 to +37 volt power supply with optional constant current output.

Figure 12.7: Variable -2 to -37 volt power supply.

Figure 12.8: Using a voltage divider for reducing the size of the output signal.

Figure 12.9: Using a TTL hex inverter to decrease the output signal.

Figure 12.10: Using a TTL hex inverter to increase the output signal.

Figure 12.11: Using an optoisolator to increase or reduce a noninverted output signal.

Figure 12.12: Using an optoisolator to increase or reduce an inverted output signal.

Figure 12.13: Mixing synchronization pulses for recording to tape using passive components.

Figure 12.14: Mixing synchronization pulses for recording to tape using active components.

Figure 12.15: Interface for controlling the power to a mechanical tape recorder so that it only records during a classical conditioning trial and not during the intertrial interval. This saves tape and makes reviewing the data much faster.

LIST OF TABLES

Table 5.1: Standard t-scores for unit activity in each Epoch compared with baseline activity in each of 12 blocks of training. (Steinmetz)

Table 5.2: Cross-correlation between behavior (stationary variable) and unit activity (lagged variable). (Lavond)

Table 7.1: Items commonly found in the surgery room. (Lavond)

Table 11.1: Equivalent software configurations for the bin controller. (Lavond)

Table 11.2: Output data for ROM controller. (Lavond)

Table 11.3: IBM-PC edge connector contacts. (Lavond)

Table 11.4: Memory map of the hardware addresses on the interface card. (Lavond)

PREFACE

When conducting scientific research in any field, it is not sufficient to simply design thoughtful and informative experiments to explore ideas and hypotheses. The experiments must be conducted in such a manner that the data generated effectively address the ideas and hypotheses under study. Collecting good data necessitates the use of good methods, techniques, and instrumentation. Behavioral neuroscience is most certainly a field that, over the years, has required novel, inventive, and effective methods and techniques to collect data on a rather difficult subject, namely, how the brain and nervous system encode behavior.

Perhaps one of the most interesting things about the field of behavioral neuroscience is that most scientists in this field are engaged in a variety of activities—it is not always the same boring routine. The rule, not the exception, in this field is that investigators are trained in a variety of techniques and skills. This work requires knowledge of skills in such diverse techniques as surgery, animal training, basic electronics, computer programming, statistics, and histology, as well as having a good theoretical background knowledge of the relevant literature and the creativity and logic necessary to design and execute critical experiments. One does not have to be an expert in all of these skills, and conversely not all skills require an expert. For example, it is our experience that an undergraduate can be trained in a short time to perform stereotaxic surgery, fabricate electrodes, and analyze behavioral and neural data, the province of a technician. Knowing what kind of surgery and where it is to be performed, knowing what type of electrode to use, or which data analysis technique is best is the province of a scientist. Not everyone needs to write their own computer programs if they are available from another source and if they are affordable. Indeed, it is very common for behavioral neuroscientists to now purchase "off-the-shelf" computer-based data acquisition and analysis systems, these providing canned programs and routines specific to the application needed by the user.

Overall, the authors have not gone the "off-the-shelf" route when procuring equipment and techniques for our research. Those who know us well know that we generally dislike "technical authority" and that it grates on us to be dependent upon others, particularly when it makes us ignorant of the underlying assumptions and operations of the equipment. Too often we have seen students and their professors who are ignorant about their own experiments, sometimes relying on someone else's word or historical usage ('this is the way it has always been done') without knowing what they are really doing. Resourcefulness has been a key to our own independence. This typically means that we have had to learn about a variety of things, like electronics and computer programming, which really do not interest us much.

Lavond's first real experience with electronics came in an Experimental Psychology laboratory class when he was an undergraduate at the University of Santa Clara, a class taught by Dr. Wendel Goesling. This laboratory was equipped mainly with old relays and switches that had been discarded ("loaned") by Dr. Nancy Daunton and her colleagues at the nearby NASA Ames Research Center. Through this course, Lavond recognized early the importance of learning electronics and programming, both for the flexibility of experimentation and because of the tremendous savings in cost over "store-bought" equipment. A year later Lavond was the TA for that same course and had to set up and repair much of the experimental stations with no assistance. The value of this knowledge was reinforced and expanded later in a graduate course Lavond took at Ohio State University on semiconductor electronics (taught by Dr. Gary Berntson) and hardware programming through connections on a patch board (co-taught by Dr. David Hothersall) and encouraged by fellow graduate student Mike Walker. Virtually everything else was self-taught through hobbyist-level books and magazines concerned with electronics and programming and occasionally through professional engineering data books and articles that appeared in psychology or neuroscience journals. Lavond does not engage in electronic design and construction or computer programming at home for fun and entertainment, and he does not even like computer games. He believes that computers are most useful as word processors and for electronic mail (a major advance over other forms of communications), for running and analyzing experiments, for computing statistics, and for creating figures. That is, computers are useful tools and have no intrinsic interest in themselves for Lavond.

Steinmetz has a similar story to tell. His initial exposure to electronics and instrumentation was as an undergraduate and then a Master's degree student at Central Michigan University. He quickly became familiar with the use of relays and timers to control equipment used in basic human memory and basic psychobiological experiments conducted under the watchful eyes and with the technical assistance of Drs. Terry Libkumen, Michael Kent, and H. Kieth Rodewald. More electronic expertise and an introduction to the use of computers in research were provided by Dr. Michael Patterson, who served as Steinmetz's Ph.D. mentor at Ohio University. However, Steinmetz's real education in electronics and computer use was under the direction of Lavond: The co-authors spent four years together as postdoctoral fellows in Dr. Richard Thompson's laboratory at Stanford University and it was during this time that we developed a mutual interest (and expertise) in a variety of methods and techniques used in behavioral neuroscience research. While Steinmetz has never had a formal course in electronics, he has had some formal computer programming classes and experience. Similar to Lavond, he finds electronics and computer programming to be neither exciting nor interesting, but rather they are necessities in making grant dollars stretch.

A little history as to how this book came to be is in order here. Over the years, the authors have willingly taught scores of students, postdoctoral fellows and technicians the self-taught knowledge they gleaned from interaction with other researchers, from our own experience and reading, and frankly, from pursuing a lot of solutions that simply did not work. We have been most useful, as you might guess, to those individuals who are interested in carrying on this tradition of self-help; not all are interested in these skills and we can empathize. Often, we have found ourselves being the source of "how-do-I-do-this" kinds of questions that can only be solved by those who are intimately familiar with behavioral neuroscience research. For the last several years, we have toyed with the idea of compiling our experiences into a book that might be useful to individuals in the field. That might take away some of the pressure from us. Indeed, Lavond has suggested such a book to representatives of several publishing companies over the last several years. No one seemed interested in the project until Michael Williams of Kluwer Academic Publishers agreed to pursue the idea. It helped that Lavond had already created a large number of circuit diagrams using the Freehand drawing program and had also created a brain atlas using Photoshop and Freehand software applications. Originally thought of as a general-purpose methods book, Williams suggested the title *Handbook of Classical Conditioning*, which was readily accepted because it reminded Lavond of the excellent "handbook" series for sports published by Knopf (e.g., *Handbook of Skiing, Handbook of Tennis, Handbook of Sailing*, etc.), where one can actually learn something useful about these activities by reading about them. The style of these books is logical, pictorial, and is targeted to a wide audience, from the beginner to the advanced. Steinmetz readily agreed to join this venture and together, the authors undertook the task of collecting and compiling a lot of information about methods, techniques and instrumentation, much of which we had already written down for our own purposes, some of which was specifically written for this book.

The overarching purpose of this handbook is to create from nearly nothing a laboratory in which experiments that relate brain function with behavior can be performed (i.e., a general purpose behavioral neuroscience laboratory). Since early in our research careers, both authors had the general attitude and approach to research methods and instrumentation of "what can

I do if I get a job somewhere with little or no facilities and little or no resources." By trading our time, we reasoned that we could save ourselves hundreds and perhaps thousands of dollars. It can be pointed out that if one has the research funds, a great deal of time and effort can be saved if the equipment is commercially obtained. This is a legitimate position. However, we still believe that a real advantage of doing it yourself, when possible, is that *you know exactly how the pieces of equipment operate, how the program works, or how the calculations and analyses are performed,* etc. In the long run, we believe that this is the best approach. It has certainly worked for us!

This book is therefore a compilation of methods, techniques and instrumentation that have proven useful for the practical conduct of classical eyeblink conditioning experiments, a paradigm that has been effectively used over the years to study the neural bases of learning and memory (see Steinmetz, 2000, Steinmetz, Gluck and Solomon, 2001, or Woodruff-Pak and Steinmetz, 2000a and b, for reviews of this literature; see Thompson, 2000, and Lavond and Kanzawa, 2001, for historical accounts of the discovery). We decided to base the book on classical eyeblink conditioning because that is what we know best. However, this book should prove useful for anyone in the field of behavioral neuroscience who conducts experiments that involve the collection of behavioral and/or neural data. The general methods and equipment used to deliver stimuli and record data are somewhat universal; eyeblink classical conditioning is but one application of these rather universal methods. Indeed, in several places throughout the book, we extensively discuss the application of our methods for paradigms other than classical eyeblink conditioning.

The book is organized into 12 chapters. Chapter 1 provides an overview of the classical conditioning paradigm including a summary of the types of classical conditioning often used and parametric considerations. Chapter 2 is concerned with the delivery of stimuli during behavioral experiments, with emphasis on the delivery of the conditioned and unconditioned stimuli used in classical conditioning. In Chapter 3, measuring behavioral responses is covered along with behavioral data analysis and number-crunching methods. Chapter 4 is concerned with basic neural recording techniques including fabrication of head-stages and electrodes, amplifying and filtering signals, and discriminating action potentials. Methods for analyzing behavioral and neuronal data are presented in Chapter 5. Summaries of some other commonly used behavioral paradigms and procedures, such as fear conditioning and spatial learning, are provided in Chapter 6 while Chapter 7 covers basic surgical methods. Chapter 8 presents information concerning lesioning techniques, ranging from permanent lesions to temporary inactivation of neural tissue, and Chapter 9 provides details concerning the

Preface

use of brain stimulation in behavioral neuroscience research. Basic histological techniques used for the processing of brain tissue to verify lesion or electrode placements are the topic of Chapter 10. Chapter 11 presents information about the computer control of behavioral and neural experiments while Chapter 12 discusses other electronics that have proven useful for behavioral neuroscience experiments. An Appendix is also part of this book with sections that list suppliers we have found useful, and a small rabbit brain atlas that was created by Lavond. The accompanying CD includes the rabbit brain atlas, electronic versions of all figures that appear in the book (except copyrighted figures), and source codes and program descriptions for the Forth and C++ experimental control systems that have been designed and used by the authors in their laboratories.

We hope that you find the information in this book useful. The authors have found great enjoyment and fulfillment in conducting behavioral neuroscience research over the last several years. We hope the information that appears in this book will be used to provide you with the same.

ACKNOWLEDGMENTS

We thank Michael Williams of Kluwer Academic Publishers for pursuing our book proposal. Michael is also credited with coming up with the title. We also thank Mary Panarelli of Kluwer Academic Publishers for her many editorial assists.

We give special thanks to Kristina Fallenius and the Nobel Foundation for permission to use their photograph of Pavlov (Figure 1.1). In his normal research Lavond had not seen many pictures of Pavlov in print and would not have recognized Pavlov in a lineup of famous scientists and ne'er do wells. Having a picture of Pavlov is a pleasure.

We also thank Richard F. Thompson for allowing us to use the historical photograph of early human eyeblink conditioning (Figure 1.2) that was given to him by Ernest Hilgard. We also have to thank Dick for the many opportunities in his lab that were essential components of the 'how to' for classical conditioning that the reader will see in this volume. We hope that some of these opportunities, like making equipment and programming computers, benefited Dick as well. We thank Dick for all the support he has given us throughout our many years of association.

A number of ideas and circuits that we have used throughout the years have come from others. Without meaning to exclude anyone, we would like to specifically acknowledge the original contributions of Robert Clark, Gil Case, Kathleen Chambers, Unja Hayes, Tapani Korhonen, Ami Makena and Paul Solomon.

Lavond would like to specifically acknowledge his co-author, Joe Steinmetz, whose publishing experience with Kluwer and whose scientific contributions to this book were immense. Joe was principally responsible for including other paradigms in this collection and for the sections on stimulation, while contributing significantly to all other aspects of this book. At the same time, Steinmetz would like to acknowledge his co-author, Dave Lavond. Dave's extensive knowledge of electronics and methodology, not to mention his great skills as a behavioral neuroscience researcher and scholar, have greatly benefitted Steinmetz over the 20 years they have been associated with each other.

Many of the figures, especially circuit diagrams, were made by Lavond for his own records and dissemination. Some figures, like the different ordered filters, were made specfically for the electronics course he developed and taught at USC (PSYCH 499: Special Topics: Laboratory Electronics for Psychologists). Other figures were made specifically for this book, principally explanatory figures like the example of 'bad trials' and measurements during the trial.

Chapter 1

THE CLASSICAL CONDITIONING PARADIGM

Throughout this book, we will use classical conditioning as the main behavioral procedure to illustrate methods and techniques commonly used in behavioral neuroscience research. It therefore seems appropriate to begin this book with a thorough discussion of classical conditioning procedures. Because classical eyeblink conditioning has become the most frequently used classical conditioning procedure, much of our discussion will be centered on this procedure. Readers who are already familiar with this behavioral paradigm might want to skip the first part of this chapter and move on to the material on setting up classical conditioning experiments that appears later in the chapter.

IVAN PAVLOV AND CLASSICAL CONDITIONING

As all introductory psychology students know, the origins of the procedure we now call classical conditioning (or Pavlovian conditioning) is credited to Ivan Petrovich Pavlov (see Figure 1.1). Pavlov, the son of a Russian priest, initially began his formal studies in a seminary in Riazan, but left the seminary before his studies were completed. In 1870, he entered the natural sciences section of the Faculty of Physics and Mathematics at St. Petersburg University. After completing his undergraduate work, he went on to earn a degree in medicine, actually becoming director of a new experimental laboratory associated with an internal medicine clinic before he completed his doctoral thesis (which, incidentally, was on the effects of several drugs on cardiovascular function). After several years of employment in a variety of rather non-descript, low-level positions, Pavlov was appointed as a professor at St. Petersburg's Military-Medical Academy in 1890. It was in this laboratory that he established a research program that eventually made him famous the world over.

In the 1890s, Pavlov's interests were centered around the functions of the digestive system. He conducted a series of carefully designed and

Figure 1.1: Ivan Petrovich Pavlov (1849-1936). Photograph courtesy of and copyright © The Nobel Foundation 1904. For further information see *http://www.nobel.se/medicine/ laureates/1904/.*

technically sophisticated experiments that explored the effects of a variety of stimuli, presented to various parts of the body, on the generation of digestive secretions. These elegant studies won Pavlov the Nobel Prize for physiology in 1904.

For our discussion here, however, we are most interested in Pavlov's lasting contributions to the study of learning and memory—his concept of the *conditioned reflex*, which was first mentioned publicly in his 1904 Nobel address. In the course of his experiments on digestive physiology, Pavlov and his colleagues noticed a very interesting phenomenon. Pavlov had developed a surgical technique that allowed him to collect digestive secretions through fistulae that were placed in dogs. Pavlov and his associates first noted that when food or acid was placed in a dog's mouth, the dog's digestive secretions would increase. They also noted that eventually the dogs would secrete digestive juices when they saw the food, or even when they caught sight of the animal caretaker approaching them with the food. Pavlov's assistants began referring to these cue-based digestive secretions as "psychic secretions" because they seemed to be generated by the thought of the food alone. Stephan Wolfsohn and Anton Snarsky, who both worked in Pavlov's laboratory, more formally studied the generation of these "psychic secretions" by closely examining the stimulus conditions that produced the reflexive and cue-elicited digestive secretions. Eventually, Pavlov called the food or acid an *unconditioned stimulus* (the UCS) and the reflexive response that resulted an *unconditioned response* (the UCR). The stimulus that was initially neutral (e.g., the sight of the caretaker or food, other visual stimuli, bells, etc.), but when paired with the UCS became capable of producing digestive secretions was called the *conditioned stimulus* (the CS). The "psychic secretions" were eventually called the *conditioned response* (the CR). This type of simple associative learning has become known as classical (or Pavlovian) conditioning. Using this basic procedure, Pavlov conducted a variety of experiments that covered topics such as generalization and differentiation (more commonly known as discrimination), "experimental neurosis," and even hypotheses and theories concerning how the brain was thought to encode the conditioning stimuli and responses (see Fancher, 1979; Furedy, 1992; Grigorian, 1974; and Windholz, 1991, for biographies on Pavlov).

Interestingly, as Furedy (1992) has pointed out in his brief biography of Pavlov, no systematic reports of the prototypic dog salivation experiments for which Pavlov is best known can be found in the literature. Indeed, reports of Pavlov's findings are largely based on case studies involving relatively imprecise stimulus delivery techniques and largely anecdotal and widely varying response measurement procedures. In addition, his preparations were reported to be difficult to set up and work with, often requiring months of adaptation and subsequent training time for the animal subjects. Given the technology available at the time these experiments were conducted, however, Pavlov's experiments and approach should be considered innovative and novel for the time. His insistence on direct observation of behavior was particularly novel and noteworthy.

It seems rather ironic that since Pavlov began studying classical conditioning, few, if any, scientists actually conducted classical conditioning experiments using methods and techniques that resemble those used by Pavlov. While the study of the conditioned reflex became very popular in the twentieth century and Pavlovian terms such as CS, US, CR and UR became commonplace in the language of experimental psychologists and other scientists, these experimentalists did not use conditioning procedures that resembled those used by Pavlov. Instead, other classical conditioning procedures were developed, which afforded better control of the stimuli used as the US and CS, and more reliable (and easier) measures of the UR and CR. Classical eyeblink conditioning is one of these procedures. The development and use of this paradigm will be detailed next.

A BRIEF HISTORY OF CLASSICAL EYEBLINK CONDITIONING

Human Classical Eyeblink Conditioning Studies

Early in the twentieth century, researchers shifted from studying the classical conditioning of the relatively slow, difficult-to-measure, autonomic nervous system responses, to studying the classical conditioning of faster, more accessible, somatic nervous system responses. Instead of studying digestive reflexes such as gastric secretions, the focus shifted to studying skeletal muscle responses. Classical conditioning of eyelid movement (hereafter referred to as classical eyeblink conditioning) was one of the somatic conditioning paradigms that was developed. Interestingly, the early eyeblink conditioning studies, for the most part, used humans as subjects.

The earliest published study of classical eyeblink conditioning appeared in a German physiology journal at the end of the nineteenth century (Zwaardemaker & Lans, 1899) — note that this is well before Pavlov's public speech for the Nobel in which he first mentioned his experiments. This was followed by another eyeblink conditioning paper published about ten years later in a German journal (Weiss, 1910). Cason appears to be the first American to publish on classical eyeblink conditioning—he published four papers between 1922 and 1925 that appeared in the *Journal of Experimental*

Chapter 1

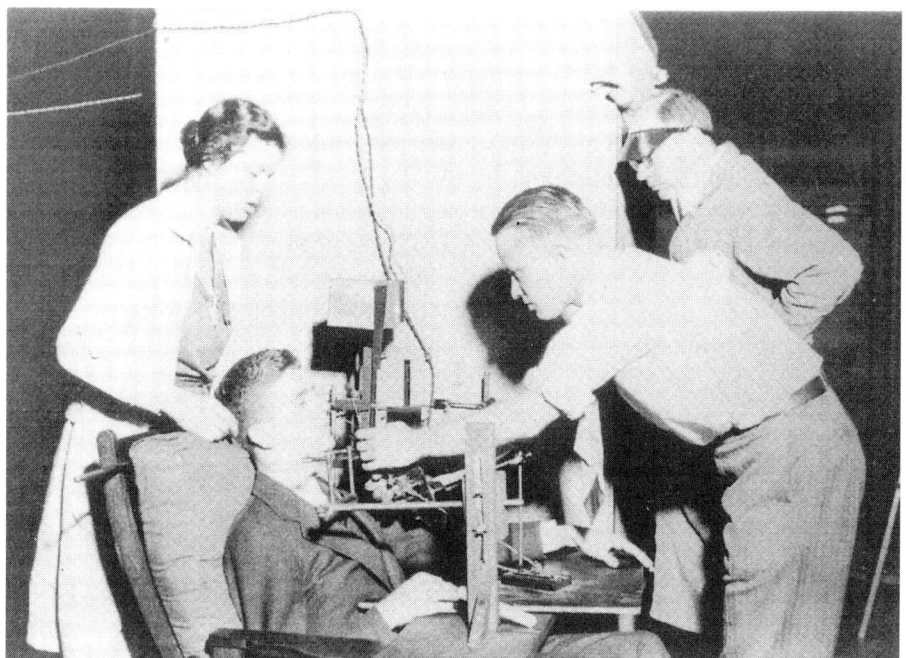

Figure 1.2: Early human eyeblink classical conditioning experiment. Graduate student Ernest Hilgard is the subject (seated) while Clark Hull (wearing the visor) looks on as Walter C. Shipley adjusts the automated wooden face-slapper (the unconditioned stimulus) with the help of postdoctoral student Helen Peak. Photograph courtesy of the collection of Richard F. Thompson.

Psychology and the *American Journal of Psychology*. Ernest Hilgard was among the first investigators to use standard conditioning techniques with well controlled stimulus and response measurement procedures (e.g., Hilgard, 1931, 1933; Hilgard & Marquis, 1936). His work established a close correspondence between human and nonhuman eyelid conditioning and also suggested that there were common neural mechanisms across mammalian species for encoding the learning. Figure 1.2 shows an early human classical eyeblink conditioning experiment.

During the last century, eyeblink conditioning was widely adopted as a behavioral paradigm useful for studying human learning and memory. Isadore Gormezano has compiled a comprehensive bibliography of human eyeblink conditioning studies that covers the years 1899 to 1985. This bibliography was published in its entirety as an appendix in Woodruff-Pak and Steinmetz's (2000a) collection of human eyeblink conditioning papers, *Eyeblink Classical Conditioning, Volume I: Human Applications*. A total of 507 citations can be found in the bibliography. Examination of the bibliography shows that 62 papers were published in the 1930s, 36 papers were published in the 1940s, 70 papers in the 1950s, 228 papers in the 1960s, 70 papers in the 1970s, and 17 papers through the first half of the 1980s. The precipitous drop in human eyeblink conditioning articles in the 1970s is noteworthy. It is likely that a large part of this drop was due to the development of animal models of eyeblink conditioning, most notably the rabbit preparation, growing interest in more cognitive human research, and growing interest in pursuing studies aimed at delineating the neural bases of simple associative learning, which was advanced significantly by the use of animal models. In the 1990s, however, there was a resurgence of interest in human eyeblink conditioning and the number of published studies has recently increased dramatically (see Woodruff-Pak and Steinmetz, 2000a).

Non-Human Classical Eyeblink Conditioning Studies

While studies of eyeblink conditioning involving human subjects advanced the understanding of basic associative learning and memory processes, since the early 1960s most eyeblink conditioning studies have used non-human subjects. The switch to non-human preparations was caused by several factors. First, measuring human eyeblink responses was relatively difficult, not only for technical reasons, but because the spontaneous blinking rate of humans is relatively high. Second, several cognitive factors are known to affect the rate and level of conditioning, such as how instructions are delivered and interpreted and how arousal and awareness influence conditioned responding. Third, interest in studying how the brain encodes associative learning grew significantly in the last few decades; prior to the advent of fMRI and other imaging techniques, humans did not make ideal subjects for studies of the neurobiology of eyeblink conditioning. Fourth, elegant animal preparations, such as the rabbit nictitating membrane/eyelid conditioning that was developed by Gormezano became available (Gormezano et al., 1962). A number of researchers across the country adopted these procedures thus creating a huge database of useful information concerning various features of this type of associative learning.

Without a doubt, the gold standard paradigm for studying the classical conditioning of somatic responses has been the rabbit NM/eyelid conditioning procedure that was developed largely by Gormezano and colleagues (e.g., Gormezano et al., 1962). However, a variety of other mammalian species have been used in classical conditioning experiments including cats (e.g., Wickens, Nield, Tuber & Wickens, 1973; Penttonen & Korhonen, 1991), dogs (e.g., Cassady, 1996) and rats (Green, Rogers, Goodlett, & Steinmetz, 2000). The conditioning of a number of response systems other than the eyeblink response has been studied, such as limb movements as exemplified by leg flexion (Cassady, Cole, Thompson, & Weinberger, 1973). Use of the rat for eyeblink conditioning experiments actually predates the use of the rabbit (e.g., Biel & Wickens, 1941; Hughes & Schlosberg, 1938) most likely because the rat was the preferred experimental subject in many other types of behavioral experiments. Indeed, for a variety of reasons, in the 1990s, a growing number of investigators began using the rat once again as a subject for eyeblink conditioning experiments. We will take up special technical issues and requirements concerning use of the rat in eyeblink conditioning experiments in later chapters in this volume.

The rabbit proved an ideal subject for classical eyeblink conditioning experiments (see Romano & Patterson, 1987, for review). In brief, the rabbit tolerates restraint well thus making it relatively easy to present stimuli and record responses during training. The spontaneous blink rate of the rabbit is quite low, thus few eyeblinks that can be classified as neither CR or UR are present. Compared to other species, rabbits were relatively inexpensive to purchase although animal care costs can be more expensive than some other species. Finally, the size of the rabbit brain is a bit larger than rats and other rodents, thus making it a bit easier to conduct studies aimed at delineating the neural substrates of classical eyeblink conditioning. Because the focus of this book is on methods and techniques and is not meant to be a comprehensive review of the eyeblink conditioning literature, suffice it to say at this point that a huge body of literature concerning the behavioral and neural correlates of classical eyeblink conditioning exists. Several reviews and edited books that summarize this literature are available (e.g., Gormezano, Kehoe and Marshall, 1983; Steinmetz, Gluck, & Solomon, 2001; Woodruff-Pak & Steinmetz, 2000b); we refer interested readers to these sources. We are interested, however, in describing many of the methods and techniques that have been used in this massive literature. We believe that these methods and techniques can be easily applied to a variety of behavioral neuroscience studies. Before presenting this information, we present here some basic information about general classical conditioning procedures, including several issues that must be considered when setting up and conducting these experiments.

THE LANGUAGE OF CLASSICAL CONDITIONING

Like all well established procedures and paradigms, classical conditioning has a common language that is spoken by researchers who use the procedures. We will use this language throughout this volume so it is important that we present it at this time.

The labeling of the stimuli used in classical conditioning goes back to Pavlov's writings. In his famous experiments, which have been recounted anecdotally for many years, Pavlov paired two stimuli to produce the learned responses; food or a chemical substance that reliably elicited digestive secretions and a light, tone, or other stimulus that initially may have produced an orienting response, but did not elicit digestive secretions. Pavlov referred to the food or chemical substance as an *unconditional stimulus* because no prior training was needed for it to be effective in eliciting digestive secretions, and he referred to the initially neutral light or tone as the *conditional stimulus* because its ability to elicit digestive secretions was dependent on pairing with the unconditional stimulus. The reflexive response elicited by the unconditional stimulus was called an *unconditional response* (because it was always produced by the food or chemical) while the learned response was called a *conditional response* (because it was training-dependent). Interestingly, due to an error in translating Pavlov's writings from Russian to English, we now do not use the original terms used by Pavlov. Instead, we know these terms as the uncondition*ed* stimulus, the condition*ed* stimulus, the unconditon*ed* response and the condition*ed* response; these are most often abbreviated as the *US, CS, UR* and *CR*, respectively.

As detailed below, the timing of the delivery of the CS and US used in classical conditioning is critical. Thus, terms are used to describe the time interval between stimuli and the time interval between discrete presentations (i.e., trials) of CS-US pairings. The *interstimulus interval* or the *ISI* refers to the interval of time between the onset of the CS and the onset of the US. In eyeblink conditioning this period is commonly 150 to 3000 msec in duration. Sometimes, and perhaps more accurately, this interval is referred to as the CS-US interval or the CS-US asynchrony. The *intertrial interval* or *ITI* refers to the length of time between discrete trials and usually ranges from 20 sec to several minutes. Rabbits and rats (and humans, for that matter) do not condition with ITIs less than 10 seconds. Typically, the duration of this interval varies in a session of training in a random or pseudorandom fashion to prevent subjects from anticipating the exact spacing of trials.

VARIATIONS OF THE CLASSICAL CONDITIONING PROCEDURE

It is common to think of classical conditioning as a single associative learning procedure. In actuality, classical conditioning includes a variety of related procedures that are loosely based on Pavlov's original conditioning methods. In general, we refer to classical conditioning as classical *defense* conditioning if the US is a negative (i.e., aversive) stimulus, including fear conditioning, and as classical *reward* conditioning if the US is a positive (i.e., appetitive) stimulus. Traditionally, classical conditioning was differentiated from instrumental conditioning on the basis of two basic classical conditioning requirements: (a) whether or not the US presented is independent of CR occurrence and (b) the CR must be similar in form to the UR, that is, the response measured as the CR must be in the same effector system as the UR. In the latter half of the twentieth century, the definition of classical conditioning was broadened to include any paradigm meeting only the requirement that a CS and US be presented independent of the target response, in some instances even ignoring the requirement that the UR and CR be similar in form and direction. This has extended the use of the term "classical conditioning" from the CS-CR procedure described by Pavlov to other stimulus-stimulus (S-S) paradigms that include CS-instrumental response (CS-IR) procedures such as conditioned suppression and classical-instrumental transfer procedures, and sign-tracking or autoshaping procedures. In addition, some discriminative approach and avoidance procedures have been commonly classified as classical conditioning procedures.

TYPES OF CLASSICAL CONDITIONING PROCEDURES

The most commonly used simple classical conditioning procedures, broadly defined are:

1) **Short-delay Conditioning**. This is perhaps the most frequently used classical conditioning procedure. In this procedure, the start of the US is *delayed* relative to the start of the CS. For classical eyeblink conditioning studies, ISIs in the range of 100 to 750 msec are typically considered as short-delay procedures, and conditioning does not occur with ISIs less than 50 msec. For other types of classical conditioning, delays of up to 1 min may be considered as short-delay conditioning. In this procedure the CS may completely overlap the US or the CS may terminate at some point before US offset.

2) **Long-delay Conditioning**. In this procedure, the start of the US is still *delayed* relative to the start of the CS, but ISIs are longer than in the short-delay procedure (i.e., from 750 msec to as long as 3000 msec for eyeblink conditioning, but as long as 5-10 min in some other classical conditioning paradigm). As in short-delay conditioning, the CS may completely overlap the US or the CS may terminate at some point before US offset.

3) **Trace Conditioning**. This procedure involves no overlap in the presentation of the CS and the US. Instead, the CS is presented, a period of time is

allowed to elapse during which no stimuli are presented, and then the US is presented. The gap between CS offset and US onset is called the *trace interval*. Typically in eyeblink conditioning, ISIs that are similar to either the short- or long-delay conditioning procedures are used (e.g., a 750 msec ISI which is composed of a 250 msec CS followed by a 500 msec trace period may be used).

4) **Simultaneous Conditioning**. In this procedure, the CS and US are presented at the same time. Classically conditioned eyeblink responses cannot be obtained with this procedure and it is sometimes used as a control procedure when stimulus exposure, without CR formation, is desired.

5) **Backward Conditioning**. The US is presented before the CS in this procedure. This type of training does not produce overt classically conditioned eyeblink responses. Like simultaneous conditioning, it is sometimes used as a control procedure when stimulus exposure, without learning, is desired.

6) **Temporal Conditioning**. In this procedure, the US is presented at regularly timed interval. In essence the background can be considered the CS for this procedure and CR acquisition is dependent on correct timing of the interval between US presentations.

7) **Unpaired Conditioning**. This procedure involves presenting the CS and US on separate trials. In the *explicitly unpaired* conditioning procedure the CS and US are presented far enough apart such that there is no chance for associations to form. Typically in eyeblink conditioning studies, this interval of time would be 5 sec or greater. In the *random control* unpaired conditioning procedure, the periods of time between CS and US presentations are randomly determined and vary greatly. Like the explicitly unpaired conditioning situation, this random control procedure involves presenting the CS and US on separate trials, the same number of times as would be presented in a paired CS-US training procedure. However, because the interval of time between CS and US varies randomly in the random control procedure, it is possible that associative learning can take place on those trials when the CS-US interval is within the conditioning range (e.g., for eyeblink conditioning, that would be ISIs of 100 msec to about 3000 msec). In practice *pseudorandom controls* are given because randomization tables and computer algorithms are often not truly random, but they are close. The unpaired conditioning control procedures assist in ruling out non-associative explanations of CR formation, such as sensitization (general non-associative increases in responding to the CS) and pseudoconditioning (CR-like responses caused by exposing the subject to the UR).

8) **CS-Alone Extinction**. Often, acquisition training is followed by CS-alone extinction training, during which the CS is presented in the absence of the US. Eventually, the number of CRs is reduced to pre-training levels as the CR is extinguished.

VARIATIONS OF CLASSICAL CONDITIONING PROCEDURES

In addition to the variations of simple classical conditioning described above, a number of variations of the classical conditioning procedures have been devised to study inhibitory learning and higher-order conditioning effects. Here are brief descriptions of a few of these higher-order classical conditioning procedures:

1) **Classical Discrimination/Reversal Conditioning**. In this procedure two CSs and one US are typically used. The CSs may be in the same modality (such as using pure tones of different frequency) or they may be in different modalities (such as using an auditory CS and a visual CS). For classical discrimination conditioning, one of the CSs is designated as the CS+ and its presentation is always followed by the presentation of the US. The second CS is designated as the CS- and its presentation is never followed by the presentation of the US. After several presentations of the CS+ and CS- trials, the subject learns to discriminate between the CS+ and CS- trials such that CRs are observed only on the CS+ trials. During reversal training the CS+ and CS- are reversed such that the CS initially designated as the CS+ becomes the CS- and the CS initially designated the CS- becomes the CS+. Subjects eventually learn to suppress responding to the initial CS+ (now the CS-) and show CRs to the initial CS- (now the CS+).

2) **Classical ISI Discrimination Conditioning**. This is a discrimination procedure whereby two different CSs (in the same or different modalities) are used to signal two different ISIs. For example, a high-frequency tone may be presented 1000 msec before a US while a low-frequency tone is presented 500 msec before a US. Using this technique, subjects can learn to perform CRs that are appropriately timed for the two distinct CSs that are presented.

3) **Latent Inhibition Conditioning**. In this procedure, a CS is presented several times before paired CS-US training is begun. The pre-exposure of the subject to the CS before paired training retards the rate of CR acquisition

relative to subjects that are not CS pre-exposed but are placed in the conditioning environment without stimulus presentation for the same amount of time before paired training.

4) **Conditioned Inhibition Conditioning**. There are a variety of ways in which conditioned inhibition training is accomplished. In eyeblink conditioning experiments, three phases of conditioning are typically used. In the first phase a CS (the CS+) is paired with a US until asymptotic CR levels are reached. During the second phase, CS+/US trials are continued, but interspersed with trials on which the CS+ is presented in compound with the second CS but not with the US (i.e., CS+/CS- trials). Typically, subjects show eyeblink CRs on CS+/US trials but suppression of responding on the CS+/CS- trials. Inhibitory learning is verified during a "retardation test" that is typically presented as the third phase of conditioning, During this phase the previous CS- is presented in paired fashion with the US. If conditioned inhibition occurs, the rate of acquisition to the previously compounded CS should be retarded relative to control subjects that do not receive the second phase of conditioning.

5) **Blocking Effects**. This higher-order conditioning procedure also involves three phases of conditioning. In Phase 1, a CS (CS1) is paired with a US. In Phase 2 of training, CS1 is presented in compound with a new CS (CS2) and the compound is paired with the US. In Phase 3, the paired CS2-US trials are presented. Blocking is measured as a retardation in the rate of learning to CS2 during Phase 3 of training relative to subjects that did not receive CS2 training in compound with CS1 (i.e., acquisition to CS2 was *blocked* during compound training because CRs had already formed to the CS1 before the compound training was begun).

6) **Second Order Conditioning**. For this procedure one CS (CS1) is paired with a US in an initial phases of training until CRs are established. During a second phase of conditioning a new CS (CS2) is presented in paired fashion with CS1 (it, in essence, has become the US). The CS2 eventually comes to elicit CRs similar to the original US.

7) **Sensory Preconditioning**. In this type of conditioning, a CS2 is initially paired with a CS1 in the absence of a US. Then, CS1 is paired with a US and this produces a CR. In subsequent sessions, the CS2 is also found to elicit a CR (or, at the very least conditioning to CS2 is facilitated relative to control subjects), even though CS2 was never paired with the US.

8) **Conditional Discrimination (Occasion Setting)**. In this procedure, on

some trials a "modulator" or "occasion setting" stimulus is presented before or in the presence of a CS that is followed by a US. On other trials, the CS is presented in the absence of the modulator stimulus and is not followed by a US. For example, a light CS is followed by a US when presented together with white noise that is turned on before the trial, but the light is not followed by a US when presented in the absence of the white noise. The subject learns to respond on those trials containing the occasion setter or modulator stimulus.

SOME IMPORTANT CONSIDERATIONS FOR DESIGNING AND IMPLEMENTING METHODS USED IN CLASSICAL CONDITIONING EXPERIMENTS

Now that we have presented a brief history and overview of classical conditioning and have discussed variations of the basic excitatory conditioning procedure, we can now turn out attention toward more technical and practical matters that must be considered when setting up classical conditioning experiments. Basically, in this section we discuss what factors must be considered when designing and implementing techniques for delivering stimuli and recording responses during classical conditioning studies (as exemplified by eyeblink conditioning). We have found it very useful to approach this problem from a computer programming point of view, mostly because this approach is very logical and also very practical—in behavioral neuroscience, computers are now used extensively, if not exclusively, for stimulus delivery and response recording. We consider here how the stimuli are delivered, how the responses are measured, how conditioned responses are determined, and what trials should and should not be included in final analyses of the behavioral data. We thought it would be helpful to identify those variables that must be defined when programming and setting up classical conditioning experiments. Thus, throughout this section programming variables that are used by Lavond in his Forth-based runtime system are identified and placed ***in italicized bold*** in the following sections. By showing what important variables must be defined and used in programming a classically eyeblink conditioning experiment, we hope the reader will get a sense of the complexity and control needed for these experiments. A summary of the variables typically used in the programs that control these behavioral experiments is available as a dictionary on the CD from the authors.

Figure 1.3 shows what is actually going on during a single short-delay classical conditioning trial. At a given point in time, a CS is turned on, presented for a period of time, then turned off. Also, at a set period of time

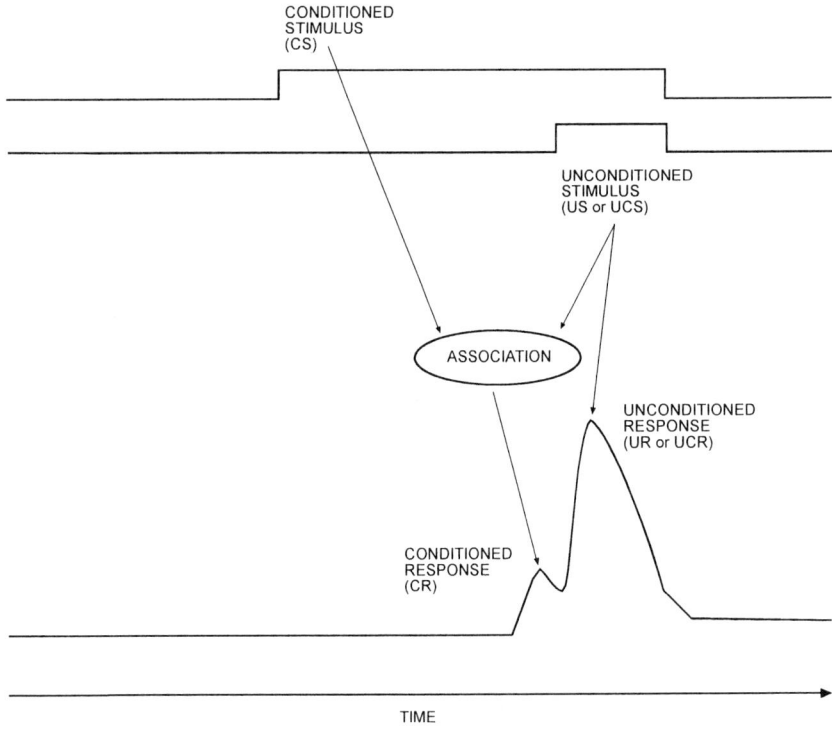

Figure 1.3: Relationship of timing of the stimuli with the behavioral response during classical conditioning.

after CS onset, the US is turned on, presented for a period of time, and then turned off. In the example shown in Figure 1.3, the CS and the US co-terminate. One important function of a computer or other control device is to control the delivery of the stimuli used during conditioning, including maintaining the proper temporal relationship between stimuli and between discrete trials. This is typically accomplished by sending control signals to peripheral equipment designed to generate the necessary CSs and USs used during training. Figure 1.3 also shows a response recorded during presentation of the conditioning stimuli. Because this is a representation of an eyeblink conditioning trial, the upward deflection of the behavioral trace seen in the figure is an analog representation of eyelid closure (i.e., the eyeblink response); the higher the trace, the more complete the eyeblink. A clear CR and UR are shown in Figure 1.3; the CR is recorded in the period between CS and US onsets while the UR is recorded after US onset and before the end of the trial. It is obvious from Figure 1.3 that the computer serves an important second function during classical conditioning experiments: The computer records data that is input to it from peripheral devices used to monitor the behavior under study. The remainder of this chapter will discuss those factors that must be considered in stimulus delivery and response recording. More details concerning the technology involved in stimulus delivery and response recording will be presented later in this book (see Chapters 2 and 3).

Timing Relationships Present on Classical Conditioning Trials

The timing relationships between the conditioned and unconditioned stimuli are commonly presented as timing diagrams. In Figure 1.4 we show the onset, duration and offset for the short-delay classical conditioning procedure. In this example, an initial 250 msec period is used to collect information about movements in the baseline time period. This initial 250 msec sampling period serves as a baseline period for collecting information about the relative activity of the subject just before trial onset. For the restrained rabbit preparation, we typically worry only about the appearance of spontaneous eyeblinks in this period. For studies using freely-moving rats and humans as subjects, however, other movement related changes in the behavioral baseline might be detected. Subsequent changes in response levels (e.g., degree of eyelid closure) are calculated relative to the eyelid position recorded during this pre-CS baseline period. In the example shown in Figure 1.4, the CS has an onset at 250 msec after the beginning of the trial (***toneon***). The CS lasts for a total of 350 msec before it terminates (***toneoff*** at 600 msec from the beginning of the trial). The US begins 250 msec after the onset of the CS (***shkon*** at 500 msec after the beginning of the trial). The US lasts for 100 msec and terminates at the same time as the CS (***shkoff*** at 600 msec from the beginning of the trial). Accompanying the timing diagrams is a cartoon illustrating the relationship of a learned behavioral response on a paired CS-US trial. The CR occurs between the times of CS onset and US onset, i.e., during the 250 msec ISI. We will consider a more formal definition of the CR later in this chapter. The UR is defined as the behavior that occurs after the US onset. Here the US period is defined as the 250 msec that follows the US onset. The trial terminates at this point, or 750 msec after the onset of the trial (***pts***). Three features found in Figure 1.4 define the short-delay classical conditioning procedure; the CS onset precedes US onset, the US overlaps with the CS in the later period, and both the CS and US terminate together.

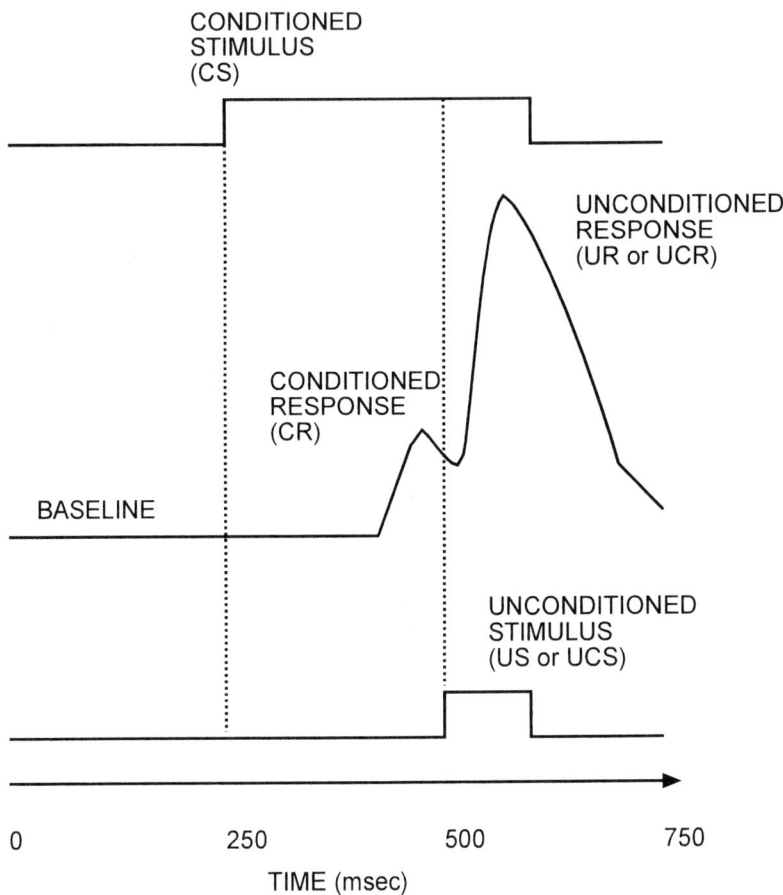

Figure 1.4: Relationship of timing of the stimuli during classical conditioning using the delay conditioning paradigm.

A few additional comments concerning trial timing relationships should be mentioned at this time. First, the ISI can actually be a range of values. An ISI of 50 msec is too short for eyeblink conditioning. An ISI of 250 msec is optimal for training rabbits and rats in this procedure. Longer ISIs of up to 3000 msec can work for training rats and rabbits but the rate of learning and the level of conditioning achieved is typically less than when optimal training timing parameters are used. For humans a 500 msec ISI seems optimal; longer ISIs may result in voluntary responding.

Second, while our example uses one-quarter second intervals (the CS starts at 250 msec, the US starts at 500 msec, and the trial terminates at 750 msec from the beginning of the trial), programming and analysis often require slightly different values for computational convenience usually associated with the rate of data collection. For example, collecting data at a rate of 250 Hz yields a bin width (***bin***) of one data point every 4 msec. This bin width is not evenly divisible into a 250 msec period (250 divided by 4 equals 62.5 data points or number of bins). For most computer systems and programming purposes, it is only possible to collect a whole number of bins, not a fractional number of bins (i.e., 62 or 63 bins and not 62.5 bins). Collecting 63 bins yields a 252 msec period (63 times 4 msec), which is close to the goal of 250 msec. Similarly, collecting at a rate of 3 msec for every bin width yields a fractional number for 250 msec (250 divided by 3 msec bins equals 83.3 data points). A period of 248 msec using 3 msec bins is also close enough to the goal of 250 msec. If one wants 250 msec periods exactly, then one should collect the data using 5 or 10 msec bins. The former will give 50 data points, the latter will give 25 data points. One of the questions in selecting the rate of collection, then, is how many data points (the number of bins which depends upon the bin width) one wishes to collect: 3 msec bins collects 83 data points in 248 msec, 4 msec bins collects 63 data points in 252 msec, 5 msec bins collects 50 data points in 250 msec, and 10 msec bins collects 25 data points in 250 msec. For delay conditioning as described here, the total number of data points collected for 4 msec bins would be 189 (***pts***) and the rate of data collection would be 250 Hz (i.e., 250 data points would be collected per second). Your choice of bin-size will likely depend on the resolution of data collection you desire and the amount of storage space you have available. In general, the lower the bin size, the greater the number of data points collected, and thus the greater the resolution for response sampling. The trade-off, however, is that the more data collected, the more storage needed in the computer to collect each trial. Not too many years ago, this was a huge problem. However, newer microcomputers have loads of storage space available so this is no longer a limiting issue.

Third, the values ***shkon*** and ***shkoff*** (shock onset and shock offset) are used to describe the onset and offset times desired when shock is used as a US. When an air puff is used as the US, the corresponding values are ***airon*** and ***airoff***, which include a correction factor for the amount of time it takes the air solenoid to open and for air to travel down the tubing to the eye. (Unlike shocks, air does not travel at the speed of light). The correction factor is determined empirically by measuring the latency between the timing signal being sent from the computer to when air actually arrives at the end of tubing positioned near the eye of the subject. The value ***shkon*** represents the time when shock would be used as a US, *and also* represents the actual time when the air arrives at the eye. For this reason we refer to ***shkon*** rather than ***airon*** in this discussion.

The trace classical conditioning procedure is schematized in Figure 1.5. As with the short-delay paradigm, an initial 250 msec period is used to collect information about movements in the baseline time period. The CS has an onset at 250 msec after the beginning of the trial (***toneon***). Unlike the delay paradigm, the CS for trace conditioning lasts for a total of 250 msec after which time it terminates (***toneoff*** at 500 msec from the beginning of the trial). The key feature of trace conditioning is that after CS offset, there is a period during which no stimuli are delivered (called the trace interval), which lasts in our example for 250 msec (from 500 msec to 750 msec from the beginning of the trial). The US begins 500 msec after the onset of the CS (***shkon*** at 750 msec after the beginning of the trial) and lasts for 100 msec, terminating at 850 msec from the beginning of the trial (***shkoff***). Accompanying the timing diagrams is a cartoon illustrating the relationship of a learned behavioral response on a paired CS-US trial. The CR could occur anywhere between the times of CS onset and US onset, i.e., during the 500 msec ISI. Here, the CR for trace conditioning usually occurs during the trace interval when no stimulus is present. The definition of a CR is the same as with delay conditioning (see below). Similar to short-delay conditioning, the UR is the behavior that occurs after the US onset. Here the US period is defined as the 250 msec following the US onset. The trial terminates at this point, or 1000 msec after the onset of the trial (***pts***). Three features are shown in Figure 1.5 that define this trial as a trace classical conditioning trial; the CS onset and offset occur entirely before the US, there is an interval when no stimulus is present, and the US occurs by itself. The key difference between delay and trace is that in delay conditioning the CS and US overlap, whereas in trace the CS and US do not overlap and there is a gap between them, requiring that the subject remember the previous CS in order to make an association with the US.

Our colleague Diana Woodruff-Pak at Temple University has found that the differences in rates and level of conditioning seen when delay and trace paradigms are compared are dependent not only on whether or not the CS and the US overlap, but also that the length of the CS is a critical feature. In Figure 1.6 we show the onset, duration and offset for a long-delay classical conditioning procedure. This paradigm has overlapping CS and US as with the previous delay paradigm, but like the trace paradigm the CS and the CS-US intervals are longer. As in the short-delay and trace procedures, the initial 250 msec period is used to collect information about movements in the baseline time period. The CS has an onset at 250 msec after the begin-

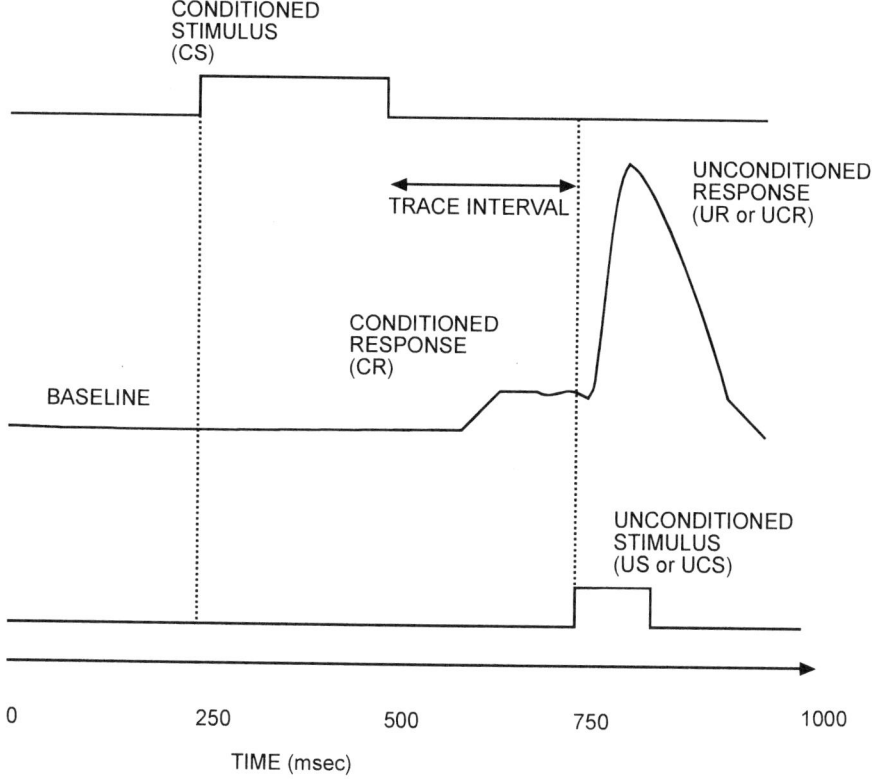

Trace Paradigm

Figure 1.5: Relationship of timing of the stimuli during classical conditioning using the trace conditioning paradigm.

ning of the trial (***toneon***) and lasts for a total of 850 msec, after which it terminates (***toneoff*** at 1100 msec from the beginning of the trial). The interstimulus interval between the CS onset and the US onset is 750 msec. The US begins 750 msec after the onset of the CS (***shkon*** at 1000 msec after the beginning of the trial) and lasts for 100 msec, terminating at 1100 msec from the beginning of the trial (***shkoff***). The US period lasts from US onset at 1000 msec after the beginning of the trial until the end of the trial 250 msec later (***pts*** at 1250 msec from the beginning of the trial). Accompanying the timing diagrams is a cartoon illustrating the relationship of a learned behavioral response on a paired CS-US trial. The CR could occur anywhere between the CS onset and the US onset (i.e., during the 750 msec ISI). The

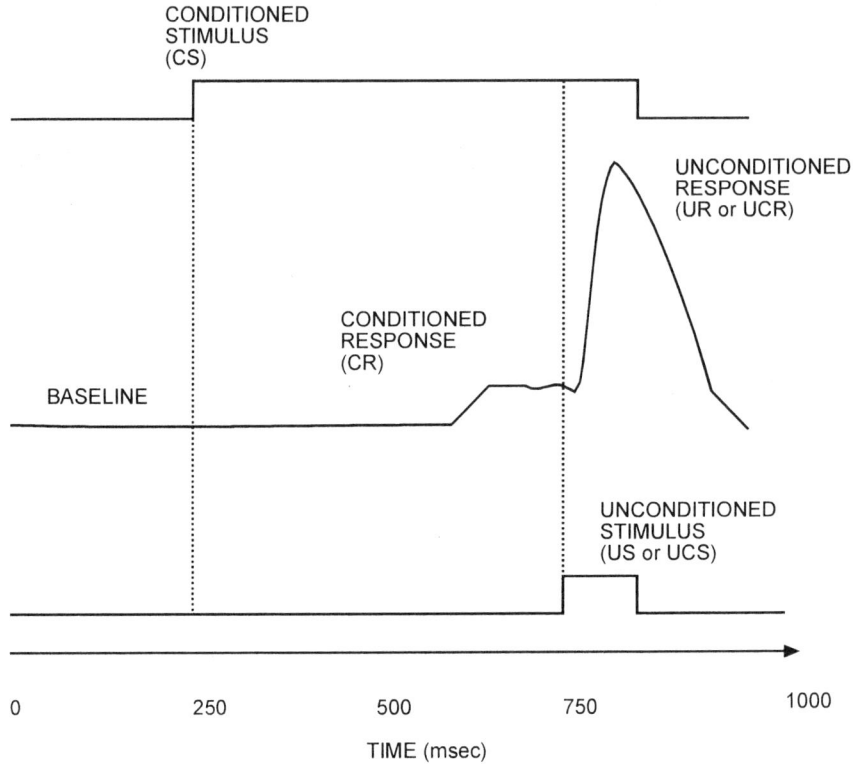

Long Delay Paradigm

Figure 1.6: Relationship of timing of the stimuli during classical conditioning using the long delay paradigm.

definition of a CR is the same as in short-delay and trace conditioning (see below). The UR is the behavior that occurs after the US onset. As with delay and trace conditioning, the US period for the long delay paradigm is defined as the 250 msec following the US onset. The characteristic features of long delay classical conditioning depicted in Figure 1.6 are: the CS onset precedes US onset by a relatively long interval (i.e., > 750 msec), the US overlaps with the CS in the later period, and both the CS and US terminate together.

The Development of Conditioned Responses: A Predictable and Systematic Process

It is instructive to briefly review the normal progression of the developing CR as training proceeds from the state where the subject knows nothing about the CS-US relationship, through the point where the subject shows evidence of developing an association between the CS and the US, to the point where the subject has learned the association very well. To make this more concrete, we need to describe and comment on how training proceeds during a typical classical eyeblink conditioning experiment.

Rabbits take an average of 130 trials before learning the association. Rats take about the same number of trials while humans take 25-50 trials to learn the association. Conditioning humans requires some special considerations. Human subjects easily get bored and sleepy during eyeblink conditioning so background tasks are often necessary to keep them awake and alert. These background tasks can affect the rate of conditioning (e.g., Papka, Ivry, & Woodruff-Pak, 1995; Tracy, Ghose, Stecher, McFall & Steinmetz, 1999) and it appears that the manner and rate of learning of some classical conditioning procedures for human subjects may depend on whether or not they can articulate the relationship between the CS and US (e.g., Clark & Squire, 1998, 2000). These factors have to be considered when describing the rate of learning in human subjects.

For rabbits and rats, we typically give a total of 100-120 trials (*#trls*) per daily training session. The interval between trials, the intertrial interval (ITI), typically averages 30 seconds (20-40 second range, pseudorandomly distributed; *range* and *minimum*), making the training session (or "run") last a little less than one hour. Investigators who use longer ITIs typically present fewer trials to avoid longer sessions of training. We now describe here how a typical training session, composed of 120 trials, is constructed for subsequent data analysis and computations.

The 120 trials that might be given in a session are usually subdivided into 12 blocks (*#blks*), each containing 10 trials per block (*#t/blk*). In a typical experiment in the Steinmetz laboratory, for example, each block consists of the following sequence: one tone/CS alone test trial (*tone trial* or *t&*), four tone-air puff/CS-US paired trials (*tone-air trial* or *t&a*), one air/US alone test trial (*air trial* or *a&*), and four tone-air puff/CS-US paired trials (*tone-air trial* or *t&a*). This sequencing yields a total of 12 tone/CS alone test trials, 12 air puff/US alone test trials, and 96 tone-air puff/CS-US paired trials per session (run), making up the total of 120 trials. Training typically proceeds over a number of days, which may or may not be consecutive. Training for five days might involve training once each day Monday through

Friday. When the authors were graduate students and postdocs, there seemed to be an obsession with making sure that training was given for seven days a week. It is our experience, however, that little harm seems to occur by allowing breaks for weekends and holidays. Training usually proceeds until a preset learning criterion is reached. We discuss this issue later in this chapter.

Figure 1.7 shows a number of hypothetical trials at successive points during classical conditioning for the delay paradigm. The hypothetical examples here represent individual trials on which nictitating membrane/eyeblink responses to tone CS and air puff US were measured during the classical conditioning of rabbits. At the beginning of training the subject shows no response during the CS-US interstimulus interval but does show a reflexive response (UR) to the US. Often the initial size of the UR is small as in this example. In some instances, the rabbit makes no response whatsoever in the first few trials. As training continues, the UR gets "better," i.e., more consistent and larger, as the rabbit develops pairing specific reflex facilitation. The reflex facilitation phenomenon is described by Don Weisz and other researchers (e.g., Weisz, Harden, & Xiang, 1992). In reflex facilitation, the UR is facilitated by the presence of the CS when compared with the size of the UR as measured on UR alone test trials.

When learning initially takes place it is first apparent on CS alone test trials. The initial learned response is typically small and occurs during the period when the US would have occurred if a paired trial had been delivered. With continued paired training, the CR onset begins to occur before US onset and therefore the CR becomes apparent on paired CS-US trials. For this reason, when scoring CRs, one needs to take into consideration the type of trial that is being delivered (*types*) as well as when in training one is looking. Traditionally, the entire post-CS onset period is examined for CRs on CS-alone trials while on paired CS-US trials, only the interval between CS and US onset is examined for CRs. Figure 1.7 provides a summary of how the CR typically developed on paired trials. As training progresses even further, the CR latency gets shorter, moving the CR closer and closer to CS onset. As this happens the trailing CR gets progressively larger.

On paired trials the first conditioned responses are small and occur just before the onset of the US. As training continues, the CR latency gets progressively shorter, and the CR moves closer to the onset of the CS. In very well trained rabbits, the CR latency can get as short as 80 msec; an average onset latency in well-trained rabbits is 130 msec. At the same time, the size of the CR gets larger. Initially, there are two peaks to the eyeblink, one for the CR and one for the UR. As training continues, the two responses tend to merge such that the CR peaks at the time of US onset. The size of the response may approach the limit of our calibrated system (25.5 mm);

however, some rabbits never develop CRs larger than 5 mm or so. Theoretically, the learned response (CR) has an adaptive function in protecting the eye from the US. Studies have shown, however, that aiming the air hose at the temporal (posterior) eye, using clips to hold open the eye, or using shock as the US, makes it difficult if not impossible for the subject to avoid the air puff with an eyeblink. Nevertheless, the subjects still learn. Thus CRs will form even though the learned response is not adaptive for avoiding the US.

Although we discuss the criteria for learning next, it is worth mentioning here that most rabbits (and rats) show very consistent responding. A single CR does not make learning—it could have been an accidental, spontaneous response that occurred during the CS-US interval. The criteria for learning include the idea that there must be some consistent evidence of CRs. It is common for rabbits to show CRs on greater than 96% of the trials, some animals achieving 100% consistently from day to day. Conversely, it is sometimes the case that a high responding rabbit will, all of a sudden, show a poor or even no CR on an occasional trial. Often, but not always, this dropped CR can be attributed to the rabbit's distraction or struggling. Occasionally with very young rabbits they seem to get bored and fall asleep (not unlike a lot of undergraduates we have classically conditioned!).

Although not an issue in rabbit eyeblink conditioning studies, other species, including rats and humans, show reflexive eyeblink responses to the presentation of the CS, especially when tone or light CSs are used. These responses have been called "alpha" responses, and are not associative in nature. Continued presentation of the CS sometimes reduces the amplitude of these alpha responses and some researchers have included CS-alone presentations before paired training to eliminate these responses. If this technique is used, one should be aware that latent inhibition effects can be produced by pre-exposing the subjects to the CS. The existence of alpha responses shortens the potential observation window in the CS-US interval that is available for scoring CRs because the initial 50–80 msec, when the alpha responses typically occur, must be ignored when scoring for CRs.

Measuring the Behaviors that Occur During Eyeblink Conditioning

Behavioral measurements are usually made during three periods; the PreCS (baseline) period, the CS period, and the US period as illustrated in Figure 1.8 (*measure*). In our example of short-delay conditioning, the preCS period lasts from the beginning of the trial to the 250 msec mark when the CS is turned on, the CS period lasts from CS onset until US onset, and the US period lasts from US onset until the end of the trial. On CS alone test trials the scan for learned responses traditionally begins at CS onset and lasts until the end of the trial, thus including both the CS and US periods (as

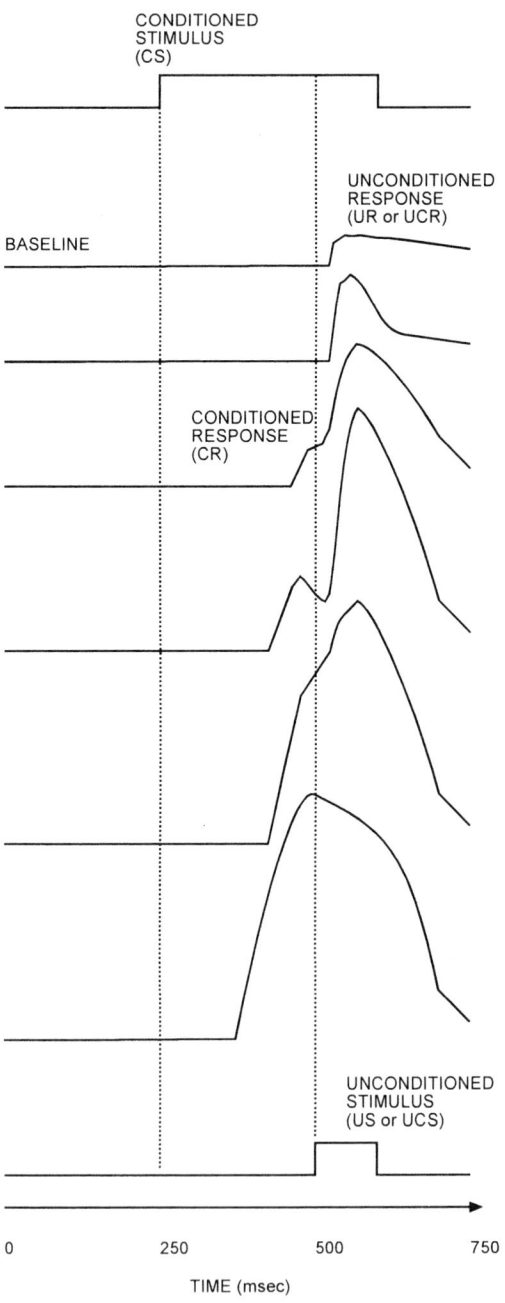

Progression of Learning

defined on paired CS-US trials). Thus, the CR scoring period for CS-alone trials is typically twice as long as the scoring period for paired trials, and this factor should be taken into consideration when CS-alone and paired trials are compared directly.

During the preCS period the average eyelid position is calculated and stored (***bln***). This measure is used as a baseline for determining when and how much the rabbit makes learned and reflexive movements in the CS and US periods. It is also used to determine whether the rabbit is making spontaneous movements that would either interfere with an accurate baseline measurement or cause the calculation of false CRs.

The CS period is scanned from CS onset to US onset (the ISI) for the first instance of a learned response (***onset***). Since the time when behavioral responses were recorded on polygraph paper, the definition of a CR has been set as the first time there is a movement in the CS-US interval that is equal to or greater than 0.5 mm (***critlevel***) above the calculated pre-CS baseline value for the position of the eyelid. The 0.5 mm criterion was actually empirically established; it was the smallest incremental change from baseline that Gormezano and his colleagues could detect using the polygraph recording devices available at the time when the rabbit eyeblink conditioning preparation was developed. This standard CR criterion has endured for nearly 40 years, however, recently some investigators have started to use other criteria for establishing precisely when a CR is detectable (e.g., a change in eyelid position that is two standard deviations or greater than the baseline level). Note that these techniques are somewhat arbitrary. As long as a criterion is properly reported and consistently applied, however, comparisons across-studies are still possible.

The time corresponding to the position during the trials period where the beginning of the CR is found is defined as the CR onset latency, that is, the time difference between CS onset and when the criterion level was met. In some laboratories the definition of the latency of the CR involves determining when the criterion level has been met, then moving backwards towards the CS onset until the response reaches the baseline level, and using that corresponding time as the latency. The choice of which method to use is one of those decisions that is at the discretion of the investigator. What is most important is that the criterion be stated and that it is consistently made. If there are no CRs then there is no latency to the CR (and the latency is typically set to the maximum value of the trial, which is the difference between the end of the trial and CS onset). The CS period is also scanned for the largest deviation for eyelid closure (as opposed to opening) from base-

Figure 1.7 (opposite page): Hypothetical development of learned responses during classical conditioning using a delay paradigm.

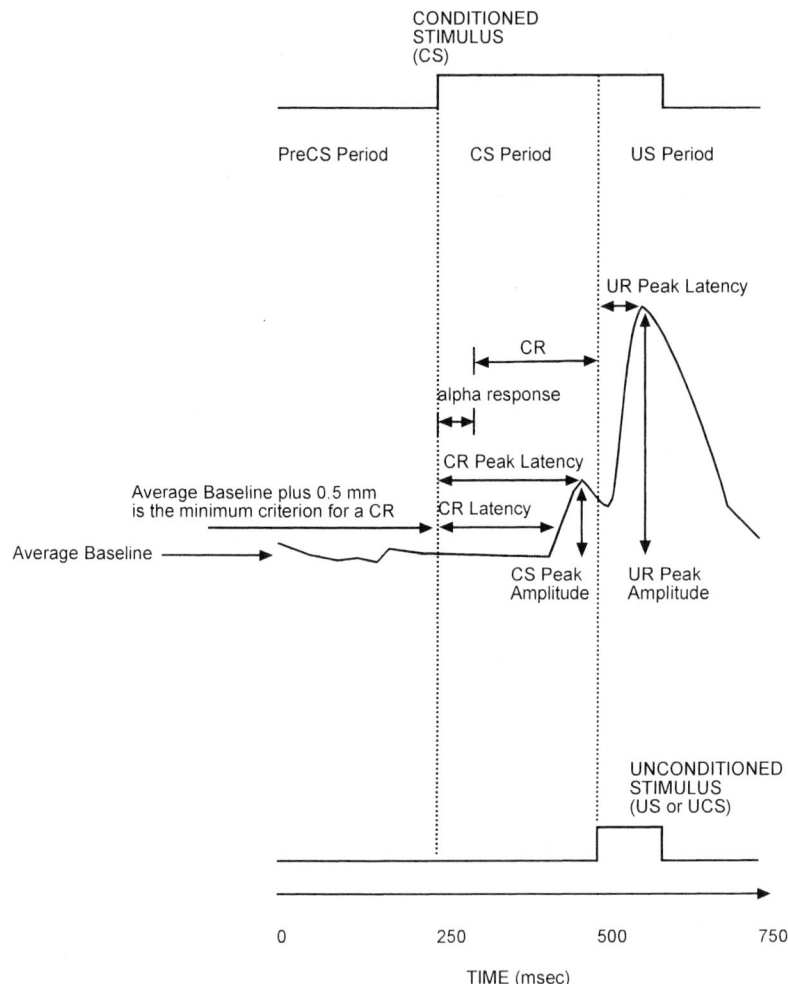

Behavioral Measurements

Figure 1.8: Definitions of the behavioral measurements during PreCS, CS and US periods for delay classical conditioning. Note that the CS and US are both present at the beginning of the US period, and that there are no stimuli present during the latter part of the US period. The latency and peak amplitudes during the CS and US periods hold the greatest interest.

line (*cs-meas*). This value is the peak amplitude, and the corresponding time is the peak latency (from CS onset). Note that the latency is the *first time* that the maximum value has been reached. Thus, if the the maximum value has been reached and remains there (or the eye opens and returns to the maximum value) then it is the first value that is reported as the latency. If there is a CR then the peak amplitude is at least 0.5 mm (*critlevel*) and the latency is at least the value of the latency for a CR. As originally specified, if there is no CR but some movement (below *critlevel*) then a zero will be reported. It is assumed that any movement less than criterion level represents noise in the recording system. An argument could be made for including the values between 0 and the criterion level. If there is no response, then traditionally the peak amplitude is set to zero and the peak latency is set to the maximum of the CS-US interstimulus interval (here 250 msec). These three measures—latency to the learned response, amplitude of the peak response, and latency to the peak response—constitute the data normally collected for the CS period.

The statistics that describe characteristics of the CR deserve further discussion. As originally conceived, trials on which CRs are detected are averaged for each block of trials and for the whole training session by averaging with all the trials (CR and nonCR trials). As a measure of the average activity in the CS period this is fine. As a measure of the average CR *when it occurs* this method is biased because it averages in nonCR trials as zero with CR trials of at least 0.5 mm. In our summary programs, we created the function *christy* (for Christy Logan who requested this function) that calculates CRs by averaging only trials that show a CR. Similarly, the summary program function *dragana* (for Dragana Ivkovich who requested this function) only averages trials on which either a CR or a UR is detected. The creation of special functions that guide data analyses illustrates a very important point: Sometimes the specific experiment that is being conducted (or at least the preferences of the experimenter) dictates the specific manner in which data are collected, summarized, and analyzed. Control programs used to deliver stimuli and record and summarize data should be maximally flexible to accommodate these individual preferences.

Responses that occur at the beginning of the CS period may not be conditioned responses (see Figures 1.8 and 1.9). Responses that begin during the pre-CS period and spill over into the CS period, for example, are treated as *bad trials* (see below). Responses that occur after CS onset but are faster than a reflexive response (i.e., less than about 25 msec, which is the time that it takes to react to a puff of air to the eye) also are considered to be bad trials. Reflexive eyeblinks to the onset of the CS, for example blinking to a light, are not considered to be bad trials but neither are they considered to be CRs, as they occur without any training. The criterion for these 'alpha responses' (as opposed to 'beta responses' with longer latencies which are considered to be valid CRs) could exclude any response whose latency is less than, say, 80 msec after CS onset. The onset of well trained CRs (beta responses) average about 135 msec when a 250 ISI is used, with the earliest response onsets observed at no less than 80 msec. An argument could be made for using a 100 msec cutoff point, as these response latencies are also rare, if the goal is to exclude any possible alpha responses.

The US period is scanned for the onset latency (*onset*) and peak amplitude of the UR and their corresponding latencies (*ucs-meas*). As with the peak amplitude and latency in the CS period, the values reported for the US period are the first time that the maximum has been reached. If the maximum value remains or is repeated (the eyelid opens then closes again to the same maximum value) then it is the first value that is reported as the latency.

We used to calculate the "(positive) area under the curve" of responses recorded in the CS and US periods as a measure of CR acquisition and performance (i.e., a CR magnitude measure). We have rarely used this measure, however, because we have discovered that there is a problem in using this computation when comparing different studies that employed different bin widths when collecting the behavioral data. Bin widths must be judiciously chosen when using this CR magnitude computation. To use this measure, one should not simply add up the response amplitudes for each bin (as we originally did). Rather, one must weight the bins by the bin width before adding the values together.

Establishing and Measuring Learning Criteria

The detection of a movement that could be considered to be a CR is *not* evidence that learning has actually taken place. Although the number of spontaneous eyeblinks typically seen in rabbits is fairly low, these spontaneous eyeblinks could appear accidentally in the CS period and these would be counted as CRs. Additional criteria are needed as evidence of learning. We normally use two criteria for training, one a learning criterion and the other a performance criterion.

In the eyeblink conditioning literature, learning (***criterion***) is often defined as the first trial on which a the rabbit makes eight CRs on nine consecutive trials (***outof***), a rate of 91.9% responding. We do not know the exact reason or precise origin of this criterion as opposed to, say, using nine CRs in ten consecutive trials, a rate of 90% responding, just as we do not know why nine rather than ten trials per block were often used in many eyeblink conditioning studies published in the 1970s and 1980s (we suspect this had something to do with the rather limited computer capabilities available

Chapter 1

at the time). Suffice it to say that some experimental procedures survive as seemingly arbitrary creatures of historical usage. What is important here is the idea that there is evidence for some consistency in responding before one declares that learning has taken place.

There are several features about using a learning criterion that need to be raised. First, the criterion trials must all be trials on which a CR could occur, such as CS alone test trials and paired CS-US trials, but not on US alone test trials where there is no possibility of a CR—the subjects would have to be psychic to reach a learning criterion that was based solely on US alone trials. Second, bad trials (see next section) are also not counted in the criterion calculation. Third, the missed trial in the eight out of nine criterion could occur at any point in the string of CRs, including the very beginning (implied) or the end (actual) of the string. This has consequences for the fourth point, which is that it is not generally well understood that learning is deemed to have taken place at the point where consistent responding begins—the animal learns before the point where the eight out of nine CRs occurs—and the criterion trials themselves are actually evidence of performance, not learning. For this reason, the performance trials should be subtracted. Another way of thinking about it is that the animal learned on the last trial *before* the performance trials (even if the last trial was not a CR). This feature leads to some interesting calculations. For example, in acquisition training, if the first eight trials of a session are CRs then the rabbit actually learned on the last trial of the previous day's training. If this is retention, then eight CRs in a row at the very beginning of testing demonstrates that there is perfect retention. Since the criterion is subtracted from the performance, with eight CRs in a row this results in a calculation of -1, meaning the previous day (in acquisition) or perfect remembrance (in retention). The negative and zero numbers that can be calculated this way appear to bother some researchers, so they report the last trial used to reach the performance criterion as the trials to criterion. In fact, the former method is the correct one, but in practice the difference of nine trials between the two methods does not amount to much. Again, it is much more important that the method be stated and consistently applied.

A performance criterion can be required once learning has taken place. This may take many forms. The eight consecutive CRs are a performance measure. More commonly, attaining a daily performance measure is required, for example, "x percent consistent responding." Many rabbits perform at perfect (100%) or near perfect (96-100%) daily performance. For difficult lesion studies we have used 70% daily performance. Informal observations have shown that, on average, the learning criterion is met on days with 30-40% daily performance. This makes us suspicious when we have seen studies report "learning" when criterion (eight CRs on nine consecutive trials) has not been met and the daily performance is as low as 20%.

Instead of trying to attain a particular level of performance one typical scenario is to train rabbits until they reach criterion, finish training that day, and give one more full day of training. Lavond's favorite protocol is to train for five days regardless of the performance criterion, provided the animals at least reach the learning criterion, and Steinmetz prefers to train for 8-10 days. Most rabbits learn in about 130 trials, or two days of training, so five days results in well-trained animals. We train rabbits for specific numbers of sessions for data analysis reasons as much as for any other reason; we have comparable amounts of data on all the subjects and do not have to resort to choosing between "Vincentizing" the data (Vincent, 1912) or only using a subset of the data. The decision to run each animal an equal number of sessions or to run each animal to a specific learning criterion depends on the hypotheses of the study being conducted. If your hypothesis testing is best served by making sure that all subjects are given equal number of training trials or equal amount of exposure to the conditioning stimuli, then it would be best to run all subjects an equal number of sessions. If your hypothesis testing is best served by ensuring that all subjects have learned the response to an equal degree, then it would be best to run all subjects to a preset learning (performance) criteria. Given that subjects learn at different rates, this may mean that a variety of different numbers of trials and sessions may be delivered to subjects in the experiment.

For simple short-delay conditioning, animals that do not learn (reach criterion) in five to ten days might be excluded from the study. Other procedures, such as long-delay, trace, or higher order conditioning may require more training. There are many extraneous reasons why a subject may not learn. There might be equipment breakdown. The person training the subject may be new and inexperienced. The personality of the subject may not be conducive to learning (e.g., if it is too anxious or aggressive). Sometimes "batches" of rabbits or rats are delivered and none of them seem to learn, suggesting perhaps a common heritable conditioning or a common ailment. If the study involves surgery around the interpositus nucleus of the cerebellum or another critical brain structure (to implant an electrode or a cannula, for example) then damage may prevent learning. The chore here is to balance out the desires to continue training, to avoid dropping subjects and needlessly biasing the data, and to avoid wasting time on animals that will never learn or learn only poorly. This decision to exclude is not taken lightly, for it is usually made after a considerable investment of time that will be wasted.

The Saga of Bad Trials — When do I Discard Individual Trials from Further Analysis?

In classical conditioning, the presentation of the stimuli (CS and US) is independent of the behavior of the animal. Even if the animal is moving around, the tone and the air puff will still be delivered. In an operant situation, however, the animal's behavior alters the contingencies. For example, an avoidance situation can be set up where, if a conditioning subject blinks after the tone onset but before the air onset then the air will not be delivered. In that situation the subject has some control over the presentation of the US, making it an instrumental training situation. But in traditional classical conditioning, the subject has no such control; no matter what the subject does both stimuli will be delivered. Nevertheless, the recorded behavioral data may be very noisy and therefore confusing. To average this behavior with normal trials could potentially obscure the results, principally by altering the baseline and creating false CRs. Recognizing and removing these "bad trials" is an important consideration.

Figure 1.9 illustrates several examples of bad trials and how to detect most of them for exclusion. The criteria for exclusion are illustrated in the bottom example. The first task is to identify instances where the baseline activity is unacceptable. After first calculating the average baseline, responses that exceed the baseline by 0.7 mm (*bada/d*) in either direction (opening or closing the eyelids) in the 160 msec (*baseln*) before CS onset (i.e., not considering the entire preCS baseline period) are excluded from analysis. These criteria allow for some baseline activity, particularly at the beginning of the preCS period, but remove trials where the movement occurs immediately prior to the CS period. Notice that the criterion of 0.7 mm means that activity that is barely a CR (0.5 mm, *critlevel*) is not excluded. Also notice that, unlike the criterion for a conditioned response that must be an eye closure, the bad trial criterion counts movement in either direction as a problem. If neural unit activity is also being monitored, then discriminated counts in the preCS period (*pcs-units*) that exceed some level (typically 50 or 100 counts, *badun*) also may cause the trial to be excluded. The second task is to identify instances where responses occur in the CS period that would count as CRs but which cannot possibly be CRs, which works out to be any CR-like movement (equals or exceeds 0.5 mm, *critlevel*) within the first 25 msec of the CS period (*badtime*). The argument is that it takes 20-25 msec to generate a reflex to the air puff US, so a voluntary response could not possibly occur in that time window. Recall that the shortest latency for a CR is about 80 msec. As a result of this exclusion, movement only really counts as a CR if it occurs between CS onset (*toneon*) plus *badtime* (25

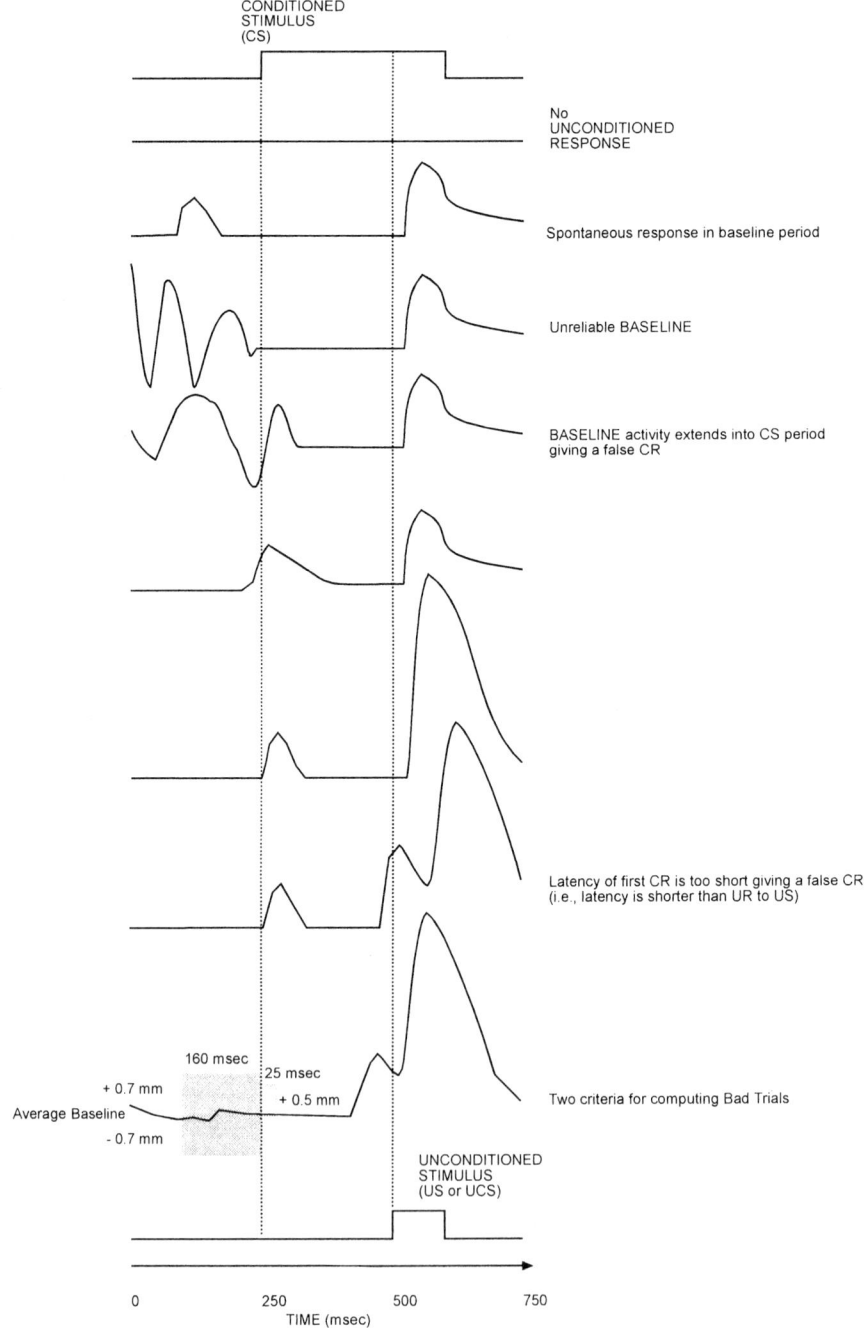

Bad Trials

msec) and US onset (***shkon***). In practice, about 5% of trials, on average, are excluded from analyses by these criteria for bad trials. This number fits well with Gormezano's observations for rabbits (Gormezano, Kehoe & Marshall, 1983). We get nervous when a rabbit's excessive behaviors (i.e., spontaneous eyeblinks or movements in the restraint box) cause a loss of 10% or more of the data. If excessive numbers of bad trials are present, we either doubt the data or conclude that our criteria for exclusion are too stringent. When conditioning rats and humans that have significantly higher spontaneous blink rates, sometimes 10-20% of the trials will be discarded because of eyelid movements during the preCS period.

Look now at the examples of bad trials shown in Figure 1.9. The top example shows a completely flat line. It could be considered a bad trial but would not be detected by the criteria just outlined. This example is a bad trial because there *should* be an unconditioned response to the air puff. It could be that this trial occurred early in training (as mentioned above sometimes there are no reflexes initially), or perhaps the apparatus for measuring the response is disconnected (mechanically or electrically) or is broken. Only experimenter vigilance will detect this problem and this illustrates the importance of the experimenter closely monitoring the subject during the experiment. The job of the experimenter is not to alter the data, but to ensure that good data—meaning believable, reliable, and replicable data—are measured.

The next two examples are cartoons showing activity in the baseline period (preCS period), which would significantly affect the calculation of the baseline. It is worth noting, however, that sometimes spontaneous blinks can be captured in the baseline period, and this may affect interpretation of subsequent data. For example, we would be less worried about the quality of the data if a subject that does not show any CRs and has very poor URs throws a great spontaneous response in the baseline period, indicating that the apparatus is okay and the animal can make good responses.

The next two examples also show abnormal baseline activity. This activity bleeds into the CS period and therefore could be mistaken for CRs. These are excluded from analysis for two reasons: eyelid movement in the baseline period was relatively high and a short latency response was found in the CS period. The final two examples do not show any abnormal activity in the preCS period but the response latencies in the CS period are too short to be voluntary or learned responses. Notice that in one instance a trial is excluded because eyelid movement was detected only within the first 25 msec. The second trial is excluded because of a short-latency response even though

Figure 1.9 (opposite page): Examples of behavioral responses that could be considered to be 'bad trials' that the experimenter might consider excluding from analysis.

it also appears to have a valid CR just before US onset. Both trials would be excluded from analysis. Of course, one could create more sophisticated criteria for exclusion and these criteria could attempt to make finer discriminations between CR trials and nonCR movement trials. Nevertheless, we typically use just the criteria for bad trials we described here. An excessive number of exclusions are not usually seen using our criteria (remember there is about 5% exclusion on average), the computation is simple (which means that it does not take a lot of time to calculate), the criteria are consistently applied by the computer, and the criteria are arbitrary in the sense that they do not bias the data by arbitrarily including or excluding CRs, as a human observer might.

SUMMARY

In this chapter, we provided some historical background concerning the classical conditioning paradigm and also described the language of classical conditioning, as well as several variations of the basic Pavlovian procedure. We will use the language of classical conditioning throughout the book, and use variations of the conditioning procedure to illustrate methodological points that should be of interest to a variety of behavioral neuroscientists, whether or not they use classical conditioning procedures in their own laboratories. Readers should therefore become familiar with basic classical conditioning terminology to maximally benefit from this book. We also laid out the important technical and methodological points that must be considered when setting up a classical conditioning experiment (e.g., stimulus delivery, response definition, trial inclusion or exclusion criteria). We will build on this information in subsequent chapters. As a beginning, we turn to an in-depth presentation of stimulus delivery in classical conditioning experiments in the next chapter.

Chapter 2

THE DELIVERY OF STIMULI

INTRODUCTION

Many behavioral neuroscience experiments require that stimuli be presented in a highly controlled and precise manner. The proper delivery of stimuli presents a variety of challenges for the experimenter. For example, the delivery of stimuli must often be precisely timed. Also, it is often desirable or necessary to manipulate various parameters of the stimuli including stimulus intensity, duration, and frequency. And, it is important that the backgrounds against which the stimuli are presented are held constant so that the relationship between the saliencies of the stimuli and the backgrounds are relatively constant. In this chapter, we use classical conditioning examples to discuss the important features and methods associated with stimulus control. Indeed, classical conditioning experiments provide excellent examples for discussing the variety of important issues that relate to the presentation of stimuli in an experiment.

As we detailed in Chapter 1, two basic stimuli are used in classical conditioning, the conditioned stimulus (CS) and the unconditioned stimulus (US). For most classical conditioning experiments, the CS is chosen because it is thought to be neutral with respect to the reflex that is being conditioned. That is, the presentation of the CS typically does not produce a reflexive movement prior to training. There are differences, however, in the relative "neutrality" of the CS. In classical eyeblink conditioning, for example, tones, lights, or tactile stimulation are typically used as CSs. When presented to rabbits, these CSs rarely produce eyelid movements. However, when these CSs are presented to humans or rats, a clear orienting or startle response to the stimuli can sometimes be seen. Also, in reduced and invertebrate preparations (such as classical spinal conditioning and the classical conditioning of *Aplysia* and *Hermissenda*), a clear response to the CS, often called an "alpha response," is present prior to conditioning. Although the term is rarely used now, a "beta response" is what we normally call the conditioned response — a response that develops as a consequence of training.

In all preparations, however, the US is chosen because it reliably produces the reflexive movement that is to be conditioned. Typically mild air puffs and mild electrical shocks are used as USs for classical eyeblink conditioning.

Regardless of which specific CS or US is chosen for a classical conditioning experiment, it is very important that the delivery of these stimuli be reliable and constant—the presentation of the stimuli should be accomplished using techniques that ensure that it is received by the subject similarly on each trial. In this chapter, we review the techniques we use to deliver the stimuli used in classical conditioning experiments. We concentrate in this chapter on peripheral equipment and circuits that we have found useful for generating and delivering experimental stimuli. Some aspects of the delivery of stimuli are under control of computer and timing equipment. These aspects of stimulus control and presentation will be covered in Chapter 11.

THE CONDITIONED STIMULUS (CS)

Auditory Stimuli

For a variety of reasons, auditory stimuli are used more often than other types of stimuli as signals that convey when events will occur in behavioral experiments. This may be due to several factors. For example, auditory stimuli are relatively easy to deliver and are easy to manipulate from a parametric standpoint (e.g., intensity, frequency, duration, etc.). In classical conditioning, acoustic CSs are employed much more often than other types of CSs and rates of conditioning seem faster when acoustic CSs are used relative to other types of CSs. Intense auditory stimuli are used preferentially over visual stimuli to elicit startle reflexes, and auditory stimuli are second only to visual stimuli for experiments designed to examine the involvement of the brain in sensation, perception, and attention. Also, auditory stimuli are often used in studies of evoked brain activity, such as when either single or multiple unit neuronal activity is assessed or when gross EEG activity is analyzed. We present information here on how to create and deliver auditory stimuli for behavioral experiments.

There are various *types* of auditory stimuli used in behavioral experiments including pure tones (simple sine waves), complex tones (combined sine waves), white noise (all frequencies equally represented within a range), pink noise (all frequencies within a range biased to human hearing), and clicks (short bursts of higher frequency noise). Besides the type of waveform used, it is necessary to describe and define two other parameters that are important for describing an auditory stimulus: the *frequency* (pitch) and the *intensity* (amplitude or volume) of the auditory signal. Electronic circuitry used for delivering an auditory stimulus must take into consideration these three parameters (type, intensity, frequency). In addition, the *timing* of the delivery of the auditory stimulus and its relation to other stimuli or behavior must be considered, where timing involves the determination of the onset and offset of the auditory stimulus and its resulting duration.

The auditory stimulus could be used as a conditioned stimulus for classical conditioning or as an unconditioned stimulus used to elicit startle reflexes. The circuitry described in this section could be used for either purpose. The difference in its application would be primarily in the relative intensity and delivery of the signal. For convenience, in this chapter we will use classical conditioning to illustrate how control of auditory stimuli is achieved. In our example, single frequency tones are the most commonly used neutral stimulus (the CS) for classical conditioning. For our typical eyeblink conditioning experiment, the auditory stimulus is a 1000 Hz, 85 dB SPL tone, where Hz (Hertz) refers to frequency in cycles per second, dB refers to intensity in decibels, and SPL refers to sound pressure level. The controlling circuitry typically delivers a trial consisting of a 350 msec tone that co-terminates with an air puff unconditioned stimulus. Time between trials (the intertrial intervals) typically range from 20–40 sec, averaging 30 seconds for a session. Usually, 100-120 trials are delivered daily, thus requiring the presentation of 100-120 auditory CSs.

Figure 2.1 provides a schematic of the elements that are needed for delivering an auditory stimulus during a behavioral experiment. These elements include a *source* for the auditory stimulus, a *gate* connected to a control source (a controller or computer) that allows the stimulus to be expressed, a power *amplifier*, and a *speaker*. The auditory source is shown here as a sine wave generator for a pure tone (pitch) although it could be a white noise or pink noise source or even a combination of selected pure tones. A control signal coming from the controller or computer determines when the tone is delivered in relation to other stimuli or the collection of behavioral and neural data, by opening an electronic gate that effects delivery of the tone stimulus and shapes the signal. The power amplifier and speaker are adjusted for loudness. A sound pressure meter is normally used for calibrating the desired loudness of the output. An *attenuator* may also be interposed in the system to decrease the sound intensity in easily determined steps. Traditionally, the attenuator is placed between the power amplifier and the speaker, although as indicated in Figure 2.1, it can be located in other places within the circuit.

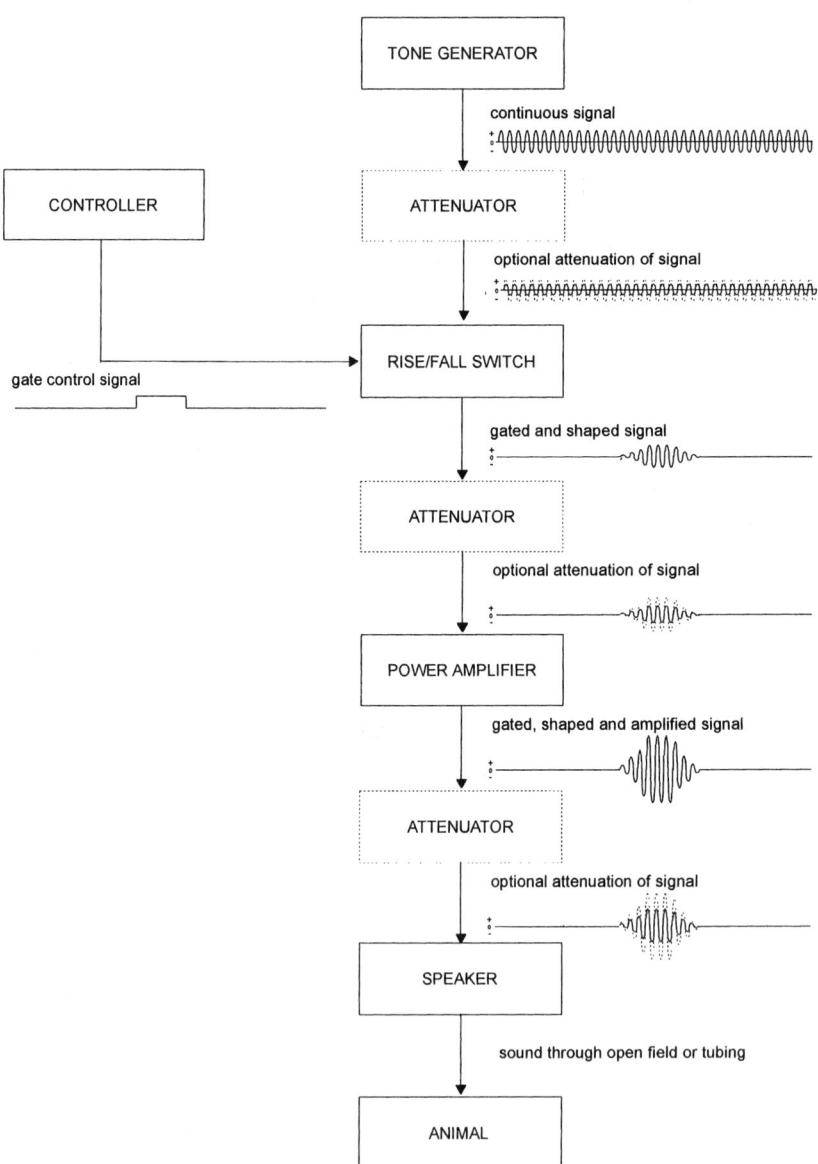

Figure 2.1: Relationship of elements to deliver a tone stimulus.

Chapter 2

The Morse Code Circuit

We first describe in Figure 2.2 a fixed frequency tone circuit that we refer to as the "Morse code circuit" because it was designed for practicing Morse code keying. Do not let the title put you off—this circuit is fully capable of being used in classical conditioning and other experiments that involve presentation of an auditory stimulus. The Morse code circuit is a simple, but complete, circuit that incorporates the vital elements of Figure 2.1, those elements being the auditory source, gate, amplifier and speaker. Important in the construction of any of these auditory controllers is the incorporation of a gate element. The gate element used in the Morse Code Circuit is realized here by an electronic switch, which Ham radio operators have designed to minimize distortion. The electronic switch gradually turns the tone on and off, avoiding the sudden onset and offset of the stimulus, which introduces high frequency noise on the transitions between states (i.e., an audible clicking sound at the beginning and end of the tone). This type of circuit is also known as a rise/fall circuit, which refers to its ability to gradually turn on and off to eliminate the clicking sound. Almost all of the components of the Morse code circuit presented in Figure 2.2 can be purchased from a local Radio Shack or electronic supply store. One critical exception is the 3080 transconductance operational amplifier[1] (which can, however, be purchased from Jameco, Newark, or Allied Electronics distributors). This circuit could be used for classical conditioning experiments with little or no modification. Its major limitation is the relative volume it can deliver, but it is more than adequate when using an earphone or headset or when additional amplification is used. Following is an examination of the individual components of the Morse code circuit.

The Auditory Source

The first stage of Figure 2.2 is a common 741 operational amplifier configured as a twin-T oscillator design (this is a common operational amplifier feedback design; see Oleksy, 1981, for example). The frequency output of this circuit is determined by the selection of resistors[2] and capacitors[3]. Halving the values of the resistors or capacitors doubles the frequency. Fine-tuned adjustments of the frequency (rate) and distortion of the sine wave is made by adjusting a 20 kΩ potentiometer[4]. Here the 741 operational amplifier is set to 1000 Hz. The amplitude of the signal is close to the maximum allowed by the 741 operational amplifier and its supply voltage.

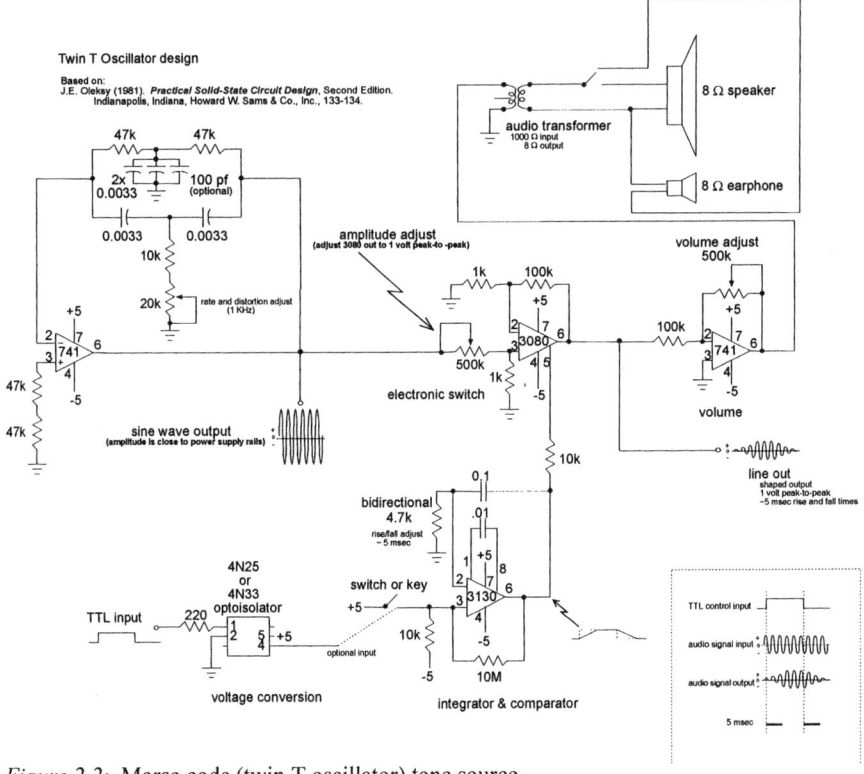

Figure 2.2: Morse code (twin T oscillator) tone source.

The Rise/Fall Switch

The key to eliminating the audible click that can be heard when a tone is abruptly delivered is finding a way to systematically vary the size of the signal in a manner that gradually turns it on and off. This gradual transition avoids creating large voltage transitions that create the clicking noise at the beginning and end of the auditory stimulus. We can envision experiments where this feature would not be critical, where one might not care if this clicking noise artifact is present or not. But if one wants to deliver a specific frequency of tone and to have the subject react to that particular frequency and not another frequency (as in discrimination training or in evoked potential experiments) then this issue becomes extremely important.

Ham radio enthusiasts have created various rise/fall circuits for applying Morse code to the airwaves without distortion or transition artifacts. In a simple form, a rise/fall switch can be built by adding an integrator (a resistor and capacitor), to the base of a transistor[5], to vary the working current. To achieve conductances for both positive and negative signals, we used two transistors in parallel, an NPN and a PNP transistor. This design was an integral part of a 6502 microcontroller[6] that we originally designed to control delivery of stimuli for classical conditioning experiments (see Chapter 11). Previous to the use of this 6502 controller, a ROM-based controller[7] was employed and this system used a 3900 current mirror in the circuit. The 3080 transconductance operational amplifier in the design presented here allows for varying the output proportional to the amount of current on the control input (which is pin 5 on the amplifier). Here we present a version of a rise/fall gate using the 3080.

The 3080 inputs can only handle small voltages, therefore the incoming signal from the auditory source from the twin-T oscillator must be reduced with a voltage divider (the 500 kΩ variable resistor and 1 kΩ resistors cause the signal to be reduced down to 1/500th of the original size). This attenuation is later compensated by a gain stage (the 500 kΩ variable resistor for volume adjust). One can think of the 3080 as acting like a very large resistor when it is turned off. When a small current (2 mA maximum) is applied to pin 5, however, the 3080 acts as if it was a very small resistor. The amount of apparent resistance is proportional to the amount of current on pin 5. In Figure 2.2 the current delivered to pin 5 is determined by the combined comparator and integrator configuration of the 3130 operational amplifier. Pressing the switch (or Morse code keyer) or using the TTL interface[8] with an optoisolator[9], causes the 3130 operational amplifier to switch states (the comparator function) with the filtering characteristics determined by the 0.1 μF capacitor and the 4.7 kΩ resistor. This configuration results in about a 5 msec rise or fall to fully turn on or off the tone (a 5 msec bi-directional control seems to work quite nicely in removing audible clicks on tone presentation). Because this is an integrator design, the rate of the rise and fall are not linear, but rather are exponential.

The rise/fall switch gradually turns on and off the tone when commanded by an on/off signal coming from a controller (or computer). While the purpose of gradually changing the output is to prevent sudden increases or decreases of the tone to the speaker, which causes sudden changes that contain high frequency components resulting in a clicking or popping sound at the onset and offset of the tone, there is a consequence to using this circuit. The penalty for controlling the on/off transition is a 5 msec delay (i.e., 5 wavelengths for a 1 KHz tone) in reaching the full intensity or zero intensity. For typical classical conditioning experiments, the total duration of the conditioned tone stimulus is 350 to 500 msec, so a 5 msec delay at the beginning and end is not much of a penalty for this stimulus control feature.

Transconductance operational amplifiers like the 3080 are very useful for a number of switching designs where the input signal (like the tone) is a varying AC-like signal. They could be used, for example, to switch on and off the input to a recording amplifier when collecting data, something one might want to do when a stimulus artifact would swamp the amplifier. Later we will see how transconductance operational amplifiers (e.g., the quad 13600) are used to create a variable filter. This design is an improvement over filters that have fixed cutoff values or that allow selection among a few cutoff values.

A Note on Calibrating the Auditory Signal to 1 V Peak-to-Peak

For the purposes of calibrating signals, often a 1 V peak-to-peak signal (i.e., 1 KHz sine wave) is put into an element like the rise/fall switch and the switch's output is adjusted to generate 1 volt peak-to-peak output when the switch is fully on. In the Morse code circuit shown in Figure 2.2, the input signal is actually a function of the 741 operational amplifier and its power supply, rather than being 1 V peak-to-peak. However, there is nothing to prevent a user from calibrating the output of the 3080 transconductance operational amplifier by making adjustments using the 500 kΩ potentiometer that forms a voltage divider with the 1 kΩ fixed resistor at pin 3 of the 3080 (i.e., the noninverting input pin). That way, if one taps into the output of the 3080 for some other purpose, a known signal level is available. The "line out" of this electronic equipment is a 1 V peak-to-peak industry standard, which means that the line out is compatible with "line in" of commercially available equipment. This feature, for example, allows one to record Morse code practice directly onto a standard tape recorder. More likely in the present context, tones of various durations and patterns of on/off (but having the same intensity) could be recorded to tape for playing back during an experiment.

Amplification and Speaker/Earphone

Achieving gain to produce volume amplification is also very important in achieving control of an auditory stimulus used in behavioral experiments. Following the rise/fall stage in Figure 2.2 is a gain stage that uses a 741 operational amplifier. This amplification could serve two purposes: it could be used to compensate for the attenuation required to use the 3080 operational amplifier, but more likely, it would be used to adjust the volume to the speaker. Here the volume adjust circuit uses a 500 kΩ feedback resistor with a 100 kΩ input resistor, meaning that the maximum gain is 5 times (i.e., 500 kΩ divided by 100 kΩ), which is more than enough amplification for this application. We have used a linear taper 500 kΩ resistor because it was easy to find. An audio taper would be a better choice, but this component can be relatively difficult to find as a trim resistor (Radio Shack, for example, has audio taper potentiometers). The major reason for preferring an audio taper resistor is that this type of resistor appears to increase the loudness proportionally to the amount of turning of the potentiometer; this makes for better human engineering.

The 741 operational amplifier is not a particularly good choice as a power amplifier to drive the speaker. Because of its inherent output resistance the 741 operational amplifier does not deliver enough current to induce the magnetic field necessary to drive the speaker diaphragm. Normally, an audio amplifier would be used instead. Radio Shack and other electronic distributors usually sell amplifier chips that can be easily used, but these usually require the trouble of adding a heat sink. Alternatively, the problem with the 741 operational amplifier can be thought of as an impedance mismatch. In Figure 2.2, we use an audio transformer (1000 Ω to 8 Ω) to overcome this interfacing problem. In the figure, either an 8 Ω speaker can be used for a free field tone or an 8 Ω earphone can be used as a way to restrict the tone to one ear. (It should be noted that unless the earphone is somehow sealed inside the external ear canal, the sound of the tone may be detected by the other ear. This could be problematic for experiments that require the presentation of auditory stimuli to one ear).

Another way to create a confined tone source would be to locate the speaker outside of the conditioning chamber, surrounded by insulation, with a funnel placed over the speaker and tubing attached to the spigot of the funnel. The tubing travels into the conditioning chamber and ends near the ear. Tapani Korhonen at the University of Jyvaskyla, Finland, uses such a system. The advantage of this system is that tone artifacts caused by electromagnetic radiation that are conducted through the wires and speaker are effectively eliminated in neural recordings. We have incorporated this design in some of our experiments. The only real problem we have encountered is being able to sufficiently insulate the speaker so that the whole lab is not disturbed by the auditory stimuli.

A Note on Calibrating the Sound Intensity

The intensity of the stimulus experienced by the subject must be calibrated. If you have colleagues who conduct auditory experiments, sometimes it may be possible to borrow a relatively expensive B&K sound pressure meter for very precise measurements. Otherwise, we have found that a relatively inexpensive analog sound meter, such as those sold by Radio Shack for home stereo enthusiasts, works reasonably well. During calibration, we use the "slow" settings. The "A" weighting is used for guessing the optimal level heard by humans and would be good for calibrating music (this setting takes into consideration the frequency range and bias of human hearing). The "C" weighting is used for calibrating equipment and should probably be used for calibrating the equipment described here.

A critical feature in calibration is the distance of the meter from the sound source. Like radiation, sound falls off in proportion to the square of the distance from the source. Calculations are not necessary, however: For "free field" auditory presentations (i.e., when using a speaker in a confined space) just place the meter where the ear or eardrum would be relative to the source, and adjust the amplification until the desired readings of decibels are achieved. For most of our work, the speaker is placed 10 cm in front of the subject's ears and the loudness is adjusted to 85 to 92 dB SPL. For confined delivery systems i.e., when using earphones or tubing that is placed inside the external auditory meatus) one has to measure, or at least make a good estimate of, the distance to the ear drum; to be accurate the meter must also mimic (or actually be inside) the confined space for this measurement to be taken. Louder intensities for studying startle reflexes would require a more powerful amplifier and speaker than those described here.

Cost

At the time of publication of this book, the Morse circuit described here cost less than $20 for the circuit board, chips, sockets, capacitors, and resistors. Connectors, switches and a box will add to the price, easily doubling it. The circuit also needs a power supply. Since this project will most likely be made in conjunction with other circuits (for air puff delivery, for unit discrimination, etc.), the box and power supply can be shared, thus cutting expenses.

Summary

The Morse code circuit described here provides a simple yet complete and relatively sophisticated tone source for behavioral neuroscience experiments. The Morse code circuit includes the individual critical elements of a tone delivery system depicted in Figure 2.1; it includes the source, gate, amplifier, attenuator and speaker, and their relations to each other. In the remainder of this chapter we focus on individual elements of Figure 2.1 rather than on a completed circuit.

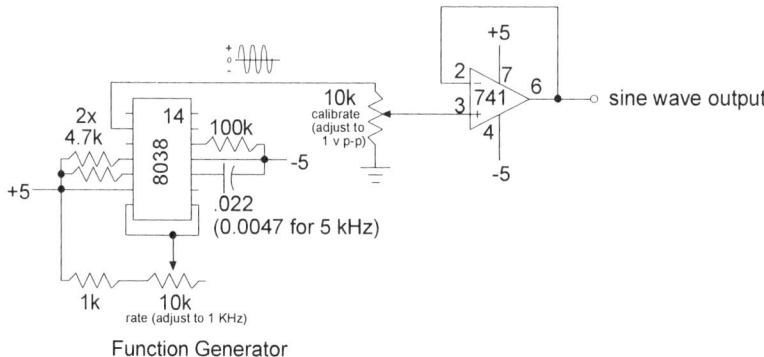

Figure 2.3: Versatile tone source.

A Versatile Tone Circuit

Figure 2.3 contrasts the fixed auditory source used in the previous Morse code circuit with a more general tone generator that we refer to here as the "versatile tone circuit." The heart of the versatile circuit is a specialized chip, a 8038 precision waveform generator[10], for setting the tone frequency. This chip is available from electronic distributors such Jameco, Allied and Newark. The 8038 waveform generator can easily be tuned to the desired frequency with a single variable resistor. Changing a single capacitor of the 8038 can be used to make large changes in the tuned frequency. In contrast, the twin-T design used in the Morse code circuit is not so easily tuned, having multiple resistors and capacitors that would have to be changed simultaneously. The XR2206 is another popular frequency generating chip that we have used.

Tuning Circuit

The versatile nature of the 8038 precision waveform generator allows for a more complex design for the programmable selection of several frequencies. We call this next circuit the "tuning circuit" because it has been used successfully in experiments designed to gather "tuning curve" information concomitantly with neural recordings. A tuning curve simply plots the evoked reactivity of a single nerve cell against a series of auditory frequencies presented as a series of stimuli. Figures 2.4a and b show the tuning circuit that uses a single 8038 with a selection of resistors and capacitors to allow choosing among the several frequencies (500 Hz, 1k, 2K, 3K . . . . . 20K Hz) by entering the selection into the 74154 address decoder[11] that is part of the circuit. The address decoder selects the correct combination of resistors and capacitors by switching relays attached to the 8038. The relays hook up different sets of capacitors and resistors as needed to achieve the desired frequency. In our use, the 74154 address decoder is connected to the first four bits of a port on our computer interface board. The operation of the port in selecting the appropriate relays is described in the text that accompanies Figures 2.4a and b (see also the CD for a discussion of the software control of this device).

In this application, the frequency must be calibrated for each selection that is made. The resistors in this design are variable (15 or 20 multiturn) and individually adjusted to the desired frequency. If this was a commercial product, the axles on the variable resistors would be glued or painted once they were properly adjusted. To adjust the frequency, the sine wave can be displayed on an oscilloscope and/or measured with a frequency meter. For calibration, we prefer to use a frequency meter because it tends to be more discriminating. Either method will require a decision about how close the frequency should be to the ideal. For example, the ideal desired frequency may be 10 kHz. Instead of achieving exactly 10.000 kHz on the meter by adjusting the variable resistor, however, the closest you may be able to achieve might be 10.973 kHz. The point here is that the desired frequency you may end up achieving is +/- 1% (or whatever criterion you can tolerate).

White Noise Circuits

Sometimes a white noise stimulus is desired rather than a pure tone as the auditory stimulus. White noise may be used as a CS for classical conditioning, but often it is used in these experiments as a masking stimulus to mask external noises that may be present during conditioning. White noise is very frequently used in startle experiments. Figure 2.5 shows several white noise generators. Unfortunately, the best source for white noise in the figure uses a single chip that is no longer readily available (S2688 or MM5837), but might still be found in old supplies, combined with a simple voltage follower circuit that includes a 741 operational amplifier. Some large-scale integrated circuits with sound generators that are incorporated into computers may have a simulated white noise function.

The other three circuits in the figure are from sources (identified in the figure) that all work on the principle of amplifying the inherent noise in a p-n junction, for example, at the base of a transistor. Our experience is that these circuits work to varying degrees of success and with varying ease of use. Alternatively, several popular white noise sound generators are now

Chapter 2

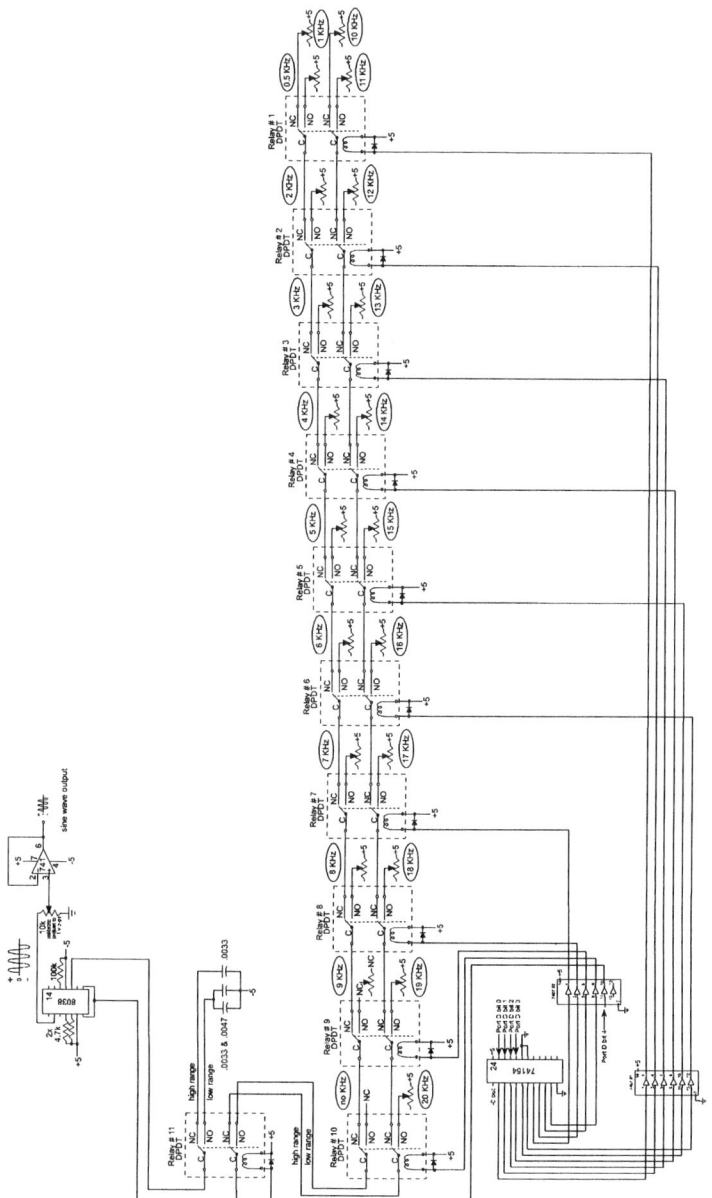

Figure 2.4a: Circuit for programmable frequency selection.

Notes for Tuning Curve Circuit

This circuit allows for selection by computer of the frequency of the tone to be tested.
Relays 1 - 10 are used to select timing resistors for the 8038 timing chip. Half of each DPDT relay is used for the low range of frequencies (0.5 KHz - 9 HHz) and the other half is used for the high range of frequencies (10 KHz - 20 KHz). The range depends upon the state of Relay 11. In the table below, NO is normally open, NC is normally closed.

Relay # & state	Var. Resistor	Frequency
first side of each relay ...		
1 NO	2 KOhms	0.5 KHz
1 NC	2	1. (default frquency if all relays off)
2 NO	2	2.
3 NO	5	3.
4 NO	10	4.
5 NO	10	5.
6 NO	20	6.
7 NO	100	7.
8 NO	100	8.
9 NO	none	9.
10 NO	none	9. (not used in programming)
second side of each relay ...		
1 NO	20	10.
1 NC	20	11.
2 NO	20	12.
3 NO	20	13.
4 NO	20	14.
5 NO	100	15.
6 NO	100	16.
7 NO	200	17.
8 NO	200	18.
9 NO	200	19.
10 NO	200	20.

Relay 11 is used to select timing capacitors to choose the range of frequencies and to direct which half of Relays 1 - 10 are used for the timing resistors.

Figure 2.4b: Notes for programmable frequency selection.

commercially available, such as devices used to generate the sound of the ocean or gentle rainfall to induce sleep, and these devices can be quite easily adapted for experimental purposes. One easy and very inexpensive source of white noise is to tune an old FM radio to a frequency between stations. Strictly speaking, FM or diode/transistor noise (as in the circuits of Figure 2.5) is a type of white noise called Gaussian, where unlike a true white noise source, the amplitudes are represented by a normal distribution over the frequency range (see Durrant and Lovrinic, 1984).

Other Sound Sources

Other sources for sounds are available but usually these cannot be sources for a gated rise/fall switch as the speaker (or transducer for sound) is an integral component. This applies to certain buzzers for doorbells, to piezo electric sound devices, and even the clicking sound of a relay turning on and off. Nevertheless, creative solutions for controllable sounds can be found and may be appropriate in certain circumstances (e.g., when a novel environment of sounds is used to make a distinctive setting).

Rise/Fall Switch

The rise/fall switch circuit shown in Figure 2.6 is the basis for the rise/fall switch used in the Morse code circuit depicted in Figure 2.2. The features of the design that appears in Figure 2.6 are the 3080 transconductance operational amplifier used to gate the signal, the 3130 operational amplifier configured as an integrator to modify the conductance of the 3080, and the digital control circuit consisting of the optoisolator and the 741 operational amplifier configured as a comparator. In the Morse code circuit the integrator and comparator functions were combined. The link between the comparator and the integrator can be a simple variable resistor that yields a bi-directional shaping of the audio signal (i.e., adjusting the resistor to give 5 msec rise and fall). Alternatively, by using oppositely directed diodes[12], the rise and fall times can be adjusted independently. We use the simpler bi-directional configuration in our experiments.

Note that for calibrating the rise/fall circuit, a 1 V peak-to-peak sine wave is put into the "input buffer." The intervening 3080 reduces the signal by 100 times (through the 100 KΩ and 1 KΩ resistors that make a voltage divider on the non-inverting input, pin 3, of the 3080). The "output buffer" is adjusted to return the signal to the original 1 V peak-to-peak size. By performing this calibration, any new incoming signal will be reproduced accurately at the output.

There are potentially some creative uses for rise/fall switch beyond simply turning a tone on and off. For example, placed into a circuit for neural recordings, the rise/fall switch could be used to turn off the input to an electrophysiological amplifier during stimulation, thus preventing saturation of the recording. As you will see shortly below, by modifying the current on pin 5 of the 3080, this rise/fall switch can also be used as an attenuator feature in the auditory stimulus control circuit (Figure 2.9).

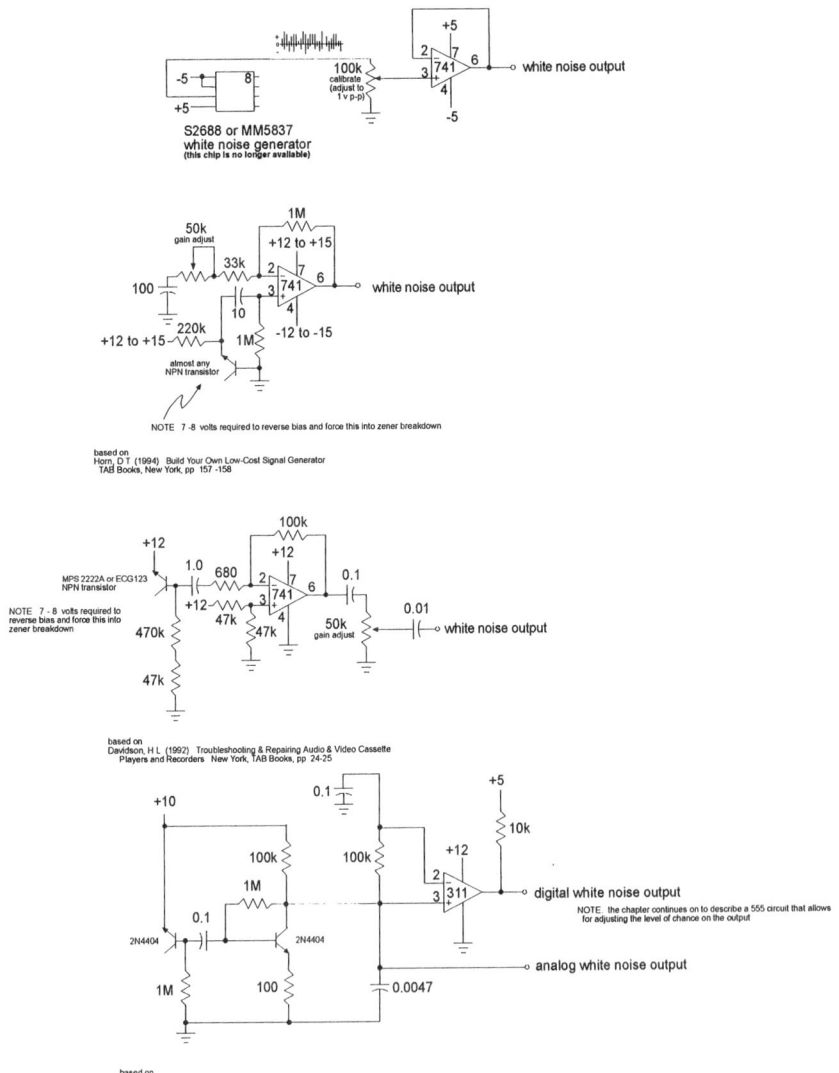

Figure 2.5: White noise circuits.

Chapter 2

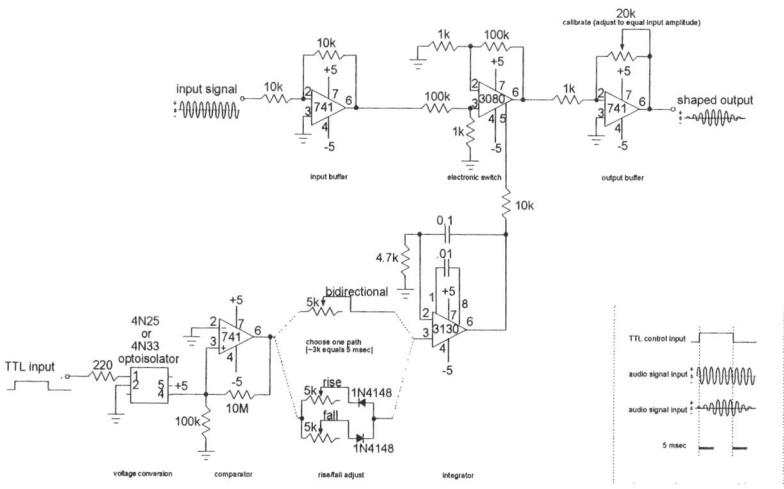

Figure 2.6: Basic rise/fall switch to control the onset and offset of auditory stimuli.

Attenuator

For most serious auditory experiments one would set the intensity (loudness) of the auditory stimulus at the maximum value the equipment can handle and then attenuate that signal for actual use in the experiments. For this purpose, an attenuator, with easily selectable decrements in decibels, is desired. The circuit in Figure 2.7 shows one basic design where optoisolator devices control relays that allow for the switching in or out of resistor networks. These resistor networks cause signal attenuation while having a 600 Ω output resistance, which is suitable for applying to an amplifier or speaker. The available attenuation values are powers of two: -0.5, -1, -2, -4, -8, -16, -32 and -64 decibels, and can be added together. In our application, the optoisolator inputs are connected to a port of our interface, and controlled by software (see CD).

Two variations of the attenuator are given in Figures 2.8 and 2.9. In Figure 2.8, a 741 operational amplifier is configured as a summing operational amplifier. The theory of operation works like this: The audio signal is split and simultaneously fed into all branches of the input adder for the maximum signal output of the 741 operational amplifier. The resistor relationships were chosen such that they apply the positive value of the decibels in gain. Thus, the resistor associated with the -2 dB pathway actually adds +2 dB of signal to the output of the 741 operational amplifier. Each of the inputs can be turned off by affecting the corresponding inverter to the appropriate gate of the 4066 analog switch[13]. By selecting the -2 dB input, what actually happens is that the +2 dB of signal is now removed from the output of the 741 operational amplifier, thus achieving a 2 dB loss. The only precaution is that the 4066 quad bilateral switches, used instead of relays, have a typical ON resistance of 120 Ω at 10 V (the absolute power supply of +/- 5 V) that contributes to the resistor value. The 120 Ω must be subtracted from the calculated resistor value (e.g., 794328 - 120 = 794208 for the -2 dB pathway). For the larger resistors, this 120 Ω is not appreciable, but for the -64 dB pathway where the calculated resistance is only 631 Ω, the 120 Ω should be taken into account. Use of the -64 dB pathway is not advised for a different reason, however: the gain associated with this pathway is so large that it is difficult to calibrate the attenuator as this swamps out all other signals. As shown in the "Table of Attenuator Values" in Figure 2.8, the +64 dB pathway contributes about 1,500 times the output signal that the +1 dB pathway does. Try discriminating +1 dB or +64 dB on an oscilloscope and you will appreciate the problem of scaling and calibration!

For calibration, 1 V peak-to-peak signal is placed on the input. The 1 MΩ variable resistor R2 is adjusted so that the output is 1 V peak-to-peak when all the attenuator inputs are off (i.e., grounded). In essence, with all the attenuator inputs off, each decibel pathway adds in the associated gain. The output is therefore the accumulation of each of the pathways. If the -64 dB pathway has been omitted, as we have advised, then the resulting accumulation of 0.5 + 1.0 + 2.0 + 4.0 + 8.0 + 16.0 + 32.0 is 63 dB on the output. The effect of the 1 MΩ feedback resistor (R2) is to attenuate the combined signal so it can be adjusted to 1 V peak-to-peak. As with the circuit in Figure 2.7, the inputs to this adder circuit of Figure 2.8 can be connected to an available port of an interface, and controlled by the software (see CD).

The second variation of an attenuator is shown in Figure 2.9. This is a modification of the rise/fall switch depicted in Figure 2.6. Here, the amount of current allowed through the controlling pin 5 of the 3080 is modified to achieve attenuation. The more current that is sent to pin 5, the greater the output of the 3080 transconductance operational amplifier. Proportionately less current on pin 5 causes proportionately less signal on the output. The modification of the current on pin 5 is achieved by the size of the associated resistors. For example, to create a 2 dB loss (attenuation) of the signal, there needs to be a reduction of the controlling signal by approximately 20.6%.

This reduction is calculated using the formula given in Figure 2.8 for V_{out}. The formula is $V_{out} = V_{in} \times antilog\,(y\,dB/20)$. If $V_{in} = 1$ V peak-to-peak, then the formula reduces to $V_{out} = V_{in} \times antilog\,(dB/20)$. Note that in the calculation for loss, dB must be a negative number. The calculation of V_{out} yields the proportion of 1 x remaining (0.7943 V) or about 20.6% loss,

Chapter 2

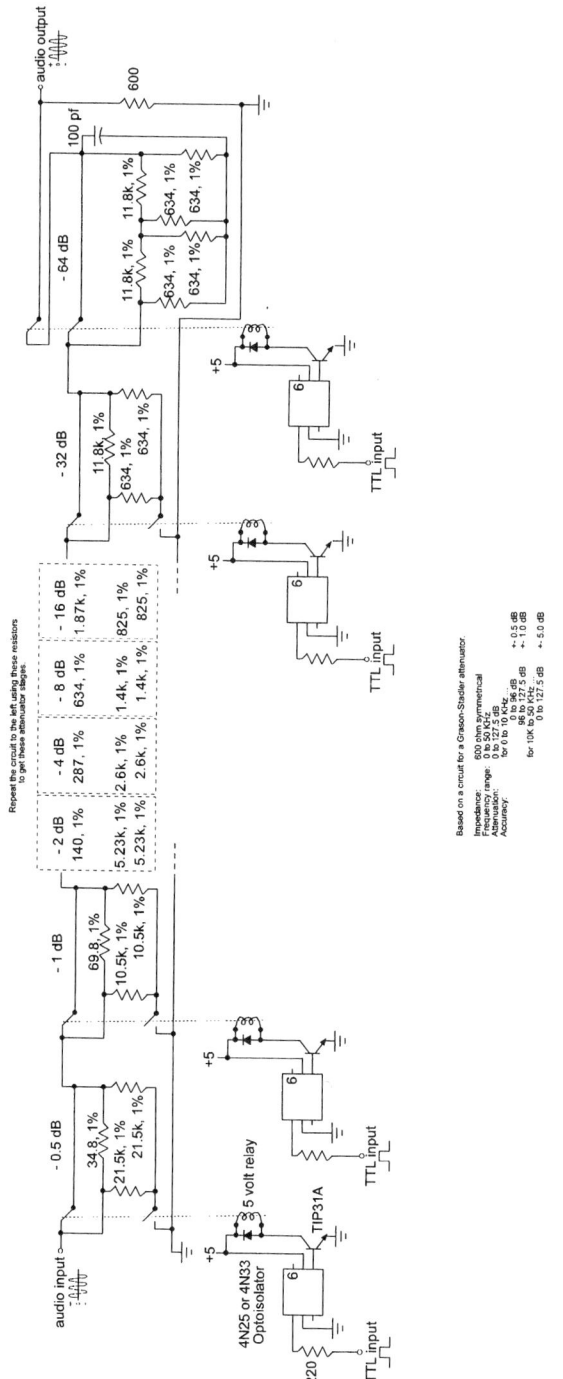

Figure 2.7: Basic attenuator circuit.

from the calculation of 100% times the difference of 1.0000 V minus 0.7943 V. The 20,567 Ω resistor achieves this proportional loss of current to pin 5. If the associated switch on the 4066 is activated, however, the 20,567 Ω resistor is bypassed (shorted out) with the 120 Ω ON resistance of the 4066, effectively causing no loss of current to pin 5. Thus, under normal circumstances with no attenuation, all the resistors associated with the attenuation function are shorted out. Only when the -2 dB pathway is chosen is the short through the 4066 turned off, thus placing the 20,567 resistor into the current path for pin 5. The gate inputs to the 4066 are connected, as in the other attenuators, to a port of the controller interface, and controlled by the software.

Power Amplification and Speaker

In most instances, a speaker needs a power amplifier to drive it. Radio Shack and other dealers sell audio amplifier chips that can be the center of powerful amplifiers. We have found that this is usually too much trouble, as complete power amplifiers can be readily and cheaply bought from Radio Shack and other sources. We have used a Radio Shack SA-10 stereo amplifier with good success, however, that amplifier seems to be no longer available. Radio Shack has other alternatives such as the Optimus SA-155. (Because we have not tried this we cannot vouch for it, but we see no reason why it should not work well.)

The speaker we have found very useful is the Radio Shack 3" midrange tweeter (#40-1289A), which has fairly good characteristics over a range of frequencies. Their #40-1282 midrange speaker (40 watt) probably would be a good substitute, as would be the #40-1218 tweeter (150 watt). Realistically most experiments do not need a range of frequencies if a single pure tone is being used, which is most often the case, and if white noise is being used then it is usually only important that many frequencies can be heard, not necessarily all of them. In short, just about any speaker will do if it is loud enough at the frequency of interest for the experiment. The main specification of concern is that the power amplifier and the speaker are matched in terms of wattage. The speaker should be able to handle the wattage of the power amplifier to avoid damage to the speaker. If the amplifier is a little underpowered for the speaker that is fine; being considerably underpowered will result in very faint sounds. The equipment for audiology experiments may be more critical although the quality of the equipment available now far surpasses the quality and price that would have been available years ago.

Chapter 2

Theory of Operation

The input signal is normally routed to ALL of the summing inputs so that the output signal is actually the sum of all the the possible dB. By giving an input (logic 1, +5) to the inverter, an input is actually turned OFF to the summing inputs. Thus, if the -16 dB input is activated with logic 1 (+5) then the +16 dB signal is REMOVED from the summing circuit, and therefore the output is reduced by 16 dB.

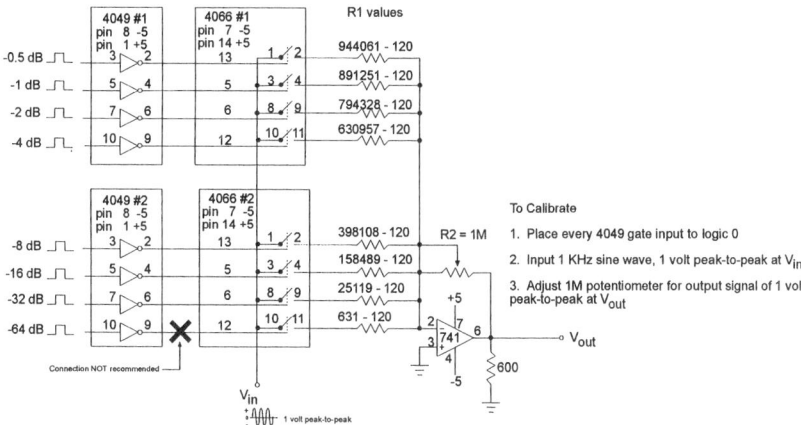

Table of Attenuator Values

dB	V_{out}	R1 (ohms)
+64.0	1584.8932	630.96
+32.0	39.8107	25118.86
+16.0	6.3096	158489.32
+8.0	2.5119	398107.56
+4.0	1.5849	630957.34
+2.0	1.2590	794328.49
+1.0	1.1221	891250.90
+0.5	1.0593	944060.87

Formulas and Calculations:

$$dB = 20 \, (\log_{10} (V_{out} / V_{in}))$$

$V_{out} = V_{in}(\text{antilog} (dB / 20))$;
if $V_{in}=1$ volt then $V_{out} = V_{in}(\text{antilog} (dB / 20))$

Gain of the op amp = R2 / R1;
if R2 = 1M then R1 = 1M / V_{out}

Figure 2.8: Attenuator circuit based on an op amp used as a mixer.

Buying Off-the-Shelf

It is possible, of course, to purchase most of the components shown in Figure 2.1 from a variety of commercial vendors. Indeed, many of the experimental set-ups in the Steinmetz laboratory have been assembled with components purchased off-the-shelf. We have a number of standard frequency generators to create pure tones, such as the Fordham AG-260 and the much older GR Model 2C-2MC oscillator. In addition, tone, white noise and pink noise generators can be purchased from companies that design and sell modular control systems such as Coulbourn, Lafayette Instruments and Med Associates, Inc. The major advantages of using these audio generators are that they can be set for a wide range of frequencies and typically contain built-in step attenuators. The major disadvantage, of course, is price: these audio generators can range in price for $200 - $800. Audio switches with shaped rise/time control can also be purchased commercially. For example, Med Associates, Inc., as part of their rack-mounted modular interface system, has the ANL 913 Shaped Rise/Time Audio Switch that is very versatile and easy to use. This device is used to gate the auditory stimulus on and off and in our application is controlled by a signal that arises from a microcomputer (see Chapter 11). It is relatively expensive, however, at $300–$400. Finally, as described above, a number of power amplifiers are available to boost the audio signal for presentation including several amplifiers sold by Radio Shack (e.g., the SA-10, STA-19, SA-155 Optimus and MPA-25 Realistic amplifiers). We have also used a variety of general purpose, high quality stereo amplifiers for this purpose, finding these at auctions, inventory sell-outs, and a variety of other inexpensive places.

Visual Stimuli

Light is probably the second most commonly used CS for classical conditioning experiments. There are at least a couple of major reasons that light has not been employed more often as a CS in classical conditioning experiments. First, the most popular classical conditioning procedure has been eyeblink conditioning and the most popular experimental subject for these experiments has been the albino rabbit. It appears that the rate of conditioning in the albino rabbit is significantly slower when a light CS is applied relative to the rate of conditioning observed when an auditory stimulus is used. This may be due to the relatively poor development of the visual system in the albino rabbit. Indeed, it is our experience that classical eyeblink conditioning in rats and humans may proceed at similar rates when tones and lights are used as CSs. Second, rabbits, rats, and humans all show marked alpha responses (short latency, reflexive blinks) when lights are used as CSs and this could confound the interpretation of when and how well conditioning occurs. While intense tones can also evoke alpha respondes, (in fact, both light and tone can be used as USs), tone CS intensities can be reduced well below the level that produces alpha responses.

In a lot of ways, it actually may be easier to present a light as a CS than to present auditory CSs because less equipment is needed to present a light CS. There are a number of relatively simple circuits available for con-

Chapter 2

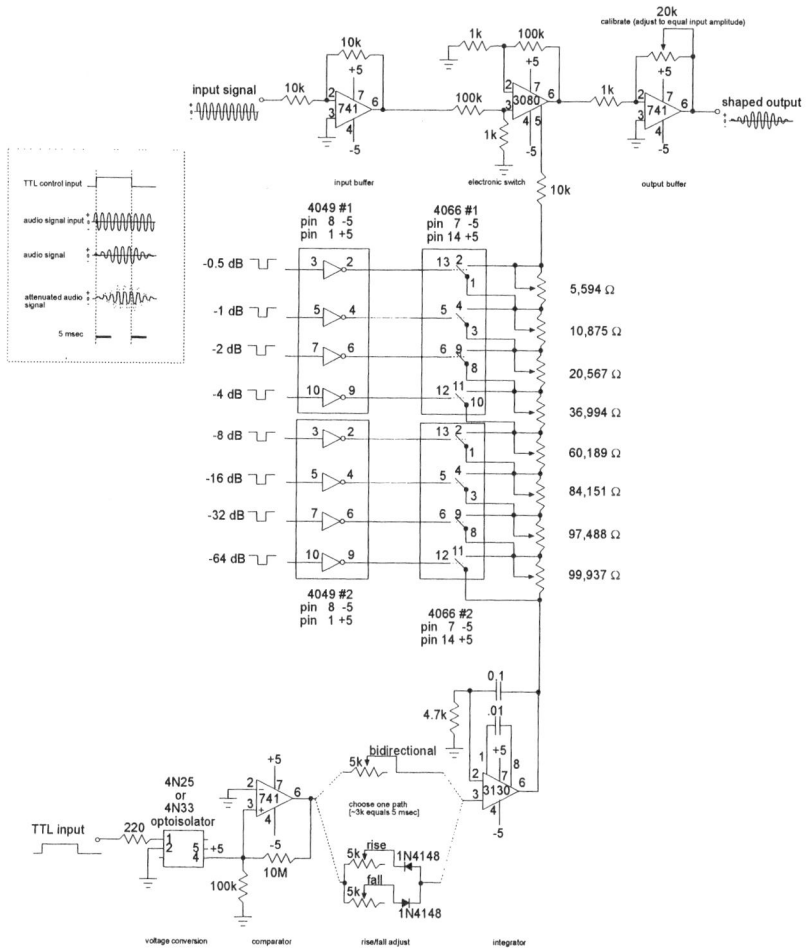

Figure 2.9: Attenuator based on rise/fall switch.

trolling a light. The circuitry we have used for delivering light stimuli are seen in Figures 2.10 and 2.11. We have used an optoisolator between the controller/computer and a larger voltage source to turn on light CS. Here the light is a small incandescent light bulb, typically used for flashlights, and is powered with 12 V batteries. The optoisolator isolates (protects) the controller/computer from this larger voltage and acts as a voltage translator from TTL levels to 12 V (see Chapter 12 on interface circuits). In both circuits the output of the optoisolator drives a power transistor. In one of the circuits (Figure 2.10) this transistor operates a 12 V relay that has contacts that switch the bulb on. Since the relay may make auditory or electrical (switching) noise, in the second circuit (Figure 2.11) a transistor is used to directly drive the bulb. To protect the transistor, the working current is limited by a 10 Ω, 10 W resistor.

Since incandescent lights take a measurable amount of time to illuminate once the current is applied, both circuits are designed to pass a small amount of current through the bulb when the light is in the off-state. When they are off, the bulbs actually radiate a small glow. This small current warms the element of the bulb so that it turns on more quickly when operated. This glow is achieved by having a small amount of current pass through four 1 kΩ resistors in parallel. The reason for using multiple resistors is to split the current through each resistor, limiting the current through each resistor to their 1/4 W rating (the current through each resistor is one fourth the total current, or 0.012 A; the wattage for each resistor is 0.012 A times 12 V, or 0.144 W). For rabbits, we place the light bulb either at the end of the air hose assembly used to deliver the air puff US (see section beginning on page 61 in this chapter on delivery of the US), or at the edge of the restrainer in front of the eye (remember that rabbit eyes face sideways). The Steinmetz laboratory has found that the rate of eyeblink conditioning increases significantly when a flashing light CS is used instead of a constant light CS. Figure 2.12 depicts the control circuit used to generate a flashing light. Other investigators have used commercially available light sources such as the Grass Photic stimulator. These are relatively easy to gate on and off and provide very good photic stimulus control.

Other light sources are possible. One common CS is to manipulate the house lights and this is sufficient for training rats. The relay design here can switch the house lights using the Common and Normally Closed contacts of a relay so that the house lights are off during the intertrial interval, on during the CS presentation. Alternatively, the house lights could be kept on during the intertrial interval and turned off as the CS presentation, using Common and Normally Open contacts. Extreme caution should be exercised when designing systems that switch line voltages.

Another source of light stimuli are LEDs. We have not used these as light sources for CSs, but know of a few instances where others have attempted with limited success to use illumination from an LED or groups of LEDs as the CS. Care should be taken so that red color LEDs are not used when training albino rabbits since these rabbits have difficulty processing red light.

As an aside, notice that we have used the relay circuit depicted in Figure 2.10 as a general solution for interfacing various pieces of equipment with stimulus controllers. The same circuit could be used to gate on a shock or stimulation as the CS or US (see discussion below). The relay contacts

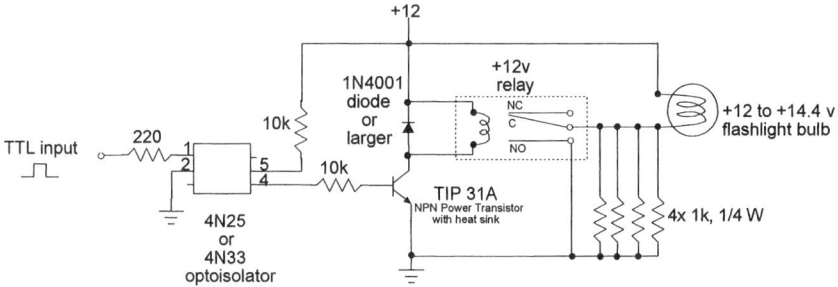

Figure 2.10: Relay control of light stimulus.

are often used to engage remote controls that are activated by merely shorting two leads on the input (i.e., require a simple switch closure), as required for the remote control of devices like tape recorders and VCRs. Some older models of Grass stimulators also have remote control functions that are activated by shorting the input leads as do some other commercial shockers and stimulators. More solutions for interfacing control circuits can be found in Chapter 12.

Tactile Stimuli

Although not used nearly as often as auditory and visual CSs, some investigators have successfully used tactile stimuli as CSs for classical conditioning experiments. A very simple method for delivering a tactile CS has been used by Solomon and his colleagues at Williams College (e.g., Solomon, Lewis, LoTurco, Steinmetz & Thompson, 1986). They simply placed a small audio speaker on the backs of rabbits that were sitting in Plexiglas restraint boxes. A train of 5 Hz stimulation delivered to the speaker makes it vibrate, thus delivering a vibratory tactile stimulation. Stimulus control can be achieved with the same interfacing equipment described above for experiments involving auditory stimuli. Some care must be taken, however, to make sure that there are no auditory components associated with delivery of the tactile stimulus, because these auditory stimuli could be used as cues during the behavioral experiment.

Other devices could also be used to deliver tactile stimuli such as battery-powered vibrators or massagers. Controlling the onset and offset of these other types of tactile stimulators is fairly straightforward, using gating circuits designed to provide timed DC current to the device. However, it is difficult to control the intensity and frequency of these inexpensive tactile stimulators. Some more refined (and more expensive) tactile stimulators are

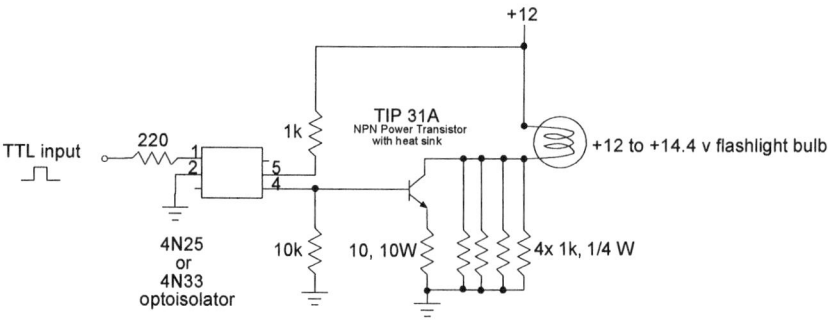

Figure 2.11: Transistor control of light stimulus.

commercially available. Tactile stimulators that produce sensations of vibrations, taps and pressure have been described including von Frey hairs, solenoids, electromagnetic shakers, piezoceramic benders and arrays as well as air puffs. Reviews of these more sophisticated tactile stimulus delivery systems have been written by Craig and Sherrick (1982) and Cholewiak and Wollowitz (1992).

Taste Stimuli

Conditioned taste aversion is a major paradigm for studying the properties of association and their neural bases. This paradigm involves the application of a novel taste, for example sucrose delivered through a water bottle, as the conditioned stimulus that is paired with illness, for example gastrointestinal nausea created by intraperitoneal injection of lithium chloride (LiCl), as the unconditioned stimulus. The behavioral measure is the degree of avoidance of the novel taste and how long it takes to extinguish. The conditioned taste aversion (CTA) paradigm typically requires one or a very few number of pairings of the stimuli, which are often separated by significant interstimulus time intervals. Although these properties are not unique to CTA, we can more easily imagine the ecological validity of learning to avoid tastes that make us ill than the relevance in the wild of taking a hundred trials to learn to blink to a tone. Conditioned taste aversion has the added advantage that it is simple to implement: sugar, bottle, water, syringe with needle, LiCl.

We ourselves have little experience with the paradigm, however we sometimes assist or collaborate with Kathleen Chambers and her students. Here in Figure 2.13 we share the method developed by her graduate student

Chapter 2

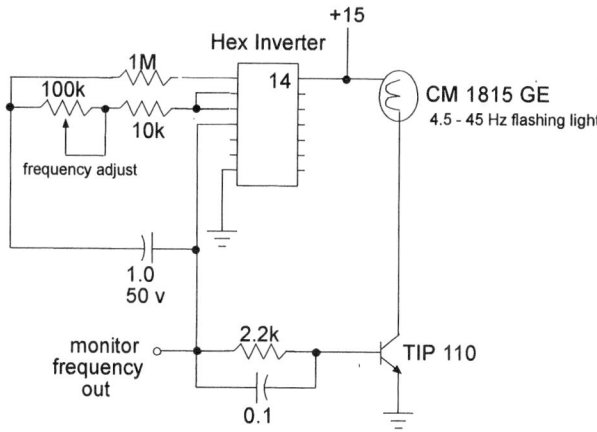

Figure 2.12: Flashing light circuit.

Unja Hayes to deliver sucrose-water directly into the mouth of rats through a fistula so that orofacial reactions about hedonic properties could be quantified from videotape.

As shown in Figure 2.13, the fistula consists of a short length of PE-100 tubing that passes through the rat's cheek. During the experiment, an injection needle passes through this tubing to allow direct injection into the mouth. The tubing is permanently held in place by two Teflon washers, constructed using a hole punch, placed on either side of the cheek. The washers are held in place by heating and flaring the ends of the tubing using a surgical cautery. The figure shows the steps in construction and application of the fistula. Note that the fistula is placed while the rat is anesthetized, using a 19-gauge needle on the tubing that passes from the inside of the cheek to the outside. The needle and excess tubing are cut off and the outside end is flared over the outer washer.

Intracranial Electrical Stimulation

Over the last several years, direct electrical stimulation of the brain has been used as an effective CS for eyeblink conditioning and has also been used in a variety of other behavioral neuroscience experiments (e.g., Steinmetz, 1990; Steinmetz, Lavond & Thompson, 1989). In these experiments, brain stimulation has been successfully substituted for peripheral CSs, such as lights and tones, and used effectively to trace critical CS pathways in the brain. We take up the special issue of electrical stimulation of

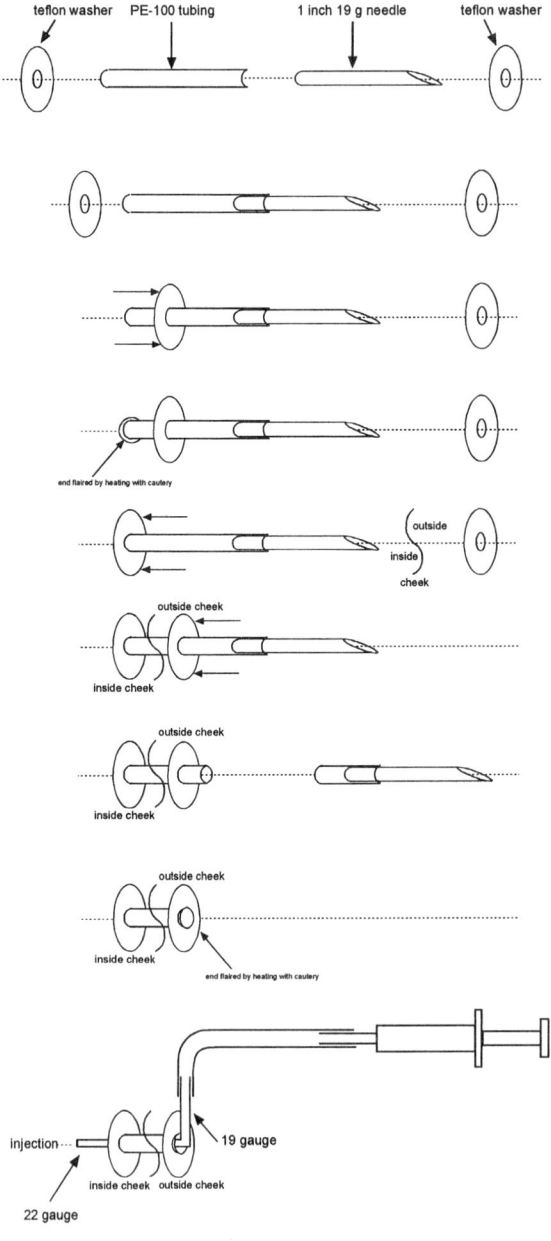

Figure 2.13: Hayes' fistula for delivering conditioned stimulus in taste aversion experiments.

Chapter 2   63

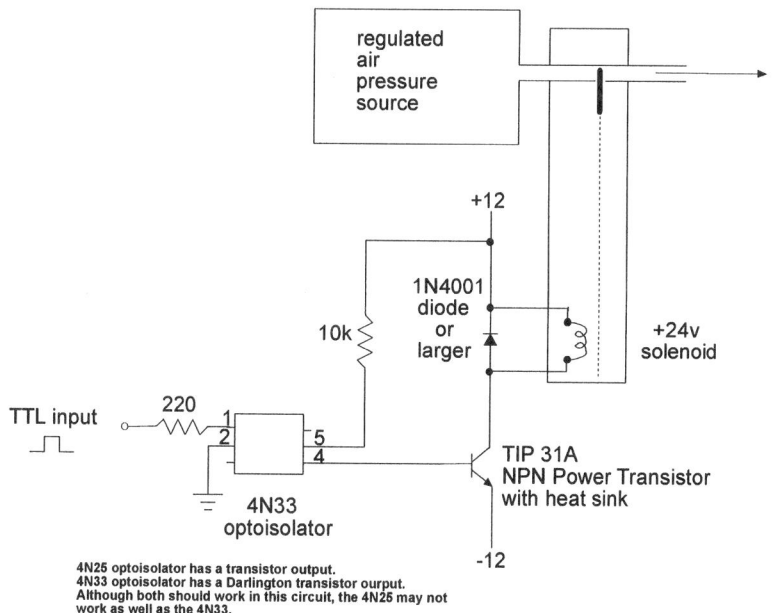

Figure 2.14: Interface to solenoid for air puff circuit.

the brain in Chapter 9. Suffice it to say that it is best to use commercially available brain stimulators because these devices allow several parameters to be adjusted such as stimulation intensity as well as the frequency and duration of pulses that are delivered. The methods for gating these commercially available stimulators on and off depend on the make and model of the stimulator being used. Some require a simple switch (contact) closure while others require the application of a control voltage to gating contacts. See Chapter 12 for a discussion of common gating circuits used to control these kinds of devices.

THE UNCONDITIONED STIMULUS (US)

The unconditioned stimulus for classical eyeblink conditioning is most often a puff of air to the eye or a mild shock delivered to the face or to a leg or forepaw. These stimuli reliably cause reflexive eye blinks and limb flexions that are paired with the CSs used during conditioning. Some advantages for using an air puff US can be cited. For example, air puff USs are more acceptable for use with humans, so the use of air puffs in non-human

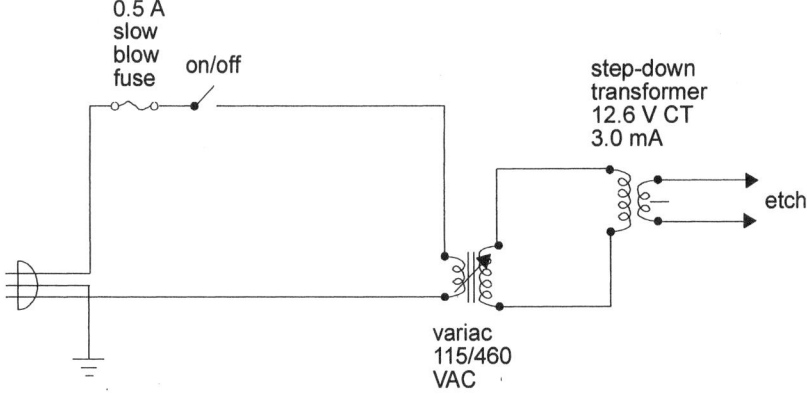

Figure 2.15: Basic shocker circuit. We recommend that the user purchase commercially available shock sources for safety.

animal studies allow a direct between-species comparison of conditioning. Also, the presentation of air puffs does not generate electrical artifacts in neural (and behavioral) records thus it is quite easy to directly measure neural or muscular electrical activity during conditioning. The chief disadvantages of using air puff USs are that the air puff requires more bulky equipment to deliver than a simple shocker, and is much harder to parametrically manipulate (e.g., it is fairly difficult to control air puff intensity). Here, we discuss the apparatus and control circuits necessary for using air puff and shock as the unconditioned stimuli.

The Delivery of Air Puffs

Assuming that a source of pressurized air is available, the electronics necessary for delivering the US are very simple. Figure 2.14 shows an air source that is controlled by a solenoid, which, in turn, is activated by a transistor and an optoisolator. The advantage of the optoisolator as will be described later in the book (see Chapter 12 on interface circuits) is that it allows easy interfacing of TTL signals from a controller or computer to affect the power supply necessary to drive the solenoid. In our system, the solenoid is a General Valve +24 volt DC solenoid (Part number 9-179-900), which we power with a +12/-12 V supply. The solenoid is attached with Teflon male-to-tubing fittings (Part number 11-12-125-2) which can be connected with any 1/4 inch (outer diameter)/0.170 inch (inner diameter) vinyl tubing from a local hardware store. The output of the optoisolator controls a power transistor (TIP 31A) set up to act as a sink. A diode is placed backwards and directly on or as close as possible to the coil of the solenoid to dissipate backward electromagnetic force when the magnetic field collapses as the solenoid is turned off. That is, the diode protects the transistor and the optoisolator output. The optoisolator protects the computer equipment.

Several sources of air have been used. In the laboratory, large tanks of pressurized air can be rented. The newly filled pressure of these tanks is about 2,000 psi. For safety reasons, air pressure is controlled using a two-stage regulator. The first stage drops the pressure from 2000 psi to about 50 psi while a second pressure regulator allows the output of the solenoid to deliver 0 – 30 psi. For most experiments, the second-stage pressure regulator is adjusted to 3 psi (2.1 N/cm^2, 2.1 nbar, 210 mPa, 157.5 mm Hg, 0.21 atm) at the source. Measuring the actual pressure as it is delivered to the eye has been problematic.

In the Lavond laboratory at the University of Southern California, the pressured air line is piped into the laboratory, as the building air supply is used as the source for the air puff US. The building pressure is somewhere in the range of 40-60 psi. A high-pressure bicycle hose is attached with a hose clamp to the standard chemistry laboratory valve, which in turn attaches to an acetylene torch regulator (Victor Equipment Company Model Number SR 260A, 400 psi input gauge, adjustable 0-30 psi regulated output and gauge) that was purchased from a welding supply company. Since the output is at a relatively low pressure, vinyl tubing leads can be used from the regulator to the solenoid. After 10 years of use the tubing was replaced because over the years the pressure stretched the vinyl tubing enough that it leaked at the fittings.

Portable systems for eyeblink classical conditioning are popular for researchers who need to visit their human subjects at off-campus locations. One system uses a small tank on wheels, similar to portable units used by out patients who carry their own oxygen at home. Another system was created for Diana Woodruff-Pak at Temple University that includes a small reservoir for holding the air and a pump to set the pressure. Lavond has used a third system in a laboratory where access to the building air supply is difficult. In this system he filled a metal air tank sold at auto supply stores that is meant to hold air for filling the tires on a car. This could be filled by the air supply at a gas station (the intended location), by a hand or electric pump, or by air drawn from the building air supply as Lavond does. Attached to this tank was placed a regulator that supplies the solenoid. The air lasts for several training sessions before needing to be refilled.

Electrical Shock and Stimulation

Figure 2.15 shows a 60 cycle shocker design based on a device made for us by UCLA Psychology Technical Services. It turns out that 60 Hz is a good frequency for neural and somatic stimulation, explaining in part why electrocution from defective household appliances is a danger. The key feature of Figure 2.15 is the variac[14], or variable AC transformer, which allows for a higher voltage. The higher voltage is necessary to effectively make a constant current output by compensating for the resistance of the animal and electrodes, as discussed in Chapter 8 where a lesion maker circuit is described. Higher voltage means that small changes in resistance have little effect on the overall current output.

A few design features of Figure 2.15 are noteworthy. First, while we are interested in the actual current that is delivered as a shock, we can use a voltmeter across a 1 kΩ resistor to indirectly measure the current. The reason for this choice is that ammeters are more difficult to find than voltmeters. Here, 1 V translates into 1 mA of current. The same trick is used in our lesion maker with a digital liquid crystal display. By selecting the decimal place the display can be used to read milliamperes without any confusion. For greater accuracy, measure several resistors with a meter and select the one nearest to being exactly 1 kΩ, or use a 1% 1 kΩ resistor.

Second, a calibration position allows for setting the current before applying the current to the animal. Here we approximate the animal and electrode resistance with a 22 kΩ resistor. Any resistor around this value would be fine. A 22 kΩ resistor is a commonly found value. Since this is a constant current design any variation in the actual resistances (the electrode and the animal) will have relatively little effect on the output current, so choosing a calibration resistor that is about the right size as the actual resistances will work just as well.

Third, both shock leads are switched through a relay, disconnecting the animal entirely from the shocker when shock is not applied. This protects against accidentally shocking the animal with one active lead through another source (e.g., through recording electrodes that might be implanted). Note also that when the relay is not actively giving a shock that both electrodes are shorted out to each other. This is another protective measure against accidental shocks. It has the added feature that the tendency for electrodes to build up charge and become polarized is dissipated by shorting the electrodes out to themselves. Again, this feature is not critical here, because the shock source is alternating current (AC), which prevents polarization. But the feature should be kept in mind for the design of direct current (DC) stimulation devices, which can cause electrode polarization (see Electrical Stimulation, below).

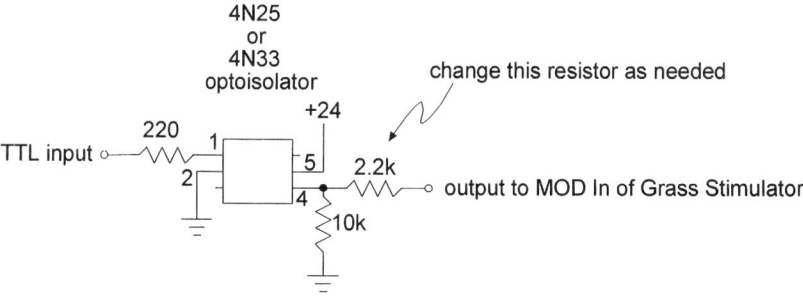

Figure 2.16: Interface to control stimulator.

Fourth, we have added an etching feature. This is a method for making fine-tipped electrodes by electrically removing the metal from the electrode shaft. The leads from the step-down transformer are attached to a carbon rod (scavanged from a dry cell battery) and to a stainless steel insect pin. Both of these are placed into a weak acid solution (10% HCl) and current is passed to etch the metal away. More details concerning the electrode etching technique are provided in Chapter 4.

The fifth feature of note in the 60 Hz shocker of Figure 2.15 is the built-in power supply and TTL interface. The power supply here is overkill. UCLA Technical Services used a simpler design with a wall adaptor (120 VAC input, +6 VDC output) instead, built directly into the shocker.

Shock generators are one device that the researcher should consider purchasing from a reputable manufacturer rather than making himself or herself. This is particularly true if waveforms other than 60 Hz sine waves are desired. The problem for the researcher then is to interface the control or operate of the shocker to individualized controller/computers. One of the versatile optoisolators circuits (see Chapter 12) can be very useful in this respect if the shocker does not already accept a TTL input. Modern equipment usually comes with TTL connections. Older shockers may require an optoisolator interface.

Electrical Stimulation

Electrical stimulators can be used to generate shock USs and for directly stimulating muscles and brain areas involved in reflex generation. John Moore at the University of Massachusetts, Amherst, for example, has very successfully used single pulse stimulation as a US, essentially eliminating the shock artifact in neural recordings that he and his colleagues have made concomitantly with behavioral training. Commercial stimulators are available with either a standard TTL interface, which is directly compatible with the controllers and computer interfaces described later in this book or are designed to interact with a compatible family of equipment from the business source, which may require one of the versatile optoisolators circuits described later (see Chapter 12). In addition, older stimulation equipment may be available. We have found the circuit in Figure 2.16 to be useful in controlling older Grass stimulators (e.g., SD 9) that were made before having TTL compatibility. The MOD inputs to the Grass stimulator required something more than +15 volts and less than +30 volts to operate the stimulator remotely. The simple optoisolator circuit of Figure 2.16 works well and has the added benefit that the controller/computer is optically isolated from the Grass stimulator.

We have not made our own stimulators, in part because stimulators have been available to us and primarily because of the amount of trouble associated with realizing all the options, i.e., chips, knobs and switches to select the frequency, pulse width, polarity, train, single versus dual pulses, remote versus local control, you-name-it. Safety for the experimental subject is also an issue and the commercial stimulators provide a well-controlled and well-designed source of electrical stimulation that can be safely applied to biological tissue. If we were to build our own, however, we would combine aspects of the lesion maker (Chapter 8), which is a battery powered and constant current design, with one of the controllers found in Chapter 11. By adding an optoisolator circuit (Chapter 12) we could incorporate both TTL and remote control interfaces.

One feature of the shocker (Figure 2.15) to keep in mind when designing a DC stimulator, as mentioned above, is to short out the electrodes between stimulation. This procedure prevents polarization of the electrodes. By polarization we mean that charge builds up on the electrodes just as it does on the plates of a capacitor. The effect of polarization, like a capacitor, is that it becomes more and more difficult to pass the current. By shorting the electrodes their capacitor-like charge is dissipated. Alternatively, one stimulator design reversed the charge on the electrodes between stimulation pulses to prevent polarization (Berntson, Ault & Walker, 1977). Insuring that the electrodes are properly grounded between stimulation pulses may also be considered.

CONCLUSIONS AND SUMMARY

In this chapter, we have presented a variety of methods, techniques and circuits that we have found useful for controlling the delivery of stimuli in behavioral neuroscience experiments. While we concentrated on stimulus delivery techniques used during classical conditioning experiments, the same methods, circuits and equipment could be used to deliver discrete stimuli during most any other behavioral experiment, such as those described in Chapter 6. It is our experience that a relatively few specialized electronic circuits are needed to interface between the computer controllers, such as those described in Chapter 11, and the pieces of equipment that generate the desired stimuli. However, we do remind the reader that the quality of the data collected may hinge to a large degree on the quality of the stimuli that are delivered and the precision in timing and parametric manipulations that are possible.

NOTES

[1] Operational Amplifier. An electronic device used for amplification, the operational amplifier or "op amp" is a general device that can be used for a variety of standard analog applications. An op amp is an integrated circuit, as opposed to a discrete device like a transistor or resistor. We normally use op amps for amplification, filtering, comparing, buffering and mixing (adding) signals.

[2] Resistor. A discrete (single purpose, separate) electronic device that impedes the flow of current.

[3] Capacitor. A discrete (single purpose, separate) electronic device that stores charge and resists changes in current flow.

[4] Potentiometer. Like a resistor, this is a discrete (single purpose, separate) electronic device that impedes the flow of current. The potentiometer has the trait that the impedance value can be changed by the user, normally by rotating an axle.

[5] Transistor. A discrete (single purpose, separate) electronic device that amplifies an analog signal but can also be used for digital switching. Multiple transistors are used to design operational amplifiers.

[6] 6502 Micro Controller. A controller is a device that incorporates several components: a processor, memory, and interfacing circuits. The processor interprets commands (software) and might store data. The 6502 microprocessor was one of the first devices used as the basis for primitive computers, the AIM-65 and KIM-1 for example, and later became the brain behind the first popular personal computers, the Commodore VIC-20 and Apple II computers for example. The Z80 microprocessor and the 6800 microprocessor were its main rivals.

[7] ROM-based Controller. A device where a ROM (Read Only Memory) stores a collection of serial instructions which, when activated sequentially, effect a series of commands. For example, timing for delay classical conditioning can be effected by commanding no output (baseline period), tone (the CS period) and air puff (US period) for a trial.

[8] TTL interface. A standard digital signal having either an on or off state. The on state is a

voltage between +2.4 and +5.0 volts, and the off state is a voltage between 0 and +0.8 volts. Voltages in between +0.8 and +2.4 are undefined. Voltages outside these ranges, i.e., below 0 volts (negative voltages) or about +5.0 volts, will damage TTL device.

[9] Optoisolator. A discrete (single purpose, separate) electronic device which has an input and output that are electronically isolated but spanned by an optical (light) signal. The input is usually an LED (light emitting diode) and the output is commonly a phototransistor (a transistor that detects light). The big advantage of an optoisolator is that devices attached to the input and output are electrically isolated, so that anything that "goes wrong" on one side or the other will not destroy the other side. This is a good way to protect an expensive computer from laboratory equipment. It is also a good way to protect a subject from the possibility of electrocution. It is also a good way to eliminate electrical artifacts in neural recordings caused by stimulation.

[10] Precision Waveform Generator. An integrated circuit used to create electronic signals.

[11] Address Decoder. An integrated circuit used in digital applications (e.g., computers) to select a particular device, where different devices have different addresses.

[12] Diode. A discrete (single purpose, separate) electronic device that allows current to flow in one direction but not in the reverse direction.

[13] Analog Switch. An integrated circuit that selects among alternative paths for an analog signal.

[14] Variac. A discrete (single purpose, separate) electronic device that acts like a variable transformer.

Chapter 3

MEASURING BEHAVIORAL RESPONSES

INTRODUCTION

In the last chapter, we reviewed methods and techniques that we have found useful for controlling the delivery of stimuli during behavioral neuroscience experiments. As we pointed out in Chapter 2, typically the major reason for presenting stimuli to subjects in an experiment is to elicit behavioral responses that are under observation. Indeed, normally the primary goal of *behavioral* neuroscience experiments is to study changes in *behavior* that occur due to a manipulation of the external (or for that matter, the internal) environment of the experimental subject. Just as it is vital that stimuli be presented in a well-controlled and precise manner, it is vital that behavior be measured with accuracy and reliability or the results of the experiment are meaningless.

In strictly behavioral experiments, behavior is typically observed in an attempt to advance our understanding of the laws and principles that guide behavior. The underlying biological processes responsible for generating the behavior are not of concern or interest to the behaviorist. In behavioral neuroscience experiments, however, the behaviors that are measured are thought to reflect underlying activity in the nervous system. The behaviors that are observed can be as simple and discrete as the movement of a single muscle, such as might be observed in habituation, sensitization or simple conditioning experiments, or as complex as a verbal utterance, the generation of a written list, or a multi-joint reaching movement. For behavioral neuroscience experiments, the selection of the behavior to observe is normally dependent on the neural process that is of interest. For example, to study frontal cortex function, relatively complex cognitive tasks that tap into executive functioning have typically been used. To study spinal cord function, simpler reflex behaviors have been observed. Regardless of the level or complexity of the behavior under observation, however, it is vital that good behavioral measurement techniques be used—the accuracy of the behavior under observation can only be as good as the accuracy of the techniques used to measure the behavior.

For the most part, in this chapter we will present and discuss methods and techniques used to measure behavioral responses in classical conditioning experiments. As was the case for the delivery of stimuli covered in Chapter 2, classical conditioning experiments afford a great opportunity for presenting information concerning response measurements. In classical conditioning experiments, the US reliably elicits a reflexive response, the UR. Eventually, with enough CS-US pairings, the CS comes to elicit the same or a similar behavioral response, the CR. As we discussed in Chapter 1, a major advantage of the classical conditioning paradigm is that the experimenter knows precisely when a response will occur with reference to the presentation of stimuli. Moreover, the response to be measured is relatively discrete, such as the closure of an eyelid, the flexion of a forepaw or a hindlimb, or a change in heart rate or skin conductance. In this chapter, we will present issues related to the successful measurement of behavioral responses in the context of what we know best, classical conditioning and a few other simple learning procedures. However, many of the issues we discuss with relation to recording these relatively simple behaviors certainly apply to situations where more complex behavioral responses are being measured.

MEASURING EYEBLINKS

By far most of our experience has been in measuring eyeblink behaviors during classical eyeblink conditioning experiments. We begin this chapter with a discussion of several ways in which eyeblink can be measured. It should be noted, however, that some of these recording techniques could be used when measuring other types of behaviors. We point out these variations at several places in this chapter.

The eyeblink reflex is a discrete, somatic muscle response that is evoked by presentation of the US during classical eyeblink conditioning. The response can readily be learned to a tone or light CS after several CS-US pairings. Actually, the presentation of the US (either air puff or periorbital shock, see Chapter 2) produces a constellation of responses. In rabbits, for example, the presentation of an air puff US causes a passive retraction of the eyeball via activation of the retractor bulbi muscle, which in turn causes a passive displacement of the nictitating membrane (the "third eyelid") across the cornea of the eye. At the same time, through activation of the obicularis oculi muscle, the upper and lower eyelid close over the eyeball and, dependent on the intensity of the US, some other facial muscles may be activated simultaneously. In other species, such as the cat, the same constellation of responses can be seen although the control of the nictitating membrane is through activation of the levator palpebrae muscles that actively retract the membrane. In still other species, such as rats and humans with less well-developed nictitating membranes, the US typically causes external eyelid movements along with the activation of facial muscles. The CR that is formed to CS-US pairings may include part, or all, of the constellation of responses that are activated by the US, depending on the relative intensities of the CS and US.

To appreciate the complexity of the response it is worth remembering the passive nature of the nictitating membrane response in the case of rabbits. In rabbits the retractor bulbus muscle pulls the eyeball back into the eye socket, displacing Hardner's gland which, in turn, pushes the nasally located nictitating membrane out and over the eyeball towards the ears (i.e., temporally). The inward movement of the eyeball can be easily observed. Any mechanism that pulls the eyeball into the socket will cause nictitating membrane extension. Simultaneous activity in the external eye muscles controlled by cranial nerves III, IV and VI, for example, contribute to the nictitating membrane response by pulling the eyeball into the socket in addition to the activity of the retractor bulbus muscle. In contrast, as noted above, the nictitating membrane response in cats is an active process where the membrane is pulled across the eyeball.

The nictitating membrane of the rabbit has been used for decades as the measure of the conditioned and unconditioned responses in classical eyelid conditioning (Deaux and Gormezano, 1963; Gormezano 1966, 1972). Originally, the nictitating membrane movements were transduced by a mini-torque potentiometer. The output of the potentiometer was put into a polygraph and calibrated so that 0.5 mm of membrane movement were recorded as 1.0 mm of pen deflection (a gain of 2). When the classical eyelid conditioning procedure was computerized, the analog signal originating from the potentiometer that signaled the eyelid movement was sent to an A/D (analog to digital) converter[1], which in turn, changed the continuous wave signal from the potentiometer into numbers reflecting the magnitude of the signal (we will take this issue up later in this chapter).

Gormezano and colleagues found measuring nictitating membrane movements in rabbits, their species of choice, to be very easy. Among the major reasons for using the nictitating membrane response was that the external eyelids could be held open with eyelid clips, yet the membrane was still free to sweep across the eye. This actually solved two problems, one practical and one theoretical. The movement of the nictitating member keeps the eyeball moist, thus preventing drying of the eye (which could be problematic when an air puff US is used). In addition, the nictitating membrane does not sweep totally across the eyeball, thus totally preventing an air puff from reaching the eyeball on trials when a CR is evident. Thus, it has

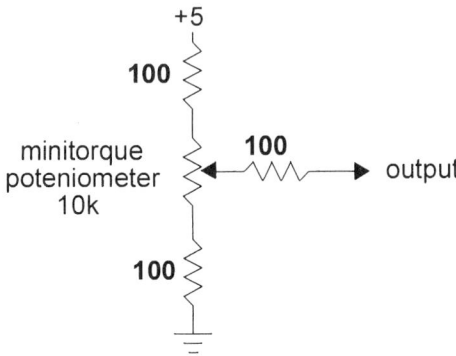

Minitorque Potentiometer

Figure 3.1: Simple eyeblink transducer using a minitorque potentiometer and current-limiting resistors for protection against excessive currents.

been argued that classical conditioning of the nictitating membrane response is not an instrumental conditioning situation, where the animal's behavior can prevent or lessen the impact of the arrival of the unconditioned stimulus. Others have minimized the instrumental nature of the conditioning procedure by using a shock US instead of an air puff US, arguing that it is impossible for the subject to avoid US presentation with a shock. While the theoretical arguments concerning the instrumental versus classical nature of this paradigm are well beyond the scope of this book, we briefly take up this issue again at the end of this chapter. The reader should be aware, however, that these issues have been discussed in the literature and may be important when considering response measurement strategies (e.g., see Gormezano et al., 1983).

A growing number of investigators have abandoned the measurement of nictitating membrane movement during eyeblink conditioning in favor of measuring movements of the external eyelids. There are several reasons for this. First, there is very good evidence that movement of the nictitating membrane and external eyelids are highly correlated and thus highly interchangeable (e.g., Lavond, Logan, Sohn, Garner & Kanzawa, 1990; McCormick, Lavond & Thompson, 1982). Second, aside from the final common pathway for response generation, the brain substrates that underlie conditioning of these two responses systems are extremely similar, if not identical. Third, there is evidence that nictitating membrane control is different across species; for example, nictitating membrane movements are passively controlled in the rabbit and actively controlled in the cat. Fourth, while the rabbit was the favored species for eyeblink conditioning studies in the 1960s through the 1980s, eyeblink conditioning in other species, most notably the rat and human, has recently become more popular. These and other species do not have nictitating membranes that are amenable to response measurement. We thus cover techniques here that are useful for measuring nictitating membrane responses as well as external eyelid movements.

Torque Measuring Devices

Minitorque Potentiometers

The original method for measuring eyelid movement in rabbits was to use a rotary minitorque potentiometer. A lever is attached to the axle of the potentiometer, which when displaced causes rotation of the potentiometer wiper between two poles, one set at ground and the other set at a positive voltage (+5 V in Figure 3.1). The potentiometer acts as a variable voltage divider, with the wiper separating the relative amount of voltage drop across the two halves of the potentiometer. The voltage drop over the resistive material between ground and the wiper is then used to measure the position of the eyelid. Because the wiper can substantially reduce the amount of resistance, causing a large amount of current flow when the wiper is near the positive pole, the small resistive material or the small wire leads attaching to the resistive material inside the potentiometer can sometimes burn up. Since minitorque potentiometers are relatively expensive ($245 the last time we looked (February, 2001), Subminiature Electronics, 950 W. Kershaw, Ogden, Utah 84401-3467), to prevent their damage we have added 100 Ω resistors to all the leads as the manufacturer recommended (see Figure 3.1). We suspect that these leads must be made more fragile today. It used to be that these potentiometers would last until the resistive material wore out from excessive use; newer potentiometers that we have purchased seem unusually prone to damage caused by burnout of the leads despite our best some of our efforts to protect them.

In setting up the potentiometer for measurement, it is best to have the axle sweep somewhere in the middle of the resistive material. Some researchers like to vary this point to keep the resistive material from wearing out. Since the potentiometer axles rotate freely through 360°, one must be careful that the sweep does not go through the 360° to 0° transition point (or back) while trying to measure the eyelid movement. This place on the potentiometer is commonly called the 'flip-point' or 'dead-spot.' Finally, depending upon the polarity of the power supplied to the potentiometer (the two power supply leads can be set up for clockwise or counterclockwise rotation) the voltage drop between the wiper and the ground can get larger or smaller as the eyelid closes. This polarity can be switched by simply interchanging the power leads.

The potentiometer is calibrated by moving the lever (and therefore the axle) over a distance of 1.0 cm, and then associating this movement with the voltage change of the divider. When looking for a minitorque potentiometer, we have used a 10 KΩ device with 0.5% linearity. There is also a parameter called 'smoothness' that is not generally publicized. Smoothness is a measure of how consistent the linearity is from position to position; this is particularly important if the potentiometer is to be used in an x-y micrometer stage, for example. Our advice is to get the smoothest potentiometer you can afford. We will describe how the potentiometer device is secured to the eyelid to measure eyelid closure after we describe a second kind of potentiometer that is increasingly being used, the polarized light potentiometer.

Polarized Light Potentiometers

A reasonable alternative to purchasing minitorque potentiometers may be to construct one's own transducing device. Paul Solomon from Williams College, for example, has built a rotary potentiometer that uses polarized plates to detect eyelid movements (see Figure 3.2). The principle of operation of this device is quite simple: By shining a light through two polarized plates, one fixed into position and the other free to rotate around an axle, different amounts of light pass through the plates depending on how their polarization fields line up. A photodetector device measures the amount of light that passes through the plates. We use a reverse biased high-output LED as the visible light source and a cadmium sulfide photocell as the light detector (both obtained from Radio Shack or other suppliers). Be forewarned that it is tempting to use the available infrared emitter/detector pairs but these will not work—the heat produced by these is absorbed by the plastic polarized plates (Edmund Scientific) that are cut, shaped and placed between the LED and detector. The hardest part about the construction of this device is machining a nearly frictionless axle (19 g stainless steel tubing or 0.016 inch brass tubing from a hardware store) and housing (e.g., PVC tubing, caps or connectors; 35 mm film cans; etc.) for the unit. We have found, however, that these can all be constructed with hand tools (a Dremel motor tool is very handy).

As with the minitorque potentiometer, the axle of the polarized po-

Chapter 3

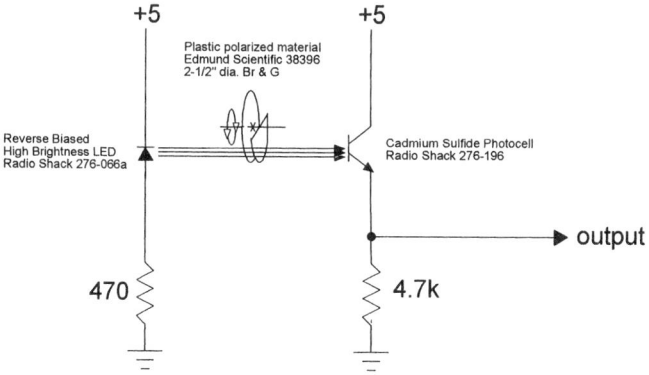

Polarized Light Potentiometer

Figure 3.2: Eyeblink transducer using the relative registration of polarized plates to vary the amount of transmitted light. Based on a design by Solomon.

tentiometer is set so that the range of eyelid movement sweeps between the positions when the polarized plates are perfectly aligned (allowing maximal light through) to when they are at 90 degrees to each other (allowing the least amount of light through). The transition from light to dark is presumably not linear. Rather, we believe it approximates a sinusoidal function. However, these devices are still useful because over the middle range, a sine wave is very close to being linear. The polarity of the polarized potentiometer can be understood as, for example, a clockwise rotation that results in either a darkening or a lightening of the relationships of the plates. To reverse the polarity of the polarized potentiometer, one just has to rotate the axle 45 degrees. Finally, to calibrate the polarized potentiometer, a lever attached to the axle is moved a distance of 1.0 cm and the relationship between voltage change and distance can be established.

Linkage

Ultimately, both the minitorque potentiometer and the polarized potentiometer must somehow be attached to the nictitating membrane or external eyelid to transduce the eyelid movements. At least two ways have commonly been used. In our labs, we sew a tiny loop of 6-0 nylon suture to the outer surface of the nictitating membrane or upper external eyelid (for further information, see Chapter 7 and Figure 7.7). A second length of 6-0 suture, with a hook secured to one end, is then attached to a counterbalanced lever that is attached to the axle of the potentiometer. The easiest way to secure the hook to the suture is to fashion a hook from the swagged, cutting edge needle that is already attached to the suture (by filing off the needle's point and clipping and bending the needle into a hook shape). During training, the hook is secured to the loop of suture that was placed in the eyelid, thus connecting the eyelid to the arm of the potentiometer via the length of suture. Since this suture is a flexible attachment, slack in its length can cause inaccurate readings. A counterbalance weight that is placed on the potentiometer arm is used to maintain tension on the eyelid loop. Excessive length of the suture can be wound around a spring as indicated below.

Other laboratories have used direct coupling systems to attach the potentiometer lever arm to the eyelid. In this method, a stiff stainless steel wire (or tubing) extends from the lever and is attached to the eyelid loop by a hook in the wire. Coupling with the rigid wire has several advantages; there is no slack in the movement (other than at the eyelid loop), there is no inertia or momentum associated with the counterbalance, and the coupling is very lightweight so it does not contribute much to the measurement of the eyeblink. In Figure 3.3 we show a method for achieving rigid coupling of the potentiometer arm to the loop of nylon suture placed in the nictitating membrane. A number of researchers have described this method (for example, Berny Schreurs, West Virginia University). Our description is based on the method used by Tapani Korhonen (University of Jyväskylä, Finland). A large hook is placed at the end of the potentiometer arm. Onto this hook is attached a lightweight (e.g., 23-30 gauge) stainless steel tubing that extends to the eyelid suture, forming a hinge. Small eyes are soldered onto the extension at different positions so that the best one can be chosen to achieve a right angle to the potentiometer arm at rest. At the other end this extension is shaped into a hook that attaches to the eyelid suture. In order for these systems to work, the connection to the eyelid suture must be secure without poking the eyelid or eye, and the joint between the potentiometer arm and the extension must be free moving with little play. The Korhonen design works reasonably well, depending primarily upon gravity and the angle of the bends. Other designs have fabricated a pivoting hinge between the potentiometer arm and the extension, and a spring-loaded hooking mechanism to attach to the eyelid suture (a miniature of the design of electronic test leads).

At this point it is worth spending some time to discuss the loop of suture that is placed into the eyelid. The nylon loop is placed into the nictitating membrane or in the upper external eyelid at the end of surgery when the rabbit is still anesthetized or in an awake animal with the eye anesthe-

Chapter 3

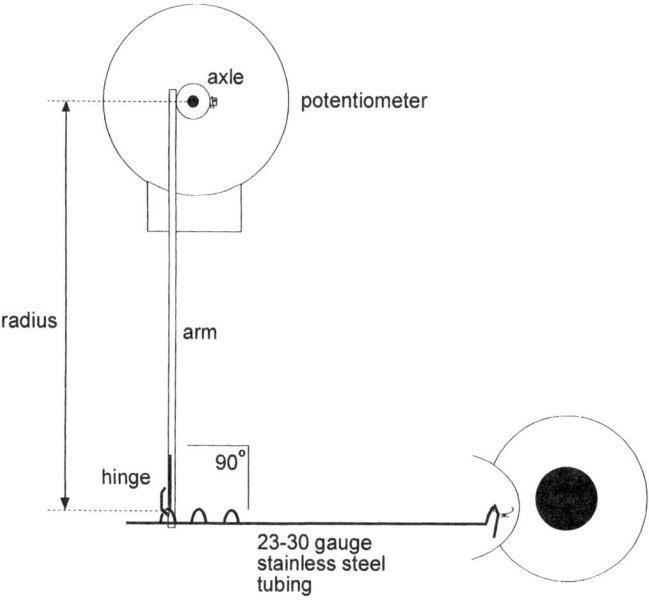

Figure 3.3: Rigid linkage of the nictitating membrane to the arm of the potentiometer.

tized with a local ophthalmic anesthetic. Various tools have been used to gauge the size of the loop (e.g., the wooden end of a cotton-tip applicator, the shank of a surgical drill bit); but by far the most popular tool is the smooth shaft of an injection needle. Injection needles have the advantage that they come in a variety of standard sizes from which to choose. Ideally, the size of the loop should be as small as possible. We typically use 14-gauge spinal tap needles as the largest size, and occasionally use 21-gauge injection needles for the smallest loops, with 19-gauge being a popular choice. If the loop is made too small, it can be very difficult to attach or remove the potentiometer linkage. If the loop is made too large is could potentially irritate the cornea of the eye and adds unnecessary slack to the system. Care should be taken to make sure that the loop is placed on the outer surface of the eyelid and not all the way through the eyelid as sutures located on the inside of the eyelid will invariably contact the cornea causing undue irritation. One of the problems we have frequency encountered with this nylon loop is that the knot can come undone, often either because the slippery nylon suture has not been tied tightly or because the ends of the suture have been cut too close to the knot, requiring that another loop replace the lost suture (usually a local ophthalmic anesthetic but occasionally a general anesthetic is used). The reader should see Chapter 7 on surgery methods. There, we have provided a rather detailed discussion of knot tying as it relates to surgery and general experimental procedures.

Finally, we close this section on potentiometers with a story on how well established methods can become "laboratory lore" and experimenters can lose sight of the real reason a particular method, tool, or technique is used in a laboratory. Michael Patterson, who was Steinmetz's Ph.D. mentor, loves to tell the following story: Patterson was a graduate student of Isidore Gormezano at the University of Iowa and was involved with many of the technological and methodological developments associated with rabbit eyeblink conditioning. Indeed, it was Patterson who introduced the classical eyeblink conditioning paradigm to Richard Thompson and his colleagues in the late 1960s at the University of California, Irvine, when Patterson was a postdoctoral fellow in Thompson's laboratory. In the Gormezano laboratory, and later in the Thompson laboratory, it became common practice to place a spring from an ordinary ballpoint pen at the end of the lever arm of the potentiometer such that the arm pierced the two ends of the spring, making it arch and rotate freely around the arm. The spring had a very simple purpose: The end of the suture connecting the eyelid to the arm could be wrapped around the spring and arm; by rotating the spring around the arm, the length of the suture could be shortened or lengthened and thus the position of the arm of the potentiometer could be easily adjusted for individual rabbits. In the late 1970s, Patterson returned to Thompson's laboratory at Irvine for a brief sabbatical and noted with some glee that his spring "invention" was still part of the potentiometers being used by the graduate and undergraduate students in the laboratory. He quickly noted, however, that few of the students actually moved the spring to adjust the length of the suture, and, in fact, sometimes used other techniques to make this adjustment (like clipping the suture and retying it to the potentiometer arm). To his chagrin, upon quizzing the students as to the purpose of the spring on the arm of the potentiometer, he heard a number of explanations including, "it is a counterbalance for the counterbalance" or "it provides further stability for the arm of the potentiometer" or, better yet, "gee, I don't know." Patterson typically concludes this story by noting that the students appeared to be truly amazed when the real purpose of the spring was revealed, typically stating, "what a useful device!" When later quizzed by Steinmetz, Lavond was one of the exceptions who understood the purpose of the spring. Don Weisz had trained Lavond in its use.

Reflected Infrared Light

Another method for measuring eyeblinks is to bounce a light off the eyeball and measure the amount of reflected light. One means to do this is to use an infrared light source and detector. This basic strategy has been described earlier (March, Hoffman and Stitt, 1979). This method has the advantage that it is not as sensitive to interference from ambient light levels as a visible light detector. The infrared system has the disadvantage, however, that the infrared spectrum is heat, and heat can get uncomfortable on the eyeball. The solutions are to use a small amount of infrared light (which can become insensitive to movement) or to adopt a strategy where the infrared light is quickly turned on and off so that the time of heating is reduced. This latter strategy requires that the eye position be measured only when the light source is on. The circuitry required to do this gets complex because of the rapid switching on and off and the integration required to compensate for the off periods, both of which can introduce electrical switching noise into the measurement circuit.

Figure 3.4 shows Paul Solomon's reflected infrared light circuit for measuring eyeblinks. The essence of the operation of this system is to turn the infrared light on for brief periods of time to avoid heat damage to the eye. By careful electronic choreography of the timing and by filtering to extend sampling a continuous sample can be simulated. The only disadvantage that we have encountered with this strategy is that the high rate of switching potentially introduces high frequency noise, but this can be limited by shielding.

LED/Phototransistor Eyeblink Device

John Disterhoft and his colleagues at Northwestern University have described a device for detecting eyeblinks that uses reflected light (Thompson et al., 1994). The apparatus consists of two parts; an optoelectronic sensor device that converts an eyeblink into an electrical signal, and an electronic circuit that amplifies, buffers and filters the voltage output of the sensor device (see Figure 3.5).

The sensor device is an Optek OPB704 (TRW Electronics) which contains a light-emitting diode (LED)[2] and a phototransistor[3] that are packaged together in a plastic container. The theory of operation of this device is conceptually quite simple: The LED emits a focused beam of low-intensity infrared light that is reflected back from the surface of the cornea of the eye. The amount of reflected light is converted to a DC signal by the phototransistor. The relative translucent surface of the cornea reflects a smaller amount of light than the relatively more opaque nictitating membrane and/or outer eyelids. The phototransistor output converts the varying amounts of reflected light into a range of voltages that correspond to the degree of eyelid

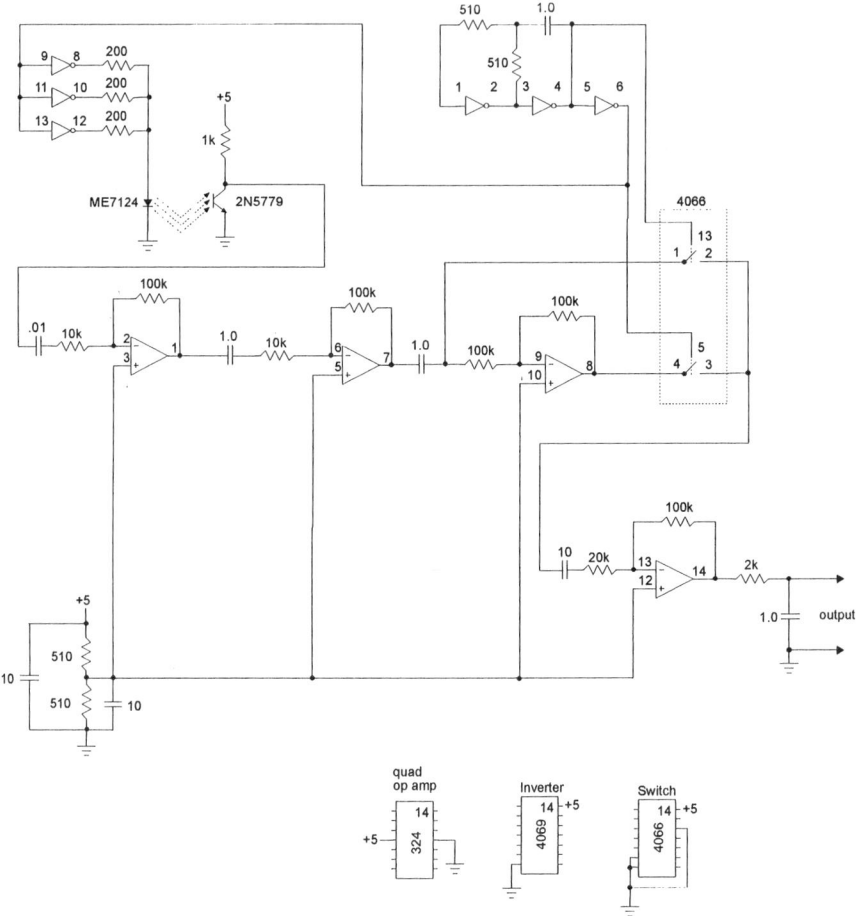

Figure 3.4: Solomon's intermittent infrared circuit for measuring eye blinks.

closure that is present. The output of the phototransistor is then amplified and filtered in a blink detector circuit (see Figure 3.5).

There are some obvious advantages of this system. The system is non-invasive; no sutures have to be placed in the eyelid as is the case for the potentiometers described above and no recording electrodes have to be placed in the eyelid musculature as is the case for EMG methods described below. Also, devices such as these have proven useful for human eyelid conditioning experiments; these devices can be mounted in glasses or goggles very easily. The system is very durable and reliable, and relatively inexpensive to build and maintain. The system operates well in darkness and under dim illumination conditions. Some disadvantages of the system include difficulties in calibrating the output of the device, difficulties in con-

Chapter 3

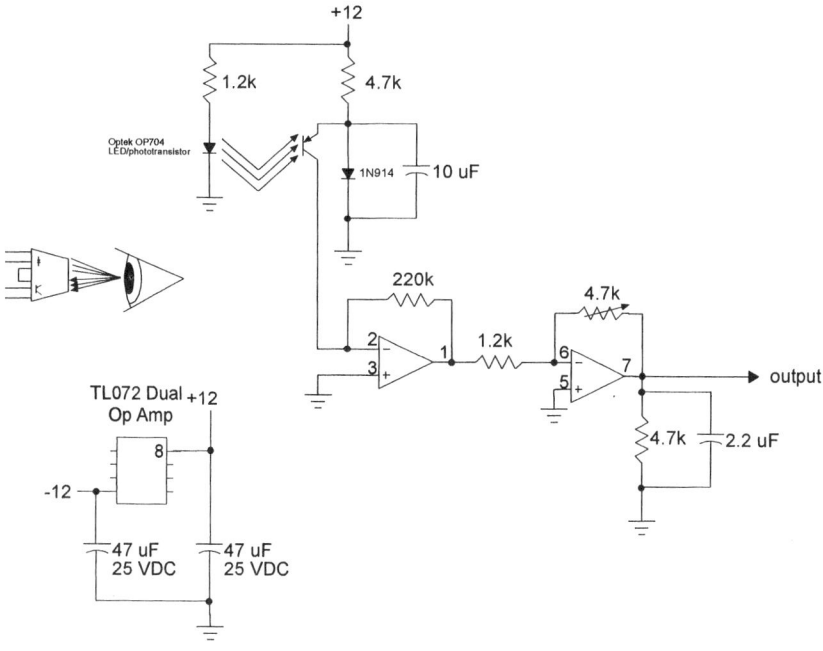

Figure 3.5: Disterhoft's reflected light circuit for measuring eye blinks.

sistently positioning the device at an optimal distance from the cornea (which is about 4 mm), and issues concerning linearity of the output. Also, this device cannot be used when training with a light CS if the light CS is delivered near the detection device (as described in Chapter 2). All in all, however, this device has proven to be a good alternative to the potentiometers described above.

Reflected Visible Light

Bob Clark, a former graduate student of Lavond's, designed the system shown in Figure 3.6 that uses visible light with an emitter/detector pair. This device uses a continuous light source. The visible light system avoids the switching and integrating problems inherent in the infrared system. However, with the visible light system one needs to be careful about the influence of ambient light on the measurement.

The Clark visible light system uses an integrated package housing both the emitter and detector. An advantage of this arrangement is that the emitter and detectors are already aligned mechanically. The photodetector

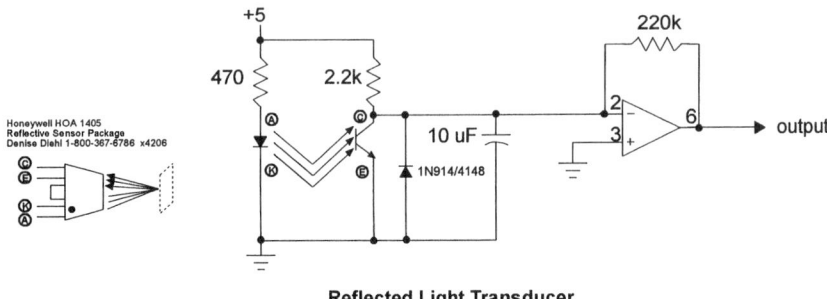

Figure 3.6: Clark's reflected light circuit for measuring eye blinks.

output is a current that is converted to a voltage by the operational amplifier configuration. The value of the 220 KΩ feedback resistor can be changed to alter the conversion relationship. Using a 3130 or 3160 operational amplifier allows for a single-ended supply whose output can take full advantage of the operational amplifier's power supply range.

The infrared and visible light reflecting systems described above can be used in humans and restrained animals. But, it is more difficult to calibrate these systems than the potentiometers discussed above. One way to calibrate human eyelid movement is to measure the normal distance between the eyelids when the eyes are open, and to relate this information to trials on which it is known that the eye was fully closed. A computer can then go back and calculate the distance of an intermediate response.

EMG Techniques

Another way to go about measuring eyelid movement is to monitor changes in the physiology of the muscles that are involved in generating the eyelid movement via electromyographic (EMG) measurements. Recording EMG activity is somewhat straightforward. Very thin wires are typically placed in the eyelid musculature. Different techniques are used for rabbits, rats, and humans.

For rabbits, after applying a local anesthetic to the skin over the muscle from which recordings are made (typically the obicularis oculi muscle), we have inserted a small-gauged needle through the skin and a small portion of the muscle and then threaded a stainless steel wire through the needle. While holding one end of the wire, the guide needle is then pulled out leaving the wire in place. The wire is then made into a small loop, the ends wrapped together, and a gold amphenol recording pin crimped onto the wrapped ends of the wire. Some laboratories have then dabbed a little non-toxic, non-irritating glue (such as tennis shoe glue) onto the wire and base of the gold pin to protect the wire and prevent movement. During training, leads are connected to the amphenol pins and the signal amplified (x100) and band-pass filtered. We have found the A-M Systems Model 1700 (four-channel) or Model 1800 (two-channel) AC amplifier to be particularly useful for amplifying the EMG signal.

Recording EMG from freely moving rats is a bit more difficult. Because they are not restrained, leads connected to externally accessed EMG electrodes are easily displaced or destroyed. Therefore, we typically route very fine wires under the skin from an acrylic headstage cemented to the skull. The EMG recording wires are constructed of two strands of ultra-thin (0.003") Teflon-coated stainless steel wires. The wires have exposed, non-insulated ends that are bent into small hooks that are positioned in the orbicularis oculi muscle. In the acrylic headstage, the EMG leads are anchored in a plug assembly that mates with a commutator connector (see Chapter 4 for a complete discussion of recording in awake, freely-moving rats). We use the A-M Systems Model 1700 or Model 1800 AC Amplifier to boost the EMG signal recorded from the muscle.

We have used surface electrodes to record EMG activity from human subjects during eyeblink conditioning experiments. The electrodes are 4 mm silver-silver chloride (Ag/AgCl) disks. The active electrode is placed directly below the lower corner of the eye and a reference electrode is placed approximately 10 mm lateral to the active lead. A clip electrode is normally attached to an earlobe and serves as the ground point for the recordings. The EMG signal is inputted to a Grass P-15 Differential Amplifier, amplifier 1000x, and band-pass filtered at 100 to 1000 Hz. We use standard electrode gel to improve contact between the electrode and the skin surface. Care must also be taken to remove make-up and thoroughly cleanse the skin with alcohol before applying the electrode gel and electrodes.

After recording and amplifying the EMG signal, there are at least two ways in which the signal can be further processed. Lavond prefers to treat the spikes recorded in the EMG record as neural unit activity; the activity is amplified, spikes are discriminated using time/amplitude criteria, and then the discriminated spikes are then collected into time bins, as detailed in Chapters 4 and 5. Steinmetz prefers to use EMG integration techniques to create an analog record of the EMG activity. There are certain advantages to the use of EMG integration, the most important one being that the resulting continuous, analog output is similar in form to eyeblink data collected with potentiometers and reflectance devices. Software already exists for analyzing the continuous, integrated data. There are some disadvantages as well to using EMG integration techniques, the chief one being that the integration of the EMG signal introduces a time delay in the process. If the delay is known, then most researchers are not worried because the delay can be factored into the analysis. But, the fact remains that the delay is a function of the slope of the integrated activity, therefore it is not a single delay but a delay that can vary from slope to slope. Choice of the equipment used for integration can also be important; some integrating devices, such as the BAK Model ABI-1 AC Bridge Integrator, are designed to minimize the delay that is introduced into the process.

Figure 3.7 shows circuitry that can be used to convert the raw EMG signal into a analog, nictitating membrane-like, DC signal. This process involves four operations: rectification, inversion, mixing and integration. In our design, rectification is carried out by operational amplifiers that are configured as positive and negative superdiodes and inversion is accomplished by differentiation in a design that also incorporates mixing of the rectified signals. The final stage involves an operational amplifier low pass filter for integration. The superdiode design is an improvement over ones which use oppositely arranged diodes to separate out the positive and negative spikes, where the forward voltage drop of a simple diode, typically 0.5 V, cuts out the signal closest to ground. The combined mixer/differentiator design simplifies the design by avoiding a separate inversion stage. As indicated in the figure, the gain of the positive superdiode must be adjusted by a simple potentiometer. To do this calibration, a sine wave can be inputted to simulate an EMG signal and the output of the mixer/differentiator adjusted until the peaks are of equal value. The final stage, the integrator, is a maximally flat, 20 Hz, low pass filter of a second order (meaning that there is a 12 dB/octave drop off of the filtering characteristics). Please note that the signals used as examples in the figures are cartoons meant to illustrate the principles of the circuit and are not precisely the output you should expect.

CONVERTING EYEBLINK MEASUREMENTS TO COMPUTER-COMPATIBLE INPUT

After recording the movement of the eyelid, the signal from one of the eyelid transducers described above is often treated further by amplification, polarity switching, or additional filtering. Originally, the output of the minitorque potentiometer was low pass filtered through a large capacitor and resistor and then amplified on a Grass polygraph from which measurements were taken by hand. The filtering was applied so that each trial would start at about the same baseline level although there may have been some movement or eyeball adjustments during the interval between trials. For example,

Chapter 3

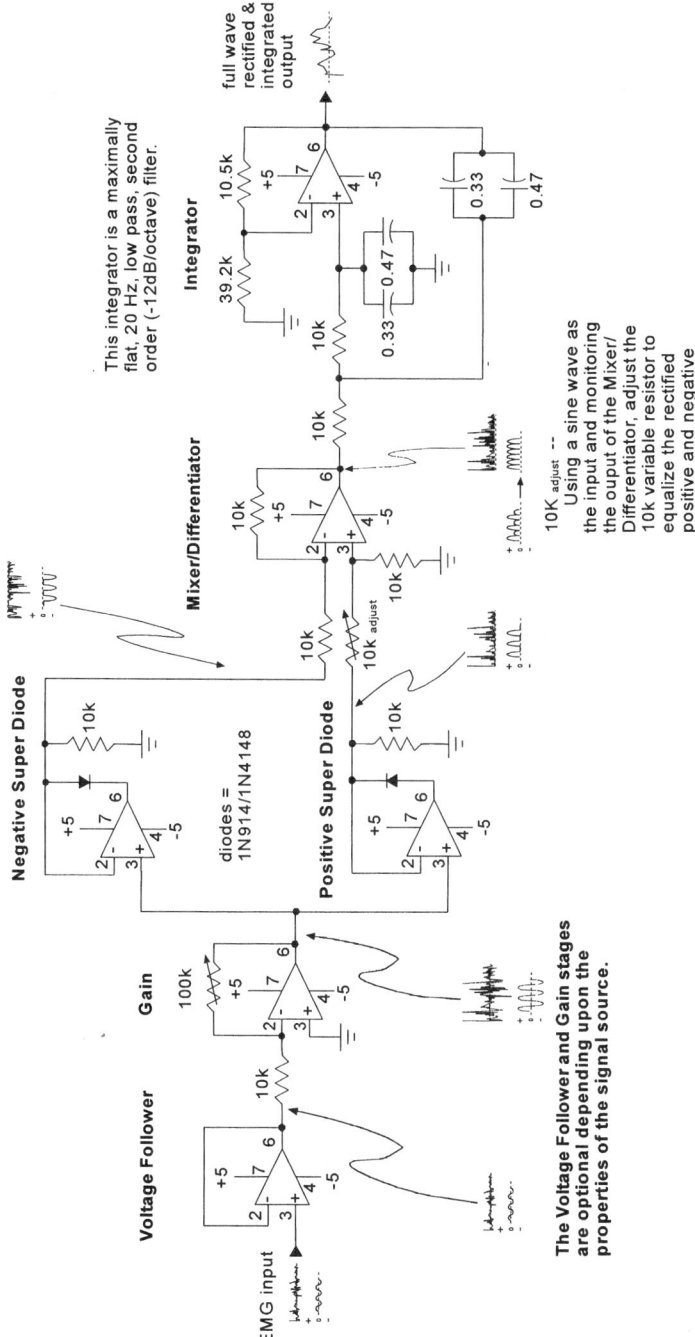

Figure 3.7: Circuit for converting eyelid EMG into a continuous wave, nictitating membrane-like signal.

the baseline of the record self-adjusts for a rabbit that holds its eye partially closed during the intertrial interval. The amplification was made as part of the calibration procedure, so that moving the lever through 1.0 cm caused a known amount of movement on the polygraph paper, which the experimenter could then go back and measure with a ruler, or use the lines on the polygraph paper. Having done this ourselves, we cannot express how thankful we are that mini- and microcomputers were developed. All of these considerations have been important for developing circuits that we generally used to refer to as the "NM [nictitating membrane] box" or "NM circuit" (so named for the original container that held the filtering capacitor and resistor). The main function of this circuit is to condition the signal from the transducer by amplifying, filtering, inverting, and adjusting the baseline of the eyelid measurement.

The circuit we use is based on a circuit that was published to automatically correct for ambient light levels at the receiver end of a fiber optic system (Myers, 1982). Our version, shown in Figure 3.8, starts with a voltage follower for isolation and impedance matching then uses the core of the Myers design for adjusting the gain, the baseline compensation, and the filtering/integrating of the signal. With the values shown, it takes about 11 seconds for the output to return to the baseline level after a full range deflection of the potentiometer. This is approximately the rate of self-adjustment we found for the original resistor-capacitor in the NM box. This value works reasonably well for dampening small changes during the intertrial interval while still being sensitive and faithful to rapid changes during the trial. This stage also has a gain/calibration adjustment. We adjust the gain so that 1.0 cm of movement of the lever attached to the axle of the potentiometer results in a change of 1.96 volts (5 V times 10 mm/cm divided by 25.5 mm). In other words, the +5 V supply of the circuit is calibrated for an 8-bit A/D converter (ranging from 0 to 255) so that each number of the A/D converter corresponds to 0.1 mm of movement; thus, our potentiometer and NM box can is calibrated to measure up to 25.5 mm of movement (i.e., well beyond the normal blinking range of rabbits). As a result of this calibration, there is no computer time required to compute the movement in millimeters—the A/D value is directly read as tenths of a millimeter, which is the defined precision of our measurements. Obviously, greater precision could be achieved with a more restricted voltage range, with greater amplification, or with a 12-bit A/D converter, for example, but we do not really see the point of that.

The next stage is optional. It is used to invert the signal before it goes to an A/D converter or oscilloscope. As noted above, inversion of the polarity can be achieved as well by reversing the power on the minitorque potentiometer or by rotating the axle of the polarized potentiometer by 45 degrees. In our software designs (See Chapter 11) we actually assume that

Chapter 3	89

eyelid closure yields A/D values that get smaller, rather than what one would normally expect, i.e., that eyelid closure would cause larger A/D values. The reason for this has to do with computation time and how points are displayed on a computer monitor. Whereas we normally would think from our experience with Cartesian coordinates that the x,y coordinate 0,0 would be represented by the lower left hand corner of the monitor, in reality computers represent the coordinate 0,0 in the upper left hand corner. It turns out, as a handy trick of programming, that inverting the A/D numbers (i.e., subtracting the A/D value from 255, the maximum value of our 8-bit A/D converter) actually converts the monitor screen from the upper left corner to the lower left corner. If we set up our potentiometer or signal conditioning circuitry to electronically invert the signal, however, then no computation is necessary and the collected A/D values can be directly displayed on the screen. This trick allows for faster display of the data since it involves no computation. It also turns out that having inverted numbers has little consequence for subsequent computations of amplitude and latency of the eyelid responses.

One of the criticisms of signal conditioning of the eyelid response is that the circuitry for filtering the intertrial events also applies the filters to the eyelid behavior during the trial itself. Although this distortion is really pretty small, it is nevertheless present. A new feature in our design is the addition of a sample and hold feature (Figure 3.8). During the interval between trials when the sample and hold circuit is given a TTL logic 0 (off), that is, when the 4066 switch is closed so the circuit is in sampling mode, the filtering characteristics are fed back and affect the output. But when a TTL logic 1 (on) signal is applied during a trial, the feedback of the intertrial baseline is held and the filtering is effectively disconnected from the circuit (the 4066 switch is shown in the open position which "holds" the baseline). Thus, there is active (filtered) coupling between trials, but direct (unfiltered) coupling during the trial. This new feature has the best of both worlds: baseline compensation due to eyelid movements between trials but direct measurement of eyelid movements during the trial. By holding with a TTL logic 1 (on) signal for sampling all the time, this NM box reduces to a direct measurement system with no filtering characteristics at all, yet it is still calibrated because the gain function is still in effect. A manual switch disconnecting the feedback would also result in a direct coupled system all the time, as would removing the feedback operational amplifier altogether, and the voltage divider for baseline adjust should be added to pin 3 of the gain/calibration operational amplifier.

A final note about the design advantages of our system is that the output is restricted to the voltage supply, here a single-ended supply between 0 and +5 V. By using 3130 or 3160 operational amplifiers whose outputs can cover the range of its power supply, the output of our circuit is also in the range of 0 to +5 V. The 3130 and 3160 operational amplifiers are unlike many popular operational amplifiers, like the general purpose 741 for example, whose outputs can only approach within a few volts of either side of the power supply. What this means for us is that the A/D converter on our PC-based interface boards can also be powered by the usual +5 V TTL supply, and that as another result, the A/D converter will use its full range of conversion values of 0 to 255 units (where a 0 V input equals a value of 0, and a +5 V input equals a value of 255).

A word of caution about data acquisition equipment like ours which depends upon calibration of the potentiometer and other measuring devices for accurate measurements. More truthfully, it is our software, not the hardware, which makes an assumption about the meaning of the voltage changes coming from the potentiometer or other recording devices. We calibrate the potentiometer arm by having it traverse through a known distance and relate this distance to the A/D converted values. By careful adjustment of the amplification of the output of the potentiometer we calibrate the system so that 1 A/D value equals 0.1 millimeter resolution of movement. Practically speaking, this is accomplished by moving the potentiometer arm (at the point where the suture would be attached) through 1.00 centimeter and calibrating the potentiometer's amplification until the corresponding number of A/D units is 100. The great advantage of this method is that there are no calculations needed to translate the A/D values into movement during the experiment. To accomplish this calibration, we construct guides to limit the excursion of the potentiometer arm to 1.00 centimeter. This means that the guide must be 1.00 centimeter plus the thickness of the potentiometer arm because we want to measure 1.00 centimeter from one edge. In other words, we want to be consistent in measuring from the leading edge of the potentiometer arm between the start and end points through 1.00 centimeter, which is accomplished by negating the thickness of the potentiometer arm. Note that this calibration is only true if we are consistent in measuring the excursion of the potentiometer arm at the exact point where the nylon suture or the rigid extension would be attached during conditioning. Great care must be taken in making sure that we consistently choose the calibrated radius of the potentiometer arm. If the radius of potentiometer arm (the distance from the suture attachment to the potentiometer axle) is not consistent, then the same eye movement would cause different amounts of rotation of the potentiometer axle, thus yielding different and incorrect measures from the radius where the calibration had taken place.

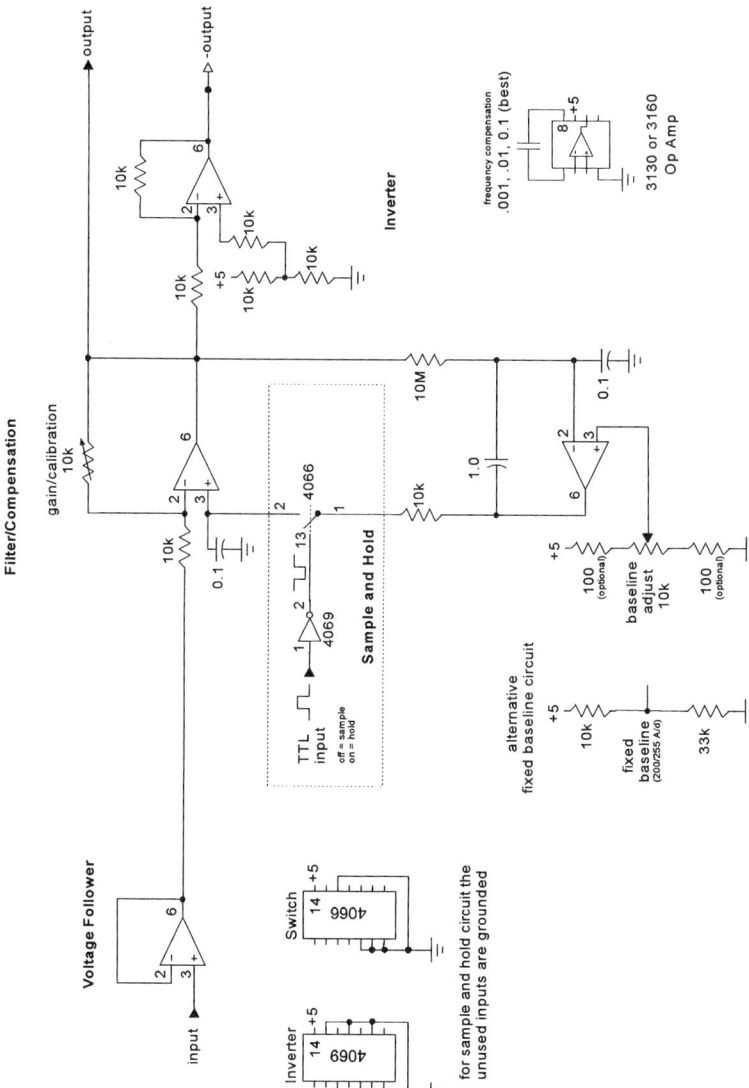

Figure 3.8: "NM box" or "NM circuit" to automatically filter, adjust the baseline and amplify the eye blink into a computer-compatible signal. This circuit allows for continuous or sampled filtering for classical or operant eye blink conditioning, respectively.

MEASURING OTHER SOMATIC RESPONSES

One of the most attractive features of using the rabbit in classical eyeblink conditioning experiments is that making behavioral measurements, using devices such as potentiometers, reflected light sources, and EMG recording equipment, is relatively easy and straightforward. The restrained rabbit sits quietly in the box thus affording easy access to the eyelid for behavioral measurements. Things are a bit more tricky when using the freely moving rat for classical eyeblink conditioning experiments. Using bulky devices such as potentiometers and reflected lights systems are quite difficult, leaving EMG recording via electrodes that are implanted subdermally as one of the few alternatives available.

Measuring behavioral responses from other somatic response systems can also be tricky, but not impossible. Measuring EMG activity might be the easiest method. EMG electrodes can be implanted into just about any muscle, using surface electrodes for restrained animals or subdermally routed electrodes for freely moving animals, and electrical activity of the muscle of interest can be recorded. Similar to eyeblink measurements, the EMG signal can be amplified and treated as a spike record or integrated to produce a smooth analog record that can be processed easily by A/D converters on computer interface boards.

Potentiometers can also be used to record behaviors from other response systems. For example, several years ago, Patterson and his colleagues used a potentiometer-based system to record hindlimb flexion responses from restrained cats. This same system can be used for rabbits and rats, as well. The animal is placed in an elevated sling or Plexiglas restrainer that contains hole through which the hindlimbs (or forelimbs, for that matter) can protrude. A small loop of suture is placed through the skin of the limb. Alternatively, if non-invasive techniques are desired, a loose knot or Velcro strap can be placed around the limb whose movement is to be monitored. A potentiometer is mounted onto a base under the elevated sling or restraint box and a length of suture is connected between the arm of the counterbalanced potentiometer and limb of the subject. Flexion (or extension) of the limb causes movement of the potentiometer arm, which in turn rotates the axle of the potentiometer, thus recording the movement of the limb. Steve Berry at Miami University has also used potentiometers to measure jaw movements in restrained rabbits. The lead to a counterbalanced potentiometer is simply connected to a loop of suture placed in the skin overlying the jawbone. Jaw movements are thus monitored by movements of the potentiometer arm.

Another indirect method for measuring whole body movements involves the use of stabilimeter platforms placed under the restraint boxes (for rabbits) or floors of cages (for freely moving animals such as rats or mice). Stabilimeter platforms can be easily fashioned from two Plexiglas sheets with springs mounted between the plates on the four corners and in the middle (see Figure 3.9). The basic idea of this device is that movements in the restraint box or cage placed on the platform will differentially depress the springs thus moving the two sheets of Plexiglas toward each other. For example, a mild leg shock delivered to a rabbit will produce a leg flexion that causes a downward push on the platform. A potentiometer attached to the platform is an easy way to monitor the relative movement of the platform during the response, although the Lavond laboratory has tried this technique and found it is not sensitive enough to pick up and measure subtle movements. Other types of activity monitoring platforms are commercially available from a number of sources (e.g., Columbus Instruments, Lafayette Instruments, and Med Associates, Inc.).

Finally, force transducers can be used in place of potentiometers to measure movements. Force transducers come in a variety of sizes and can transduce a variety of ranges of forces. These are available from a variety of commercial sources (such as Grass Instruments, Stoelting, and World Precisions Instruments). These can be directly attached to limbs of restrained animals or they can be attached to the stabilimeter platforms described above.

MEASURING HEART RATE AND OTHER AUTONOMIC RESPONSES

A number of investigators have measured autonomic responses during classical conditioning and other types of behavioral neuroscience experiments. Autonomic conditioning is relatively rapid, seems to precede several different types of somatic conditioning, and is extremely robust. The three most popular autonomic responses that have been measured during conditioning experiments are heart rate, respiration, and skin conductance.

Recording Heart Beats

Figure 3.10 shows a circuit we have used for heart rate recordings in rabbits during classical conditioning experiments. A TL074 quad operational amplifier is used to create a differential amplifier with high common mode rejection. The active leads are placed on either side of the chest cavity of the restrained rabbit and the ground is placed on a distant point, like the leg, for a reference. We used Ag/AgCl cup electrodes and electrode paste to

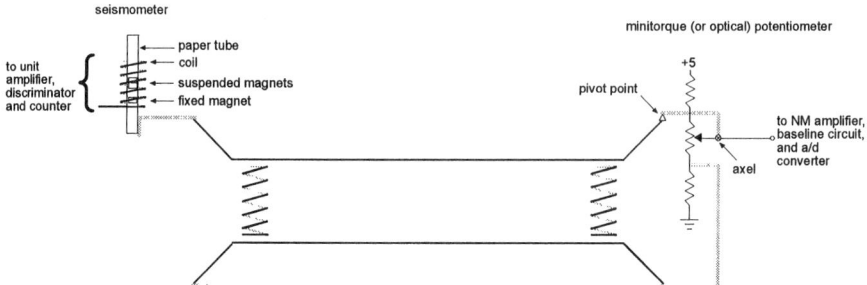

Figure 3.9: Simple stabilimeter (jiggle stand) for measuring gross movements.

record the heart rate response. The cups were held in place by plastic clip leads that grabbed a small area of shaved skin. Subcutaneous electrodes (loops of stainless steel wire placed through the skin) were also effective, but there is a risk of infection when these are used for long-term chronic experiments. Filtering on the input side (AC, or Active Coupled) and output side removes DC potentials. A separate 741 operational amplifier is used for signal gain. An additional 741 operational amplifier is configured as a second order, Dual Component VCVS low pass filter to filter out frequencies above about 70 Hz. From this filter the heart beat output can be monitored. Normal rabbit heart rate baseline activity averages about 200 beats per minute. The final stage is a voltage height comparator that is used to give a TTL level output signal for each heart beat. A timer could be added after this stage to provide a uniform pulse duration for each heart beat. For our heart rate experiments, the output of the filter was amplified and recorded on a Grass polygraph, along with signals on another polygraph channel that indicate the onset of the CS and the US. By comparing the stimulus signals with the heart rate, the first 10 heart beats before CS onset (baseline) and the first 10 heart beats after CS onset (conditioned response period) can be measured to determine habituation and conditioning.

By routing the TTL output signal for each heart beat into the interface card (See Chapter 11) a computer program could be written that measures interbeat intervals (i.e., heart rate). For example, the heart beat TTL signal could be read by the bit on an 8255 Peripheral Programmable Interface (PPI) chip[4]. An interval timer could then be activated until another heart beat is detected. The interval difference between the two beats could then be measured. This period could then be converted into a frequency per second (Frequency = 1/Period), and then converted into beats per minute by multiplying the frequency by 60. Alternatively, the interval difference could be used as an index that is referenced to a table that has already calculated the beats-per-minute for any given interval. Either way, the interval timer could be reset after each reading, so that the heart rate is continuously deter-

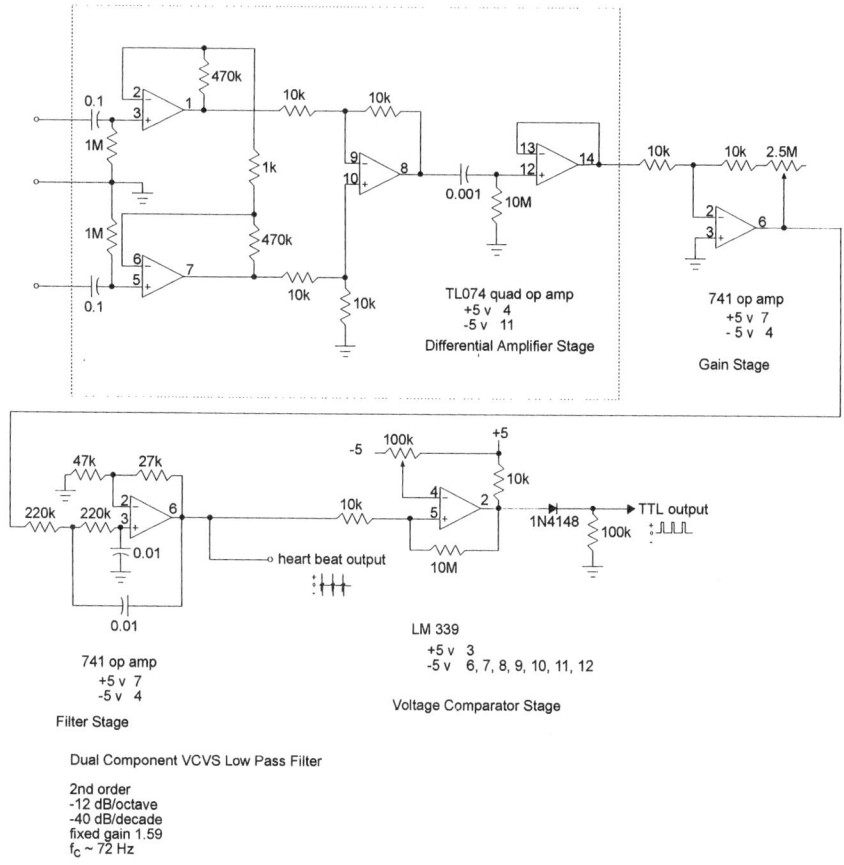

Figure 3.10: Differential recording heart rate amplifier.

mined, beat by beat. Using this arrangement, a type of cardiotach can be fairly easily implemented.

There are other ways to measure the number and spacing of heart beats. For example, the output of the recording electrodes can be amplified and filtered and then sent to a window discriminator/comparator used for spike detection. Heart beats can then be detected and discriminated as discrete spikes when they break a preset voltage level and then the output of the discriminator can be routed to the circuitry described just above for counting. Alternatively, the amplified and filtered signal could be input directly to an A/D channel on a computer interface and heart beats counted using software routines.

Recording heart rate activity from freely moving rats can be a bit difficult, but certainly not impossible. We have used two methods. Two small Ag/AgCl electrodes can be implanted under the skin over the chest muscle with a third ground electrode place in a muscle in the scapular region. The leads from the electrodes are run under the skin and then attached to a plug assembly cemented into an acrylic headstage. A swivel and tether device can then be plugged into the headstage (see below) and the electrocardiographic (ECG) signal can be routed through the swivel to amplifiers and recording devices. We have found that this system works well for a few days and that rats show no signs of discomfort or irritation caused by the subdermally routed electrode leads. However, if more extended training sessions are necessary, the leads to the recording electrodes often break or wear out with movements of the animal. Linda Rorick, a graduate student in the Steinmetz laboratory, has come up with a non-invasive alternative to the implanted ECG electrodes. She has fashioned a stylish vest for rats that contains the recording and ground electrodes on the inner surface of the vest (see Figure 3.11). Leads from the recording electrodes exit the back of the vest and can be connected directly to the swivel and tether system without creating a headstage and plug assembly. The rats seem to quickly adapt to wearing the vest and this recording method has yielded reliable data concerning heart rate.

Recording Respiration Responses

Another popular autonomic response that can be recorded is respiration. Perhaps the easiest way to record respiration is to use respiratory transducers that monitor changes in force due to the expansion and contraction of the chest cavity that occurs with respiration. The output of the transducer is typically routed to an amplifier and filter which converts the respiratory signal to a DC signal. The signal can then be input to an A/D channel, counted by a spike discriminator, or shaped and measured in a variety of other ways. Respiratory transducers are available from a variety of commercial sources for a variety of species, ranging from mice to humans (e.g., Grass Instruments, Stoelting, World Precision Instruments).

Recording Skin Conductance Responses

There is a very large literature on the use of skin conductance measures to assess changes in autonomics during learning and emotional states, including studies of fear conditioning (see Chapter 6). Most of this literature involves the use of humans as subjects although some studies involving rats

Chapter 3

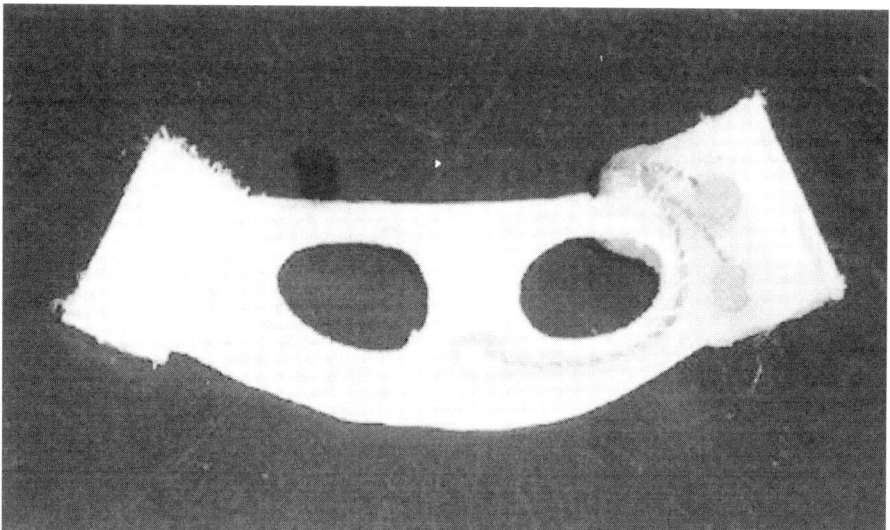

Figure 3.11: Rorick's stylish rat-sized vest for recording heart rate.

have been conducted. For human experiments, the skin conductance response is typically recorded via specialized bipolar finger electrodes that contain Ag-AgCl electrodes (6-8 mm in diameter) attached by Velcro straps to the middle phalanges of the subjects' third and fourth digits. Standard electrode gel is used as an electrolyte. The electrical signal that is recorded by the electrodes is typically amplified and filtered through a 90-1000 Hz bandpass using standard biological amplifiers available from a variety of commercial sources (e.g., Coulbourn Equipment, Lafayette Instruments, Biopac, and Stoelting). It is a bit more difficult to record skin conductance responses from rats, but not impossible. Some investigators have coated the rat's paws with electrode gel, placed the rat in a box that contains a floor composed of metal bars, and recorded changes in skin conductance via paw contacts made with the floor grid.

DRINKOMETERS OR LICKOMETERS

For the most part, the type of classical conditioning that we have normally observed in our laboratory has involved the modification of a reflex. But there also exist paradigms for which classical conditioning is used to suppress responding, as in fear conditioning (e.g., Mahoney & Ayres, 1976). For example, one could train an animal to associate footshock US with a tone CS in one experimental context, then test the animal in a different experimental context where the animal seems relaxed and is engaged in an appetitive activity like drinking water. When the tone CS is presented in this second context the animal will stop drinking. This suppression of drinking is thought to reflect the extent to which the animal is fearful of the tone. The lickometer or drinkometer circuits described here for detecting contact such as drinking can be used in conjunction with our classical conditioning programs and the computer interface (Chapter 11) for experiments on conditioned fear. The circuit is not confined to detecting liquid appetitive behaviors, as any contact (for example, touch with the forepaws) can be detected.

Figures 3.12 and 3.13 show two simple circuits for detecting licking (or touching). Note that both circuits actually use the animal as part of the circuit and that the detector is the same in both circuits. Here, a quad comparator[5], the LM339, can be used to detect the behaviors of four different animals. The comparators are referenced (the inverting or negative input) to a voltage that is half way between the power supply (created by the voltage divider using the two 1 MΩ resistors). The input to the comparator from the animal (the noninverting or positive input) also involves a voltage divider. This voltage divider consists of a 10 MΩ resistor to the positive power supply in series with the animal's resistance to ground. For testing purposes a 22 kΩ resistor and momentary switch substitute for the animal. Contact of the animal (or switch) brings the voltage low on the noninverting comparator input, below the reference value on the inverting comparator input, causing the comparator output to go from high to low. By reversing the roles of the 10 MΩ resistor and the animal (resistor and switch) on the noninverting input, the output of the comparator will go from low to high with contact. The current passing through the animal during contact is about 0.5 µA, too small a current for the animal to notice.

The two figures differ slightly after the comparator in the way the detected contact is treated. In Figure 3.12, the output of the comparator activates one of the timers on the 4528 CMOS dual timer[6]. The negative input is used on the timer. The output of the timer is a positive pulse of about 150 msec in duration, which can be used by a counter (for example, this can go to one of the counters on the 8253 of the computer interface, see Chapter 11). In Figure 3.13, the timer output is directed into a flip-flop[7] created from two gates on the 4001 CMOS quad NOR gate[8]. The flip-flop will remember a contact until it is reset by the positive pulse, therefore the flip-flop is only good at remembering a single event.

A computer program that continually polls the condition of the flip-flop (e.g., using the 'trigger' or 'unused' inputs to the 8255 of the computer interface, see Chapter 11) and continually resets the flip-flop (e.g., using the 'pulse' or 'unused' outputs from the 8255 of the computer interface, see Chapter 11) could detect and count the contacts. Indeed, we have designed such a system for Kathleen Chambers and her student Unja Hayes at the

University of Southern California to monitor liquid consumption in conditioned taste aversion experiments. Our design allows up to 40 animals at a time to be monitored using multiple comparator circuits and flip-flops with our computer interface and programming. Their circuit is made even simpler, however, by eliminating the need for the timer (4528).

Figure 3.13 shows the circuit modified for this simpler requirement. Here, the negative-going comparator output goes to both the timer and the flip-flop. Either the timer or the flip-flop of the circuit can be eliminated from the design as desired. The timer output could substitute for discriminated neural units and counted using the 8253 as described above. The flip-flop could be used with just the comparator for conditioned taste aversion experiments as indicated. The flip-flop is reset with a high-to-low pulse. Note that the flip-flop in Figure 3.13 uses a 4011 CMOS quad NAND gate[9], whereas the flip-flop in Figure 3.12 uses a 4001 CMOS quad NOR gate. The difference is that the 4001 accepts and returns low-to-high (positive logic) signals, whereas the 4011 accepts and returns high-to-low (negative logic) signals. Figure 3.12 could easily be modified so that the comparator yields a low-to-high signal (as indicated above, reverse the roles of the 10 MΩ and animal on the noninverting input) in which case a 4001 would be used as the flip-flop (not shown).

There is one final note. Although Figures 3.12 and 3.13 show separate voltage dividers for each comparator, a single divider could be used as the input for all four comparators. Instead of using a voltage divider, an improvement would be to create a single voltage source using a 1 KΩ resistor to +5 V in series with a 2.4 V zener diode to ground. The point between the resistor and the zener is a +2.4 V source that could be used on the reference inputs to the comparators.

LEVER PRESSING

B. F. Skinner popularized the use of lever (or key) pressing in operant learning situations and this behavior has been monitored in a variety of species including mice, rats, pigeons, cats, dogs and even humans. Lever presses are very discrete responses that can be repeated at relatively high rates. Lever pressing behaviors are relatively easy to shape, easy to measure, and easy to quantify. It should therefore not be surprising that this simple behavioral response has become a staple in behavioral neuroscience experiments. For example, Steinmetz and colleagues have used this response in a variety of experiments designed to compare appetitive and aversive motivated learning in rats (Steinmetz, Logue & Miller, 1993; See Chapter 6 for details).

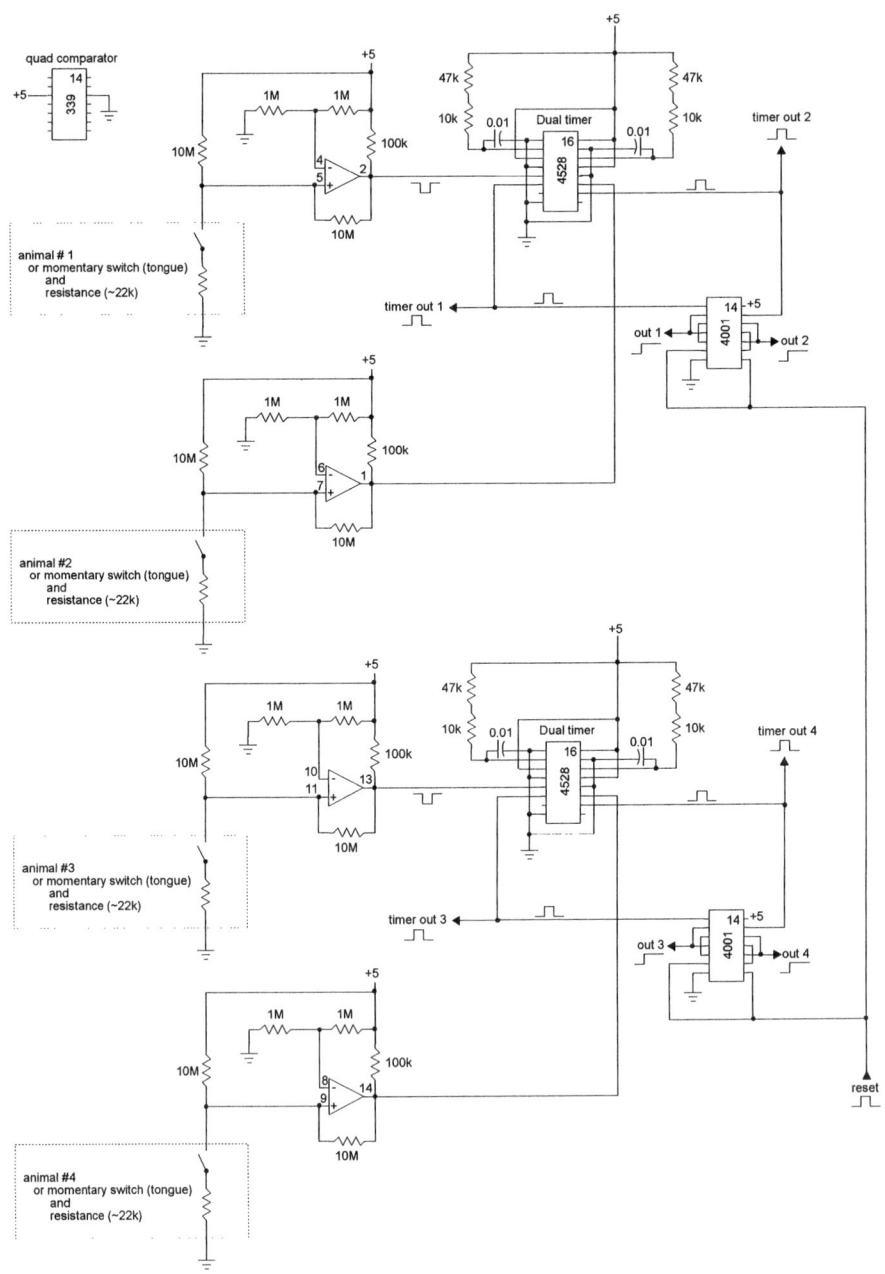

Figure 3.12: Latched drinkometer (contact) detection with positive reset.

Measuring a lever press requires nothing more than measuring a switch closure and the circuit depicted in Figure 3.14 provides a very simple way to generate a voltage pulse that mirrors the lever press made by the animal. Typically, we measure the lever press by either routing a 0 to +5 V pulse to a TTL-compatible I/O chip (such as the 8255, see Chapter 11) or by routing the switch closure to an A/D Channel on an interface board (see Chapter 11) and treating the switch closure as an analog response. The simplest way to accomplish this is to gate +5 V to the A/D input channel whenever the lever is depressed, using relay or optoisolator circuits (described in Chapter 12). At the A/D, each bar-press response is seen as a change from 0 V to 5 V. Most investigators treat the lever press response as a simple yes/no or on/off event, thereby missing valuable information that can be obtained by more careful, detailed measurements that may be afforded by reading the signal as an analog response. There is much information in the relative timing and duration of individual lever presses in addition to the obvious information obtained on lever pressing frequency. For example, motor impairments can be seen as delayed onset latencies or perhaps increased level pressing durations. All of this information can be gleaned by careful analysis of the lever press responses made during a session.

VIDEOTAPING AS A BEHAVIORAL MEASURE

Videotaping is a very valuable method for measuring behavior and has been applied in a variety of ways. It is frequently used in naturalistic and semi-naturalistic observation situations when other recording devices would be intrusive or not practical. We will not attempt to review techniques associated with the use of videotaping as a behavioral measure as there are other sources of information available (e.g., much of the infant development literature uses videotaping for data collection and this primary literature can be consulted for tips on using this technique). We will present a couple of interesting uses of this technique in behavioral neuroscience experiments, though, because they illustrate the usefulness (and requirements) of videotaping in measuring behavior.

Mark Stanton and his colleagues at the Environmental Protection Agency, Research Triangle, North Carolina, have conducted an interesting series of developmental studies involving eyeblink conditioning in very young infants. Using human infants in research poses some very special challenges: For example, they squirm, they are generally not cooperative, they cannot follow directions, and they sleep a lot. Stanton and colleagues have overcome many of the obstacles inherent in infant studies and have successfully classically conditioned the eyeblink response of infants using

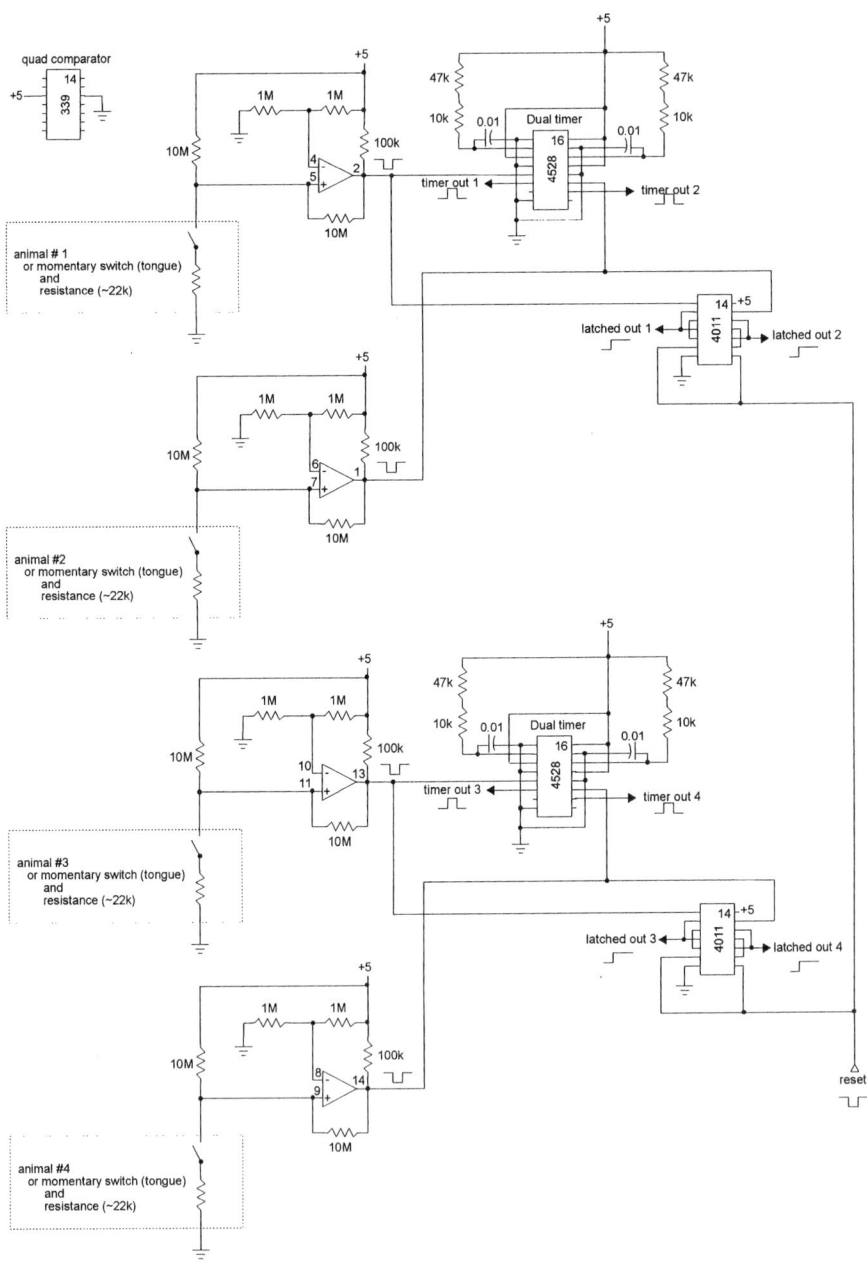

Figure 3.13: Timed or latched drinkometer (contact) detection with negative reset.

Chapter 3

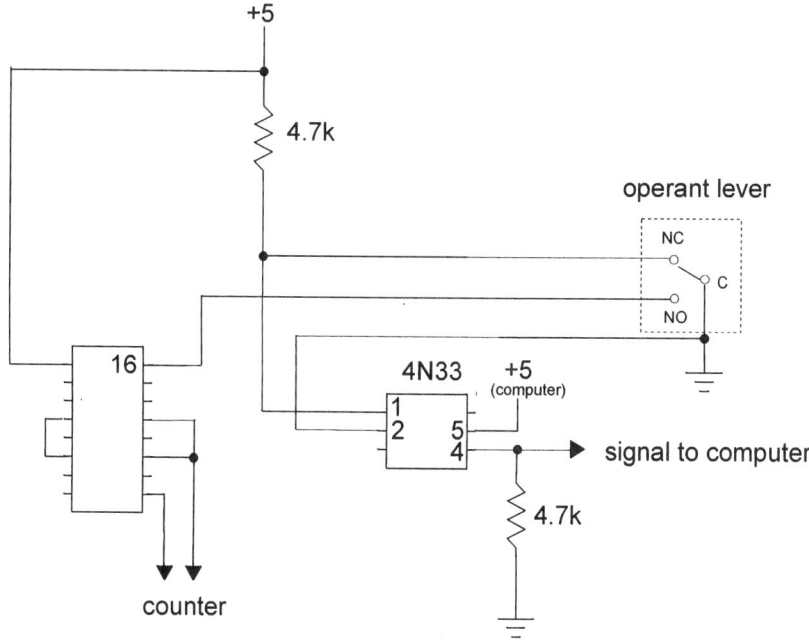

Figure 3.14: Circuit for detecting lever pressing as an analog rather than digital event.

some rather simple procedures (e.g., Ivkovich, Eckerman, Krasnegor & Stanton, 2000). In short, during conditioning the infant is entertained by an experimenter, seated in front of the infant, who presents a variety of colorful objects to the young subject. During the presentation of trials, the infant is seated on his or her parent's lap. Tone CSs are presented through speakers located on either side of the infant and an air puff US is presented via a small nozzle that is affixed to a soft headband placed around the infant's head. While it is possible to record EMG activity from the region of the eye, this is not necessary because eyeblink activity can be scored from videotapes made during the experiment. To accomplish this, Stanton and colleagues have used two video cameras, one placed to the front and right side of the infant and a second focused on a signal box that uses LEDS to indicate when the CS and US are presented. The first video camera catches the eyeblinking behavior of the infant while the second camera captures the timing of the stimulus events. A video splitter is used to display the output of both cameras on one monitor and the split screen video can be analyzed off-line, frame by frame, for the timing and magnitude of blinking.

The Steinmetz laboratory has used videotaping methods to score freezing behavior during fear conditioning experiments. In this type of learning procedure, a tone or light CS is presented just before a mild footshock US. We have recorded tone-elicited accelerations in heart rate during this procedure as a measure of fear learning but have also measured behavioral freezing as another index of fear. We simply position a relatively high-resolution video camera in front of the operant box that contains the rat being conditioned. If a tone CS is used, the audio channel of the camera can be activated during training so CS presentations can be detected. Alternatively, LEDs that indicate when the CS and US are turned on can be placed in front of the chamber and therefore caught simultaneously on tape. Individual training sessions are taped and the tapes are analyzed offline for signs of freezing behavior using well established criteria (e.g., Anagnostaras, Maren & Fanselow, 1996). We have found this videotaping technique for scoring freezing to be very reliable and easy to use. In addition, this technique can be used to score a variety of other behaviors (such as rearing, chewing, grooming, and general activity) when used in conjunction with other behavioral manipulations.

Depending on the resolution required, this method can be relatively inexpensive to set up, requiring very little other equipment to effectively collect data. The choice of a camera is very important. Its lens system should have enough resolution to pick up the desired behavior when the camera is located in the desired location. Relative illumination that is available is another factor (low- versus high-light conditions). Researchers now have several recording formats from which to choose, including digital, that have advantages and disadvantages, especially related to the devices that are required for subsequent playback. Perhaps the best way to make a decision concerning videotaping equipment is to locate individuals who use videotaping techniques for data collection purposes and consult with them concerning the proper equipment to be used for the application you have in mind. The Steinmetz laboratory has used VHS and Hi-8 videotaping and have found both methods to be useful.

ANIMAL RESTRAINT ISSUES

Among the many advantages that Gormezano recognized for using rabbits in classical conditioning of the eyelid response was that they accepted restraint very well. This acceptance of restraint has a distinct advantage for neural and behavioral measurements—few artifacts are introduced into the neural and behavioral measures during recording. How long will restrained rabbits sit quietly? We do not know the limits. We do know that when we began working in Thompson's laboratory that the intertrial interval averaged 60 seconds and that 117 trials were typically given. This means

Chapter 3

that each training session lasted about two hours. Sometime in the early 1980s, Dave McCormick, among others in the laboratory, began using a shorter interval that averaged 30 seconds between trials, which meant that twice as many rabbits could be run in the same interval (i.e., sessions lasted about 1 hr for each rabbit). Many in the field have adopted this shorter intertrial interval.

Not all animals tolerate restraint well. In fact, the rabbit, perhaps due to its strong, innate tendency to freeze in fearful situations, can be considered rather unique with regard to acceptance of restraint. Many species, including rats and mice, show significant increases in corticosteroid levels during even mild episodes of restraint and these elevated hormone levels could significantly affect behavioral and neural responding. Some care must therefore be exercised in the interpretation of data obtained from animals that show this strong reaction to restraint.

Plexiglas Restraint Boxes

The standard restrainer for rabbits is constructed from Plexiglas and a version of this restraint box is shown in Figure 3.15. The basic design for this restrainer was created in the early 1960s by John Waltke, an instrument maker at Indiana University. The first boxes were built for Isidore Gormezano, a faculty member at Indiana University at the time, and this basic design is still used today by a number of researchers. Figure 3.15 shows individual pieces of the restrainer with their approximate dimensions as well as a fully assembled restrainer. The dimensions given in the figure work reasonably well for rabbits that weigh up to 3 kg. The unit is constructed of 1 in thick Plexiglas in one of the models that we have, making the restrainer quite heavy. Thinner pieces of plastic could be used as well. The pieces can be assembled with screws and/or glued together. Two pieces are not fixed to the restrainer, the stock and the back plate. These pieces are moveable to adjust for differences in the sizes of different rabbits. As shown, the stock is held in place by a screw and metal extension piece that yield a pressure fit in position. We have seen different designs for positioning the back plate, one using side rails that the back plate slides along and into position, the other using springs and pins that fit along a track with several holes. In addition to these six pieces of Plexiglas, three pieces of 1/16 inch aluminum plates are associated with the stock. The two pieces marked 'outer' and 'inner" are placed along either side of the stock to act as guides for the stock as it slides up and down the front piece. The ears lay on top of the 'outer' plate, covered with thin closed (preferred) or thick open cell foam rubber to add padding comfort. A hinge attaches the third aluminum plate to the front wall piece. This third plate may also be padded (ours is padded with half inch thick, closed cell foam). The ears are held between the clamp and 'outer' plates by a hinge on one side and a locking mechanism (e.g., a screw) on the other side. The hinge and locking mechanism are not shown in the drawing. In a more recent design, we replaced the hinge and latch with a piece of 2 inch wide tubular webbing bought at a climbing store, about 13 inches long with Velcro hooks we sewed on both ends. A metal plate is placed inside the webbing. Complementary Velcro loops are glued onto restrainer.

The restrainer can be fitted with a removable top (not shown) made of thin Plexiglas to enclose the rabbit's body. Often pieces of foam rubber are used in the spaces between the rabbit and the top plate. Other designs include a metal back plate bent at 90 degrees to form a top plate that extends forward to cover the back of the rabbit. Instead of a back plate, Lavond uses layers of closed cell foam rubber that are held in place with 1 in flat nylon webbing. The webbing is attached to the restrainer with Velcro.

The advantages of the plastic restrainer are that it is a) relatively quick and easy to place the rabbit in or out of the restrainer, b) the restrainer is relatively secure given its mass, c) it is easy to clean when rabbits make a mess (and believe us, they can and do), and d) the head and eye are easily accessible for conditioning experiments. By cutting out a rectangular area from the bottom plate and placing the restrainer off the floor via legs attached to the bottom of the restrainer, the hind limbs can be pulled through and leg flexion experiments can be done as well.

Patterson and colleagues have described a plastic box restraint system for cats (Romano, Steinmetz, & Patterson, 1980) and this box has been successfully used in classical conditioning experiments involving the eyeblink and hindlimb flexion responses. As anyone who has had a cat for a pet knows, cats do not tolerate restraint particularly well. Nonetheless, cats required little adaptation to this system and appeared to tolerate the restraint well. The Plexiglas box used in this restraint system is very similar to the box used for rabbits (see Figure 3.16). A major difference is that a chin plate extends from the front of the box on which the cat's head can rest. Nylon releasable cable ties are used to secure the head to the chin plate along with a small bite-bar. A hinged lid covers the back of the animal and paint rollers are placed between the animal's back and the lid to increase comfort. This restraint system alleviates the problem of lengthy adaptation sessions; the cats accept this restraint for as longs as 2 hours with little or no struggling after a single adaptation session of 1 hour.

Some investigators have conducted classical eye blink conditioning studies using restrained rats. Nestor Schmajuk and colleagues (Duke University) have described a restrained rat preparation (Schmajuk & Christiansen, 1990). During surgery, a 13 mm stainless steel bolt is ce-

Chapter 3

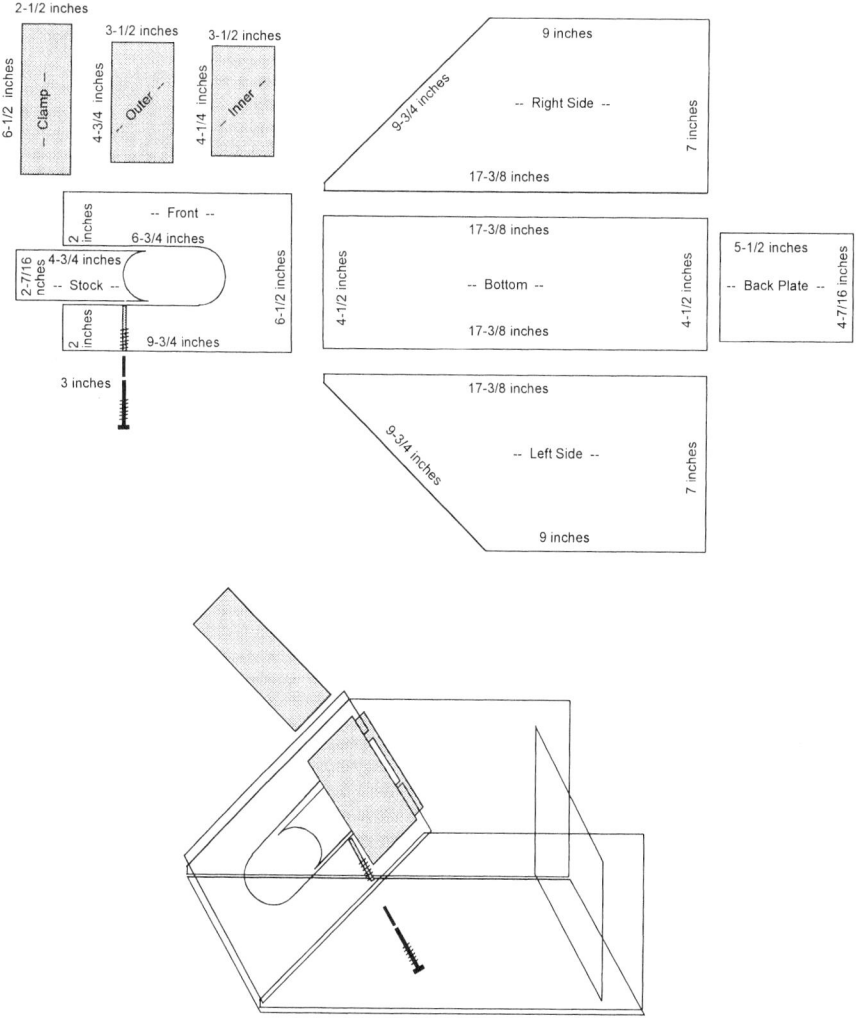

Figure 3.15: Plexiglas rabbit restrainer.

mented in an upright position to the skull of the rat using dental acrylic. About 6-7 mm of the threaded shaft of the bolt is left exposed. During training, the rats are restrained in a 108 x 48 mm standard restraining (injection) cage, which is available from several commercial sources (e.g., Fisher Scientific, Harvard Apparatus; see Figure 3.17). The cage is constructed from clear acrylic with slots that can fit a tailgate, which, when in place prevents the rat from backing out of the device. The bottom has a wide slot to allow waste to drop out and a hole is located on the top of the box over the area where the rat's head is located. A small metal sheet with a hole drilled in it is anchored over the hole and accepts the bolt implanted during surgery. Securing the bolt with a nut to the metal sheet restrains the rat in the acrylic tube. Air puff USs can be delivered via a plastic tube that is positioned through a hole in the side of the acrylic restraint tube. Because the rat is restrained, a number of different devices could be used to monitor eyeblinks or other behaviors (including potentiometers, reflectance devices and EMG methods). The major advantage of this system is that it is quite easy to deliver stimuli and record behavioral as well as neural data. The major disadvantage of the system, however, is that restraint is known to cause elevation of stress hormones in rats and this could affect the behavioral and neural results that are obtained.

Wraps, Bags, Slings and Hammocks

Other methods for restraint include wrapping the animal in a cloth. Gil Case in Thompson's laboratory worked out and championed the wrap method of restraint for rabbits based upon a veterinary "cat bag." The wrapping technique has been used previously for temporary restraint in rodents to give injections or take blood samples. Figure 3.18 shows the approximate dimensions for a wrap that can be used for very small and very large rabbits. Darts are cut into the cloth at the end where the rabbit's head exits the wrap. The darts are there only to help in the folding. Here the darts are 3 in equilateral triangles, but the shape and dimensions are not critical.

The lower schematic in Figure 3.18 shows the position of the rabbit and the order for folding the wrap. The rabbit is placed with the head near the darts, the left and right sides of the cloth are brought around the back and the rear end folded up and on top where they are fixed in placed (for example, with safety pins or with Velcro fasteners). The front paws are placed under the fourth fold and then the left and right corners are brought up behind the ears and fixed in place. Safety pins seem to work better here than Velcro.

The advantage of the wrap is that it appears to be more comfortable for the rabbit. The disadvantages are that the rabbit can still move around despite being in the wrap, it is more difficult to clean when the rabbit makes a mess, and it is hotter than the plastic restrainer. This last problem is worse when a thicker felt-type material is used. Felt material was introduced because Velcro hooks readily attach to it, making felt a convenient way of cinching up the wrap, but this has to be weighed against the added weight and heat. We have not adopted this method ourselves. To make this system practical, Case found that the head must be immobilized to prevent the animal from moving around, just as Disterhoft had previously shown.

Chapter 3

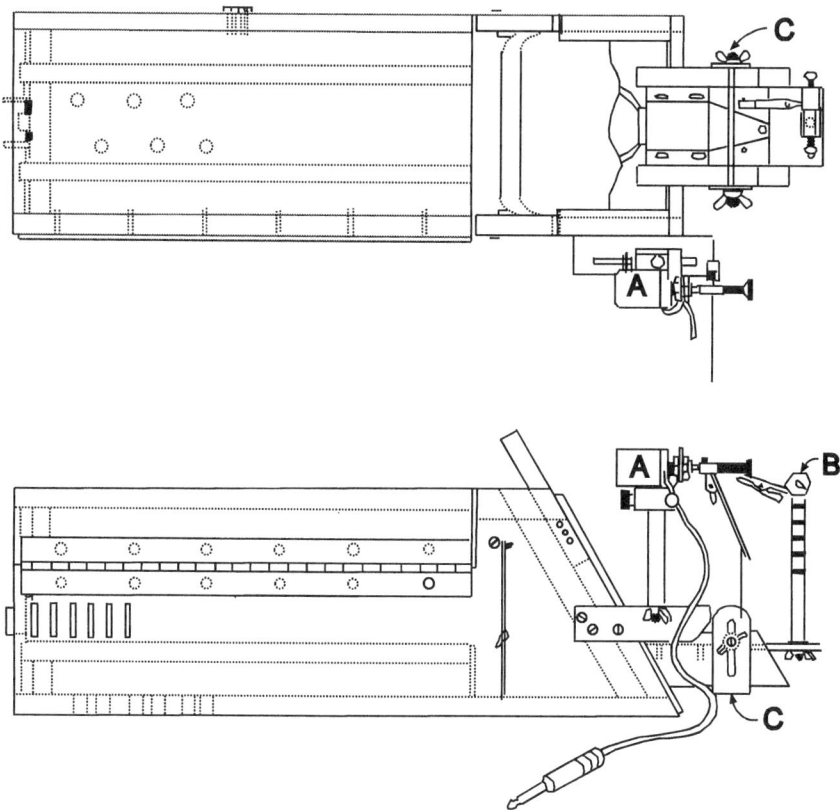

Figure 3.16: Plexiglas at restrainer.

For several years, Disterhoft and his colleagues have used a canvas bag in conjunction with a stereotaxic frame to restrain rabbits (e.g., Disterhoft, Kwan, & Lo, 1977). Prior to training, during surgery, four No. 6-32 x ½ inch nylon machine screws are positioned in dental acrylic that are cemented to the skull. For training, the rabbit is placed in a canvas bag that is secured around the neck and behind its hind legs. The four nylon machine screws fit into a ¼ x 4 inch steel bar that is attached to a Baltimore Model 'L' stereotaxic instrument, which in essence restrains the rabbit. The Steinmetz laboratory has used a similar device that involves implanting two head bolts during surgery and securing the rabbit in a stainless steel support system that incorporates a standard Kopf stereotaxic frame. We have found this system to be excellent for single unit neuronal recording with moveable electrodes (see Chapter 4).

Slings and hammocks have also been used to restrain cats and dogs during experiments (e.g., Adams, 1983; Bruner, 1969; Pavlov, 1927; Wick-

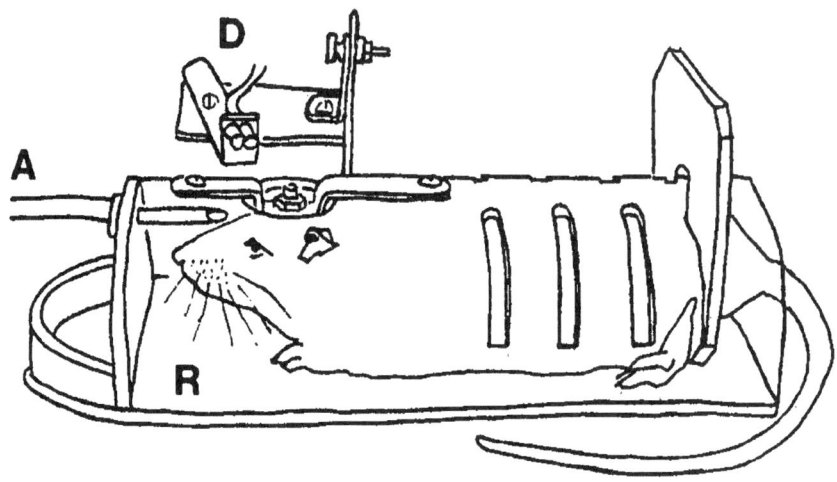

Figure 3.17: Plexiglas rat restrainer.

ens, Myers & Sullivan, 1961). In this form of restraint, the animal is restained and in some designs suspended over a platform. However, these systems have generally required extensive periods of adaptation before stimuli could be delivered without undue struggling. When Lavond was a graduate student, he weekly witnessed Dave Tuber's cats run out of their home cages in the vivaria, scamper down the hall to the Wicken's laboratory, where they jumped up onto the apparatus, waiting to be harnessed while they purred. Slings and harnesses have been used with some success for rabbit experiments because rabbits appear to adapt well to mild restraint.

Using No Restraint

Freely moving rabbits can also be studied. Michael Gabriel at the University of Illinois, for example, studies rabbits that move around in a large activity wheel. Studies of classical conditioning in freely moving rats and mice have been done in the last few years in the Thompson and Steinmetz laboratories. The traditional method for achieving relative freedom of movement when using recording and stimulation electrodes is to attach the wires by means of a commutator (such as Model 53509 available from Stoelting, Inc). Commutators (also called swivel/tether devices) can include as many as 10 channels of wiring to connect the animal with peripheral equipment. In addition, commutators can include small tubing for the delivery of drugs or other liquids. With a commutator the animal can circle in either direction as much as it likes without tangling the wires. The only real problem for classical conditioning experiments is that the commutator cannot handle well the rotation of the hose for the air puff. As a result, the US used in experiments of freely moving animals is typically stimulation/ shock to the face or periorbital region coming from wires that are routed through the commutator. The free-moving rodent is much preferred to using a restrainer or wrap as with rabbits. Our personal experience is that rats definitely do not like to be restrained.

In the Steinmetz laboratory, the wires running from the commutator to the animal are encased in either a thin, flexible metal covering or in a relatively thick, shielded, rubber insulation (see Figure 3.19). We have found that it is quite easy to prevent rats from chewing on the leads if there is little or no slack in the leads running from the commutator to the animal. In a typical rat classical eyeblink conditioning experiment in the Steinmetz laboratory, we use all 10 channels available in the commutator to connect a rat to external equipment. In surgery, an acrylic headstage with a standard 10 pin Augat socket is cemented onto the rat's skull. Attached to the 10 pins of the socket are: a connection to ground, two wires used to record EMG activity from eyelid closure musculature (see above), two wires used to deliver an electrical stimulation US to the skin around the eye, and up to four channels used to record brain signals (see Chapter 4). Another channel in the commutator is used to carry a voltage that powers a small FET circuit that is located on the end of the leads connecting the animal to the commutator (see Figure 3.20). This circuit is used as a first-stage amplifier and is designed to minimize movement artifacts that are introduced into the brain and EMG signals. We have successfully used this tether and swivel system in a variety of experiments involving freely moving rats. While the general preparation of the rat for these experiments is time consuming, we have found that the benefits of using the non-restrained rat far outweighs the cost of preparation.

Anesthesia as a Restraint Method

We also have tried the alternative to restrained and freely moving rats, i.e., trying to classically condition anesthetized rats. There are reports in the literature that this is possible, especially with adrenalin injections, which we (Lavond and Matti Mintz, Tel Aviv University) tried, but our experience is that rats learn to the extent that they are *not* anesthetized. This view is in keeping with most of the human surgical literature that shows no evidence of learning or remembering if the anesthesia is effective. Steinmetz has successfully conditioned anesthetized rats and rabbits using

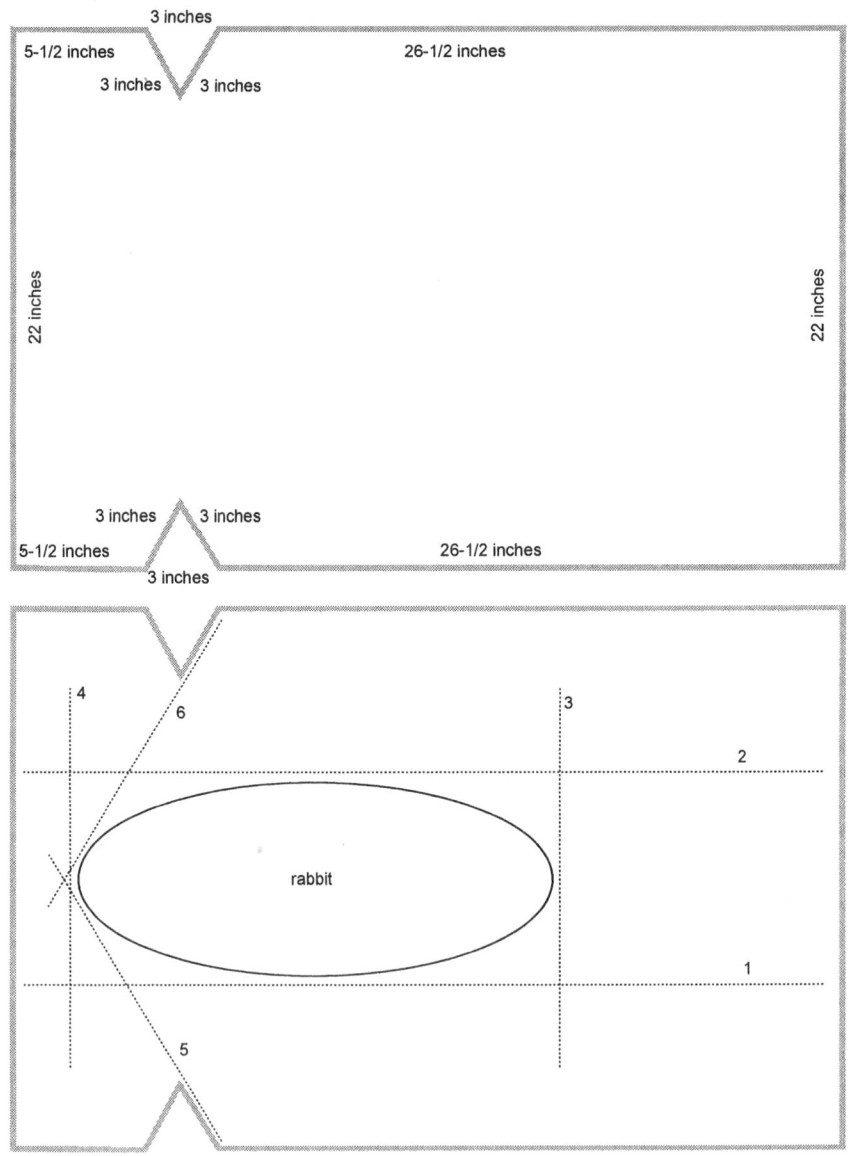

Figure 3.18: Wrap for rabbit restraint.

microstimulation as the CS and US instead of peripheral stimuli. However, further attempts to pursue this line of research were abandoned when it was evident that the anesthesia was affecting the conditioning process and the neural activity under anesthesia was somewhat different than activity recorded in the awake, behaving animal.

USE OF EYE CLIPS DURING EYEBLINK CONDITIONING

Before closing this chapter on behavioral response measurement, we would like to touch on one final issue, which may be of particular interest to classical eyeblink conditioning researchers; the issue of using eye clips to hold back the external eyelids during training.

Classical conditioning is traditionally distinguished from operant conditioning by the fact that no matter what the animal does its behavior does not alter the delivery of the CS and the US. In an operant variation of classical eyeblink conditioning, for example, if the animal learns to blink (CR) after tone onset (CS onset) and before air puff or shock onset (US onset) then the air puff or shock is not delivered (avoidance learning). Such control is not the case for "pure" classical conditioning. No matter what the animal learns, the air puff or shock US is still delivered. Another way to think about this situation is that the learned CR may not be adaptive in avoiding the US. Obviously, in eyeblink conditioning, closing the eye mitigates the effect of the UR. Two ways to avoid (or at least minimize) this operant aspect to the classical conditioning situation are to use a US that is not affected by eye closure (e.g., shock as the US instead of an air puff) or to prevent the eyelids from closing using eye clips (which Steinmetz notes can be likened to the famous scene involving Malcolm MacDowell in the Stanley Kubrick film, *A Clockwork Orange*; Lavond had to read the book for a high school class but has never seen the film).

In human experiments of classical conditioning where eyelid closure is adaptive (shocks are not used and the eyelids are not held open) no one worries much about this contamination of the paradigm. In rabbit experiments some researchers (like Lavond) make an effort to remain faithful to the paradigm described by Gormezano and colleague by using devices that hold the external eyelids open and by directing the air puff towards the undefended temporal half where the nictitating membrane cannot reach. Other researchers (like Steinmetz) do not believe that using these eyelid restraint devices make much difference in the conditioning process and do not use them in order to increase the comfort level of the subject. In fact, there is evidence that eyelid restraint interferes with extinction procedures, whereas

Figure 3.19: Commutator for recording from a freely-moving rat.

Chapter 3

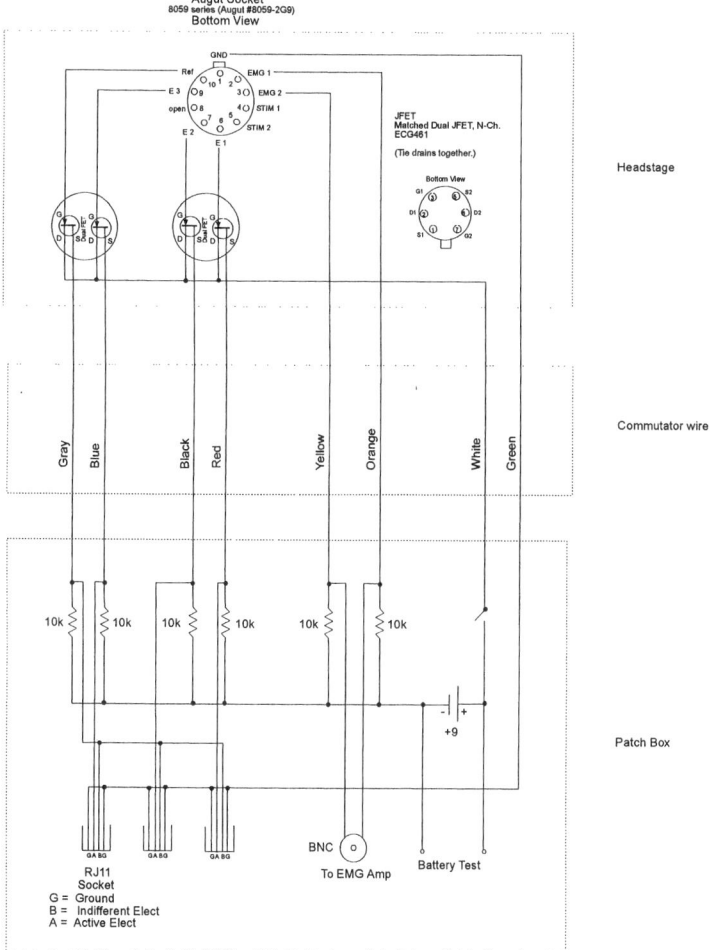

Figure 3.20: Amplifier to minimize movement artifact in the freely-moving rat.

no restraint is associated with normal extinction.

Some digression to discuss the rabbit eye is warranted at this time, and the reader may wish to consult Prince (1964) on the subject. Like most animals the rabbit has two external eyelids, just like humans. Like many animals, including snakes, frogs, sharks and cats, the rabbit has a "third eyelid," the nictitating membrane, which is a membrane found inside the external eyelids, attached to the eye socket near the nose (the nasal position) and closes by sweeping across the eye towards the sides and ear (the temporal direction). In rabbits, extension of the nictitating membrane (i.e., closure of that eyelid) is a passive phenomenon, whereas in cats the nictitating membrane is attached to a muscle that pulls the membrane across the eye, innervated by the nerve supplying the superior oblique muscle. Rats also have a nictitating membrane but it is a vestigial structure. The nice feature about rabbits is that the external eyelids can be held open yet the animal can still moisten the eyeball with sweeps of the nictitating membrane. Gormezano discovered many years ago that measuring movement of the nictitating membrane was an excellent way to monitor eyelid conditioning.

Standard Eye Clips

Figure 3.21 shows the basic design for eye clips that we have used for many years to hold open the external eyelids of rabbits. The two bands for the eye clips are made from 11 inches of common elastic. One band reaches from the upper eyelid, goes over and around the head, under the chin and up to the lower eyelid. The hooks are modifications of hook-and-eye fasteners or constructed of stainless steel wire. The second band attaches to either side of the first band and is positioned behind the rabbit's ears. The attachments are loops made in such a way that they can be slid up and down the first band for optimal positioning. The 11 inches of each band are purposely too long; the slack is taken up by using alligator clips to take a bite (a 'bite' in the sense of a knot) temporarily out of the band for a snug fit. New rabbits may resist the eye clip. Like all animals, but new rabbits especially, rabbits can squint their eyes closed at any time but this occurs especially during adaptation. Be patient. Rabbits get tired of resisting and eventually relax the eyelids. It helps to make sure that the tension on the bands is not overly tightened.

Ophthalmic Eye Clips

Tapani Korhonen (University of Jyväskylä, Finland) developed the eye clips shown in Figure 3.22, based upon the ophthalmic instrument used in human eye surgeries familiar to his wife, a surgical nurse. It is similar in design to eye clips used by Patterson and colleagues for cat eyeblink conditioning experiments. The dimensions in the figure need only be approximated. These eye clips are made from a length of stainless steel wire that acts as a spring to hold the external eyelids open. The spring end is oriented towards the ear, where it is less likely to be dislodged during training (using the eye clip shown in Figure 3.22, the rabbit would be facing towards the left). We use 21-gauge wire but different sizes could be used to achieve different tensions. For 21-gauge stainless steel wire the length is about 4 cm and the outside (open) dimensions are about 2.7 cm before placing the de-

Chapter 3   117

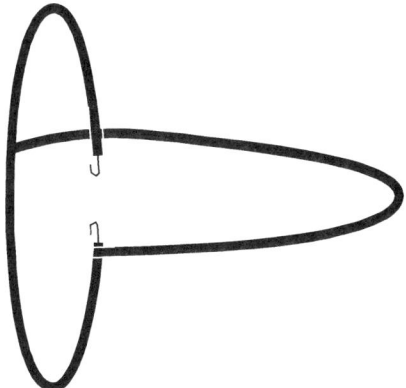

Figure 3.21: Traditional elastic bands with hooks to retract eyelids during eye blink conditioning.

vice on the eyelids. We solder the ends to each other to achieve a nice smooth finish. Acid flux is necessary for soldering stainless steel and should be washed away with soap and water. Unlike the elastic eye clips above, these ophthalmic spring eyeclips can be placed into sterilizing solution.

No Eye Clips

As mentioned above, some researchers like Steinmetz have abandoned the use of eye clips altogether. This is due to several reasons. First, the animal seems to be much more comfortable without the eye clips and no adaptation to the clips is necessary. Second, some species like rats and humans do not have nictitating membranes so eyelid conditioning must involve the measurement of movement of the external eyelids. Using eye clips can affect these measurements, especially when EMG recordings are taken. Third, it is not clear that aiming the air puff tube toward a region of the eyeball not covered by the rabbit or cat nictitating membrane completely eliminates the instrumental aspect of the training. Regardless as to where the air nozzle is pointed on the eye, the air spreads to affect the whole eye as well as part of the skin around the eye. Nictitating membrane extension at the very least protects a portion of the eyeball from receiving the air puff. Fourth, eye clips should be cleaned and sterilized between animals so that infection does not spread from one to another. Finally, the use of shock cannot absolutely rule out instrumental components to the conditioning process. The state of the musculature around the eye is very different before and after training. Indeed, the muscle is in a relative state of contraction on CR trials when a US is presented, a contractive state that is not present on non-CR trials (i.e., the US may be sensed differently by the animal).

Interestingly, whether or not eye clips are used may affect the rate of extinction. In Lavond's experience extinction training in rabbits with eye clips has been unusually difficult to achieve. Extinction training is the process where the US is withheld after initial learning with paired CS-US training. Typically, after a number of CS alone trials are presented the animal figures out that the CS no longer predicts onset of the US and gradually stops showing CRs. The learned behavior is said to have extinguished, like a flame deprived of oxygen. We know of rabbits that learned in only a few days of training but that continued to show excellent CRs after as many as 20 days of extinction (CS alone) training (e.g., Mintz, Lavond, Zhang, Yun and Thompson, 1994). A number of researchers have made similar observations. *Not* using eye clips, however, seems to yield much faster extinction. Independently, A.J. Annala and Steinmetz, among others, have found that extinction training is more reliably obtained without eye clips. It could be argued that training rabbits without eye clips is no longer classical conditioning (at least classical conditioning as defined by Gormezano and colleagues), but then none of the literature on human eye blink conditioning would be either.

HEADSTAGES AND OTHER RELATED DEVICES

The term 'headstage' is a generic term for a foundation that attaches to the head of a subject for the purpose of holding apparatus important for the conduct of the experiment. 'Headstage' is ill-defined in that it can refer to either the foundation or the accompanying apparatus. The foundation could consist only of the dental acrylic mound created during surgery to hold recording electrodes. Here we use the term headstage to refer to all the hardware that attaches both the air puff stimulus and eyelid measurement apparati to the rabbit's head during classical conditioning along with connectors used to interface the electrodes with the amplifiers and other devices used to record neuronal activity (See Chapter 4). Typically, the headstage hardware consists of two separate pieces that couple together like a socket and a plug, although those terms are rarely used. The socket-like piece is normally attached to the animal's head, either permanently as in the case of dental acrylic or temporarily as in the case of a harness.

Chapter 3

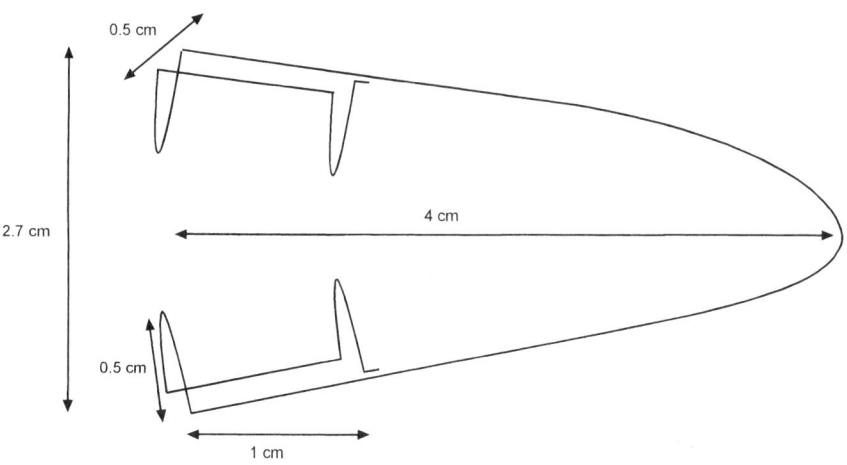

Figure 3.22: Ophthalmic eye-clips to retract eyelids during eye blink conditioning.

In addition to allowing brain amplifiers to be connected to implanted electrodes, the plug-like piece also holds the apparatus for delivering the air puff and measuring eyelid extension. When coupled, the completed headstage ensures that the air puff delivery and eyelid measuring apparatus are aimed at the eye no matter where the animal's head is pointed. There are a number of possible solutions to accomplish this positioning. The major difference between these methods is in the actual pieces used for the socket and plug. Tapani Korhonen (Finland) uses DB25 sockets and plugs (connectors associated with serial computer ports), which has the advantage that it is very sturdy when connected. DB-type connectors have been used most commonly for large animals. Originally we used ITT orange strip connectors with gold pins to construct both the socket (female pins) and plug (male pins). This construction required a separate screw to secure the connection.

In Chapter 4, we describe permanent headstage devices that are used in conjunction with neural recording. For purely behavioral studies or for situations where training occurs before surgery, then a harness holding the socket is placed over the head and face of the rabbit. We have seen harnesses made from Velcro. Our favorite harness, which we have not seen for years now, was made from a table microphone stand (the inverted V piece) and a flat piece of plastic similar in idea to the harness we describe here. The harness we describe is based on a design from Paul Solomon.

Figure 3.23 shows the parts and their assembly into a harness for the headstage socket. The pieces for the headstage are: 7-inches of number 8 copper wire shaped into an inverted U, a 3/4 by 2 inch piece of 1/16 inch aluminum bar stock, a 14 pin machined contact DIP socket with the pins cut from the underside (cut the pin but not the socket), a 3/4 inch size 4-40 stainless steel screw with washer and nut, and four small machine screws. In addition, two ordinary rubber bands are needed to attach the harness to the rabbit.

Figure 3.23 shows two views of the assembled socket headstage (left side view and front view). The pieces are put together with glue (copper wire, socket and 4-40 screw to the aluminum) or holes drilled for the small machine screws (it is not necessary to tap threads into the soft copper and aluminum). We have used either two-part, 5-minute epoxy or glue from a hot glue gun, and prefer the hot glue because it holds reasonably well, can be shaped while it cools, and is fairly easy to be removed. Two 1/4 inch holes (not shown) are drilled near either end of the flat aluminum piece to get stronger bonding. Finally, the nut and washer are used to hold the headstage plug (next section) to the harness. The nut can be tightened with a nut driver. Alternatively, wing nuts or connectors with knurled edges (our favorite) can be used for finger tightening.

The harness is attached to the animal (usually with the plug already attached) with two simple rubber bands. One rubber band attaches between the two small machine screws near the DIP socket and is stretched over and behind the ears. The ears are then clamped in the clamp of the rabbit restrainer. The second rubber band is placed on one of the small machine screws on the copper wire. The rubber band is stretched, then placed under the rabbit's chin and around to the small machine screw on the opposite leg of the copper wire. Three words of caution: make sure to stretch the rubber band before attempting to wrap the band under the chin (this prevents the irritation of pulling little hairs or whiskers); make sure the rubber band goes under the chin and not in the animal's mouth; and finally, make sure the path of the rubber band goes behind the copper wire (towards the animal) for the best security.

Besides the alternatives choices for the nut there are two other common modifications of this harness design. The first modification is to shorten the legs of the copper wire. Sometimes the rabbit pulls himself into the restrainer such that the front edge of the restrainer pushes up on the harness legs, defeating the alignment of the air puff and eyelid measurement, sometimes even dislodging the harness from the animal's head (which is not such a great thing since the eyelid measurement device is usually still attached). Shortening the legs will help to prevent these problems. The second modification is to place several screws along each leg of the copper wire. These screws can be used to adjust the tension of the rubber band for the best fit. The position we show, about 1 in from the top, seems to work well enough. A similar idea could be done for the rubber band that goes around the ears but this tension is usually not a problem.

Headstage Plug

By definition the headstage plug is a temporary connection, used intact from one rabbit to the next. Figure 3.24 shows the components of the headstage plug, indicating their points of assembly and giving different rotational views and cross sections of the components as might be helpful. The components of the headstage plug are much more complex than for the socket and harness. The components are: a 1 by 2-1/2 inch piece of 3/16 inch flat plastic used as a base with two holes for screws and a slot carved at the rear end (for accepting the screw from the headstage socket and harness); a 14 pin machined contact DIP socket with a similarly carved rear slot; a potentiometer for measuring eyelid movements (with an arm that is attached to the axle by means of a wheel collar and mounting screw used for remote controlled model planes and cars); two air puff mountings made from 1/4 inch round aluminum stock (slotted and shaped flat on one end) and tapped for screws or air nozzle, these being the greatest challenge to construct; one 4-40 screw with knurled connector to attach one air puff mounting (#1) to the plastic base; one 6/32 inch thumb screw to attach the second air puff mounting (#2) to the first airpuff mounting (#1) at one end and with a 3/16 hollow aluminum tube attached to the other end to act as the nozzle (3/4 inch long) for the airpuff. Not shown are the electrical connections to and from the potentiometer. Also not shown are the connections from the potentiometer arm to the eyelid (see below).

Figure 3.25 shows two views (left side and top views) of the assembled headstage plug. Special mention should be made that the DIP socket is glued (with hot glue or epoxy) on the underside of the plastic base. This is necessary so that the pins of the machined contacts of the socket and plug will fit together. Here the integrated DIP sockets are used just for structural purposes to hold the two halves together, with the 4-40 screw from the harness. In some designs the DIP sockets are used to make electrical connections from recording or stimulating electrodes (see Chapter 4). If this is done then the sockets are usually bought with gold plating for improved electrical connection. Voltage follower transistors or operational amplifiers can be mounted using the DIP sockets for improved performance. Figure 3.25 shows a left side view of the assembled halves of the headstage.

Rabbit Transport

A practical question that often comes up is how to transport the rabbit from the vivaria to the experimental apparatus and back. Animal carriers

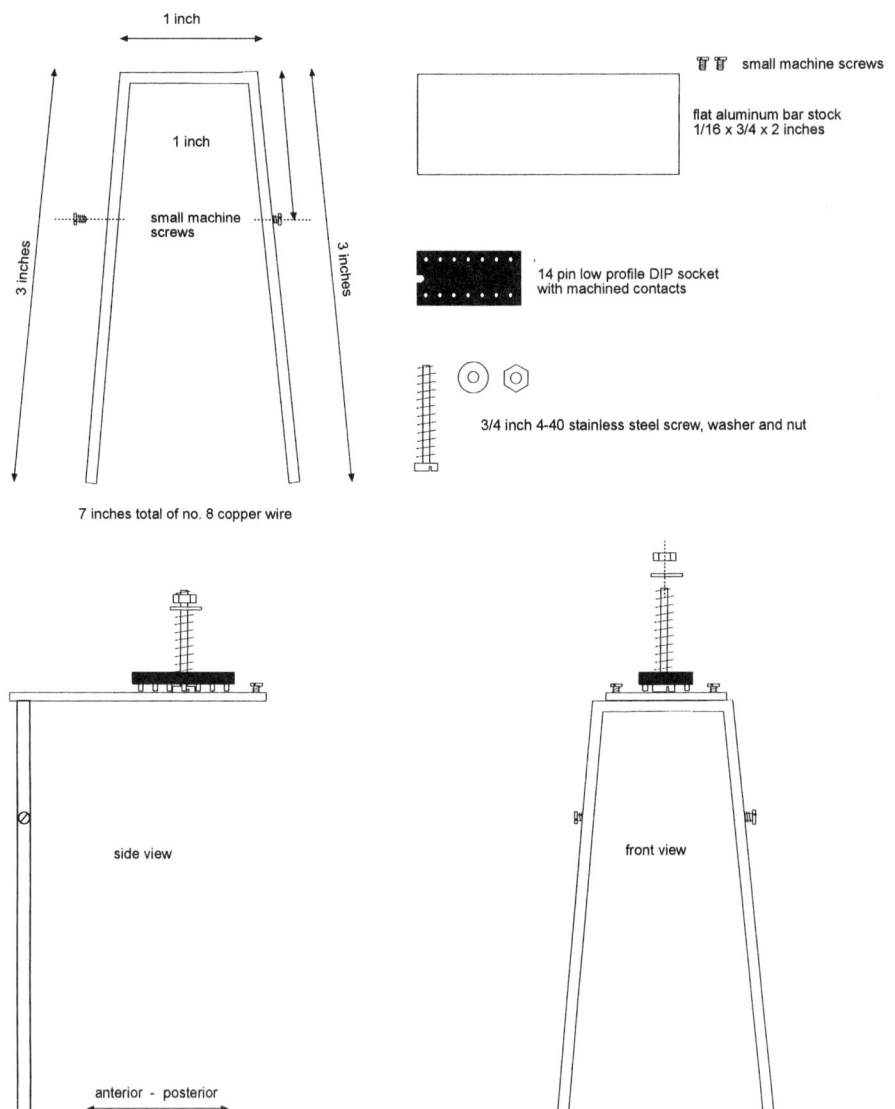

Figure 3.23: Parts and assembly of harness for attaching headstage to an unoperated rabbit.

Chapter 3

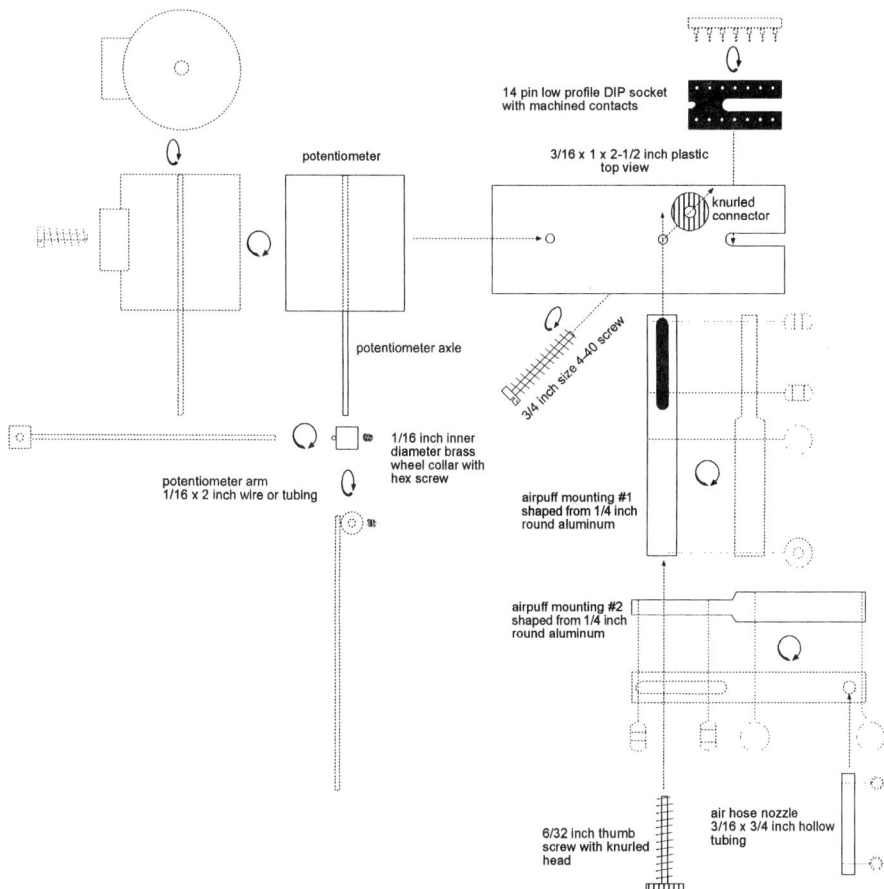

Figure 3.24: Parts and assembly headstage that holds the airpuff delivery system (unconditioned stimulus) and minitorque potentiometer for behavioral measurement.

for pets are available for dogs and cats that could be used. These are somewhat expensive and often not really suited for rabbits. Whereas a dog or cat will often jump out of the carrier door, a rabbit normally has to be lifted by the scruff of the neck and pulled out. It is difficult to pick up the rabbit in the restricted space of a pet carrier. Important design features for a rabbit carrier include top opening and containment of urine and fecal waste material.

For the most part, a plastic dishwashing pan works very well. The rabbit is placed in the pan and simply carried to the location. When placed on the floor, most rabbits will stay in the pan although they could easily jump out. Rabbits that jump out onto the floor, where they might chew on electrical cords, can be discouraged from jumping by simply placing the pan on a chair or table. Rabbits understand a cliff. We have never seen a rabbit

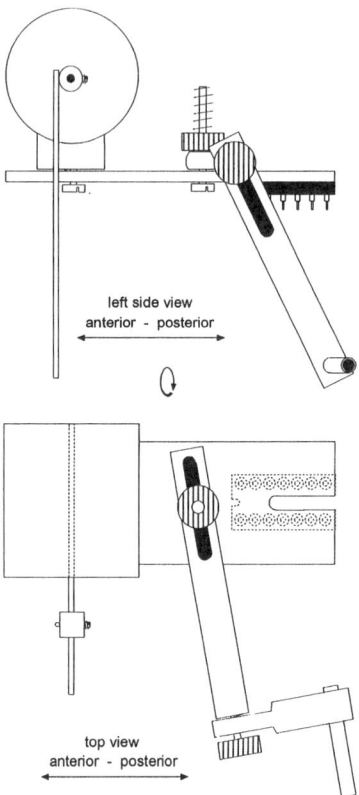

Figure 3.25: Two views of assembled headstage.

jump from a pan placed on a chair or table. Similarly, in our experience, we have not seen a rabbit jump out during transport. Most rabbits seem to enjoy the view. The few that do not like the view usually hunker down. Nevertheless, an enclosed apparatus may be desired, particularly if one wishes the rabbit to be hidden from view during transport.

We have found that a plastic cooler works very well for short duration transport (Figure 3.26). The gap along the sides between the top and bottom pieces provides adequate ventilation for short trips (less than one or two minutes). We open the top when we reach our destination, so the actual duration of confinement is very brief. Recent concerns from vivaria personnel caused us to add a total of 770 ventilation holes in the top to meet a standard of 15% openings along a major side. As shown in the figure, 385 ventilation holes are drilled into the top on both sides of the molded handle. Each hole is 1/4 inch in diameter, a small value chosen so that nothing (cannulas,

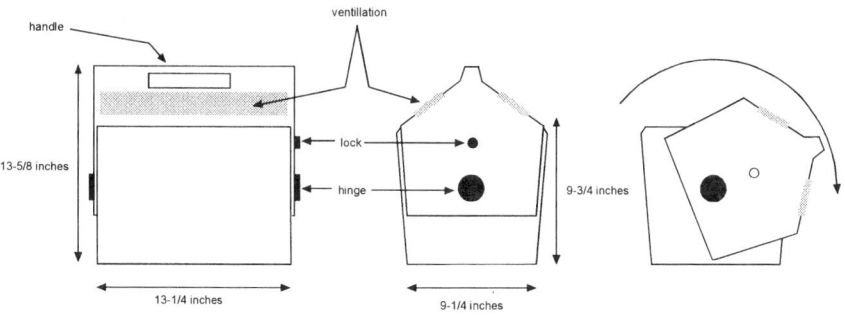

Figure 3.26: Rabbit carrier.

electrodes, sockets) on a headstage could get caught in the holes.. This yields 37.8 square inches of holes for a top that measures (liberally) 10 x 14 inches, yielding a conservative estimate of 27% opening, not taking into account the additional substantial 'leakage' between the edges of the top and bottom pieces of the cooler. The vaulted ceiling of the carrier allows for this large percentage of ventilation and provides head room for the rabbit. The holes were spaced by placing 1/4 inch wire screen over the location and drilling through every other space in the mesh. This carrier is inexpensive, simple, secure (it locks), has a convenient handle, and is easily cleaned.

SUMMARY

In this chapter we have presented some information concerning the recording of behavioral responses during behavioral neuroscience experiments. We cannot emphasize enough the importance of using careful and systematic techniques for measuring and recording behaviors. After all, the quality of the behavior one records can only be as good as the techniques used to record the behavior. In this regard, we should also note that physically healthy and emotionally happy (i.e., well-treated) animals yield reliable results. It is our experience that researchers genuinely like animals and treat them well.

NOTES

[1] A/D Converter. An electronic device that proportionately converts an analog signal into its digital representation. The number of "bits" associated with the converter indicates the precision. For example, an 8-bit A/D device converts an analog signal into numbers in the range of 0 to 255, whereas a 12-bit A/D device converts and analog signal into numbers in the rage of 0 to 4095.

[2] LED. A discrete electronic device that gives off light — the Light Emitting Diode. Here used as a light source for converting mechanical movement into an electronic signal for computer acquisition. Originally only available in colored lights, the commonly available red was not a good choice as a source for a light CS. Now LEDs are available in white which are good choices as a CS.

[3] Phototransistor. A discrete electronic device that proportionately converts light into an electronic signal. Here used as a detector for converting mechanical movement into an electronic signal for computer acquisition.

[4] 8255 PPI Chip. A large-scale integrated circuit, the Programmable Peripheral Interface chip allows a computer to activate or receive up to 24 digital signals associated with peripheral devices. In the later computer interfaces that are described, the 8255 chip is used to control tones, lights, airpuffs, shocks, etc.

[5] Quad Comparator. An integrated circuit used to convert an analog signal into a digital yes-or-no signal by comparing the analog signal to a reference value (either the analog signal is or is not above the reference value). Comes in an integrated package with four independent comparators, hence a "quad comparator."

[6] CMOS Dual Timer. An integrated circuit, the 4528 chip has two independent timers.

[7] Flip-Flop. After a capacitor, a flip-flop is the simplest of memory devices having only one of two states, yes or no (on or off; flip and flop, if you will). A flip-flop can be implemented with relays or with integrated circuits called 'gates,' in particular with NAND or NOR gates. A flip-flop is often used to debounce noisy mechanical switches.

Chapter 3

[8] NOR gate. An integrated circuit, the NOR gate is the complementary (opposite) of an OR gate in logical function. A gate typically has two inputs and one output. The output depends upon the states of the two inputs. In an OR gate, the output is a logic 1 (yes, on) if either or both of the inputs is also a logic 1. A NOR gate is the complement, where the output is a logic 1 (yes, on) only when both inputs are logic 0 (no, off).

[9] NAND gate. An integrated circuit, the NAND gate is the complementary (opposite) of an AND gate in logical function. A gate typically has two inputs and one output. The output depends upon the states of the two inputs. In an AND gate, the output is a logic 1 (yes, on) if and only if both of the inputs are also a logic 1. A NAND gate is the complement, where the output is a logic 1 (yes, on) under all conditions *except* when both inputs are logic 1 – then the output is logic 0.

Chapter 4

RECORDING NEURONAL DATA DURING CLASSICAL CONDITIONING EXPERIMENTS

INTRODUCTION

Recording and describing the activity of neurons during behavior is a major tool for trying to understand the relation between brain and behavior. Indeed, one of the major reasons that classical conditioning became a very popular behavioral paradigm for studying the neural correlates of learning and memory was that neural recording, used in conjunction with classical conditioning, was relatively simple. As we pointed out in the beginning of this book, during classical conditioning experiments the experimenter has full control of when stimuli are delivered and knows when to expect conditioned and unconditioned behaviors to occur. Further, in some classical conditioning procedures, such as eyeblink conditioning, the length of the trial is relatively short—typically on the order of 500 to 1500 msec.

The fact that stimulus/response characteristics were well known and that a relatively short sampling period was used made neuronal recording during classical conditioning easier than during operant and instrumental procedures. First, the relatively short trial lengths proved very useful as data sampling could be confined to the discrete trial periods. This can be contrasted with recording during operant tasks, where the experimental subject controls when responses are emitted and in some cases when stimuli are delivered. Neural recording during operant tasks typically required relatively long periods of sampling to capture behavior- and stimulus-related neural activity. This was especially difficult in pre-computer days and even later when computers were available but their speed and capacity were severely limited. Second, the other major advantage of using classical conditioning for neural recording experiments was that the experimenter knew exactly when stimuli were delivered and responses should have occurred. It has therefore been somewhat easier to discriminate between stimulus- and response-related activity in the brain. Interestingly, high speed processors and high memory capacity in today's computers has made neural recording during operant and instrumental paradigms relatively trivial.

In this chapter, we present methods and techniques that can be used to facilitate brain recording during classical conditioning experiments. We present information on the choice and fabrication of electrodes and cover the use of amplifiers and filters to condition the signal recorded from the brain. We also present information on calibration issues and techniques used to sort individual action potentials (i.e., spikes) during recording. Finally, we cover issues concerning surgical preparation for chronic brain recording. For some of the material presented here, we have borrowed (very liberally, we might add) from a two-part paper that was published by Gould, Sears and Steinmetz (1991a and b) in the *Kopf Carrier*. The *Kopf Carrier* is published by David Kopf Instruments, a company that makes a variety of equipment that is used in behavioral neuroscience experiments.

THEORY OF NEURAL RECORDING

Before discussing methods and techniques involved in neural recording, we present here a brief overview of the theory of neural recording. Because this chapter covers only extracellular unit recording, we will restrict our overview to this type of recording. In principle, brain recording involves lowering an insulated electrode with a very small exposed tip into a field of neurons that is capable of generating electric currents that are measured as voltage changes relative to a defined ground point. Actually, the bioelectrical signals from neurons can be picked up without even penetrating the brain via skull-mounted surface electrodes. These recordings, commonly called electroencephalograms (EEGs) or evoked response potentials (ERPs) have been useful for studying human brain activity, but lack the precision that is afforded by single- and multiple-unit recording with a microelectrode. In extracellular single- or multiple-unit recording with a microelectrode, neuronal spikes associated with the generation of action potentials are the desired data. Because spike activity is generated at the axon hillock of the neuron after the firing threshold has been exceeded, the electrode that is positioned in or near a population of neurons is atually picking up the currents generated by the spiking cells that are within the vicinity of the electrode tip. The precise form or template of the spike being recorded by the electrode depends, to a large extent, on the position of the electrode tip relative to the cell body of the neuron. Also, tip size and electrode diameter determine, to a large extent, the number of spikes that can be recorded from a given electrode tip position. Generally, larger tip sizes result in the recording of small amplitude spikes from many cells while smaller tip sizes result in the recording of large amplitude spikes from a few cells.

As you will see on reading this chapter, careful selection of electrodes, amplifiers, filters, and display devices are important for neural recording experiments. While the basic theory of recording is relatively straightforward, the quality of the data that are collected during brain recording experiments is determined by choices that are made concerning equipment used during the recording process.

ELECTRODES

The precise glass microelectrodes used for intracellular single unit recordings are too fragile for experiments in behaving animals. The brain is an organ that is buoyed in cerebrospinal fluid and constrained by flexible meninges inside the cranium. Its inertia causes the brain to remain in place while the body moves around it. Relatively speaking, the brain moves inside the head when there are head and body movements. Given the movements of the animal, a microelectrode that is luckily or skillfully placed inside a neuron has little chance of remaining in place when the brain moves around together with even the slightest head movements. An exception to this, however, is the series of studies conducted by Woody and colleagues, who studied the intracellular responses of cerebral cortex neurons during classical conditioning that involved pairing acoustic CSs with a glabellar tap US (e.g., Woody, Vassilevsky, and Engel, 1970). Because extracellular recording is much more common, we will focus here on the use of more durable metal electrodes that are placed extracellularly near a few neurons for single- or multiple-unit recordings.

We have done both single-unit and multiple-unit recording during classical conditioning experiments. These two techniques differ in the number of neurons being monitored for activity during the recording sessions. During multiple-unit recording, the activities of 3-6 neurons located near the tip of the electrode are monitored while during single-unit recording, the activity of one neuron is studied. The major determinant of whether multiple- or single-unit recording is done is the impedance of the electrode, which is determined mainly by the size and shape of the exposed tip. In general, multiple-unit recording electrodes have impedances less than 1 MΩ while single-unit recording electrodes are greater than 1 MΩ and typically more like 3-5 MΩ.

There are strengths and weaknesses of both multiple- and single-unit recording procedures. Multiple-unit recording is especially well suited for exploratory experiments in which the experimenter wishes to record from populations of neurons to get information about how populations of neurons are related to behavior. Single-unit recording is used mainly to determine the properties of individual neurons within a given population. It is impossible to determine the relative contributions of inhibitory versus excitatory patterns of action potential discharges to the population response with multiple-unit recording; this is quite easy to do using single-unit recording techniques. However, single-unit recording may lead to a sampling distribution bias in recording results. There is a tendency to record the activities of the largest and most active neurons with this technique because only the most isolated neurons, with respect to signal-to-noise ratios, are recorded with this technique. Recently, a compromise recording technique has been introduced: Multiple-unit recordings are made and single-unit response patterns are established from the multiple-unit records using wave-form template algorithms and techniques (see below).

For multiple-unit recording in rabbits and rats, we typically construct extracellular microelectrodes from stainless steel insect pins (size 00 or 0 insect pins available from Ward's Biological Supplies). The electrodes may be used directly out of the package or they may be further shaped by etching or grinding (see below). The electrodes are then coated with a plastic resin insulating material (Epoxylite #6001-M electrode insulator, Epoxylite Corp, Westerville, OH). The insect pins are prepared for insulation by clipping off their nylon heads and inserting the clipped ends into the flat surface of a 1" cork that has a long, ¼" diameter bolt protruding from the opposite side. Care is taken to align the pins perpendicular to the cork surface making sure that they do not touch each other. Insulation is placed on the electrodes by inverting the cork so that the pin points are oriented downward, and slowly dipping the pins into a wide-mouth jar of Epoxylite. We have successfully used two methods for lowering and raising the electrodes: (1) The cork's bolt can be clamped in a Kopf electrode manipulator that is mounted on a stereotaxic frame and the cork assembly lowered in and then slowly and evenly raised out of the Epoxylite, or (2) A slow, reversible motor can be fitted with a clamp assembly that holds the cork's bolt to lower and raise the pins. With either method, care must be taken to maintain a slow, steady rate when dipping and raising electrodes and to prevent the cork from making contact with the resin. We have found that even coats of insulation are produced when the pins are lowered slowly then immediately, but slowly, raised (i.e., not allowed to remain stationary in the Epoxylite). Typically, the electrodes are dipped 6 to 12 times and baked in an oven between dips for 3-12 hr at 99° C. The precise number of dips required depends on the age of the Epoxylite, the rate of dipping, and the amount of time the electrodes are baked between dips.

The metal tips are exposed by removing the insulation just before using the electrodes. For multiple-unit recording, this removal is accomplished by scraping the tip with a razor or scalpel blade, or by using an electrical voltage difference to blow the insulation away. At the other end of the electrode, we typically attach a stainless steel wire to the electrode by soldering to the electrode, by using a wire wrapping technique, or by using both wire wrapping and soldering. This wire is subsequently used to attach to plug devices that route the signal to external interfacing equipment.

A comment about soldering is in order. Soldering stainless steel requires using *stainless steel solder flux* for good joints; not just any flux will do. Good solder joints look like the solder has flowed onto the metal being soldered. We further test the joints by trying to mechanically bend them: Good joints are solid, bad joints flex and easily come apart. Residual flux is washed away from the electrode after soldering.

Single-unit electrodes are typically fashioned from relatively small diameter stainless steel rods or tungsten wire. To obtain an impedance of greater than 2-3 MΩ, the electrode must be etched before being insulated. Insulation cannot be removed by scraping but rather requires that a chemical or electrical process be used to expose the minute tip. Great care must be taken to not inadvertently bump the exposed tip against anything, as these electrodes are extremely fragile. Electrodes of various sizes and with various tip profiles that can be used for single-unit recording can be commercially obtained from Frederick Haer, Inc., and A-M Systems, Inc.

Almost any stainless steel source can be used for the electrodes. Insect pins are convenient because they are relatively inexpensive and about the right size and shape. Insulated insect pins can be used without having their shapes altered for lesion electrodes, stimulating electrodes, and recording electrodes intended for multiple unit electrodes. Straight stainless steel rods ranging in diameter from 75 µm to 200 µm or straightened stainless steel wire (25-100 µm) could also be used, assuming the tips have been shaped for recording. We have also recycled commercially available stainless steel microelectrodes for lesions, stimulation or multiple unit recording electrodes by removing the insulation, creating a tip by scraping with a razor or scalpel blade, and then using fine 220 grit sand paper to hone the end. Excellent electrodes can also be made from tungsten wire or rods (25-75 µm) and from platinum-iridium (80%/20%) wire (75-125 µm).

We shape the electrodes, especially the tips for single unit electrodes, by one of three methods. The first method has been a long-standing tradition in the Thompson laboratory, that of acid etching. Figure 4.1 shows the technique, where a variac is used to adjust the voltage to a 12.6 volt stepdown transformer. The electrode is attached to either lead from the transformer. A carbon rod, salvaged from the terminal of a dry cell battery, is attached to the other lead from the transformer. The electrode and carbon rod are placed into a 10% solution of hydrochloric acid and the variac adjusted (we usually set it at 1/4 to 1/2 its value). Be careful not to accidentally touch

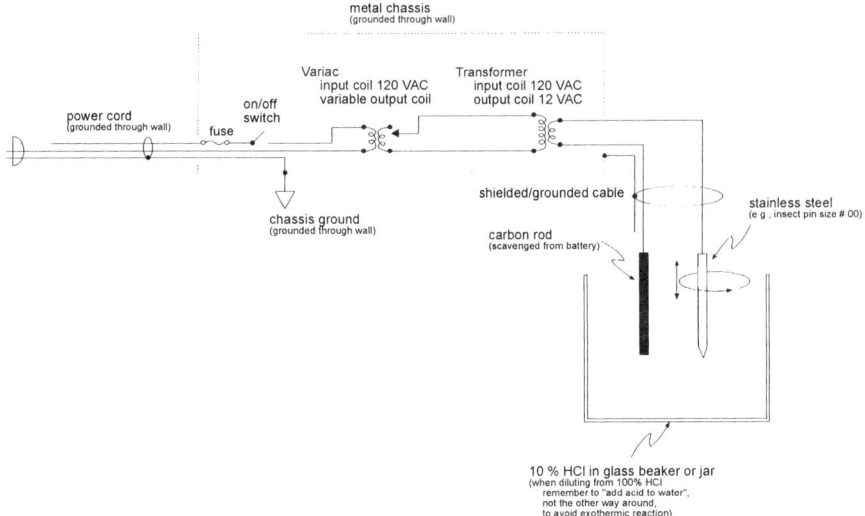

Figure 4.1: Acid etching the tips of stainless steel wire or insect pins for creating recording electrodes.

the electrode and the carbon rod or their connections. Etching can be seen indirectly by the bubbles that form around the electrode. The usual strategy is to keep the electrode moving (small circles work), otherwise the side of the electrode that is closest to the carbon rod etches away disproportionately and the electrode appears to "bend" as etching progresses. Ideally, just the tip is etched. In practice, the electrode gets skinnier and shorter. It takes about two minutes to etch each electrode. A variety of schemes to etch multiple electrodes at a time with automated circular movements and withdrawing the electrode have been tried; we have found them rarely to be successful.

The second method for shaping the electrodes was introduced to us by Tapani Korhonen (University of Jyväskylä, Finland). This method involves grinding the electrode with an electric grinding disk, while viewed under a microscope. Key to this method is having a very steady grinding wheel. This precision is accomplished by the smoothness of the motor and its bearings, by the sturdiness of the motor's mounting, and by the mass of the grinding wheel. The electrode is held in the hand and pressed against the grinding wheel. We do not have as nice a set up as Korhonen, but it works within our less discriminating level of acceptance. A simple rotary tool such as the kind available from Dremel or Sears, for two examples, works by choosing the thickest and smoothest grinding wheel. Mounts are available for the rotary tools, but we just mold children's modeling clay (*not* play dough) between the rotary tool and the microscope base or table. While not quite as good as Korhonen's electrodes, we feel that grinding is a significant

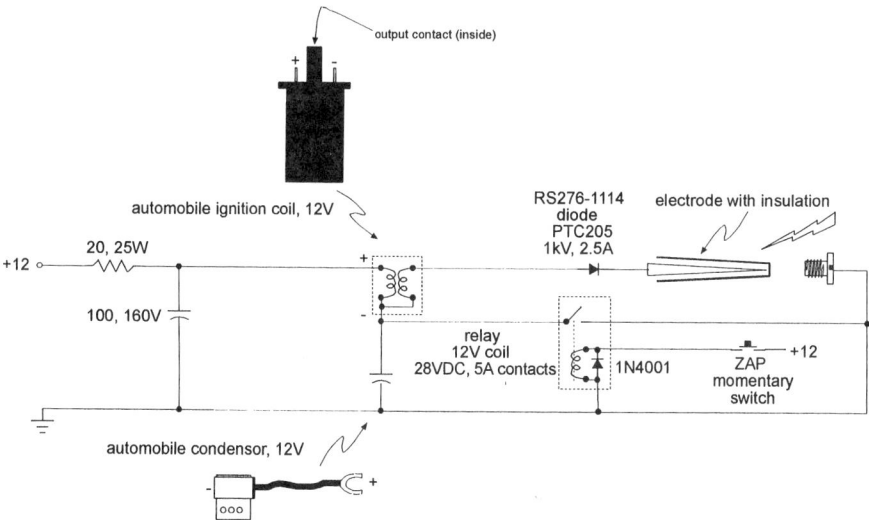

Figure 4.2: The 'zapper' for blowing off the insulation at the tip of a recording electrode.

improvement over etching, especially for electrodes in the middle to low impedance range.

Finally, it is possible to etch tungsten wire using a third method: Tungsten wire can be flame-formed to a precise tapered point using an acetylene torch. This method, though very effective, requires a great deal of trial-and-error experience to get the precise time of flame application for the tip profile that is desired.

Once the electrode shape is satisfactory and the insulation applied, the problem then becomes one of how to remove just a little of the insulation from the tip. For lesion electrodes (0.5 mm tips), stimulation electrodes (0.25 mm tips), and multiple unit electrodes (40-60 μm tips) we use a sharp scalpel blade under stereomicroscopic viewing with a steady, braced hand. For single units (20 μm tips), with a great deal of care and an equal amount of luck we can use the same method of scraping. However, for better uniformity we use a "zapper" to electrically blow off the insulation from the tip.

We do not know who originated or worked on the zapper that was used for years in the Thompson laboratory. It could have been Dave McCormick, Greg Clark, Mike Mauk, or some other equally resourceful individual. Figure 4.2 shows the circuit for the zapper. A 12 volt power supply (e.g., Radio Shack unregulated +12 VDC power supply) is substituted for a car battery used to power an automobile ignition coil and condenser (plans for a similar device were published Ciacone and Rebec, 1989). The high voltage generated between the insulated electrode and the screw separates the insulation from the electrode. In practice, the electrode and screw are viewed under a stereomicroscope, the electrode tip and the screw separated by tens of microns (the particular value is probably less important than trying to be consistent with this distance). With darkened illumination the momentary switch is closed and a spark leaps between the electrode and the screw, taking some insulation with it. Often times, a single zap is enough for a single unit electrode. Three or four zaps creates a multiple unit electrode. One of the advantages of the zapper over scraping is that one is less likely to deform the tip when zapping. With scraping, the pressure of the scalpel can ruin the tip, although sometimes it can be rehabilitated by bending it in the other direction to straighten it. The disadvantages of the zapping method are that one cannot visually tell whether insulation has actually been removed or where the insulation has been removed. If a spark is observed then it is usually safe to assume that insulation has been removed. Testing with an impedance meter is a better test that insulation has been removed and gives an indication of the amount of removal. A good, commercially available impedance tester for metal electrodes with impedances between 500 kΩ and 5 MΩ is the BAK Model IMP-1 Electrode Impedance Tester (BAK Electronics, Inc., Germantown, MD). Sometimes with an impedance meter one can tell where the insulation has been removed. It is generally true that insulation is removed from the closest part of the electrode to the screw (i.e., from the tip) but occasionally sparks have been observed from further up the shaft. It is reassuring to see the spark associated with the tip. If you miss seeing the spark, then you cannot be sure where the break in the insulation has occurred.

There are other schemes for fabricating electrodes. Stainless steel or tungsten wire or rods can be insulated with glass, which produces a very fine, well-controlled tip size (and thus electrode impedance). However, a glass-puller is required for fabricating these electrodes. Another technique is to use fine stainless steel wire that is insulated, for example, with Teflon coating. An advantage of fine wires is that they may float with the moving brain and thus maintain contact with the recorded cell or cells for longer periods of time. A disadvantage is that it may be difficult to control the position of the tip for recording. For this method to work it requires that the wire has some slack in it. Wire bundle electrodes are commercially available from a number of sources such as Frederick Haer, Inc., and NB, Inc. While relatively expensive, the use of wire bundles coupled with template-matching spike separation techniques is becoming increasingly popular. Also, for multiple recordings, two or more stainless steel or tungsten electrodes can be glued (e.g., with Superglue) together. There is a problem inherent in recording with wire bundles or multiple electrodes, however. Insertion of the electrode assembly into the brain may cause appreciable damage because of the increased diameter of the bundle or array of electrodes. One way to get around this problem is to implant a cannula and insert several fine, insulated wires through the cannula for the recordings. Care must be taken, however, to avoid breaking the insulation by scraping the electrodes against the edges of the cannula.

One other aspect of recording is deserving of mention here: Will the electrodes be chronically implanted or will moveable electrodes be employed? For chronic recordings, individual or multiple electrodes are lowered into the brain during a surgery then fixed permanently in place using dental acrylic. During behavioral training, recordings are taken from the chronically implanted electrodes. For recording with moveable electrodes, during surgery a hole is opened in the skull over the brain area from which recordings will be taken and a metal or plastic base is cemented over the opening. During subsequent recording sessions, an electrode is mounted in a micromanipulator, the micromanipulator is secured to the base that was implanted during surgery, and then the electrode is lowered into the brain. A variety of micromanipulators are available including those that are manually operated as well as those that use electronic and hydraulics to lower the electrode into the brain (e.g., those available from Frederick Haer, Inc., Kopf, Inc., and Narishige, Inc.). There are advantages and disadvantages of both approaches. Chronic electrode recordings can provide you with a snapshot of activity from a specific brain area over the course of training. However, the number of sites that can be monitored is limited since the electrode is fixed throughout training. Moveable electrodes increase the number of sites you can sample from a given subject. However, it is very unlikely that you can record from precisely the same population on a day-to-day basis, thus limiting your ability to monitor the activity of a given population across training.

AMPLIFICATION AND FILTERING

Assuming that the electrode has been placed correctly in the brain structure of interest, the brain signal must be amplified and filtered before being routed to the oscilloscopes or computer interfaces used to display and archive the unit activity. Biological amplifiers are very special devices. They are designed to accept a relatively high impedance signal and to output the signal to a relatively low impedance display or storage device. We have found a variety of amplifiers that are good for this purpose.

For many years the Thompson laboratory constructed amplifiers based on the design by Brakel and his colleagues (Brakel, Babb, Mahnke, & Verzeano, 1971). This amplifier works very well for extracellular neuronal recordings. We have found that to maximize its performance, it is best to construct it on an etched board. This amplifier has built-in filters, but to achieve the range of filtering we usually needed to add additional filters. Additional gain may also be desired. An advanced feature of this amplifier is the voltage follower transistor which sits on the anima's head near the electrodes. This transistor greatly reduces noise and artifacts in the neural record caused by animal movements.

It is sometimes advantageous to have a tunable filter, particularly so when a homemade amplifier like the Brakel amplifier is constructed. Below, we describe a variable filter we designed that is based on the LM 13600 dual transconductance operational amplifier (described in an article by Marston, 1988). Unlike most filters which are either fixed in their value or allow a limited selection of filtering, our design allows for tuning the cutoff frequency of the high and low pass filters by turning a simple potentiometer. The first operational amplifier is simply a voltage follower configuration to provide input buffering and a defined ground return path for the low pass filter. The order of the types of filters is important, first encountering a selectable high pass (low frequency) filter then a selectable low pass (high frequency) filter. A selectable notch filter (Mims, 1982) finishes the filtering that is available. The last stage is a 10-times gain stage which is sometimes handy. If none of the filters or the gain is selected then the input signal simply feeds through, buffered by the voltage follower, guaranteeing a high input impedance and a low output impedance. More on filtering appears later in this chapter

For our experiments we also use commercially available biological amplifiers. The quality, features and price of these amplifiers make them real bargains for experiments using multiple unit recordings. Lavond uses an A-M System Model 1700 4-channel amplifier, with selectable gain and filters (low and high pass filters, notch filter). Steinmetz has successfully used a number of different biological amplifiers including the A-M System Model 1700, the Haer 74-20-3 Differential, the WPI DAM-50 and ISO-DAM8A, and the Grass P15.

We should, at this point, say a few words about single-ended versus differential recording because different types of amplifiers are needed for these two different types of neural recording. For single-ended recording, an electrode (the active source) is lowered into the brain and voltage changes are measured relative to a fixed ground point (which is normally a stainless steel screw that is placed in the skull). For differential recording, two electrodes are placed in the brain: an active source, which is lowered into the population of neurons of interest, and an indifferent source, which is placed somewhere else in the brain. During differential recording, a differential amplifier is needed. A differential amplifier amplifies the "difference" signal between its two inputs (i.e., the active and indifferent inputs) relative to a defined ground point and rejects any voltage, which is common to both inputs (which includes for the most part noise). The voltage, which is common to both inputs, is referred to as the Common-Mode Voltage while the characteristic of a differential amplifier to ignore the VCM is referred to as Common-Mode Rejection. The advantage of using differential amplification is obvious—this recording configuration reduces noise. There are some drawbacks, however, of choosing this mode of recording over single-ended recording: Under some conditions the signal may be attenuated significantly and also a second electrode (i.e., the indifferent) must be used.

CALIBRATION

Amplifiers like the Brakel, which rely on an FET transistor as the voltage follower coupled to the electrode, introduce a variable amount of gain that is peculiar to each transistor. Matched FET transistors can be bought on a single chip, ensuring that each one has the same gain for each of the unit recordings. Small differences in amplification at the voltage follower or at any stage where less than precision components are used, can be magnified into significant differences. For multiple unit recordings, knowing the actual size of the spike may not be very important. For single unit recordings, however, the actual gain might be of interest, particularly when the size of the spike is used as a gauge of the unit type (e.g., Purkinje cell versus granule cell in the cerebellum). One method of determining the actual size is to calculate the product of the gain along the several stages of amplification. Another method is to empirically measure the gain by putting in a signal with a known size and measuring the ultimate effect on size, from which the gain can be calculated. This latter method has the advantage that the components can be tested to ensure that the recordings are being measured correctly. This technique has been used to solve problems as mundane as determining that a wire has been broken, identifying malfunctioning operational amplifiers or wrong filter settings, or determining the gain of the system.

The calibration circuit in Figure 4.3 creates a very small signal that can be used to test the recording apparatus from the electrode through a discriminator. An 8038 precision function generator is used to create a 1 KHz sine wave. The amplitude of this sine wave is attenuated through the potentiometer set up as a voltage divider. This signal is then buffered by a voltage follower operational amplifier having a high impedance input and a very low impedance output. A 10 mV peak-to-peak sine wave is a good size for calibrating a multiple unit recording system. Those 10 mV sine waves are very difficult to see on our oscilloscope, however, because it is at the limit of our

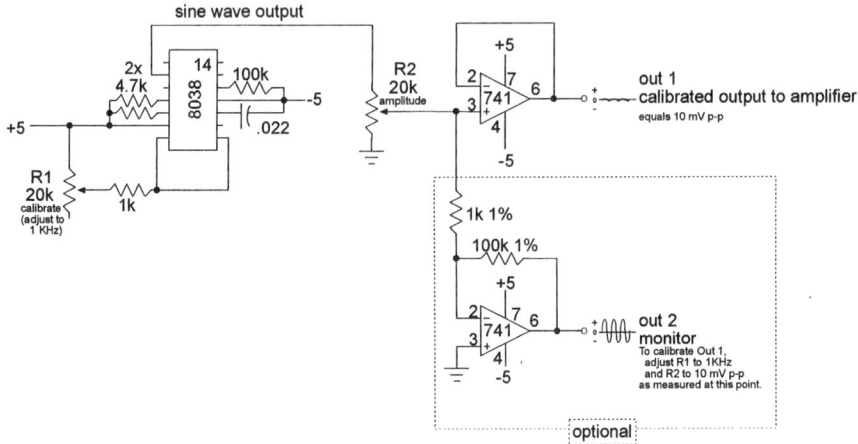

Figure 4.3: Circuit for creating a calibration signal for testing amplifiers and filters used for recording neural unit activity.

oscilloscope. The 10 mV, 1 KHz sine wave disappears into the noise of the oscilloscope's electronics. A parallel amplifier with a fixed gain of 100 (marked 'optional' in the figure) overcomes this problem by allowing the oscilloscope to operate within a more comfortable range of its amplification ability. A 10 mV signal at 'Out 1' is seen as one hundred times that amplitude as a 1 V signal at 'Out 2.' Out 2 is used to set the voltage divider to yield a 1 V peak-to-peak signal. At the same time, Out 1 will be 10 mV peak-to-peak. Out 2 is only used for the initial calibration. This 10 mV signal from Out 1 is then sent through the recording set up.

This calibration circuit will work with any dual battery system (e.g., two 9 V transistor batteries that provide ±9 volts) as long as it is set up by monitoring Out 2 each time it is used if it is being used to determine the gain of the system. Note that the discriminator which follows has a variable gain stage that needs to be taken into consideration when determining the size of the original unit signal. As the batteries wear down the amplitude of the 1 KHz signal changes proportionally. Our solution is to use batteries that are voltage-regulated to ±5 volts as seen in Chapter 12 (Power Supplies) to provide some stability of the signal size until the batteries fall below about 7 V when the supply's output can no longer provide a regulated 5 V. If the only interest is to use this calibration circuit for continuity, to check out that everything is in working order, then the actual amplitude is not important and simple batteries can be used.

FILTERS

Data signals often need to be processed before they can be analyzed. One way to process a data signal is to digitize it with a simple comparator (see below). Another way is to sample the signal at a high rate and convert the analog signal into a digital representation of its magnitude (A/D conversion). Often the data signal, whether it is behavior or neural recordings, needs processing even before it reaches the comparator or A/D converter. Most often this processing removes unwanted noise, principally 60 Hz noise, or restricts the range of frequencies data considered, as in neural activity. This section of this chapter is primarily concerned with building simple filters to create restricted bands of information for unit activity.

Before we examine the filters, it is important to consider terminology. A 'low pass filter' is a filter that allows low frequency signals through while blocking higher frequencies. A 'high frequency filter' is one that that filters out high frequencies, allowing low frequencies to pass through. Thus, a 'low pass filter' and a 'high frequency filter' refer to one and the same thing. Conversely, a 'high pass filter' and a 'low frequency filter' are exactly the same thing, allowing high frequencies through but cutting out the lower frequencies. The confusion comes when a 'low pass filter' and a 'low frequency filter' are mistaken for being the same thing—they are not, they are complementary to each other. To avoid confusion, it is recommended that one stick with one type of terminology or the other, talking about what frequencies 'pass' as opposed to which frequencies are 'filtered,' for example. Unfortunately for the reader, we will use the appropriate terms interchangeably because we think the reader should become familiar with both concepts. One additional term related to filters should be obvious. The 'cutoff frequency' defines the frequencies below or above the point that the filter affects.

Passive filters

The simplest filters use passive components (resistors and capacitors). Figure 4.4 shows the simplest high pass and low pass filters. The two types of filters are readily identified when it is recognized that capacitors pass changing frequencies (i.e., higher frequencies) while blocking direct current (DC) or slowly changing frequencies (i.e., lower frequencies). A capacitor between the input and output identifies a high pass/low frequency filter. The accompanying resistor goes to ground. If the capacitor goes to ground and the input and output are separated by a resistor, then it is a low pass/high frequency filter. Think of it this way: In the low pass/high fre-

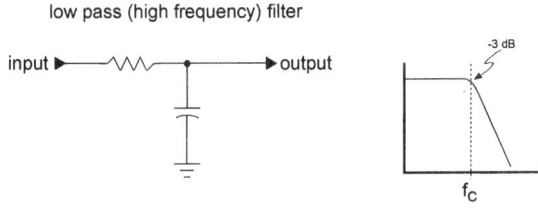

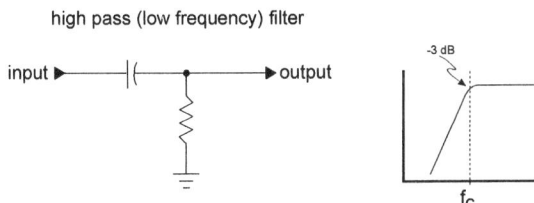

Figure 4.4: The simplest low and high pass circuits using passive resistor and capacitor components.

quency filter the capacitor is coupled to ground, allowing higher frequencies to short out to ground rather than making it through to the output.

The cutoff frequency by definition is the point where the filter begins reducing the amplitude of the input signal. By convention, the cutoff frequency is the point where the input signal has been reduced by 3 decibels (dB) of voltage, which works out to a value of 70.8% of the input amplitude. The formal definition of a dB is

$$dB = 20 \log (e_{out}/e_{in})$$

where dB stands for decibels, e_{out} is the output voltage and e_{in} is the input voltage. A 1 dB loss (negative dB) of signal between the input and output voltages means that the output is 89.1% the size of the input [-1 dB = 20 log (0.8913/1.000)]. The conventional cutoff at -3 dB calculates to 20 log (0.7079/1.000) or 0.708 the original size. Note that the input amplitude is assumed to be 1 V peak-to-peak, but other values could be used, requiring simple scaling of the output. The cutoff frequencies for both high and low pass filters is calculated by the formula

$$f_c = 1/(2\pi RC)$$

where f_c is the cutoff frequency, R is the resistance in ohms and C is the capacitance in farads. An intuitive understanding of the formula recognizes that as the resistor or the capacitor gets smaller the cutoff frequency gets higher. The corresponding period (T) is the inverse of the cutoff frequency, calculated by the formula

$$T = 1/f_c$$

using f_c from above.

A simple example of filtering is the original 'NM box' that was used in the Thompson laboratory. The NM box contained a passive high pass filter consisting of just one resistor and one capacitor. This NM box was juxtaposed between the eyeblink minitorque potentiometer (variable 25 KΩ) and the Grass polygraph which was set, as best as we can recollect, for half amplitude cutoff at 0.5 Hz high pass (only the polygraph driver was used). The NM box high pass filter used a 100 µF capacitor and a 470 KΩ resistor (i.e., the capacitor was between the input and output, and the resistor went to ground). The calculated cutoff frequency is 0.0034 Hz. In other words, the NM box was used to cut out any DC potentials on the NM signal. In this way the output of the NM box was self-adjusting for the resting position of the eyelid (baseline) between training trials.

Active Filters

The performance of passive filters is limited by the sharpness at the cutoff frequency and the rate of fall off. Improved performance is achieved using active filters created from operational amplifiers. Making active filters has become pretty routine through several publications. We routinely use Lancaster's (1975) or Berlin's (1977) books.

Figures 4.5, 4.6 and 4.7 show the designs for low and high pass active filters of 2nd order, 4th order and 6th order, respectively. The order refers to the rate of drop off (slope, sharpness) of attenuation of the affected frequencies. This slope is described as voltage (amplitude) change per octave, where an octave is a doubling (or halving) of a frequency. Both 500 Hz and 2K Hz are one octave away from 1K Hz (below and above, respectively). First order filters change by ±6 dB/octave; 2nd order by ±12 dB/octave; 4th order by ±24 dB/octave; and 6th order by ±36 dB/octave. Passive filters are 1st order; 6th order filters are very sophisticated filters. Most physiological recordings are more than adequately served with 2nd order filters.

As an illustration of designing a filter, suppose the goal is to create a 2nd order, low pass filter (Figure 4.5) with a 3.5 KHz cutoff. Each of these

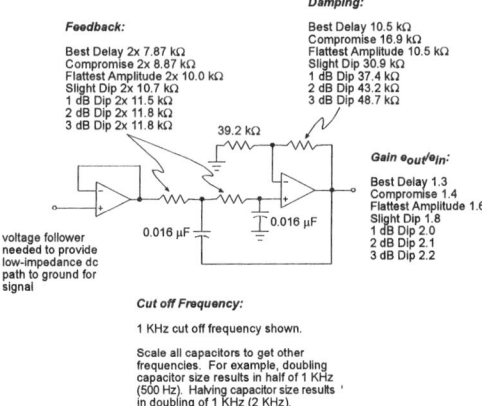

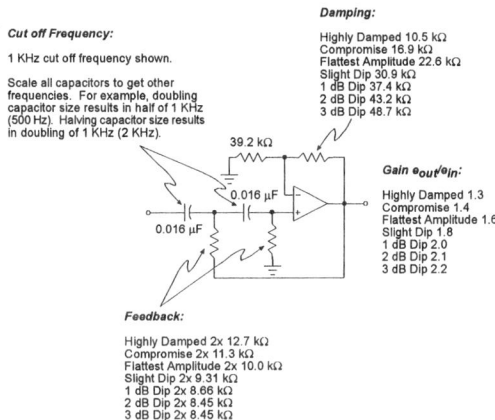

Figure 4.5: Designs for creating second order low- and high-pass filters.

filter designs (Figures 4.5, 4.6 and 4.7) is based on a cutoff frequency of 1 KHz (1000 Hz). As indicated in each of the figures, other cutoff frequencies are achieved by changing the all capacitor values by the same factor. For example, for a 2 KHz low pass/high frequency filter change all the 0.016 µf capacitors to 0.008 µf (halving the capacitors doubles the frequency). In general, to figure out the capacitor value for any frequency, divide 16 by the desired frequency (in Hz). Thus, for a cutoff frequency of 3.5 KHz the capacitor value is 0.00457 µf (16 divided by 3500).

The way the filter performs depends upon the values of the

Chapter 4   145

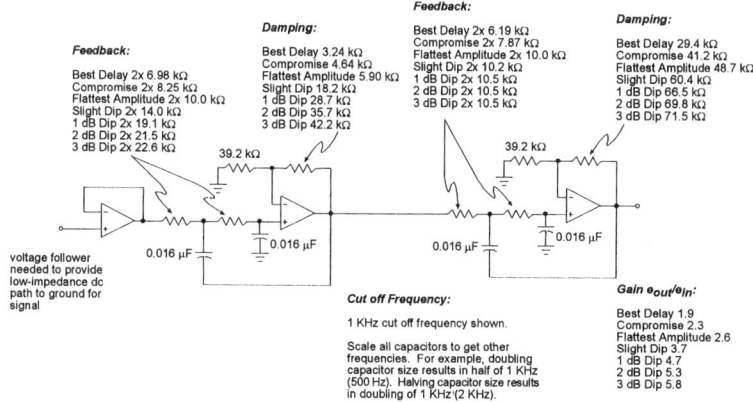

4th order, low pass filter

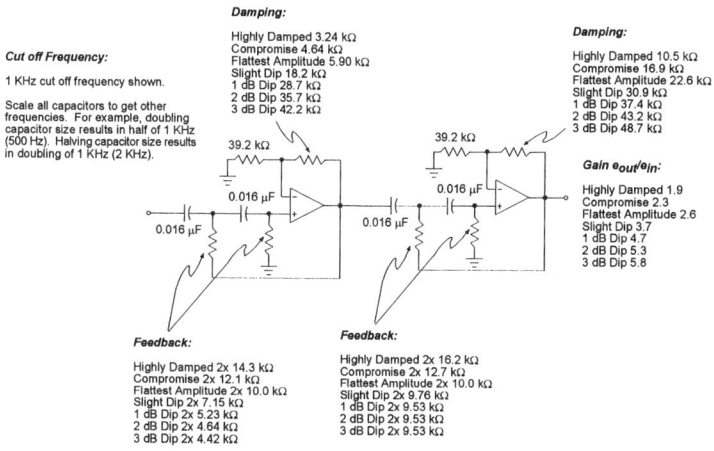

4th order, high pass filter

Figure 4.6: Designs for creating fourth order low- and high-pass filters.

'feedback' and 'damping' resistors. The 'best delay' filter, for example, waits the longest before filtering begins. At the other end of the designs, the '3 dB dip' design has large ripples causing unequal gain in the pass-band, but then has a very sharp cutoff. For most physiological work, the commonly chosen design is the 'flattest amplitude' which has reasonably fast responsiveness with the least distortion. For the flattest amplitude, the feedback resistors are each 10 kΩ (like the capacitors, there are two relevant resistors) and the damping resistor is 10.5 kΩ. The resistors should be within ±5% of these values. Besides filtering out high frequencies, the lower fre-

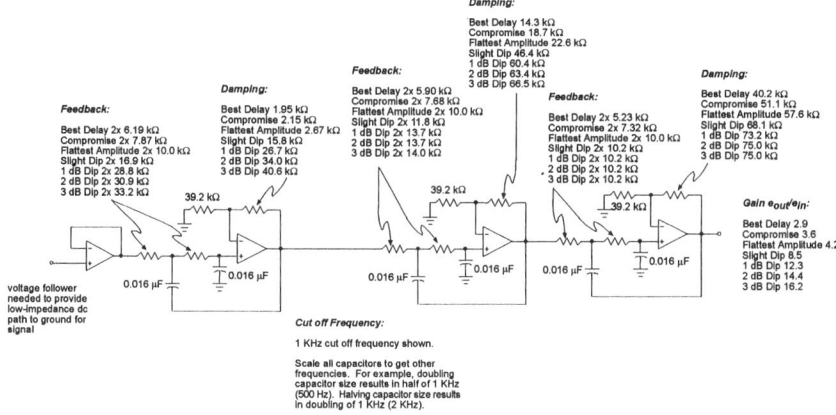

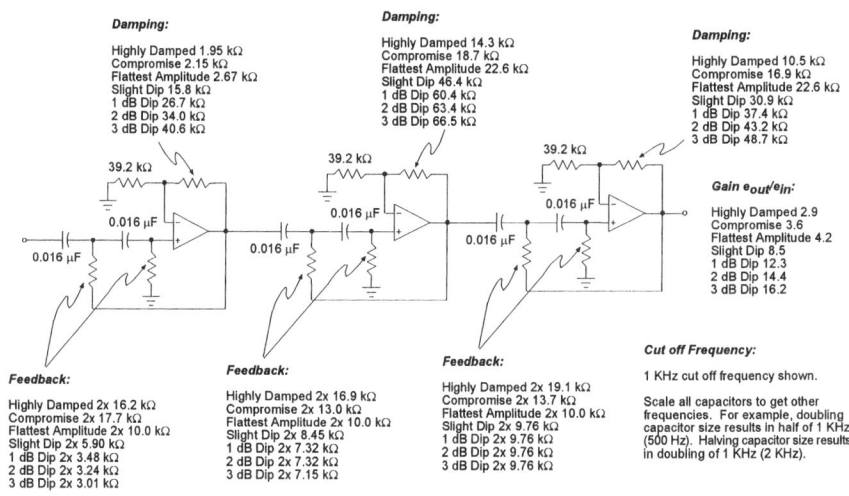

Figure 4.7: Designs for creating sixth order low- and high-pass filters.

quencies in the pass band are amplified. For the flattest amplitude design, the gain in the pass band is 1.6 times larger than the input signal (a 1 V input signal is passed as a 1.6 volt output signal).

Finally, note that the low pass and high pass filters differ in that the low pass filters require a voltage follower on the input. The voltage follower provides a necessary 'return ground' for the low pass filter designs. Without the return ground, the capacitors would charge with an input signal without ever discharging. The books by Lancaster and Berlin make note of the need for a return ground for low pass filters but do not typically include the voltage follower in the pictures of their designs, where it is annotated. The complementary high pass filters do not need the voltage follower.

Higher order filters are not constructed by simply chaining lower order filters together. Two identical 2nd order filters result in 2nd order filtering, not a 4th order filter. The higher order filters are achieved by changing the damping resistors to achieve 4th order (Figure 4.6) and 6th order (Figure 4.7) filters. In all other respects the higher order filters are the same: the cutoff frequency is manipulated the same way (changing all the capacitors by the same factor) and the feedback resistors are the same (for a given type of filter, e.g., flattest amplitude). The gains of the higher order filters is slightly different, with the gain of a 4th order low pass filter with the flattest amplitude being 2.6 times the input's amplitude, and the gain of a 6th order low pass filter with the flattest amplitude being 4.2 times the input's amplitude.

With single chips having up to four operational amplifiers, it is tempting to make higher order filters. We note however, as we did above, that 2nd order filters are more than adequate for most physiological recordings of neural unit activity.

For neural unit recordings we usually design a high pass and a low pass filter to achieve a 'pass band' between them. This ordering is important, first the high pass then the low pass, because the high pass filter provides the return path for the low pass filter. Typical frequency ranges for multiple unit band-pass recordings of unit activity are 300–3000 Hz or 500–5000 Hz. Alternatively, instead of using high and low pass filters, there are active filter designs specifically designed as band pass filters. The problem with this design is that their pass band is much narrower than these ranges for multiple units.

The filters described in this section have the disadvantage that their cutoff frequency is generally fixed to a single value because multiple capacitors would have to be changed simultaneously. Variable capacitors that are 'ganged' together (a single axle changes the capacitance of two or more capacitors) are a possible solution but they are bulky and difficult to find. An alternative solution is to use a ganged selector switch (which is slightly more common) where each position selects a pair (2nd order filter) or quad (4th order filter) of fixed capacitors, allowing for the choice of a few set frequencies. For example, a four-position dual throw rotary switch could be used to select between four filter settings on a 2nd order filter, if it could be found. For most practical purposes, however, these filters are fixed. In the next section we describe a variable filter that avoids this difficulty.

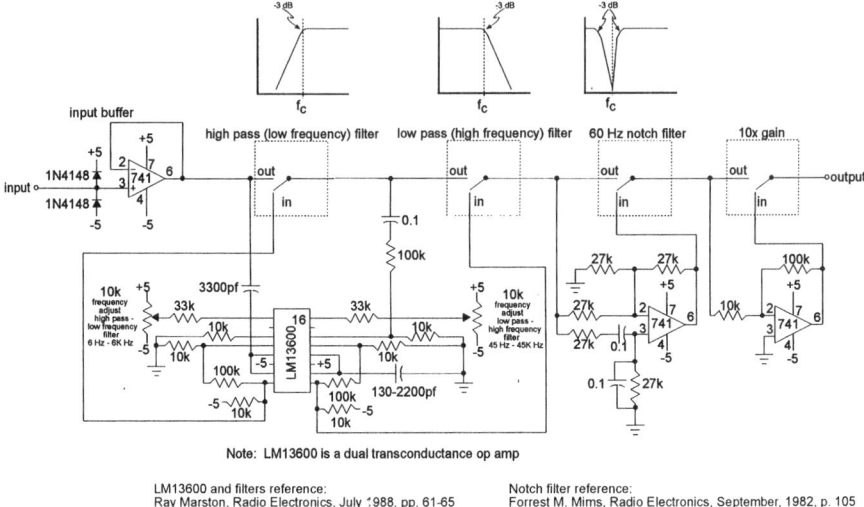

Figure 4.8: Buffering plus selective, tunable high- and low-pass filters using transconductance op amps combined with selectable 60 Hz notch filter and gain stages.

Transconductance Filters

As we noted above, we discovered an article using transconductance operational amplifiers to create filters (Marston, 1988). Unlike the filter designs in Figures 4.5, 4.6 and 4.7 described above, which require changing two, four or six capacitors simultaneously to change the cutoff frequency, the big advantage of the transconductance design is that a single variable resistor can be used to change the cutoff frequency in a linear manner.

In Figure 4.8 we show a design for a complete filter system for use with multiple unit recordings. After a buffering voltage follower, our design includes high and low pass filters based on the LM 13600 dual transconductance operational amplifier, followed by two circuits using traditional operational amplifiers to create a notch filter (Mims, 1982) and a gain stage. The notch filter removes frequencies around the cutoff frequency, in this case removing a small band of frequencies around 60 Hz. At each point in the overall filter design a switch selects in or out of one of the operational amplifier stages (four switches, one each for the high pass, low pass, notch and gain). The voltage follower acts both as a buffer so that the filter does not load the input signal, and as the return ground should the low pass filter be the first one selected. The knobs for the wipers of the variable resistors can be marked on the chassis for the filter so that the frequency can be set visually. This requires the forethought of having previously calibrated the filter by putting in known frequencies and marking the knob's position at the appropriate cutoff frequency. Alternatively, a rotary switch could be set up to select between different variable resistors, each preset to yield a desired cutoff. Either of these custom designs could be a valuable asset in your experiments. Alternatively, you might just want to build a notch filter by lifting that component out of the circuit. If you build your own biological amplifier or have another need for filtering then the circuit in Figure 4.8 may be a valuable asset. But as we noted above, although this is a nice filter design, most physiological amplifiers now come with built-in, switch-selectable filters.

Perspective

One of the most pervasive misconceptions about filters is that they completely eliminate the frequencies beyond the cutoff frequency. In fact, filters attenuate frequencies at a predictable but gradual rate (the number of decibels per octave). The closer to the cutoff frequency the less effect; the further from the cutoff frequency, the greater effect. The relationship is not linear but exponential, best described as loss per octave (doubling or halving of the frequency).

Filters reduce the amplitude of the signal, but they do not eliminate it. This can be observed by the following experiment. Observe any sine wave on an oscilloscope, for example a 60 Hz signal. Run the 60 Hz signal into an appropriate filter. Now observe that the signal on the oscilloscope appears to be a flat line. The filter worked, right? The answer is yes and no. Yes, the amplitude of the 60 Hz signal appears to be gone. But no, turning up the gain of the oscilloscope reveals that the 60 Hz signal is still there although it is much reduced in size. Further filtering (with a better 60 Hz filter or by increasing the cutoff frequency of a low pass filter) can reduce the signal even further, but turning up the gain on the oscilloscope still reveals that the 60 Hz signal is present, albeit very small. It is only when the amplitude of the signal is reduced into the noise level of the oscilloscope or the associated electronics (e.g., the power supply) that any signal is effectively eliminated, but even then this may be an illusion caused by our inability to measure it.

A second pervasive misunderstanding about filters is that they operate on frequencies, but in fact they operate on the slope of the signal (rate of voltage change per unit of time). Frequencies are just easier for us to talk about by considering ideal sine waves. But since the effect of filtering is to reduce the amplitude, and thus the slope, the result is to shift higher frequency signals to a lower frequency range. Lavond once heard a well-respected researcher find fault in a fast Fourier (FFT) analysis because some small higher frequency signals did not have much power associated with them. He advocated additional high frequency filtering, not recognizing that the current high frequency filtering had already shifted those signals into the slower frequency ranges that bothered him. The smaller slow frequency signals were really artifacts caused by reduced slopes of large fast amplitude signals.

Finally, a third misunderstanding is that filtering eliminates the cutoff frequency. That is, if the filter is set to cut off frequencies below 500 Hz, for example, then the mistaken belief is that 500 Hz is completely gone, as well as 499, 498, 497, etc. In reality, for a maximally flat filter, the frequencies just above 500 are attenuated slightly (the filter begins above the cutoff frequency), the 500 Hz frequency itself is reduced by about 30% (it is 70.8% its original size), and there is a gradual lessening of the amplitude at successively lower frequencies. There is no perfect filter that will completely eliminate the cutoff frequency, with the possible exception of some digital filtering techniques.

The cautionary warning here is to impart the idea that filtering changes the data. We have seen filters turn unit data from one form into completely different forms. Sometimes the filtering does not come from filters per se, but from the frequency response characteristics of the operational amplifiers in use. As a point of caution, it is best to avoid any filtering as much as possible. Be suspicious of your filtered data. When in doubt, systematically alter the filtering to see if an artifact has been introduced. Above all, be your own worst critic. If you cannot convince yourself of your data, no one else should believe it. Remember, a major point of science is reproducibility, to have others be able to replicate your results, so report what you can trust. As a point of practicality, specify your cutoff frequencies and know the characteristics of your filters.

ISOLATING ACTION POTENTIALS: THE USE OF WAVEFORM-MATCHING ROUTINES OR DISCRIMINATORS

There are least two ways to collect the unit data to a computer. One method involves using an analog-to-digital (A/D) converter to digitize the waveforms and store the data. This method is what we referred to above as "waveform-matching" techniques. In principle, this system works by isolating from the record those digitized waveforms that are similar to each other in amplitude, rise-time and duration (i.e., confidence intervals are established around a waveform and those that fall within the interval are counted as firings from a single neuron). There are at least two major advantages of using this type of system. First, single-unit recording results can be obtained

Chapter 4

using multiple-unit recording techniques, thus, data concerning population and single neuronal response patterns can be extracted from the same record. Second, using the proper analysis routines, the relationship in firing between units can be established from a single record. When coupled with electrode array methodology, this is a powerful technique for establishing variation and co-variation of the firing of neurons within a population. There are some disadvantages of using this method, as well. The equipment needed for this method, including amplifiers, filters, and interface circuitry is relatively expensive. Also, fast computers with very large amounts of memory are needed. While computer speed is no longer an issue (i.e., microcomputers with high-speed processors are common and relatively inexpensive), even with today's hard drives having sizes in the range of tens of gigabytes, collecting unit data by digitizing the waveforms can still consume tremendous amounts of storage space and time spent in off-line data reduction is vastly increased. The discrete nature of the trials delivered during classical conditioning experiments greatly reduce the amount of data that has to be collected, however. Steinmetz prefers the template matching methods for collecting unit data and now employs this technology in all of his neural recordings experiments (e.g., Katz & Steinmetz, 1997). The equipment for collecting spike data using this technique is commercially available from CED, Ltd. (the Spike2 system), Brainwave, Inc. and Alpha Omega Engineering, Inc. An alternative method is to recognize units based on one criterion and save only the times of the units. This requires some fast computations, but a more moderate amount of data storage.

At the extreme of data reduction is a simple discriminator, a comparator that gives a digital signal only when a neural unit spike exceeds a particular voltage height. This is the simple method we describe here, a method preferred by Lavond for collecting single and multiple unit neural data. To discriminate single units from multiple unit records Lavond uses an Alpha Omega Multispike Discriminator, whose TTL outputs can be directly fed into the unit counters for online data collection.

Figure 4.9 shows an overview of the components for a simple discriminator. A recording electrode for a single cell or multiple units is amplified and filtered by a commercially available biological amplifier. This signal is to be compared to a reference level voltage, such that every unit spike that exceeds the reference level causes a digital pulse that can be routed to a computer for counting. In support of this comparator function we have added a few features. The signal from the biological amplifier can be increased by a variable gain stage. This enhanced amplified signal can maintain its orientation or the polarity can be inverted. For an extracellularly-placed electrode, neurons that are close to the electrode will cause a signal going in the complementary direction to an intracellular electrode due to the

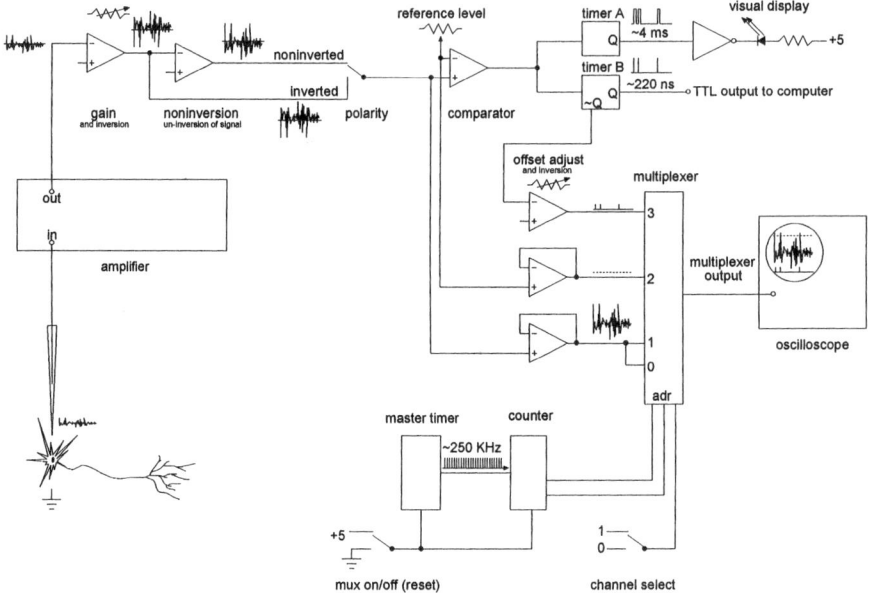

Figure 4.9: Overview of the components for recording neural unit activity and simple data analysis using a height discriminator.

direction of current flow. Thus, while we are used to seeing an intracellular spike in textbooks that goes up in a positive direction from some negative DC voltage baseline (the resting potential), an extracellular electrode placed near the cell will have a spike going in the opposite negative direction from ground potential (i.e., the DC voltage is removed). For the discriminator to work in our design, positive transitions above the reference level are required. Note that the comparator will discriminate all spikes that exceed the reference level. Different spike heights are an indication of the distance the cells may be from the recording electrode, and the characteristic spike height and width of a particular cell. For cells that are further away from the tip of the recording electrode, the polarity is just the opposite because of the direction of current flow. That is, recording from distant neurons yields positive going action potentials similar to those we are familiar with in textbook descriptions of intracellular recordings. The polarity switch allows for inverting the extracellular signal. By having this polarity feature either near or far cells can be discriminated.

After the comparator there are three additional supporting features to make use of the discriminated unit activity in Figure 4.9. The most important of these three features is the TTL level output to the computer (see Figure 4.10). This output is conditioned by a timer (Timer B) to give about a 220 nanosecond pulse for each neural spike that is detected crossing the reference level. This duration is meant to approximate the pulse width used by the Haer window discriminator, a commercially available discriminator manufactured by Frederick Haer, Inc. The second supporting feature is a longer timer (Timer A) set to light an LED for visual display (see Figure 4.10). The intensity of this LED is proportional to the amount of discriminated spike activity, providing a visual display for the experimenter when setting the reference level. The third supporting feature is a multiplexed display to an oscilloscope (see Figure 4.10). This display is an expansion of the "chop" sweeping function of an oscilloscope, whereby one beam of the oscilloscope display is rapidly switched between two different oscilloscope channels, settling long enough at each channel to display part of the data. The keys to seeing the two displays are: having the vertical transition between the two channels be so fast that they are not noticed; and settling for a short time on each channel. This technique is known as vector graphics, as opposed to the more familiar raster graphics technique that is seen on a TV screen or computer monitor. Modern oscilloscopes automatically choose this kind of display when the time base is very slow. The alternative is to "alternate" the sweeps, which modern oscilloscopes may choose, where the first sweep displays the first channel, the second sweep the second channel, then back again, with each sweep being so rapid that it is not noticed that the sweeps alternate. Older Tektronix oscilloscopes had switches to allow for manually selecting chop or alternate. The Haer window discriminator expands upon this idea by displaying three sets of data on a single chopped sweep, the reference level, the unit record, and the TTL pulses that go to the computer. Our oscilloscope display mimics these chop features, with the addition that our height discriminator has four discriminators. The Haer discriminator is a single channel device, but it has additional features like a window discriminator having three reference levels (rather than a height discriminator). Window discriminators can be used to isolate spikes that break one level but not a second or third level. In this manner, mid-sized spikes can be isolated from the record. The master clock and counter in Figure 4.10 are needed to control the chop function to give a multiplexed display.

At this time we want to digress briefly to discuss our protocol for setting the reference level of the discriminator. At the beginning of the experiment, before running trials (e.g., 108 trials per day) with fixed electrodes or at the beginning of a block of trials for a new recording location (typically, 12 blocks of 9 trials per block make up 108 trials per day) with a moveable electrode, we set the discriminator reference level and leave it. If, during the sampling the unit activity changes, we might be tempted to change the reference level (or conversely the gain or polarity of the signal;

154 *Handbook of Classical Conditioning*

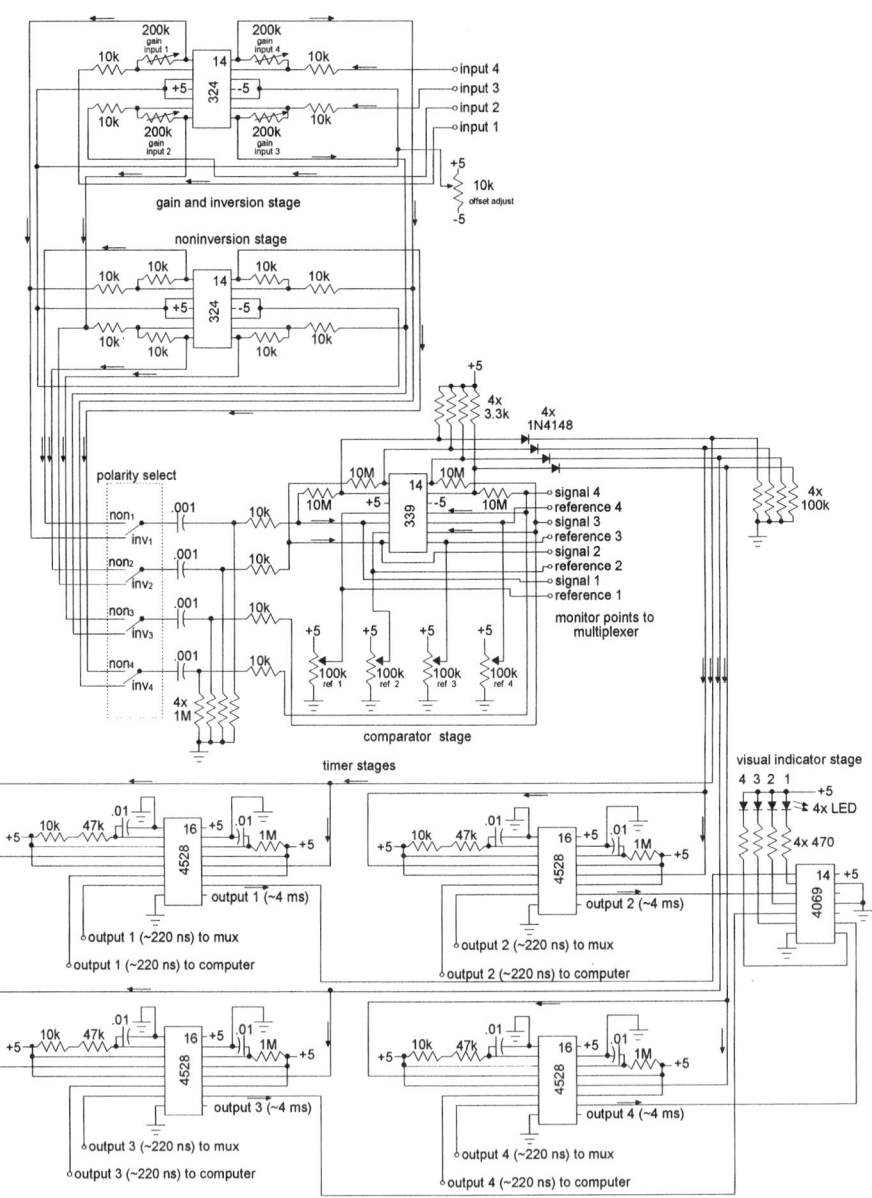

Figure 4.10: Gain, polarity and comparator functions for unit discrimination.

we record changes in our experimental notes) or to abort the testing. Special care is made *not* to change the reference level (or the gain or polarity of the signal) during a trial so that the data is not contaminated.

Figure 4.11 shows the circuitry needed for the gain, polarity and comparator functions, as well as the conditioning of the TTL signals to the computer and the LED display. The first LM324 quad operational amplifier has the dual functions of giving gain and inverting each of the four input signals. The preferred (simplest) configuration of an operational amplifier giving gain yields an inverted output. Normally this inversion does not matter for many electronic problems, but here we take advantage of it for our design. The second LM324 quad operational amplifier has unity gain (no gain) but, also being in the favored simpler configuration, inverts the inverted signal, yielding the original noninverted signal. This may seem convoluted, but anyone familiar with disinhibition in the nervous system caused by one neuron inhibiting a second inhibitory neuron should be able to relate. A polarity switch allows selection of either the noninverted or inverted signal, independently for each channel.

The LM339 is a specialized quad operational amplifier used for comparator functions. The reference level for each comparator is set by a potentiometer that varies between ground and the positive power supply (+5 V). A 10 MΩ feedback resistor from each output to its corresponding noninverting input provides for hysteresis. Each comparator output is an open collector, requiring a 3.3 kΩ pull up resistor to the positive power supply. Up to this point the chips are all powered by a dual power supply ($\pm$ 5 V). From this point on the semiconductor chips are all powered by a single ground-to-+5 V supply so that it is compatible with TTL levels that can be used by a computer. To achieve this restriction the output of each of the comparators goes through a switching diode (1N4146 or 1N914) and a pull down resistor (100 kΩ) to define ground. As a result, the signals from the comparators now are restricted from ground to +5 V of the power supply.

Each signal from the corresponding comparator is used to trigger two timers (the 4528 is a dual timer). One timer on the 4528 is set to give about a 220 nanosecond TTL level pulse to the computer each time the comparator detects a unit that crosses the reference level. The other timer on the 4528 is a longer duration (about 4 msec) to light an LED. Higher rates of detected neural unit activity cause a brighter LED display, whereas lower rates of detected neural unit activity cause a dimmer glow of the LED. This is used to give the experimenter a crude level of feedback as to the rate of unit activity when setting the reference level of the comparator.

Figure 4.11 shows our solution for multiplexing the two signals used by each comparator (the neural unit signal and the reference level) and the corresponding TTL signals sent to the computer. These three signals are

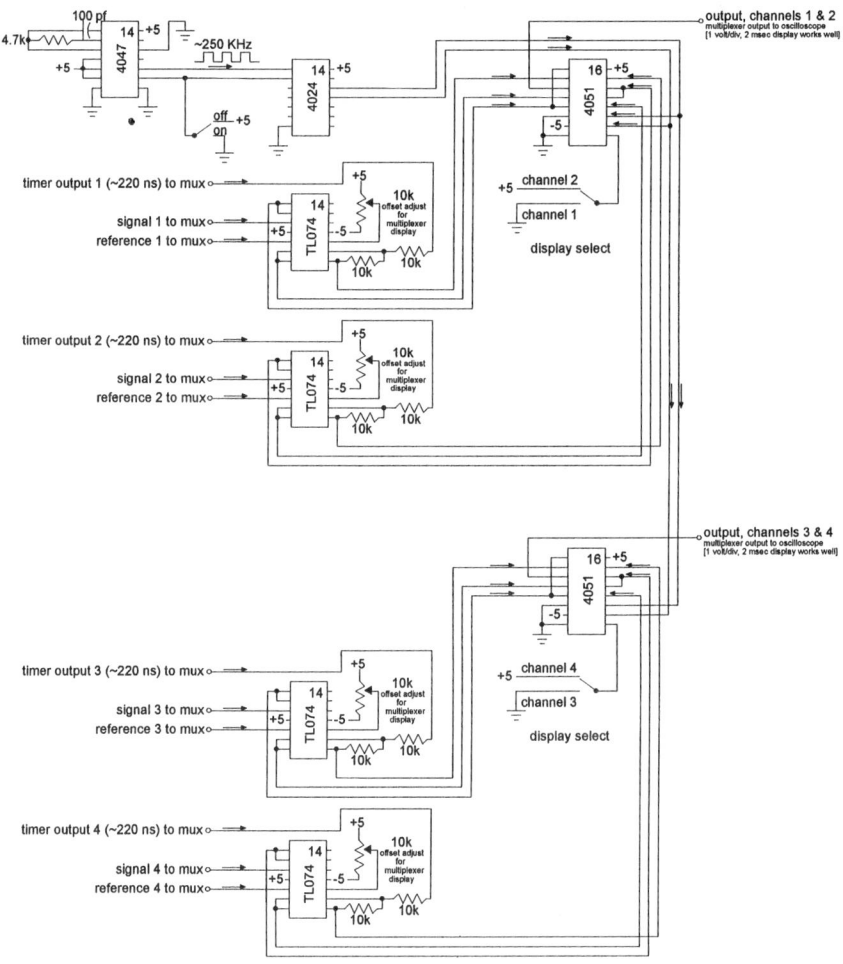

Figure 4.11: Multiplexed display for unit discrimination.

conditioned by three of the four operational amplifiers on the TL074 quad operational amplifier. Two operational amplifiers are merely voltage followers (for monitoring the comparator signals) and the third operational amplifier is used to give a DC offset to the TTL display. Unit inputs 1 and 2 share the same 4051 multiplexer, and unit inputs 3 and 4 share a second 4051 multiplexer. The 4051 is a 1-of-8 multiplexer/demultiplexer. Unlike most other multiplexers which are limited to routing digital signals, the 4051 can route an analog signal as in this application (see Lancaster, 1988). An analog signal can be routed to one of eight outputs, or one of eight analog inputs can be routed to a single output which is the way we use it. The analog output of the 4051 multiplexer is selected by the addresses created by the 4024 binary counter and the 4047 master clock. These addresses are used by all the 4051 multiplexers. Each 4051 multiplexer has an additional high address line that is selected by a switch (for example, in the upper 4051, select channel 1 or channel 2 for display). This higher address line is used to select between the first four or the second four input lines of the 4051, allowing us to select between one channel which uses the lower four inputs versus the other channel which uses the upper four inputs. Essentially, we have converted the 4051 into two data channels that share the same output. Finally, note that for any channel, two of the four inputs to the 4051 actually come from the same source, the unit recording, biasing the display to spend more time showing that information (which is more complex than the straight line for the reference level and the binary signal for the TTL display). This persistence of the unit data gives a better display.

In operation, we put the output of the upper 4051 into channel A of a dual channel oscilloscope and the output of the lower 4051 into channel B of the same oscilloscope. By setting the display select switches on the discriminator and choosing to display either channel A or channel B on the dual oscilloscope, one oscilloscope can successively display the signals related to one of four channels. In viewing the display for a channel, we use this feature to set the gain, polarity, and reference levels for each of the comparator channels. Because of the high frequency related to the clock (about 250 kHz) we turn the multiplexed display off during actual data collection.

An alternative design for this final display stage would be to use just one TL074 and one 4051. In this design, the four (or more) channels would be selected for the different inputs to the TL074, perhaps using the 4066 as a part of a scheme for a selector switch. Consideration should be given to using the fourth (currently unused) operational amplifier on the TL074, with a one-to-one output to the 4051. (The biased 4051 display for the unit record would be made by using two TL074 operational amplifiers for the same signal.) Such a design would make this a more general solution for displaying multiple data onto single oscilloscope sweep.

SURGICAL PREPARATION CONSIDERATIONS

Chapter 7 provides detailed information concerning surgical procedures that are used in classical conditioning and other behavioral neuroscience experiments. Nevertheless, a few words about headstages that are used during behavioral/neural recording experiments are in order here. Here we describe a couple of systems, one for the rabbit and one for the rat, constructed of sockets used for integrated circuits and a screw attachment.

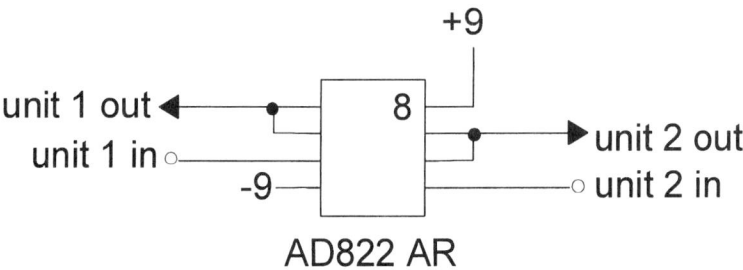

Figure 4.12: The headstage is equipped with dual op-amps configured as voltage followers to eliminate movement artifact in the neural recordings from a freely-moving animal.

During surgery, after electrodes have been positioned, they are typically held in place by dental acrylic that bonds around the electrodes and secures them to the skull. Leads that were attached to the electrode prior to implantation are typically routed to pins in a permanent headstage socket that is cemented with dental acrylic onto the skull and stainless steel screws at the end of surgery. For rabbits, the Lavond laboratory uses a 14 pin, low profile, machined contact DIP (dual-in-line-parallel) socket. Machined contact DIP sockets are different from the commonly available DIP integrated sockets in that the recesses of machined contacts are round rather than consisting of spring leaves as in the common sockets. The machined contacts are sturdier and tighter. Between the two rows of pins on the socket we also place an inverted (head side down) 3/4 inch, size 4-40 stainless steel machine screw. The socket and screw are cemented in place. This type of permanent socket arrangement is appropriate if a surgery is performed before any training. For rabbits and rats, the Steinmetz laboratory routes leads from the electrodes to pins located on an 8- or 10-pin mini-transistor socket. The electrode leads are soldered onto the pins and the socket is then carefully imbedded into the dental acrylic before it sets up. Subsequent connections to the electrodes are then made through the transistor socket via a plug that mates with the socket. For rabbit studies, the air puff nozzle and/or eyeblink detection equipment are secured to a small machine screw that is positioned in the acrylic before it sets up.

RECORDING FROM FREELY MOVING ANIMALS

One of the major reasons rabbits became the species of choice for

Chapter 4   159

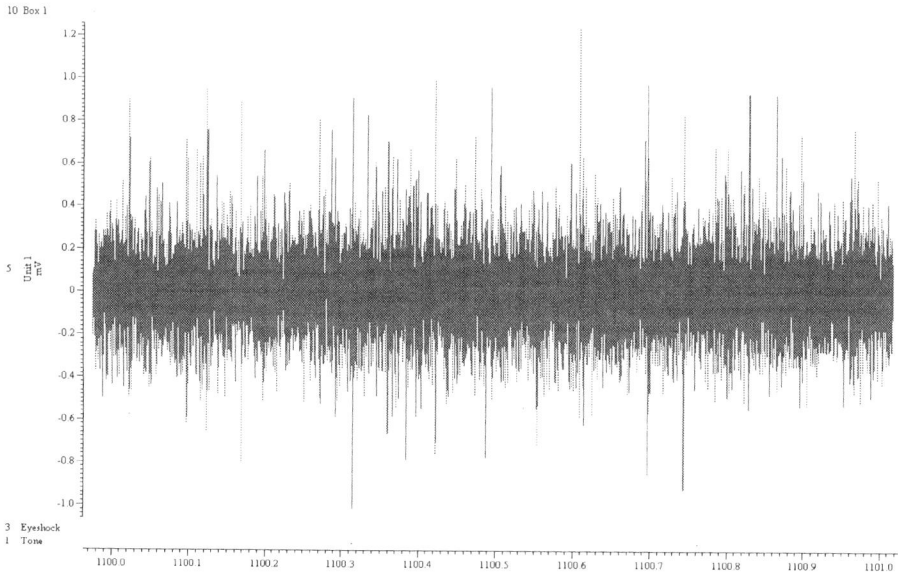

Figure 4.13: Unit activity recorded from the interpositus nucleus of a freely moving rat without having any movement artifacts.

classical conditioning experiments was that they accepted restraint well (see Chapter 1). This was an ideal situation for neural recording as stable single unit records could be obtained from restrained, yet fully awake, subjects. Recently, there has been an increasing use of rats as subjects for classical conditioning (as well as other behavioral) experiments and several laboratories, including the Steinmetz laboratory, have developed behavioral and neural recording procedures that allow the use of awake, freely moving rats as subjects.

The biggest challenge, of course, in using unrestrained animals in neural recording experiments is getting the neural signal from the recording site to the amplifier. To accomplish this, we implant recording electrodes into the desired brain locations and connect leads from the electrodes to the pins of a standard 10 pin minitransistor socket. A ground lead is run from a screw that is secured to the skull during surgery to one of the pins of the minitransistor socket. These sockets are about 1 cm in diameter and are thus not too large to be cemented to the rat's skull. A swivel/commutator device is used to connect the brain amplifier to the electrode (see Figure 3.19). The swivel is positioned above the recording box or chamber and allows free movement of the animal within the box or chamber. The leads from the swivel are encased in flexible metal or thick plastic, which protects the leads from being chewed through. Some of our set-ups use a counter-weighted arm to hold the swivel and this provides additional tension on the leads that connect the swivel to the animal, thus eliminating the opportunity for the animal to chew the leads.

The leads from the swivel are attached to a plug assembly that mates with the minitransistor socket that is cemented to the rat's skull during surgery. First-stage amplification of the signal is actually accomplished through an FET or operational amplifier that is incorporated into the plug assembly that mates with the socket in the rat's headstage. Figures 3.20 (plug assembly) and 4.12 (op amp) provide diagrams of these circuits. These first stage amplifiers are the key to obtaining good signals from the freely moving rats. They provide excellent means for picking up the relatively high impedance signal and also provide suppression of movement related artifacts in the neural record. Output from the swivel (Figure 3.19) is fed into a standard biological amplifier and filtered and then input to an oscilloscope and standard computer interface equipment. We have successfully recorded from as many as four brain sites simultaneously with these procedures and the recordings have been clean and artifact free (see Figure 4.13).

CONCLUDING REMARKS

The neurophysiological approach has been particularly successful, when combined with lesion studies, in defining the areas of the brain that are responsible for classical conditioning. For readers interested in more information about neural recording we suggest recent books by Nicolelis (1998) and Wallis (1993). There is a tendency in the field of neuroscience towards 'technique snobbery,' where a particular methodology like recordings is seen as more precise and sophisticated than lesions. This attitude ignores the severe limitations each method has by itself. Recordings, for example, can tell that a population of neurons is involved in classical conditioning but cannot tell whether the neurons are critically involved or just informed about activity elsewhere. The hippocampus comes to mind, where its unit activity models behavioral learned responses but hippocampal lesions do not prevent learning or retention of a classically conditioned response. Lesions, on the other hand, can tell whether a region is importantly involved, but cannot say in what way. Lesion of the hippocampus does not prevent classical conditioning. The weaknesses of one methodology are the strength of another, and by wisely combining multiple techniques a coherent picture of the involvement of the brain in classical conditioning has emerged.

Chapter 5

COLLECTING AND ANALYZING BEHAVIORAL AND NEURAL DATA

OVERVIEW

In this chapter, we take up the important topic of data analysis. Using classical eyeblink conditioning as a model behavioral system, in this chapter we discuss how behavioral and neural data are typically analyzed after the data have been collected. For this discussion, we assume that the behavioral and neural data have been stored to a computer in some form (more on the use of the computer to control stimuli and collect data is presented in Chapter 11). A tacit assumption we also have, of course, is that all of the equipment involved in the data collection process is working properly. After all, the quality of data stored in the computer is determined, for the most part, by the quality of the components of the equipment that records and feeds the data to the computer. In writing this chapter, we have borrowed extensively from a couple of two-part papers that Gould, Sears and Steinmetz (1991 a and b) and Katz and Steinmetz (1995 a and b) wrote for the *Kopf Carrier*, a publication that is distributed to the general neuroscience community by David Kopf Instruments (Tujunga, CA).

GETTING DATA FROM THE SUBJECT TO THE COMPUTER

Because in this chapter we are interested in behavioral and neural data analysis techniques, at this point it seems useful to provide a general overview of a typical experimental set-up we use for getting behavioral and neural data from the subject to the computer. As we have presented in previous chapters, large portions of our research efforts have involved multiple- or single-unit recording from rabbits or rats over a number of sessions of classical eyeblink conditioning. During the training sessions, eyeblinks are

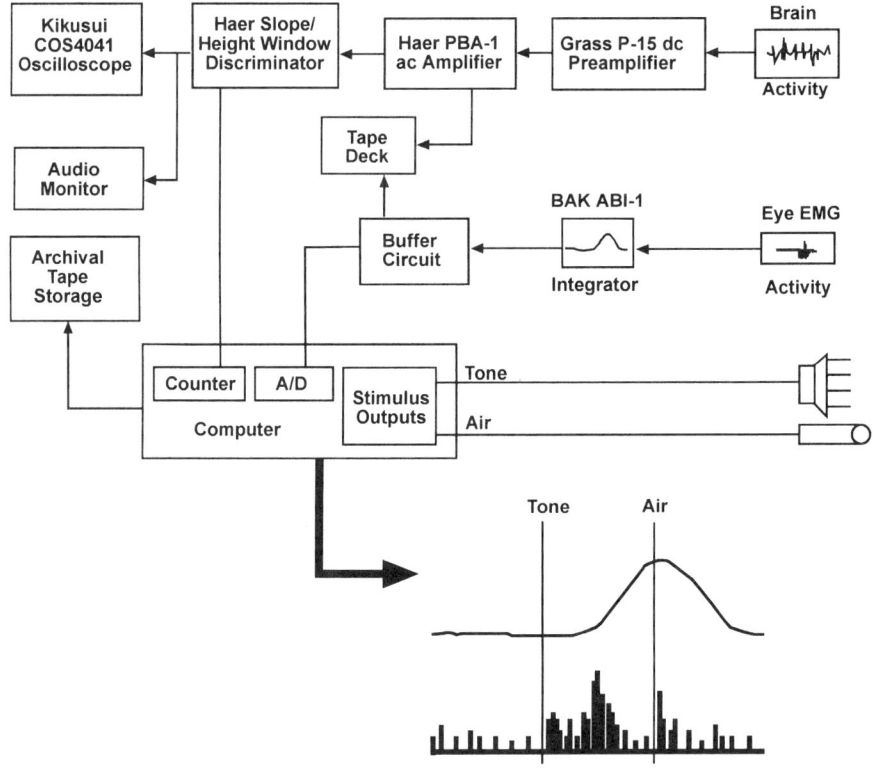

Figure 5.1: Schematic of experimental layout and data collection.

recorded using a variety of devices including minitorque potentiometers, photodiode detectors and EMG recording electrodes. Usually, neural activity is concomitantly recorded from microelectrodes that were either chronically implanted during an aseptic surgical procedure or lowered using a micromanipulator into the brain region of interest during the training session.

Figure 5.1 shows a schematic overview of the apparatus used in a typical eyeblink conditioning experiment. Equipment responsible for generating the CS (e.g., tone) and US (e.g., air puff) that are presented during training are turned on and off by a computer that controls the within- and between-trial timing of the presentation of the stimuli as well as the duration of the stimuli.

In Figure 5.1, eyeblinks are monitored using EMG recording from wire electrodes placed in the musculature around the eye. The electrical activity of these muscles is then amplified and routed to an integrator, which rectifies and converts the raw EMG signal to a smooth DC signal that represents increases in electrical activity by an increase (or decrease, depending on polarity) in voltage. We typically use a buffer circuit to scale the DC signal between 0 V and +5 V then feed the signal into one channel of an A/D converter that resides within the computer. The DC signal is also routed to an oscilloscope so that the response can be closely monitored in real time. Note that if a potentiometer or photodiode system had been used to monitor eyeblinks, the output of these devices would be handled similarly to the integrated EMG signal (i.e., the signal would be scaled to 0-5 V and input to an A/D converter).

Brain activity recorded by the microelectrodes is amplified and filtered, displayed on an oscilloscope, and depending on the method for spike counting (see Chapter 4), routed to either a discriminator or to a high-speed A/D acquisition system used for offline spike sorting. If a discriminator is used, those spikes that break the preset discriminator window are represented as TTL pulses and are input directly into counter chips that reside within the computer. If the brain signal is routed to a high-speed A/D acquisition system, the entire raw signal is digitized and stored to disk either in the computer that is responsible for delivering the stimuli or in a second computer that is synced to the stimuli-generating computer.

It is important at this point to step back, take a breath, and consider exactly what is stored in the computer at the end of each trial and at the end of the session. After each trial, the computer contains a digitized record of the behavioral response (i.e., the eyeblink) and either 1) a record of when discriminated action potentials were recorded within the trial or 2) a complete digitized record of the amplified and filtered brain activity recorded during the trial. At the end of a session, the computer contains several trials worth of digitized behavioral and neural data. An important point that is missed by many researchers is that the data stored in the computer is, at most, an approximate representation of what was picked up by the devices originally used to collect the data. The responses have been transduced, amplified, filtered, scaled, and digitized. It is far from the "raw data" that most researchers think they have. For this reason, great care must be taken in the choice of equipment and parameters that are used in the chain of devices between the subject and the computer. A potentiometer that does not provide a linear output or an EMG integrator that integrates too slowly will not faithfully represent the amplitude of the eyeblink that was generated by the subject. As we pointed out in Chapter 4, selection of levels for the high-pass and low-pass filters greatly affects the signal that is viewed and analyzed. Additionally, the rate of digitization during the conversion of the analog signal greatly determines the shape of the data that is ultimately stored in the computer: If digitization rates are too slow, the resolution of the behavioral response is relatively poor. If digitization rates are relatively fast, stray voltage signals and signal artifacts are more likely to be seen. Also, a large number of data points have to be collected and stored.

Subsequent data analysis involves further reduction of the "raw" data that is stored in the computer in digitized form. For behavioral data analysis after classical eyeblink conditioning trials have been delivered, this involves figuring out on which trials in a session conditioned responses occurred as well as a variety of properties of the behavioral responses that were recorded. For neural data analysis, this involves determining when a neuron discharged within individual trials. We will begin by discussing behavioral data analysis techniques.

THE COLLECTION AND REPRESENTATION OF BEHAVIORAL DATA

After a training session is finished, digitized behavioral data that were collected during the session typically reside in a computer in some sort of a file. The size of this file is dependent largely on three factors: 1) the length of the trial, 2) the A/D conversion (sampling) rate, and 3) the precision (i.e., resolution) of the A/D sampling. The longer the trial duration (e.g., 1000 msec versus 500 msec), the larger the behavioral data file can potentially be. The A/D conversion rate refers to the rate at which the analog signal was sampled and digitized (e.g., 10 KHz versus 20 KHz)—the faster you sample, the more data points are generated. Precision refers to the relative range of numbers that is used to numerically represent the analog input. For example, an analog signal that varies from 0 to +5V is represented as numbers that range from 0 to 255 when 8-bit precision is used. The same signal is represented as a number that ranges from 0 to 4,095 when 12-bit precision is used. All of these factors affect the size of the behavioral data file with which you must deal.

To begin our discussion of behavioral data analysis, we present the following example. Let's assume that we have recorded external eyelid responses during a classical conditioning experiment using a minitorque potentiometer with a voltage output that ranges from 0 (when the eye is open) to +5 V (when the eye is closed). We deliver 100 trials during the session and each individual trial length is 1500 msec; a 500 msec period before CS onset (which we call the preCS period), a 500 msec period between CS onset and US onset (which we call the CS period), and a 500 msec period after US onset (which we call the US period). The signal from the potentiometer is fed into an A/D device residing within a computer, which digitizes the data at a rate of 500 Hz (i.e., 500 samples per second or one sample taken every 2 msec). The digitized data are stored in a file for later analysis.

Using this example, we can get an idea of how data are represented and subsequently analyzed. First, since we used 8-bit resolution, the data file will contain numbers that range from 0 to 255. A 0 indicates that the eyelids were fully open and a 255 indicates that the eyelids were fully closed. Numbers inbetween indicate a partially closed eyelid (e.g., 128 would indicate that eyelid was approximately halfway closed, assuming a linear output from our potentiometer). For each trial, 750 data points would be collected: Remember, we digitized at 500 Hz (once every two milliseconds). So for the session we would have 75,000 data points stored (100 trials X 750 data points/trial). Because we used 8-bit precision in storing our data, we would need 600,000 bits (8-bits X 75,000 data points) of storage in the computer (it is more common to express this as Kbyte of storage—in this example, this is about 73.5 Kbytes, a tiny file by today's standards). Therefore, at the end of the session we used in our example, a file of behavioral data would be created that contains a string of 75,000 numbers with values that range from 0 to 255. In this string of numbers, every block of 750 numbers represents an individual trial.

CRUNCHING THE BEHAVIORAL DATA

An amazing amount of information can be extracted from the string of numbers that comprise the file of digitized data that represents eyelid movements recorded during the training session. Both individual trial data and session-wide data are typically extracted. With the speed and capacity of computers now available, within-trial information can be extracted in a second or two after each trial. This is quite handy for monitoring the progress of training during the session, especially using graphics that allow traces representing the analog signal to be reconstructed from the digitized array of numbers representing the behavior. Session-wide data analysis, of course, can only be performed after the session has been completed.

Within-Trial Number Crunching

At the level of an individual classical conditioning trial, the most basic question than can be answered is: Did a CR occur on this trial? In addition, however, careful analysis of the digitized eyeblink response can generate other useful information such as the timing and amplitude of the response as well as the general topography of the response. In this section, we describe the basic procedures we have used to extract all of this information for a trial, using the string of numbers that was stored in the computer during the session.

Did a CR occur on this trial?

To answer this question, we typically scan the string of numbers to find out a) if the numbers deviated from a baseline in the direction and above a level that would indicate a CR occurred, and b) if the deviation from baseline occurred between CS onset and US onset. We first need to establish a baseline. This is normally done by averaging all or a portion of the data points that were collected before CS onset (in what we call our PreCS period). Once this baseline is established, individual data points collected from the CS onset onward are scanned for numbers that exceed the baseline by a preset amount (assuming, of course, that higher numbers in our array indicate eyelid closure). A variety of methods have been used to determine what change is necessary to indicate a significant response occurred. Some use absolute levels (e.g., an increase of 10 from baseline), which can be calibrated to represent the extent of movement of the eyelids. Others look for changes after the CS onset that take into account variability in the numbers that make up the baseline and post-CS responses (e.g., look for a change of two standard deviations from the baseline).

If a significant departure from baseline is found, we must determine when the departure from baseline occurred. This is accomplished by noting where in the string of numbers the significant departure occurred then scaling this point in time. In the example we used above where a sampling rate of 500 Hz was used, a data point is created once every 2 msec for the 1500 msec duration of the trial. Therefore, 250 numbers are created for the preCS period, the next 250 numbers are created between CS onset and US onset, and 250 numbers are created after US onset. If our 'response detection' routine determines that the 300th data point exceeds our response criterion, we would conclude that a CR occurred on that trial because that data point is between the 250th and 500th data points that represent eyelid position during the CS-US interval. If our response detection routine determines that the first significant departure from baseline in our data set is the 510th data point, we would conclude that no CR occurred because CS-US interval ends at the 500th data point (when the US is presented). Assuming that a US was presented on this trial, we would likely assume that this response was a UR because it was recorded after US onset.

Good trial or bad trial?

We typically do not include trials into our session-wide data analyses on which excessive eyelid movements during the preCS period were detected. Whether or not CRs occurred on these trials is difficult to determine

Chapter 5

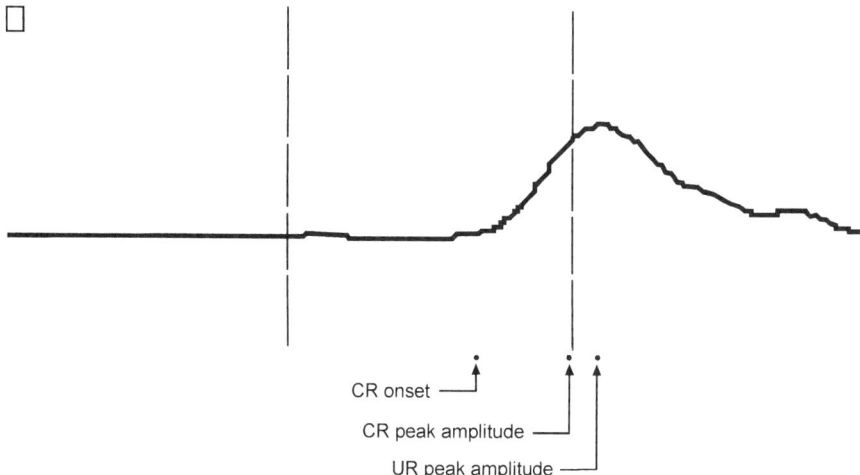

Figure 5.2: Common measurements for behavior can be made on individual trials or on an average of good trials. Here the square in the upper left indicates this is a 'block' of trials. CR onset is the first time a criterion amplitude is met (here 0.5 mm) in the proper time frame (here between 25 msec after CS onset to US onset). CR and UR peak amplitudes are the greatest values found in the CS and US periods, respectively. The display uses dots to indicate these measurements. Note that these measurements may not always be obvious in the display because the pixels on the monitor cannot display the full range of A/D values.

because it is difficult to decide what baseline should be used when examining the trial for responses (e.g., if the animal blinks just before or during CS onset the baseline would be relatively high). To determine which trials to include in the final analyses and which to discard, we examine the preCS period for excessive activity. This can be done in several ways. One way is to calculate the mean and variance of the preCS data points and reject trials on which excessive variation is found. Another way is to calculate the average for the entire preCS period then compare a few data points collected just before CS onset with the average. If a significant difference is found (either higher or lower), the trial is not included in further analyses. Like the determination of whether or not CRs and URs occurred, the reader should note that this process is a bit arbitrary.

CR and UR amplitudes

Other measurements that are often extracted from the data file are the amplitudes of the CR and UR, assuming that these responses have been found (see Figure 5.2). These measures are identified by scanning the array of data points collected after CS onset and finding the largest difference from baseline that occurs a) after CS onset but before US onset (the CR amplitude) and b) after US onset (the UR amplitude). The values that are found can be scaled to represent actual eyelid movement. For example, potentiometers can be calibrated at changes from 0 V to 5 V to represent a known distance (say, in mm) and integrated EMG responses can be calibrated to represent the relative electrical activity in the muscle (in mV). If these calibrations are known, the numbers stored in the behavioral data array can be transformed to mm (distance) or mV (electrical activity) that represent actual movement of the eyelids. The resolution of these measures is determined to a large extent by the resolution of the A/D converter that is used: a 12-bit A/D provides a larger range of numbers to represent the analog input than does an 8-bit A/D.

Onset and peak latencies for the CR and/or UR

It is possible to examine timing characteristics of the CR and UR (see Figure 5.2). We typically calculate values for the onset and peak latencies for both the CR and UR. First, the CS-US interval is scanned for the number that marks the first significant departure from baseline, as described above, and this point is defined as when CR onset latency occurred. The data point in the CS-US interval that represents when the biggest departure from baseline is defined as when the CR peak latency occurred. Next, the data points collected after US onset are scanned for those values that represent the first significant departure from baseline as well as the biggest deviation from baseline, and these are marked as when the US onset latency and the US peak latency occurred, respectively. All array positions are converted to time units (in msec). One thing to consider when calculating latency data is that on some trials, no response is detected. What latency data are entered for these trials? Different laboratories handle this situation in different ways. For example, some laboratories assign the maximum latency for onset and peak latency values. If the CS-US interval was 500 msec and no CR was detected, 500 msec is used as the onset and peak latency. Other laboratories only include those trials where CRs were found in the CR latency calculations (or URs for the UR latency calculations). Arguments for and against both methods can be raised.

CR and/or UR magnitude measures and CR/UR risetime

Some investigators measure the area under the CR and/or UR as a measurement of response magnitude. This can easily be done by calculating a response baseline then summing the differences between the baseline and each data point in the array. We have not found this measure to be particularly useful. Another measure of response topography is the response risetime. This measure indicates the relative speed and vigor of the response. It is very simply calculated by taking the difference between the CR or UR onset and peak latencies.

SESSION-WIDE NUMBER CRUNCHING

Most often, the behavioral data reported in publications are based on session-wide summaries of individual trial data. This is most certainly the case for classical eyeblink conditioning. In part, this is due to the fact that the rate of eyeblink conditioning, like other behaviors that one can monitor on a trial-by-trial basis, is somewhat variable. To get an accurate picture of the overall acquisition and performance of this type of learning, a number of trials must be combined together. However, there are a number of changes in behavioral responding that occur within a session and analyses of these changes deserve attention. As a compromise, we calculate session-wide statistics as well as statistics for blocks of 10-30 trials that are combined together.

The most widely used measures of classical conditioning are the percentage of trials on which CRs (percent CRs) were recorded and the average amplitude of the CR (CR amplitude). While percent URs are rarely reported, it is very common to see reports of UR amplitude because this provides an important measure of basic response performance capabilities. This has proven very useful and important (though not altogether free of controversy) for experiments that have used lesion techniques to study brain function. To calculate percent CRs, the total number of trials on which a CR was detected is divided by the total number of 'good' trials presented (where a good trial contains no excessive movements before CS onset). CR and UR amplitudes are simply the average of the CR and UR amplitudes calculated for 'good' trials. Zeros are averaged in when no responses are observed. There are some differences between laboratories on how trials are handled that contain deviations from baseline that do not break criterion to be counted as a response. For example, let's say that the criterion for a CR is set at 0.5 mm movement of the external eyelids. The recording equipment actually detects a 0.4 mm movement during the CS-US period of a given trial. Some laboratories enter a zero for this response when averaging while other laboratories use the 0.4 mm. This subtle difference in data handling can have a big influence on how results are reported and interpreted.

One of the strengths of classical conditioning of somatic responses, such as the eyeblink, is that response latencies and topographies can be reported. This is extremely powerful when coupled with brain research techniques such as lesioning, stimulation and recording. It is common to see reports of session-wide and block averages of the onset latencies and peak latencies for the CS and the US. Again, it is very important to know how non-response trials were included in the analysis. As described above, some laboratories include all 'good' trials in the analysis and use default values (normally the maximum latencies possible) for non-CR or non-UR trials. In this manner, the latency data gets weighted for the lack of a response. Other laboratories include only the 'good' response trials on which responses were observed into the latency average.

As covered in Chapter 3, it is also very common for research to present CS-alone trials along with paired CS-US trials during a session. Most often, these trial types are kept separate during the session-wide and block data summaries. The inclusion of CS-alone trials is very important because it is possible to see learned responses on these trials in the absence of the UR, which by definition occurs on all trials in which a US is presented. CS-alone trials are particularly useful for studying onset and peak latencies because the entire post-CS trial period is, by convention, analyzed on these trials. For example, some lesions cause a slowing of behavior; this effect might only be seen on CS-alone trials where UR are not present to mask or hide late CRs.

ANALYSIS OF OTHER BEHAVIORS

As in other parts of this book, we have concentrated on data analysis involving classical eyeblink conditioning data to illustrate how data reduction and analyses are performed. However, almost any behavior can be represented as digitized data and much of what we discussed above is applicable to these situations. For example, heart rate responding during fear conditioning can be easily digitized and heart rate and interbeat intervals calculated by finding onset latencies and peak latencies of the digitized heart beat. Also, the Steinmetz laboratory has used operant and instrumental lever-pressing paradigms to study behavior-brain relationships. To analyze lever-pressing behavior, we have the lever hooked up to gate 4-5 V into an A/D device. In this manner, we can get all of the measures mentioned above for eyeblink conditioning (including response amplitude if we use a pressure-sensitive lever and lever-press duration). We are always surprised by the number of studies reported in the literature that could, but do not, record latency and duration of lever-pressing. We have found these to be extremely valuable measures, especially in pharmacology and recording studies.

Chapter 5

ANALYZING NEURAL DATA

The coupling of electrophysiological recording with classical conditioning of the eyeblink response has produced an impressive amount of data concerning the involvement of a variety of brain structures and systems in learning and memory. When analyzing unit data, two major questions are typically asked: 1) Is there a significant change in unit activity that is related to the training stimuli (CS and US), and 2) is there a significant relationship between the unit activity and either the reflexive or the learned behavior? If the neural activity is related to the behavior, then it becomes critically important to understand how it is related. In particular, it is critical to establish whether unit activity predicts learned behavior or whether the unit activity follows the behavior (e.g., feedback). We start our discussion by reviewing how neural data are collected and stored during each trial and then we present some methods commonly used to relate unit activity to behavior.

Collection and Storage of Neural Data Using a Discriminator

In Chapter 4 we detailed how discriminators were used to collect spike activity during neural recording experiments. Typically the output of a discriminator (which is a simple TTL pulse that indicates when a spike that broke threshold was detected) is routed directly to a counter that resides within a computer. The counter accumulates events (i.e., it counts) until it is reset. In the Forth-based and C++-based systems described later in this book, we have created routines that read and reset counters that receive input from spike discriminators. Basically, these programs define cycling times for reading and resetting the counters on individual trials. For example, let's say we are presenting trials that are 1500 msec in duration and sampling our behavior at a rate of 200 Hz (one A/D conversion every 5 msec). CS and US onsets occur at 500 and 1000 msec, respectively. In a typical recording experiment, we might use the same sampling rate to read and then reset our counters (i.e., every 5 msec, we read the counter, which tells us the number of action potentials that were counted in the preceding 5 msec, then reset the counter to 0). This process is repeated continuously for the duration of the trial. A common way to look at the end result of this process is that for each trial we create a string of data points each representing the number of spike discharges counted during a 5 msec 'bin' of time. In our example, we would have a string of 200 numbers, each number representing the number of unit discharges counted for a 5 msec bin at some point during the trial. In a typical multiple-unit recording study, the numbers in this array will range from 0 (no activity recorded) to perhaps 6 or 7 (depending on how many neurons are being monitored). In a typical single-unit recording study, the numbers in this array will range between 0 and 2 (from no activity to one or two discharges of the neuron in the 5 msec period). If a session included 100 trials, there would be a total of 20,000 data points stored (100 trials X 200 points/trial). Because the magnitude of the numbers stored in the unit data array are quite small (normally less than 10), these arrays tend to be quite small in size (at best 8-bit precision is used; 4-bit is probably all that is required).

Collection and Storage of Neural Data Using Waveform Matching Algorithms

More and more laboratories have begun using waveform matching software to isolate responses from single neurons from a multiple-unit record. The Steinmetz laboratory, for example, now only uses this method for collecting and analyzing spiking data. In this method, instead of routing the amplified brain signal to a discriminator that counts spikes that break a preset amplitude level, we route the amplified brain signal to a high-speed A/D interface system (the Spike2 system manufactured by CED, Ltd). In our setup, two computers are actually used; one that generates the conditioning stimuli and timing signals for delivering trials during a session and a second that accepts the brain activity data. The A/D converter on the spike collecting interface system typically operates in the range of 100–300 kHz, depending on the number of brain channels we are monitoring. We also acquire the digitized behavioral responses through the high-speed interface system. The spike and behavioral collection system is turned on and off by signals that originate from the computer that controls the delivery of stimuli and trial and session timing. One computer could be used to control the delivery of stimuli, trials and session timing, and control the acquisition of spike and behavioral responses. However, we chose to use a two computer system because we wanted to continue to use some existing equipment and software we had (see Chapter 11).

Perhaps the greatest disadvantage of using a waveform matching system is that the data files collected for each session are quite large. For example, a single classical conditioning trial that involves the digitization of 1500 msec of brain activity at 200 kHz (8-bit resolution) generates 300,000 data points (about 146 Kbytes). Approximately 14,600 Kbytes or 14.6 Mbytes of data would be collected across a session with 100 trials. Because we often record from more than one unit channel and also concomitantly collect behavioral data, the storage requirements for a single session for a single subject can get quite large.

The major advantage, however, of using template matching procedures is that this technique has proven very powerful for reliably isolating the activity of more than one neuron from a multiple-unit record. We typically have recorded from 2–3 units per session per site using this system. Much of the analysis is done off-line after the session has been completed. The digitized data for the session are analyzed using algorithms that identify spikes on the basis of amplitude, duration, and rise- and fall-times (i.e., the session's data are examined for spikes that are similar in shape). The user sets the criteria for determining when individual waveforms are identified as being generated by a single neuron so great care must be taken in setting the parameters for inclusion. Once individual spikes have been identified, their temporal location within each individual trial is marked. In this manner, spike firing patterns are defined for individual trials and we have arrived at the exact point in unit data collection that is achieved when a discriminator is used to isolate spike activity. Note that unlike when discriminators and counters are used, it is not necessary to define a bin duration for starting and resetting counters to accumulate spike activity. However, it is very common to form bins of activity after spike isolation with the waveform matching method for further analysis of spike activity.

Further Reduction of Unit Data

No matter whether a simple discriminator or a sophisticated pattern matching system has been used to collect the data, most laboratories form peristimulus time histograms (PSTHs) from the discriminated unit data to describe and further analyze the brain activity. Figure 5.3 depicts an example of a display for a single trial, showing the A/D converted eyeblink (top), timing marks for the onset of the CS (left) and US (right), at 252 msec and 504 msec from the beginning of the trial respectively, and patterns of discriminated brain activity collected from three brain areas. The stepped drawing for the eyeblink data and the PSTHs for the unit data emphasize the binned nature of the data collection. Here each step (the associated software variable is **bin**) is 4 msec worth of time. The total trial lasts about three quarters of a second (756 msec is the software variable **trialdur**, which is equivalent to a total number of 189 bins or the software variable **pts** for each data). Units are simply counted into each of the 189 successive bins (**pts**) and displayed as a PSTH. The behavioral response from the peripheral monitoring device has been digitized once per bin (i.e., once every 4 msec). This trial data represents the 'raw' data from which further analysis and averaging is derived.

We should point out at this time, however, that these data are far from 'raw.' The unit data shows only when a spike was detected and tell us

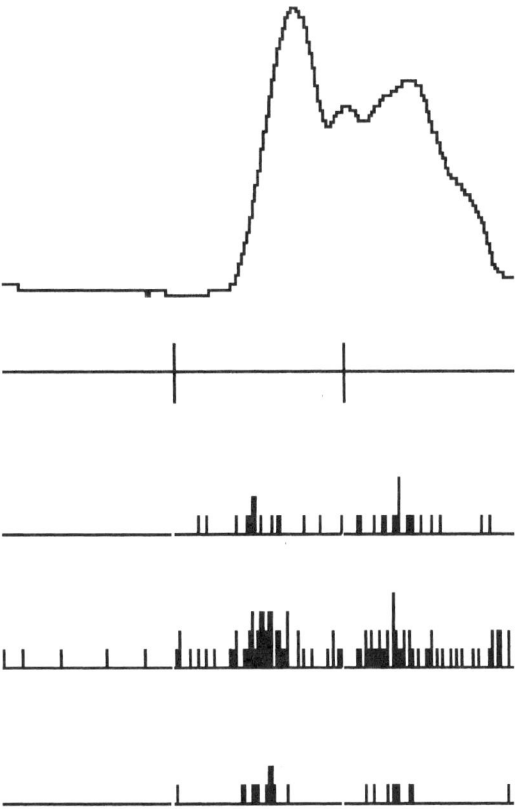

Figure 5.3: Data collected during a single trial showing eyelid position (top), timing marks (CS onset on left, US onset on right), and three levels of discriminated unit activity.

nothing about the size and shape of the action potential or even if that action potential is coming from a single neuron. Also, the eyeblink data has been transduced, amplified, filtered and digitized at this point and is not truly 'raw' data at all. Indeed, most labs usually average and smooth the behavioral response that is displayed. For example, Lavond's laboratory smoothes the digitized behavioral response using a three point moving average (the routine **smooth**) that is activated by setting a soft switch (to **filter on**). Our understanding of the purpose of filtering is to correct for small variations in the minitorque potentiometers that are used to transduce the eyeblink, i.e., to make the response look nice. This should not be confused with the use of smoothness as a technical term and property that describes how well a potentiometer moves in fine increments, of which most of us are unaware.

As we pointed out above, while there are important data that can be gleaned from a single trial, the data from several trials are normally combined for further analysis. We have described above how averages of behavioral measures are obtained. It should be noted, however, that it is possible to get an average digitized behavioral response by simply averaging together the digitized values of the behavior collected over a block or session of training. This usually provides a very good idea of the general shape of the behavioral response.

PSTHs of the unit data are formed by adding, on a bin-by-bin basis, unit activity; PSTHs are not typically formed by averaging. Typical PSTH routines simply add spike counts to each of bins, yielding an accumulation of counts over the course of several trials (i.e., either a block of trials or the whole session). Accumulation of units works better than averaging because there may be some bins that contain very few or no counts from trial to trial. Averaging would wash out their contribution entirely. The technique of simply accumulating without computing the average is taken from evoked potential research. In essence, very small changes that might occur only after a large number of trials can show up with accumulation that otherwise would become all zeros by averaging.

ANALYZING PSTHS

T-score and z-score analyses

There are several ways to further analyze the PSTHs that are formed, including a number of variations of standard score analyses that have been described (e.g., Katz & Steinmetz, 1997; Sears & Steinmetz, 1990). In this method, the activity in the bins in the baseline preCS period is compared to all the activity in the bins in the CS or the US periods. For this comparison, t- or z-scores can be calculated and probabilities determined for significant increases or decreases in activity. Z-scores were described by Thompson and colleagues (Thompson, Berger, Cevgaske, Patterson, Roemer, Teyler & Young, 1976) and incorporated as the z-score program by Lavond (see **z.scr** program). T-scores have been used in programs by Steinmetz and in the Datamunch programs (originated by A.J. Annala, subsequently enhanced by D.G. Lavond, D. Krupa, J. Tracy and D. King, in succession). In t-score programs the bins may not be collapsed, and the comparison is between successive bins in the CS period (for example) with the corresponding bin in the baseline period. By identifying when significant activity occurs a latency can be determined, and this unit latency can be compared with the latency of the behavior.

A description of how the Steinmetz laboratory uses standard score analysis may be useful here. The first step toward statistical analysis of the relationship between the eyeblink response and neural activity involves con-

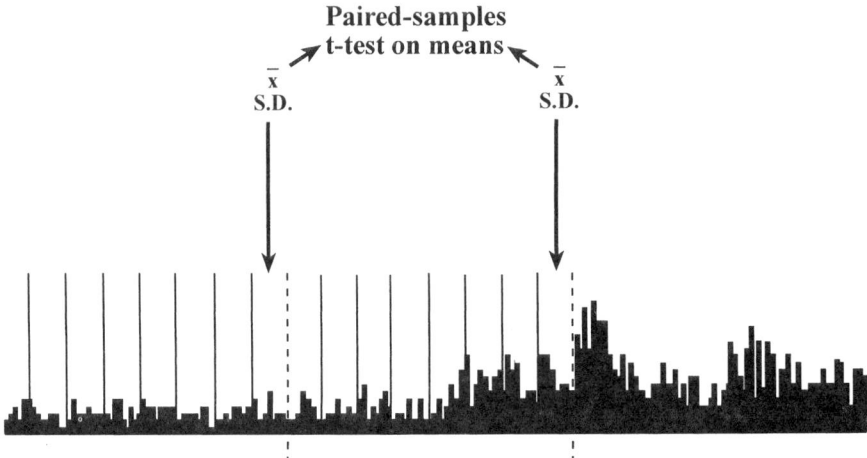

Figure 5.4: Average of a block of trials in which only 'good trials' have been included.

verting the raw number of action potentials into standard scores. To standardize the amount of neural firing, we first aggregate a block of 10 trials, summing the neural activities. Assuming that the neural pattern of activity is reasonably stable, the result is the PTSH described above that contains time periods during which bursts of unit activity are identifiable as clusters of increased amplitudes in the PTSH. We then divide each section of the preCS, CS and US periods into eight bins of the same length, each bin containing the number of action potentials recorded within that particular time window. With an equal number of bins in each period of the trial, we can compare stimulus- and response-related neural activity to the block-specific baseline. Each of the eight CS period bins and US period bins, numbered one to eight from the beginning to the end of the period, is compared to the same number bin of the pre-CS period.

Figure 5.4 illustrates this technique, showing as an example the comparison between CS Period Bin 8 and the corresponding baseline bin (i.e., PreCS Period Bin 8). A simple t-test is performed on the average numbers of action potentials for the two bins. As the two averages are calculated from the same subject and the same trials, they are not considered to be independent of one another; therefore, a paired samples t-test is the appropriate statistic (Hays, 1988). The result of the t-test is the standard scores for CS Period Bin 8. If the average number of action potentials produced in CS Period Bin 8 is significantly higher than the average number of action potentials isolated for the PreCS Period Bin 8, then the standard score for that comparison will exceed the critical t-value for the degrees of freedom of the t-test. In other words, depending on the sign of the t-value, a significant in-

Chapter 5 177

Table 5.1: Standard t-scores for unit activity in each epoch compared with baseline activity in each of 12 blocks of training.

Block	\multicolumn{16}{c}{Epoch}															
	cs1	cs2	cs3	cs4	cs5	cs6	cs7	cs8	us1	us2	us3	us4	us5	us6	us7	us8
1	4.12	X	3.14	3.54	6.55	5.71	6.79	6.71	7.21	5.75	4.39	2.86	4.04	X	3.10	X
2	X	X	2.97	2.33	X	2.41	7.06	4.11	6.47	2.22	X	X	X	1.86	X	X
3	X	X	2.76	2.49	4.02	4.04	4.94	2.58	5.08	2.34	X	X	X	X	X	X
4	2.70	X	2.34	3.50	3.15	X	2.74	5.19	4.77	X	X	X	X	X	X	X
5	X	3.28	X	2.43	X	X	X	3.12	7.58	7.80	X	2.41	X	X	-2.78	-1.93
6	X	2.53	X	2.71	4.95	5.85	5.25	9.00	4.78	3.63	X	X	X	X	X	X
7	2.38	X	2.43	3.24	3.60	3.64	4.95	4.26	5.49	X	X	2.24	X	X	X	X
8	X	X	X	5.41	X	4.02	5.76	5.45	6.97	X	X	2.04	X	2.65	X	X
9	2.45	4.02	X	X	3.07	7.52	4.09	3.63	7.96	4.62	3.15	3.19	5.32	6.25	6.62	4.71
10	X	1.94	2.73	3.35	3.52	X	3.45	3.43	5.17	3.71	X	X	X	X	X	X
11	X	6.82	X	X	X	4.87	3.61	6.98	4.65	3.55	X	X	X	2.25	X	3.05
12	2.21	3.84	1.89	2.47	3.04	9.13	3.37	3.23	7.06	5.30	3.04	X	X	X	X	X
Session Mean	1.74	2.44	1.87	2.67	2.95	4.11	4.41	4.81	6.10	3.38	1.55	1.68	1.24	1.40	1.10	0.48

crease or decrease in spiking is said to have occurred in CS Period Bin 8.

Table 5.1 displays the output of this technique for a session involving 12 blocks of training. A matrix of 192 standard scores is computed representing the level of training-related activity (across eight CS period bins and eight US period bins) for each of 12 blocks. In addition, the bottom row provides a simple average of the standardized activity for each bin, so that the data may be examined at the level of the sessions as a whole. The increased amplitude seen in the PSTH shown in Figure 5.4 has now been quantified such that its statistical significance is made plain.

It might be argued that the sheer number of comparisons represented by the matrix drive the protection level—the probability of finding a seemingly significant result by chance alone—too high. While this may be true for an examination in which the mere presence of some significant result is the aim, it does not present much of a problem when the search is for a consistent pattern of increased firing block after block. In fact, by using session averages in the analysis, we tend to err on the side of caution. We require that the overall standard scores meet the more stringent critical values used for single blocks (in our case, $df = 9$).

This test is neither perfect nor adequate for our purposes, however. For one thing, background activity in the neural single makes it difficult to detect brain-behavior relationships involving the inhibition of action potentials. The floor of noise may make an inhibition effect virtually undetectable, even if it is the case that the in-task inhibition suppresses action potentials completely. Aggregating the data across several sessions will minimize this problem, but only if great care is taken to ensure that the baseline activity remains fairly constant for all such sessions. Such aggregation, meanwhile, rests on the perhaps invalid assumption that the sought after inhibition is fairly stable across sessions.

In fact, the baseline activity itself may not be stable, but may change with training. The stability of the baseline can be tested within a session as along as amplifier gains are left untouched, by comparing the raw numbers of discriminated spikes from the first to the last blocks. Between sessions, the validity of baseline activity comparisons is difficult to support.

Another limitation of the standardization procedure is that it only tells us whether the neural activity exceeds baseline; that is, the t-test tells us whether the height of the PSTH is greater than zero at each time point. They tell us nothing about the relationship between adjacent bins, or about the distributions as a whole. The shape of the distribution remains mysterious, as does its relationship, in shape and timing, to the eyeblink.

An alternative to t-test determinations of significant changes in unit activity is the z-score statistic, which has been used in the Lavond laboratory. Figure 5.5 shows the results of a z-score analysis for a block of trials

Chapter 5

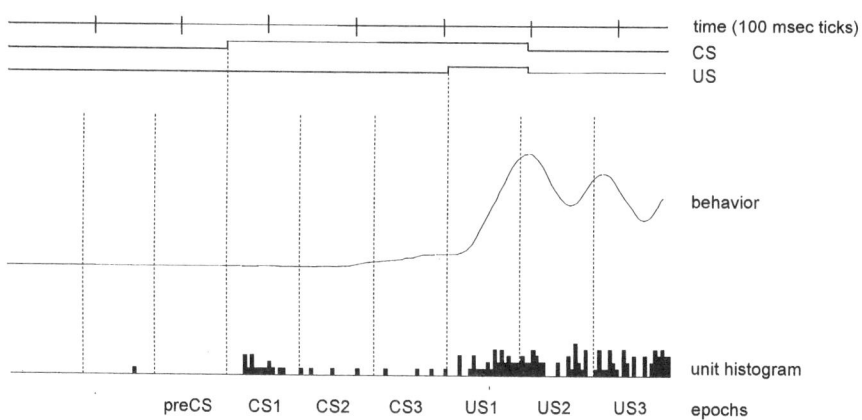

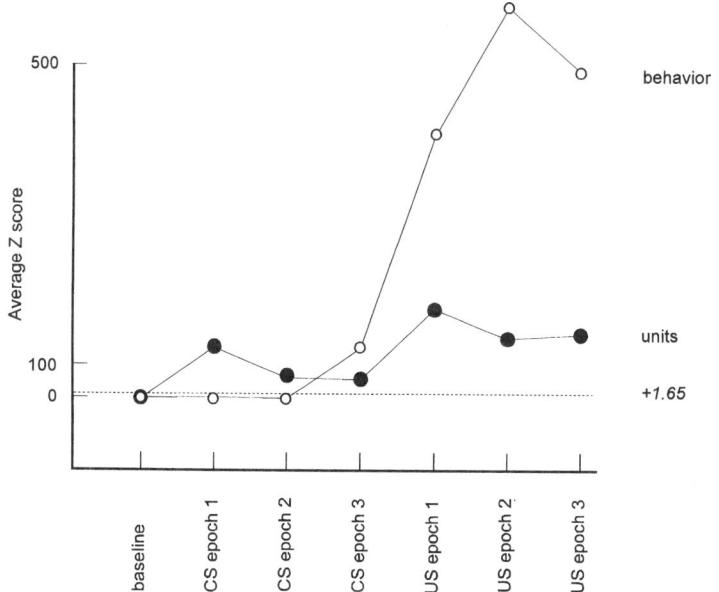

Figure 5.5: Z-score analysis of unit activity for a block of trials compared with averaged behavioral response.

on neural unit recordings taken from the pontine nuclei before and after cooling-based inactivation of the interpositus nucleus of the cerebellum. The mean baseline activity is compared with the mean activity during successive epochs of the trial (three CS epochs and three US epoches) divided by the standard error of the baseline period to compute the z-score. Activity above +1.65 z-score units is significant at the $p < 0.05$ level for a one-tailed test. The top figure shows a recording from the pontine nuclei with auditory-evoked responses (CS Epoch 1 and US Epoch 1) and significant activity related to the learned behavior (CS Epoch 3) and the reflex (US Epochs) in filled circles. The imprecision of the z-score epochs can be seen in that CS Epoch 2 could be related to stimulus-evoked or learning-related activity or both. With inactivation of the interpositus nucleus by cooling (open circles) the learning related activity (CS Epoch 3) and reflex (US Epochs 2 and 3) disappear, with auditory evoked activity still present. The bottom of the figure shows a different pontine recording that showed behavior related activity (CS Epochs 2 and 3 for learned behavior; US Epochs 2 and 3 for reflexive behavior; it cannot be ruled out that US Epoch 1 could include an auditory reaction to the white noise created by the US air puff). Inactivation of the interpositus nucleus (open circles) abolishes all significant activity.

Cross-correlation methods to establish brain-behavior relationships

Cross correlations are used as a means to associate the pattern of neural activity as the causal agent of the shape of a learned behavior. For anyone who knows about statistics, that statement may seem inconsistent with the capabilities of correlations. Most of us understand that stronger correlations suggest stronger associations. High correlations of the patterns of neural activity are said to *model* the shape behavior. However, correlations themselves do not indicate causal relationships. The association could indicate that neural activity causes the behavior (as in causation), that the behavior causes the neural activity (as in feedback), or that another neural area causes both the neural activity and the behavior (as in coincidence). An argument can be made, however, if it can be shown that the neural activity *precedes* the behavioral response suggesting (but not proving) a causal relationship. The opposite result would disprove a causal relationship. These interpretations are not perfect, but eliminating the feedback interpretation is helpful in establishing causation.

To further extend the analysis, we often directly compare the distribution of standardized neural activity across time to the distribution of eyeblink activity across time, a technique similar to that described by Berger, Latham and Thompson (1980). The data are arranged such that the height of the standardized neural distribution at each time point is paired with the height of the behavioral distribution at the same time point (i.e., bin-to-bin correspondence between neural activity and behavior). With the data set arranged in this way, the two distributions can be cross-correlated: A Pearson product-moment correlation (Hays, 1988) of the paired time points allows us to discern the similarity between the two distributions throughout the task.

The obtained r-score provides a good estimate of how similar the shapes of the distributions truly are. Obviously, significant tests may be brought to bear on the r-score.

More than this is required, however, to relate a distribution of neural activity to the shape of a behavior that is presumably caused by the neural activity. The transmission of neural commands to effectors takes time—a signal must travel down axons, navigate synapses, and induce muscle activity. It is necessary to take this delay into account when cross-correlating the two distributions. For that reason, we 'slide' the two distributions past each other, offsetting them by a set amount of time (i.e., number of bins) at a time. At each offset, we calculate a new cross-correlation on the overlapping portions of the distributions. Calculating multiple slides and correlations is known as the cross-correlation.

The value of calculating cross-correlations at various offsets is potentially large. Fisher's z-tests (Hays, 1988) reveal how significantly each correlation differs from zero. The highest z-score tells us, at long last, what the best-fitting relationship between brain and behavior is. Perhaps more importantly, the peak z-score also provides information about the time lag between neural activity and behavior. This latter measure, and its reliability across blocks, can be used as converging evidence that the recorded activity is neural activity (as opposed to artifactual noise caused by the movement) that is tightly linked to the behavior. For instance, conditioned eyeblinks and action potentials recorded from the interpositus nucleus tend to correlate as highly as $r = 0.80$ to 0.90, and sometimes even higher. These correlations cluster at an offset of approximately 30-50 msec, a biological plausible delay given the synapses and distance between the cerebellum and eye muscles. These data have provided additional strength to the argument that the cerebellum is critical for eyeblink conditioning.

Of course, time-lagged cross correlations is not without its drawbacks. The largest concern can be summarized in a single word—gain. Cross-correlation assumes that changes in the two distributions are linearly related. That is, it assumes that a change of size X in the height of the neural activity PSTH will always be associated with a change of size Y in eyeblink size. If, for instance, arithmetic increases in the number of spikes cause multiplicative changes in the size of the blink, then the resultant cross-correlation will underestimate the relationship between the two. The fact that the records (neural and behavioral) are collected by separate amplification systems compounds the problem.

Another potential limitation of the procedure has to do with the possible imprecision of the 'best' correlation judgment. As with the spike standardization t-tests, Fisher's z-tests compare the observed r-scores to an r-score of zero; they do not compare the observed r-scores at similar offsets.

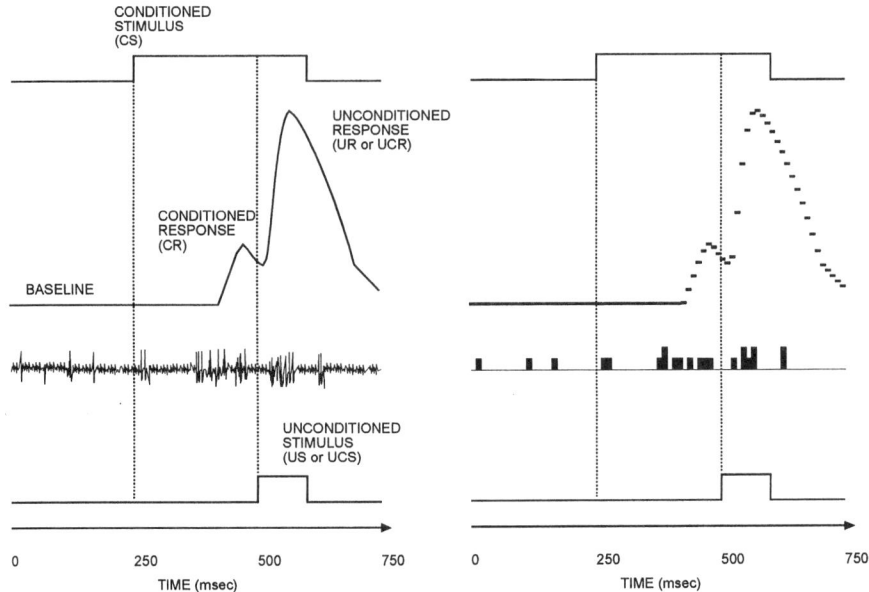

Figure 5.6: Comparison of fictitious raw behavioral and unit data (left) converted for analysis by computer (right).

It is possible that the correlations observed at adjacent time lags are not significantly different from each other. In essence, it is possible that a large confidence interval exists around the 'best offset.' As with the spike standardization procedure, however, this problem is minimal given the phenomenon observed in our preparation. If the most significant correlation is found at similar offsets for two sessions in a row, it is reasonable to ignore the confidence interval of each individual calculation.

Figure 5.6 (left side) shows fictitious behavioral and neural data corresponding to the CS and US elements of a single trial. The two data sets are converted into numbers for every bin of the trial (Figure 5.6, right side). In this example the data are collected once every 10 msec for 250 msec of each the preCS (baseline), CS and US periods, for a total of 750 msec (75 bins total equals the value of pts in the software). For each bin, the behavior is sampled and converted to a digital value corresponding to the analog position of the eyelid. For the corresponding bin, the number of counts of neural activity above a discriminator level is placed into a peristimulus time histogram. A correlation between these two data sets can be called a 'simple' or 'straight correlation,' or a 'cross correlation with no shift in the time frame' (zero time lag, see below).

Figure 5.7 (top) shows the simple correlation between the behavioral and neural data (no time lag). There are a few aspects worth particular note.

Chapter 5

First, the baseline (preCS) data is not included in the calculation of the correlation. This feature is important because it is the baseline data that is shifted so that cross correlations can be made. A good consequence is that the cross correlations will therefore always be made on the same number of pairs of data, therefore they will all have the same power. Second, it should be pointed out that while we are using a single trial's data as an illustration here, in reality cross correlations are not normally performed on single trial data, but rather on data from many trials. The eyelid data is an *average* of a block of trials (9 trials) or the whole session (108 trials), whereas the neural activity is an *accumulation* of neural activity from all the corresponding trials. Third, the number of pairs of x,y (behavior, neural data) is large, here being 50 (500 msec divided by 10 msec binwidths). The greater the number of pairs the greater the likelihood of finding a statistically significant correlation with even poor matches or 'models.' A penalty for using more precise binwidths is that the number of x,y pairs gets even larger. For example, we have used 4 msec binwidths and 252 msec epochs (versus 10 and 250 in our illustration here), yielding 126 pairs of data. A less obvious consequence is that fewer units are counted in any bin with shorter binwidths, making flatter histograms. Fourth, the range of numbers for each bin in the neural histogram is usually very much restricted in comparison to the behavioral data. The fictitious behavioral data illustrated here has a range of about 200 values (0 through 200), whereas the typical units in the illustration have only three values (0, 1 or 2 counts per bin). This disparity is partially overcome by accumulating neural activity (rather than averaging) over several trials. Nevertheless, for the typical data which is a comparison of a block of trials (9 trials) the degree to which the 'flatness of the neural data' contributes to the 'model"is often underestimated, yielding a range of only 0 to 18 in this example. With the exception of the first observation (on the exclusion of the baseline) these other problems contribute to the criticisms of the cross correlation methodology that we will discuss momentarily.

The next three illustrations in Figure 5.7 show how the neural data is shifted in time and new correlations are computed. The shifts illustrated are zero (top), 5 lags, 10 lags and 25 lags (bottom). A lag corresponds to shifting a single bin's worth of time. Therefore, 25 lags at 10 msec bins for each lag has the effect of shifting the entire 250 msec baseline into the calculation. Typically, the way cross correlations are used, each and every lag from zero through the number of lags available in the baseline period are calculated (here, 25 correlations). By the unstated rules of the game, the experimenter searches through the table to find the first time that the highest correlation is reached and, noting the number of lags, translates this into the latency difference between the neural and behavioral responses. Admittedly, experimenter bias could enter here. In effect, the highest correlation indi-

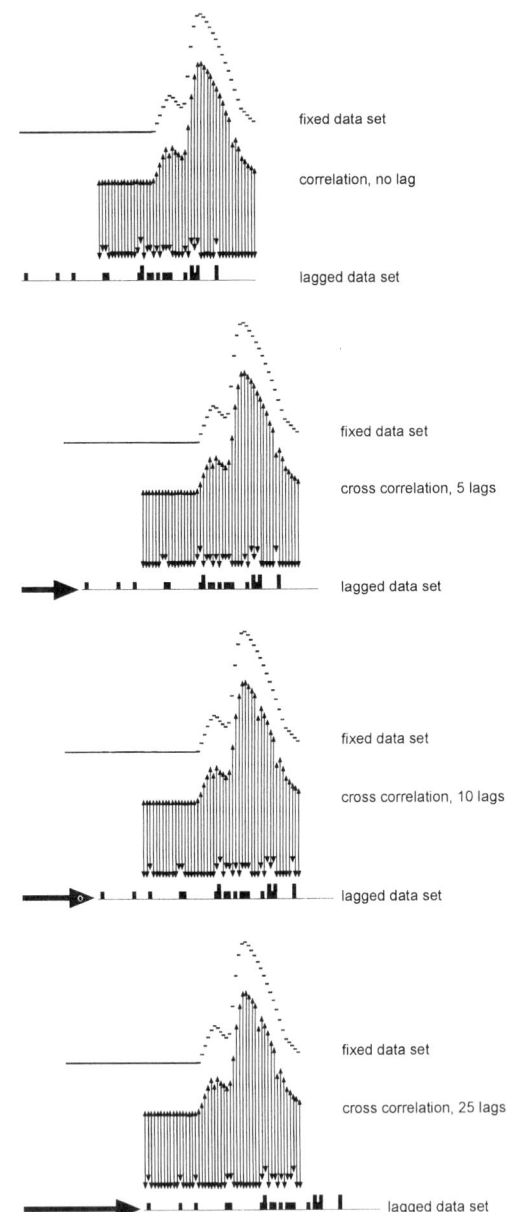

Figure 5.7: Comparison of the original behavioral and unit data (top) with successive shifting of unit activity for computation of cross-correlations.

cates the best model, and the corresponding shift indicates the latency. This assumes that the shape of the individual data is reasonable. We already noted, however, that the data that makes up the neural histogram is unusually restricted in its range. Another problem is that by shifting the neural activity we are biased towards the hypothesis that neural activity predicts the behavior. In comparison, by fixing the neural activity and shifting the behavior (not illustrated) the opposite hypothesis is examined, where neural activity reflects feedback from the motor response. In practice, the calculation for feedback is rarely computed. The cross correlation method is indifferent to the data analyzed; one could just as easily compute a cross correlation between two neural recordings or two behavioral responses. For example, Merrill, Steinmetz, Viken, and Rose (1999) used cross-correlational techniques to demonstrate the close similarities in response topography that can be seen when CRs and URs from twin subjects are compared.

A Cross-correlation program

The Summary program that is written in Forth has several menu options for viewing already collected data. These options are called Review programs, which graphically display the data on the monitor in the trial data that was displayed during its collection (Runtime). The Review programs allow more than simply viewing of individual trials, however. A common activity is to average selected trials, for example all the paired trials from a block where a drug has been applied. We are using the term 'average' loosely here: the A/D data is averaged whereas the the neural activity is accumulated, as indicated.

One of the Review programs has the option for calculating cross correlations. When this option is chosen, the file **cross.scr** is loaded (see CD). Forth as a programming language originally did not include floating point operations, but cross correlations requires floating point calculations for the Pearson r coefficient. A software implementation of floating point is loaded with the relocatable file **float.rlf** as part of the Forth implementation package. The calculations for the cross correlations themselves are straight forward. The rest of the cross correlation program involves manipulation of the fixed and lagged arrays of data to match up the x,y pairs. The cross correlation is performed on averaged data.

Some drawbacks of using the cross-correlation technique

The goal of finding neural activity which models the behavioral re-

Table 5.2: Cross-correlation between behavior (stationary variable) and unit activity (lagged variable).

Lag	r	n	msec
0	.3865	126	0
1	.4199	126	4
2	.4513	126	8
3	.4825	126	12
4	.5120	126	16
5	.5389	126	20
6	.5659	126	24
7	.5943	126	28
8	.6170	126	32
9	.6372	126	36
10	.6535	126	40
11	.6665	126	44
12	.6718	126	48
13	.6790	126	52
14	.6853	126	56
15	.6876	126	60
16	.6905	126	64
17	.6891	126	68
18	.6880	126	72
19	.6840	126	76
20	.6799	126	80
21	.6750	126	86

sponse is admirable. It is also fraught with experimenter biases and data and statistical limitations. At best, cross correlation can help to identify the direction of causation—whether unit activity predicts behavior or vice versa. But even in the best of circumstances, as illustrated in Table 5.2 from actual data, the results are often too imprecise to be of much use. One source of the imprecision is the binned nature of the data, which means that the resolution is at 4 msec intervals in this example. This resolution is probably too gross to distinguish between the latencies of two different neural areas.

Another source of imprecision is related to the calculated correlations. In the table the largest correlation is positive 0.6905 at 16 lags or 64 msec between the neural activity and the behavior. A correlation of 0.6905 is not bad as an indication of a neural model of the behavior; although correlations as high as 0.99 have been reported between neural activity and the behavior, usually they are much lower. What is disturbing, however, is that this high correlation of 0.6905 was obtained from neural data that showed a strong response to CS onset in addition to activity related to the learned response. There was no corresponding behavioral activity to CS onset yet the correlation is fairly high. Even fairly flat neural activity and behavior will yield good correlations yet no one would argue that this involves a model. In fact, *all* of the correlations in Table 1 are significant from zero at the $p < .01$ level (for $df = 100$, the critical r value is 0.256). In other words, even with no lag there is a significant correlation. The consistently large correlations that are typically seen when the behavioral and neural data are compared are likely due to the fact that the general forms of the distributions on which the correlations are made (i.e., the behavioral responses and unit PSTHs) are generally very similar: They typically show one or two areas of increased amplitude.

A further problem is encountered when asking whether there are significant differences between the correlations. For example, is the highest correlation of 0.6905 at 64 msec significantly different from 0.6876 at 60 msec or 0.6891 at 68 msec? Is the correlation 0.6905 at 64 msec significantly different than the correlations at *any* of the other time lags? In fact, it is not different than any of the correlations in the table within any reasonable time frame that would indicate whether the neural activity causes the behavior. As a result, there is no arbitrary, independent way to determine whether the latency of the model is as little as 20 msec or as great as 84 msec (the limit in the table) in this data. Here we are using the term 'arbitrary' in the statistically good sense to mean that it is unbiased.

While it is clear that there are some problems in using cross-correlational analyses to establish brain-behavior relationships, this method has proven useful over the years, especially if one keeps in mind that the overall size of the correlation at time-shift points is not the only data that should be considered when using this method. Rather, the block-to-block or session-to-session pattern and consistency of correlations, and not only the absolute size of the correlation, is important for establishing the brain-behavior relationship.

Cluster analysis

Again, the major goal in using electrophysiological techniques together with the observation of behavior is to establish causal relationships between brain function and behavior. More specifically for eyeblink conditioning, the question is whether neural spikes predict learned behavioral responses. We have seen z-score, t-score and cross correlation methods that attempt to relate neural activity to behavior. These methods are more or less successful, that characterization depending on what is available and the limits of the conclusions as well as the consistency in results.

The greatest difficulty in identifying the brain-behavior link is determining the latency of the neural activity. None of these methods seems to do as good a job at figuring the latency of neural activity as simply examining the unit histograms by eye. We are pretty good at identifying about where there are significant increases or decreases in unit activity. The difficulty comes in deciding exactly where the significant change begins which is important for determining the latency of neural activity. The larger difficulty is to make these judgments consistently and without bias.

The following represents a work in progress that tries to instantiate some of the properties of 'eyeballing' the data in a rigorous way. As with the other methods, the key data are the unit histograms of spike data, whether the source is from single units or multiple units. As discussed later, the source of the data is important for interpreting the relationships, or more correctly put, the source is important for limiting the interpretations. As it turns out, the method discussed here has certain advantages in this regard.

Recently the Lavond laboratory developed a program we call cluster analysis because it looks for aggregates of activity (or inactivity) that look like clusters in a peristimulus time histogram. Figure 5.8 illustrates the major features of this technique (the figure is for illustration purposes only; the data is imaginary). The key for this analysis is the definition of a 'cluster': A cluster is any histogram bin (a single count) or group of bins (multiple counts) that is bounded by inactivity. Figure 5.8 shows seven examples of a cluster. The frequency of the cluster is the number of unit counts for that cluster divided by the total time of the cluster, where the time includes the preceding period of inactivity and the period of unit activity. The histogram indicates the number of unit counts, and the underscores identify the cluster times for each of the seven clusters. Since it is not known when unit activity might have occurred before this data sample, there is no way of knowing what the cluster time is for the first unit activity identified as an 'indeterminate cluster.' Therefore, the frequency of the first unit activity cannot be known for certain. We further discuss this situation below. Every other activity can be defined as a cluster with its frequency and onset. The onset of the cluster (indicated by the arrow) is defined as the first bin of unit activity in the cluster.

These two measures, frequency and onset, characterize each of the clusters. The frequency gives an indication of the size of the activity. Although it appears that only increases in frequency matter, in reality significant decreases in activity (as before US onset in Figure 5.8) can also be identified. Onset is useful for determining the latency of the unit activity before or after an event. For example, clusters 2 and 5 appear to be a CS- and US-evoked neural responses. The difference in time from CS onset to cluster 2 onset is the latency of the cluster response. A latency which is too long or too short could indicate other processes. The interpretation of cluster 5 is much more complex than indicated, however, in that it could be a somatosensory response to an air puff US, or an auditory response to hearing the air puff, or motor activity related to the reflexive eyeblink. If cluster 5 remains when the air hose is directed away from the eye then it would clearly be an

Chapter 5   189

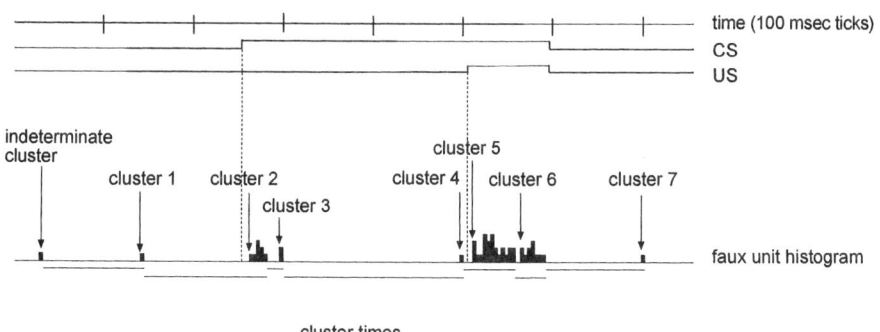

Figure 5.8: Definitions of unit clusters.

auditory evoked response. If cluster 5 is not seen with spontaneous eyeblinks then it could not be related to the motor component of the eyeblink. If the onset of cluster 5 is too long or too short then it may not be related to somatosensory information. These considerations of cluster 5 illustrate that the interpretation of the neural activity still requires the intellectual and experimental participation of the researcher.

It should be apparent that neural activity as defined by clusters can be associated with stimulus events (evoked potentials) and motor responses (causative or feedback). Similar to the results of standard score and cross-correlational analysis, when cluster analysis is used, arguments that the stimulus causes a neural response depend upon *order*, (the stimulus precedes neural activity), *appropriateness* (that the neural response occurs in a reasonable time frame), and *consistency* (the neural response is present from trial to trial). For motor responses, the latency differences between the motor response and unit activity can resolve whether the unit activity might cause the motor reflex or whether unit activity is related to feedback from the motor reflex. The criteria for a reasonable time frame and consistency also apply to either type of motor response.

The problem of determining causation becomes more tenable when looking for neural activity that predicts learned behavioral responses. Figure 5.9 shows real data from a single trial in a trained animal. In this figure the clusters have been identified in the histogram and their frequencies plotted above. The frequency graph makes it easy to see the sizes of the associated frequencies and their latencies. We have marked several of the latencies from CS onset for illustration. The first peak on the frequency graph in the CS period has a latency of 8 msec to the onset of the auditory CS—the order, appropriateness and consistency of this unit activity allow us to suggest that this is an evoked response to the CS. Note that all the latencies will be mul-

Comparing behavior with unit frequency and histogram

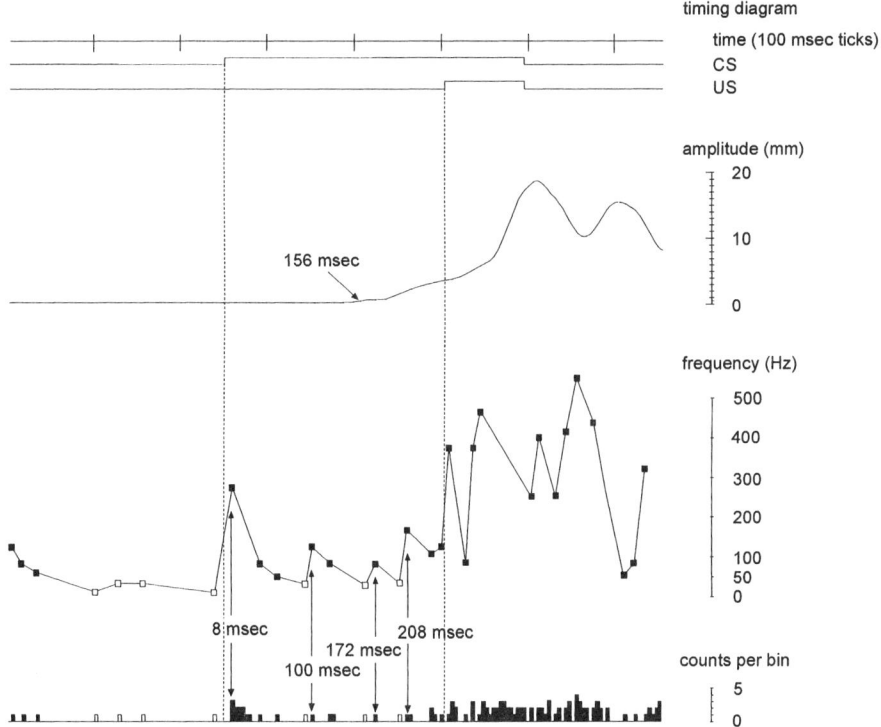

Figure 5.9: Cluster analysis for a single trial.

tiples of 4 msec, the bin-width for data collection. The difficulty comes in deciding which of the other three peaks at 100, 172 or 208 msec is related to the behavioral response, which has an onset latency (0.5 mm) at 156 msec as indicated in the behavioral plot. At 100 msec the unit activity precedes the behavior; at 172 and 208 msec the unit activity comes after the behavior. The unit activity at 172 and 208 msec could be proprioceptive, somatosensory or motor feedback from the learned response, or even a random event unrelated to the behavior.

The criteria for order, appropriateness and consistency are more difficult to achieve when looking for neural activity that predict behavior. Even for stimulus-evoked behavior, however, a relationship cannot be resolved in a single trial. Figure 5.10 shows an average of several trials, typically referred to as a block of trials. In reality, the behavior is an average and the unit histogram is an accumulation of all the unit activity in those trials. The frequency data for the individual trials is overlapped. There is

Chapter 5

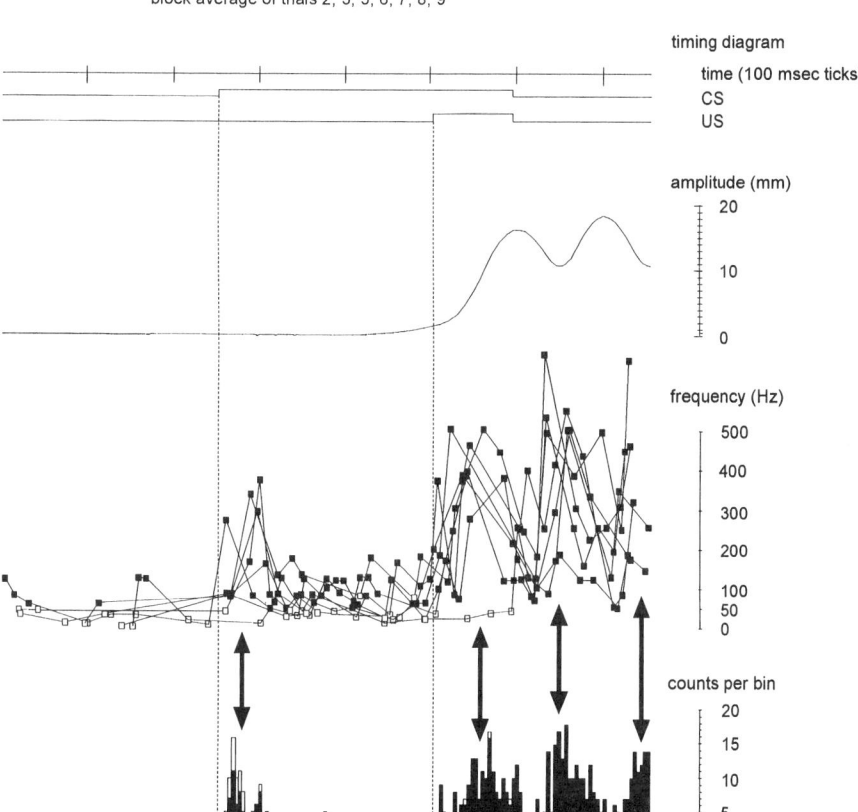

Figure 5.10: Cluster analysis for a block of trials.

some consistency of unit activity as indicated by the double arrows between the histogram and frequency plots. The information is slightly different, however, in that the peak of the histogram shows the maximum unit activity whereas the peaks of the frequency plot indicate onset of the clusters. Also noticeable is that the learned response pretty much washes out in the behavioral average as well as in the frequency plots and histogram.

To be useful a judgment must be made about the significance of the cluster frequencies. One way to do this would be to find the mean frequency and standard error in the baseline (preCS) period. Using the baseline from a whole session, the standard error gets extremely small so that small changes appear significant at two or three standard errors away. Ironically, a simpler approach is more conservative than using standard errors. In Figures 5.9 and 5.10, significance is defined as any cluster frequency that is three times the average baseline frequency. (In this example, using a criterion of twice the frequency is more than six times the standard error.) In both figures, significant frequencies and clusters are indicated by filled symbols, non-significant ones are indicated by open symbols. It can be readily seen in Figure 5.8 that raising the criteria could keep the 8 msec latency significant but cause the other identified latencies (at 100, 172 and 208 msec) to become non-significant. Alternatively, 'significance' in a lay sense can be determined by the apparent peaks in the CS period. A peak is defined as a point with zero slope bounded by negative slopes, which computationally reduces the task of finding the single points that are bordered by smaller values on either side.

However one decides which frequencies are worth investigating, the question is how to determine whether they are related to classical conditioning. For CS evoked activity, a simple calculation of the variance of the first (second, third, etc.) cluster latencies will show which peaks appear to be time-locked. Similarly, the variance of clusters in the US period can reveal US evoked clusters. The overlapped frequency plots in Figure 5.10 give an idea about time-locked variability for CS and US evoked clusters. The temptation is to compute a correlation between the CS (or US) onset and the first (second, third, etc) cluster to examine CS (US) evoked relationships, but this cannot be done because the CS (US) onset time is always the same, therefore it has zero variability, and hence a correlation cannot be computed.

Correlations can be used to compare onset times of behaviors (learned or reflexive) with onset times of any and all clusters to help establish *causation*. A high correlation satisfies the criterion for *consistency*. The relationship of the latencies (which one is shorter than the other) satisfies the criterion for *order*, which should be confirmed on a trial-by-trial basis for consistency. The order will show whether the unit cluster predicts (potentially causes), is indifferent (occurs at the same time; presumably controlled by a third factor that causes both the behavior and the cluster) or reacts (feedback) to the behavior. The difference between the onsets satisfies the criterion for *appropriateness*. If the difference is too large then the gap may be mediated by sensory events, for example. If the difference is too small, then the relationship could be coincidence or both factors could be controlled by a third agent. A similar argument can be made for the relationship of reflexive responses (URs) and clusters in the US period.

Note that in this discussion there is no inherent distinction between causation for a learned or reflexive behavior other than when it occurs, in the CS or in the US period. One might expect that the origin of learning in the nervous system would have a larger difference in onset to the behavior (appropriateness) than for a reflex, although this assumption may not be true. A different approach is to look for the earliest sign of learning related unit activity as an indication of the origin of learning. Using the cluster analysis

Chapter 5   193

technique might enable the researcher to finally distinguish the earliest latencies that could be related to learning in the nervous system. In Figure 5.9 the criteria for order and appropriateness support the role of the 100 msec cluster as being related to the onset of the learned behavior at 156 msec. More trials with conditioned responses would be needed to support this conclusion with the criterion of consistency. The 172 and 208 msec clusters which follow the behavioral onset at 156 msec suggest these clusters are related to feedback. In fact, this is the most likely conclusion since this recording is from the pontine nucleus, which has been shown in other studies to receive feedback about learning-related neural activity from the interpositus nucleus. As with the interpretation of causation, a single trial is not decisive. More trials with conditioned responses would be needed to support one conclusion or the other with the criterion of consistency.

Knowing the baseline firing frequency of a neuron is important in determining significant clusters. We considered several different approaches for calculating the baseline frequency. The obvious approach is to identify baseline clusters and determine their average frequencies. The difficulty as noted above is that the very first cluster of the session, whether it is in the baseline or CS or US period on even in the next trial or later, is indeterminate because the definition of a cluster includes the preceding time of inactivity and it is not known when a cluster last occurred before sampling began. Even when there is a lot of baseline activity this problem still exists at the beginning of each and every trial. If baseline activity is around 4 Hz it would be expected that, on average, only one spike would occur in the usual 250 msec preCS baseline period, and its frequency would be indeterminate. At least two clusters, a minimum of two spikes, would be required to get a measure of the baseline frequency. One way to get around this problem is to assume that the baselines of every trial are independent samples, which they are, and simply string them together so that the end of the baseline of trial 1 is added to the beginning of the baseline of trial 2, etc. That way, only the very first cluster of the entire session is indeterminate rather than the first cluster of every trial.

An alternative approach is more traditional: Simply accumulate all the baseline unit activity, divide by the total baseline time, and scale the frequency in Hertz. For this calculation it is not even necessary to identify clusters per se. Although it is never stated that way, this traditional method essentially assumes that baseline periods can be strung together. That realization, and the attractiveness of the simplicity of the calculation, led us to adopt this method for calculating the baseline frequency. Note that with this method there are two deeply hidden, unacknowledged or unrealized assumptions—first, that a neuron fired just before the very first baseline sample of the training session; and second, that another unit fired just after the very last baseline sample. Thus, the assumptions are that the baseline activity is bracketed by one unit count before and one unit count after the recording. We know that the latter is usually not true because we have examined the CS period of the very last trial of many sessions and a unit response does not often begin that CS period. However, the error that violating this introduces is negligible when the baseline frequency is calculated. We simply point out this implication because there is often a smugness about favored methodology, which in this instance is ignorance about hidden assumptions. Assumptions are important for understanding the limits their implications, which in this instance will slightly bias the baseline frequency towards a higher value. This is fine because most neural activity related to classical conditioning (Purkinje cells in cerebellar cortex are an exception) is excitatory, meaning that a high baseline frequency will give a conservative estimate of significant increases. In other words, the implication of the assumption here works against our hypothesis, which is the way one should operate.

A Computer Program for Cluster Analysis

Lavond and colleagues use two programs for cluster analysis. The first program is found in the file **dave.txt** or **dave.4th**, which allows one to review collected data offline and has several averaging functions. This first program gives a graphical display of individual trials, averaged blocks of trials, and frequency displays of clusters (individually or for several trials). The displays from this review program were used for Figure 5.9 (an individual trial) and Figure 5.10 (an average of several trials) in this chapter. It should be noted that Figure 5.9 is an accurate representation of the output of the **review** program (the **review**ed data was printed on a laser printer, digitally photographed with a Sony Mavica camera, traced in Photoshop, exported as Illustrator paths, then imported and rendered in Freehand). Figure 5.10 is also an output of the review program, treated the same way as Figure 5.9, except that Figure 5.10 was doctored so that the cumulative histogram included symbols for both significant and insignificant clusters derived from their respective individual trials.

The second cluster analysis program is found in the file **cluster.txt** and is evoked by **cluster**. This second program prints out pages of cluster information on a trial by trial basis. The information includes the choice of identification of the frequency and latency of every cluster, identification of significant positive and significant negative clusters as opposed to insignificant clusters, and trials to criterion based on significant clusters in the CS period. This second cluster program requires first loading in **dave.txt,** which includes the basic cluster identification routines and then loading **cluster.txt,** which includes print formatting routines and trials to criterion analysis. Although it is possible to invoke **review** after loading in **cluster.txt**, the formatting of the numerical information in the graphic display is not pleasant because of changes made necessary for the cluster print out.

The cluster program is written in Forth-83 standard using a Windows version of Forth (WinForth) by Laboratory Microsystems, rather than using the DOS version of Forth (MasterForth) by Micromotion, which was used to write the orginal Runtime and Summary programs. There are some slight differences between the two versions of Forth (for example, MasterForth uses **is** whereas WinForth uses **equ**). The obvious differences are in DOS versus Windows displays. The **review** program originates in the DOS version of the Summary program. This DOS-Forth **review** program was imported into and modified for the Windows version of Forth. We see no reason why, other than our time, we cannot go back and incorporate the essential cluster features of the Windows version back into the DOS version.

A variation of the cluster program would be to dispense with identifying clusters and simply use the frequencies of activity within each bin. This method would yield an upper limit to the possible frequencies since only a certain number of units can be counted in each bin. For discriminators with 200 microsecond pulses for each spike, for example see above and Chapter 4, the maximum number of spikes in a 4 millisecond bin is 15 (i.e., one less than 4 milliseconds divided by 200 microseconds because there must be a space between each count otherwise 16 counts would yield a continuous signal or just one count per bin). In our runtime program, **unitduration** is set to the discriminator pulse width, **bin** is the duration of each sample period, **sanity** calculates the number of valid counts within a bin and stores this in the variable **sane#**, and the subroutine **'valid** is used by the data collection subroutine **'datsr** to check and correct the count as the data is collected. The number of errors (counts over the maximum possible) is logged into the array **#errors**. In the current example, 3750 Hz would be the highest possible frequency for 15 counts in a 4 millisecond bin — a ridiculous number for single units, a wildly theoretically possible but not encountered combined frequency for multiple units. (What this indicates is that very high rates could be measured so that ceiling effects are not likely to be encountered by the hardware and software.)

Some Perspective on Cluster Analysis

The term 'cluster analysis' is unfortunate because the method is not related to the statistical method known as a cluster analysis. However, the term does accurately reflect the basic operating principle, which is to try to organize unit activity the same way that we perceive it. Other than the definition of a cluster and the treatment of the baseline as noted above, the cluster analysis program actually makes far fewer assumptions about the data than cross correlations, t-scores or z-scores. All these methods share a degree of imprecision since this is binned data. But the cluster analysis program operates on the original histogram rather than on data that has been smoothed, as in the t-score programs of Datamunch that involve addition of a sliding Gaussian curve. This data smoothing smears the onset information, adding imprecision. Because the cluster analysis program avoids this data manipulation it yields a more accurate determination of the latency. Comparing onset latencies between behavior and units is theoretically possible with cluster analysis. A great advantage of cluster analysis is that it analyzes on a trial-by-trial basis, whereas the other methods operate on pooled data from several trials. Determining trials to criterion using unit data is therefore possible with cluster analysis.

It should be noted that, with the possible exception of pattern matching (which itself may erroneously identify a unit), none of the programs distinguish between single unit data or multiple unit data. They all operate on digitized data. That acknowledgement and limitation is the responsibility of the researcher. Single units analyses are exalted in the field, a view that occasionally is unjustifiably revered in our view. An argument can be made that ensembles of multiple units should be more properly studied. Multiple units also have an advantage when sampling for any interesting activity in an area, or for looking for activity that may have been overlooked. Multiple units can also give a better idea about the relative frequencies and relatedness of activity. Single unit data can be gleaned from multiple unit recordings by pattern matching (e.g., Datawave or Spike2) or multiple criteria system (e.g., Alpha Omega discriminators) at the time of data collection. Unless the original record is saved, however, discriminated data can never recover the original recording. In the figures presented here, multiple units were recorded which show stimulus evoked and motor reflex related activity; individual trials may show activity related to learning but this is not robust in these examples. It is not known if it is the same unit or different units for these responses. If the question is to sample the activity of an area then this data is not only fine but it is preferred over single unit recordings. If the question is whether the same neuron reacts in multiple ways, then single unit recordings are essential. This is particularly true when the question is whether a single unit changes over the course of learning, or whether additional units become recruited over learning. The point is that single and multiple unit recordings have their advantages and disadvantages that should be favored depending on the experimental question.

Berthier and Moore's Algorithm for Neural Analysis

There have been a few other analysis techniques that have been described for relating unit responding to behavior. As an example, we describe a methods used by Berthier and Moore (1990) in an experiment designed to assess the involvement of the deep cerebellar nuclei in eyeblink conditioning. Discriminated pulses representing cerebellar neuronal activity on individual conditioning trials were stored in a computer. They first determined if short-latency responses to the CS occurred by generating interspike interval data across a number of trials for the 100 msec preceding and following the CS, and comparing the distributions using a two-tailed Kolmogorov-Smirnov test. A similar procedure was used to detect changes in firing before and during CR execution.

Berthier and Moore determined onset latency for neuronal firing using an edge-detecting digital filter that was similar to one described by Marr and Hildreth (1980). First, the cumulative sum of spike activity was low-pass filtered by convoluting with a Gaussian. The cumulative sum of the jth time step was equal to

$$\sum_{i=0}^{j} (g_i - g_{mean})$$

where g was an impulse sequence representing spike activity and the mean number of spikes per time step (1 msec) in the preCS period. Next, the low-passed filtered data were differentiated to get a firing frequency and then these data were passed through a third stage that calculated the second derivative of the firing frequency. The time points where firing frequency changed were then estimated by taking the zero-crossing of the output. The filter was designed as follows:

$$f(x) = D^2[D(G*g)](x),$$

where * is the convolution operator, D is the derivative operator, g is the input to be filtered, and f is the filtered output. G is the Gaussian function:

$$G(x) = [1/\sigma(2\pi)^{1/2}]exp(-x^2/(2\sigma^2)).$$

By the associative and derivative properties of convolutions, this yields,

$$f(x) = (D^3G * g)(x)$$

Completing the differentiation,

$$f(x) = [(-1/\sigma^3 (2\pi)^{1/2})(-x/\sigma^2 - 2x/\sigma^2 + x^3/\sigma^4)exp(-x^2/2\sigma^2)] * g(x).$$

Berthier and Moore sampled D^3G at 1 msec intervals and convolved with the cumulative sum of spike activity. The σ was set at 2.4 for all cells. The cumulative sum of spike activity and f were displayed on a graphics terminal and the experimenters set a criterion for the slope at the zero-crossing above which a change-point in firing was accepted as significant.

After the CR and spike-change onset times for each unit were determined, the data were analyzed further to see if the unit was showing stimulus- or movement-related firing. The activity of the unit was determined to be stimulus-related if its activity did not shift when the CR shifted in time (but rather was time-locked to CS onset). The unit was classified as movement-related if the onset of the firing time followed the onset of the behavior on a trial-by-trial basis. Further, the relative movement- versus stimulus-relatedness of the individual units was quantified using three variables: a) the difference between the stimulus onset and spike-change times (Variable x), b) the difference between spike change and the onset of the CR (Variable y), and c) the difference between CS onset and CR onset (Variable z). Finally, variance ratios, i.e., σ_y^2/σ_x^2, were calculated to further determine movement versus stimulus relatedness. A variance ratio of less than 1 indicated a movement-related neuron while a variance ratio greater than one indicated a stimulus-related neuron.

The Steinmetz laboratory has recently begun developing a slight variation of this Berthier and Moore method. So far, we have found the technique useful for separating stimulus- and movement-related responses. Our early results indicate that this method is very useful, especially when coupled with standard score and cross-correlational methods. Indeed, the answer to in-depth descriptions of brain-behavior relationships may very well involve the use of multiple approaches for analyzing behavioral and neural data. Recently, some books concerned with neuronal data analysis have been published (e.g., Nicolelis, 1998 and Rieke, Warland, de Ruyter Van Steveninck & Bialek, 1997). We direct interested readers to these books for addition information about the analysis of neuronal data.

Chapter 6

OTHER BEHAVIORAL PARADIGMS

OVERVIEW

To this point, we have concentrated much of our discussion concerning experimental methods on the use of the classical conditioning paradigm, specifically, classical eyeblink conditioning. We have pointed out in previous chapters that due to the high degree of stimulus control and the relative ease of recording responses, the classical conditioning of skeletal muscle responses has proven extremely useful for studying brain-behavior relationships. It is important to note, however, that there are a number of other behavioral paradigms that have been used to successfully study how the brain encodes learning and memory. This is especially true as relatively inexpensive, high-speed computers have become available for controlling the delivery of stimuli and recording behavioral responses. In this chapter, we review some of these other behavioral paradigms.

FEAR CONDITIONING

After classical eyeblink conditioning, the study of the brain correlates of fear conditioning has generated more data about brain-behavior relationships during learning than any other behavioral procedure. While there are subtle variations in experimental protocols, the basic procedures used during fear conditioning are relatively similar. In the basic fear conditioning experiment, a neutral stimulus such as a tone (the CS) is paired with an aversive foot-shock (the US). After a few pairings of the CS and US, the subject begins to show signs of fear to the previously neutral tone CS (i.e., fear is acquired through a Pavlovian conditioning process). Some fear conditioning experiments measure the amount of freezing or behavioral immobility in the animal as a sign of fear. However, other laboratories have measured changes in autonomic responses, such as heart rate, blood pressure, or pupillary dilation, as indices of fear. Hence, this type of learning is sometimes referred to as autonomic learning or autonomic conditioning. Studies have shown that conditional fear increases as the number of CS-US pairings increase and also as the intensity of the US increases. Interestingly, the CSs used for fear conditioning do not have to be discrete stimuli. The context in which the shock US is presented can serve as the CS for fear conditioning (i.e., the experimental subject makes an association between the context [environment or apparatus] and the foot-shock, thus subsequently demonstrating fear responses when replaced in the same contextual situation in which the foot-shock was delivered.) Researchers have exploited this feature of fear conditioning to study the processing of specific cues like a tone versus context during the acquisition and maintenance of fear conditioning (e.g., Fanselow, 2001).

The basic equipment that is used for Pavlovian fear conditioning experiments is relatively simple. For example, Maren and colleagues (e.g., Maren, 1998) use a number of identical observation chambers (30 x 24 x 21 cm) that are commercially available. These chambers have aluminum sidewalls and Plexiglas front walls and top with a hinged Plexiglas front wall. The chambers are placed in chests that are located in a brightly lit and isolated room. The floors of the chambers are comprised of 19 stainless steel rods (4 mm diameter) that are spaced 1.5 cm apart (center to center). The rods are wired to a shock source and grid scrambler for the delivery of foot shock USs. The grid scrambler is very important as subjects can very quickly learn to stand on only those bars of the same polarity (all grounds or all hots) to avoid getting shocked. Other stimuli are generated using standard equipment (see Chapter 2). Typically, background noise (e.g., 65 dB) is supplied via the motors of ventilation fans. The context for training can easily be altered by either using different shaped and painted conditioning chambers or by placing the conditioning chambers in differently decorated chests or in different rooms. The use of different odors during training also provides a powerful contextual difference. Fear conditioning has been studied in humans, rabbits, rats, rabbits, mice, cats, guinea pigs, and pigeons. The choice of response systems to monitor during the fear conditioning procedure has been determined in part by the species under study. It is important to realize, however, that the conditioned fear response is not a single learned response but rather a constellation of response changes that are seen including freezing, alterations in blood pressure, changes in regional blood flow, changes in hormone release, papillary changes, skin response changes and even hormonal changes. Researchers have typically selected the easiest response to monitor when measuring fear conditioned responses. In the rat this has included measuring increased heart rate, increased blood pressure or increased freezing. In the pigeon, heart rate increases were most frequently measured during fear conditioning. In the rabbit, heart rate deceleration has been observed as a result of pairing CSs with strong aversive USs. Given their small size, freezing seems to be the response most often measured in the mouse during fear conditioning procedures. Fear conditioning has been measured as increases in heart rate, blood pressure, and galvanic skin responses in humans, while papillary dilation has been measured in the cat. Fear conditioning can be measured rather robustly by monitoring any of these responses.

One of us (Steinmetz) has recently conducted a series of fear conditioning experiments in rats, measuring both freezing responses and heart rate changes. For these fear conditioning experiments, we have used the same C++ based software and interface system that we have used for classical eyeblink conditioning to control the timing and duration of the CS and US (see Chapters 2 and 11). Indeed, with exception of the basic interstimulus interval length (it is typically several seconds in fear conditioning versus two seconds or less in eyeblink conditioning), stimulus delivery in fear conditioning differs very little from classical eyeblink conditioning. This should not be surprising given that both are Pavlovian procedures. What is different is how responses are measured. We use videotaping technology to measure freezing, a technique that has been used by many other investigators (e.g., Desmedt, Garcia & Jaffard, 1998; Kim, Decola, Landeria-Fernandez, & Fanselow, 1991; Lattal & Abel, 2001; Schafe & LeDoux, 2000). Observers, blind to the experimental condition of the rat, typically score the rats behavior for movements at regular intervals (both during a trial and between trials). The scoring is a simple "there is movement" (no freezing) or "there is no movement" (freezing), with freezing defined as the absence of all movement except that movement related to respiration (after Blanchard & Blanchard, 1969). Normally, the amount of freezing is quantified as the percentage of observations during which freezing was observed for a session.

Measuring heart rate is a bit more complicated. As we discussed in Chapter 3, Linda Rorick in Steinmetz's laboratory designed and constructed a nifty vest, which contains heart rate recording disks electrode sewn onto its inside surface of the vest (see Figure 3.11). This vest provides a noninvasive means for monitoring heart rate activity. As you might imagine, at first it is a bit difficult getting the vest onto the rat. However, they very quickly adapt to wearing it and do not attempt to remove it. After first placing the vest on the rat, the heart rate is slightly elevated and this is likely due to mild stress. However, the heart rate quickly returns to baseline and remains there on subsequent sessions. Electrical activity from the heart is amplified a hundredfold, filtered between 100 and 500 Hz, and routed to a Spike2 waveform analysis system. Using template matching algorithms (similar to action potential detection methods), heart beats are isolated and their distribution determined in off-line analyses. We also note here that we have also conducted autonomic conditioning experiments involving humans, measuring variables such as skin conductance, heart rate and respiration. For these studies we have chosen to use commercially available devices for their expertise and liability issues. Grass Instruments, for example, sells a large line of stand-alone recording systems and devices (e.g., polygraphs and digital systems) that can be used for this purpose.

Weinberger and colleagues have studied fear conditioning in mildly restrained guinea pigs, and have measured cardiac activity using electrocardiograms (e.g., Cruikshank, Edeline & Weinberger, 1992). During surgery they fastened metal clips to both the dorsal and ventral thoracic skin surfaces from which recording leads were attached. The electrocardiogram was amplified a thousandfold, bandpass filtered between 80 and 250 Hz and sent through a voltage discriminator, which generated output pulses for each discriminated beat that could be counted by electronic counters.

Other laboratories have used a variety of methods to measure fear conditioned responses. Maren and colleagues have used load cell platforms to measure freezing behavior (e.g., Maren, 1998). The load cell platform is placed under the conditioning chamber and is used to record chamber displacement in response to the animal's movements. After calibration (especially important if more than one chamber is being used), the output of the load cell is set to a gain that optimizes the detection of freezing behavior. The amplified load cell signal is routed to a digitizer and recorded offline using commercially available software (e.g., MED Associates, Inc.). This activity monitor could easily be built by the investigator using force transducers and a low-gain amplifier, and feeding the signal into an A/D system. A variety of force transducers are commercially available with varying degrees of sensitivity. For example, Grass Instruments (Model FT10) and Stoelting (Model 56350) sell force transducers with amplifiers that are used for measuring muscle movements; these could be used to measure grosser movements, such as those related to cage and chamber movements.

Kim and colleagues have recently used automated methods to record freezing during fear conditioning and also measured ultrasonic vocalizations as a measure of emotional learning in rats (e.g., Lee, Choi, Brown & Kim, 2001). Freezing was measured using a 24 cell infrared activity monitor that was mounted on the top of the conditioning chamber. This device detected the movement of the emitted infrared (1300 nm) bodyheat image from the rats in the x-, y-, and z-axes. A computer program calculated the total time of inactivity and freezing was defined as continuous inactivity lasting greater than 3 seconds. Ultrasonic vocalizations were measured using a heterodyne bat detector (Mini-3; Noldus Information Technology, Wageningen, The Netherlands), which transforms the high frequency ultrasonic vocalizations into the audible range. The signal was then filtered and routed to a microcomputer that contained specialized analysis software that determined the onsets and offsets of vocalizations. Other investigators have used much simpler measures of freezing such as changes in gross body movements (counted as the crossing of lines drawn on the floor of the training chamber) and rearings (e.g., Avanzi, Castilho, de Andrade & Brandao, 1998). Still other investigators have used conditioned suppression of licking as a measure of conditioned fear (e.g., Mahoney & Ayres, 1976; see Chapter 3 for a discussion of devices and circuitry used to measure licking in the rat).

POTENTIATED STARTLE

Potentiated startle has been used extensively by Davis and colleagues to study the neural substrates of emotional learning and memory (Davis, Falls, Campeau & Kim, 1993). In this procedure, a loud tone that induces a startle reflex (typically a vigorous movement of the whole body of a rat) is initially presented and the startle reflex is measured. Next, a light CS is paired with a footshock for several trials. When the light CS is subsequently presented just before the acoustic stimulus, the startle reflex is potentiated (i.e., the rats jump higher when the startle eliciting stimulus is presented together with the stimulus that was previously paired with an aversive event). Importantly, before training, the light CS does not elicit the startle response nor does it become capable of eliciting the startle response when delivered in unpaired fashion with the footshock US. This behavioral procedure has proven very useful for studying the neural correlates of emotional learning and memory, especially when coupled with lesion, stimulation and pharmacological techniques. Recording neural activity during this behavioral procedure is difficult; the excessive movement artifacts during startle are tough to eliminate from the neural records.

We will not review the extensive literature concerning emotional learning and memory that has been created by Davis and his colleagues through the use of this procedure. We can, however, say a few words about methodological issues. First, the delivery of stimuli is quite simple and can be achieved using methods described in Chapter 2. Recording the response is a bit trickier. Basically, in this procedure we are interested in how high or how vigorously the rat jumps to the startle stimulus before and after the light-shock pairings. This can be done using videotaping methods, but this process is rather time consuming. Davis and colleagues have automated this recording process by using stabilimeter technology (e.g., Campeau and Davis, 1992; 1995). Their stabilimeter is an 8 x 15 x 15 cm Plexiglas and wire mesh cage suspended within a 25 x 20 x 20 cm heavy steel frame. The floor of each stabilimeter is made up of four stainless steel bars (6 mm in diameter) that are spaced 20 mm apart (center to center). Electric shocks are delivered through the metal bars. Within the steel frame, the cage was compressed between four springs above and a 5 x 5 cm rubber cylinder below, with an accelerometer (Endevco 2217E) positioned between the cage and the rubber cylinder. Movements of the animal produces a cage movement that displaces the accelerometer. The output of the accelerometer is amplified, digitized and stored in a computer for off-line analysis. Davis and colleagues have defined startle amplitude as the maximum accelerometer voltage that occurred during the first 200 msec after the onset of the startle stimulus. Typically the stabilimeters are placed within a ventilated, sound-attenuating chamber. Startle is usually induced by presenting a 105 dB, 50 msec burst of white noise delivered through a speaker that is positioned 70 cm in front of the stabilimeter. Commercially manufactured startle reflex monitoring systems are also available (e.g., the Responder-X-1 system made by Columbus Instruments, the SR-Lab Startle Response System made by San Diego Instruments, and the SOF-815 Startle Response Apparatus made by Med Associates).

Before leaving the topic of fear conditioning, we should say a few words about the use of this paradigm to study emotional learning and memory in humans. The bottom line is that it is an excellent procedure for this use. The Steinmetz laboratory, for example, has begun studying this type of learning in obsessive-compulsive individuals. Specifically, using the same equipment we use for classical eyeblink conditioning, we have studied potentiation of the eyeblink response. Subjects are seated comfortably in front of a computer screen that typically is running a moderately engaging background task. On a random schedule, a loud burst of white noise is presented and the amplitude of the resulting eyeblink recorded using an infrared eyeblink recording device (see Chapter 3). Next, 8–10 paired 1 KHz tone (or light flashes) CS and air puff US trials are delivered (a colleague of ours has successfully used a mild forearm shock instead of the air puff; we find the air puff to be equally effective, though). Potentiated eyeblinks can be seen on subsequent white noise presentation trials, thus demonstrating the acquisition of the fear-potentiated startle response. Importantly, no change in responding to the tone or light CS is seen and unpaired presentations of the CS and US do not produce potentiation. We have also recorded altered skin conductance responses and respiration during this procedure. In addition, an EMG-based startle reflex testing system is commercially available from San Diego Instruments (Model EMG-SR).

INSTRUMENTAL CONDITIONING PROCEDURES

Most of the discussion of methods and techniques in this book have centered around classical conditioning procedures such as those used during classical eyeblink conditioning, classical autonomic conditioning, and Pavlovian fear conditioning. In classical conditioning, the subject that is undergoing conditioning has no control over the delivery of the stimuli. However, there are many situations in which the organism's behaviors dictate when stimuli and other events occur. We could say that the subject's behavior is 'instrumental' in determining the events that follow. Learning that involves this type of situation (i.e., where the subject's behavior affects subsequent events) is termed "instrumental conditioning" or "operant conditioning." Over the years these terms have become synonymous. The word "operant" was introduced by Skinner (1938) to describe a class of behaviors that were different than the reflexive (i.e., respondent) behaviors that were altered with classical conditioning procedures. Operant behaviors included those behaviors "which are emitted by the organism" and "which cannot be shown to be under the control of eliciting stimuli" (Skinner, 1938, p. 19). In his excellent book "The Principles of Learning & Behavior," Michael Domjan presents a comprehensive look at instrumental/operant learning (Domjan, 2000). Interested readers are encouraged to consult this book.

Thousands of behavioral experiments using instrumental conditioning procedures have been carried out over the years and these experiments have produced an impressive array of behavioral data concerning general laws and rules that govern changes in instrumental responses with learning. As we have pointed out in previous chapters, compared to the number of studies that have used classical conditioning procedures to study the neurobiology of learning and memory, with some very notable exceptions, fewer neurobiological studies have been conducted using instrumental conditioning procedures. We have argued in earlier chapters that this is probably due to the nature of the tasks: Before the advent of high-speed computers, which have aided in the concomitant collection of behavioral and neural data, it was relatively difficult to collect data during instrumental tasks because the subject, and not the experimenter, controlled when responses occurred and when subsequent stimuli and events occurred. Recently, the high-speed computer has made this a non-issue; huge streams of data can be taken in over large periods of time. Thus, there has recently been an increasing number of reports in the literature about neural systems and structures that are involved in instrumental conditioning. While space constraints prevent us from reviewing this extensive literature, we can present some examples of how instrumental conditioning is accomplished from a methodological perspective.

One of the major considerations for designing methods and equipment to be used for instrumental conditioning experiment is whether discrete trials or free operant responding is required. Some instrumental conditioning procedures involve placing the subject in the apparatus and providing

Figure 6.1: Operant chamber or Skinner box.

one opportunity for the subject to respond (i.e., a trial). For example, when a T-maze is used, the subject is removed from the apparatus before a new trial is begun. (We will discuss maze running paradigms in a section on spatial learning that appears later in this chapter.) Other instrumental procedures are called free operant procedures because the subject can respond more than one time before a session (or trial) is terminated. Lever pressing for a food reward is an example of a free operant instrumental task; the subject might be required to press a lever 20 times to get a single food reward and 50 rewards might be given in a single session. Often discrete trial instrumental conditioning involves a behavior that is already in the repertoire of the subject (such as maze running) while free operant conditioning involves the acquisition of a behavior that has to be acquired (such as lever pressing). The reader should note that whether the instrumental situation is defined as discrete trial or free operant is not dependent on the specific response being observed. Signaled lever pressing tasks, such as the ones described below, can be considered as a type of discrete trial instrumental learning tasks, as is avoidance eyeblink conditioning.

Many instrumental conditioning experiments take place in a "Skinner box," a rectangular Plexiglas and galvanized or stainless steel box that was popularized by B. F. Skinner in his operant conditioning work (see

Figure 6.1). Several styles of boxes have been made for pigeons and rats, but they all have some commonalities. Typically the floor is fashioned from a grid of stainless steel bars that allow for the delivery of mild footshocks for aversive conditioning experiments. In addition, the boxes may be equipped with food wells, food trays, or water spouts, located in one or more of the walls of the box, into which food or liquid rewards (i.e., reinforcers) can be delivered. The box may also contain levers (also called keys or bars) mounted into the side walls or extending down from the top of the box. Responses involving these levers or keys often constitute the behaviors that are monitored during the learning and memory experiments. We have discussed in Chapters 2 and 3 how stimuli can be delivered to these boxes and how responses can be processed. However, given the potential usefulness of instrumental conditioning for neurobiological studies, we will look more carefully here at some of these behavioral procedures.

Perhaps the simplest instrumental situation is free operant responding for a food reinforcement. In these behavioral experiments, subjects are placed in the conditioning box, shaped to make a response for a food or water reward (such as lever pressing or licking), then placed on a schedule of reinforcement that sets the response requirements (e.g., a reinforcement is delivered for every five lever presses made or a reinforcement is delivered after a set period of time, etc). For many years, researchers used the cumulative recorder to measure and display the learning and performance of these operant responses. This device was created by Skinner and was initially composed of a rotating drum, which slowly feeds a roll of paper out of the recorder, and a pen, which draws a continuous line on the paper when the drum is turning. Each instrumental response made by the subject (e.g., a lever press) would vertically move the pen on the paper a set distance. At the end of the session, instrumental learning is assessed by studying the pattern seen on the cumulative recorder. The higher the pen moves vertically on the drum, the more total responses have occurred. The rate of responding can be determined by the steepness of the cumulative record. Nowadays, the drum and pen cumulative recorder has been largely replaced by the computer. It is somewhat trivial to input each behavior event as a discrete signal to a computer interface that increments a counter whenever a response is made. Graphic displays on the computer screen can then show the pattern of responding that has occurred and return rather precise measurements of response rates and timing. For example, WPI, Inc. manufactures a relatively inexpensive "paperless chart recorder" (Model Duo18), which among other things, could be used together with a personal computer as an event (cumulative) recorder.

Classical eyeblink conditioning can become an instrumental procedure with one very simple change—omit the US on trials when a CR is produced by the subject. In eyeblink classical conditioning, the performance of the CR does not determine whether or not a US is presented, the experimenter does. In eyeblink instrumental conditioning, an active avoidance procedure, performance of a CR by the subject prevents the delivery of the US. A few studies have been done of eyeblink instrumental conditioning including one by Polechar and colleagues (Polenchar, Patterson, Lavond & Thompson, 1985) that showed that lesions of the cerebellum abolished this type of instrumental learning. Indeed, there have been great theoretical debates over the years concerning whether or not classical eyeblink conditioning may have an instrumental component (see Gormezano, Kehoe & Marshall, 1983). When an air puff is used as a US, eyelid closure can prevent the reception of the aversive air puff by the eye (a potential instrumental response). As we pointed out in Chapter 3, it is for this reason that some investigators have used eye clips to keep the eyelid extended when an air puff is used as a US or have used eye shock as a US (because it is hard to see how the eye shock can be avoided by closing the eye). If you have the equipment in place for classical conditioning experiments, it is quite easy to convert from eyeblink classical conditioning to eyeblink instrumental conditioning. You can use the same equipment to generate stimuli and record responses and do one of at least two things: First, in the software, you can include a routine that checks the amplitude of the eyeblink response as each digitized data point is obtained. If the response breaks a preset amplitude criterion, you can direct the program to NOT turn on the US at its preset time. Alternatively, you can use a hardware solution that involves logic chips. Responses that break a present voltage level can be used to "cancel" signals to devices being used to gate the US on.

Steinmetz and colleagues recently developed an instrumental conditioning task that they have used to study neurobiological correlates of instrumental learning and memory in rats (Steinmetz, Logue & Miller, 1993). In particular, we were interested in comparing and contrasting appetitive and aversive learning using a within-subject approach. In this procedure, rats are trained to press a response bar during a tone presentation to either receive a food reward (appetitive signaled bar-press learning) or to avoid a mild foot shock (aversive signaled bar-press learning). Typically, tone presentations last for 3–8 sec (and in a few cases, less than 3 sec). In our typical experiment, the same rat is exposed to both the aversive and appetitive procedures. They are trained in the same chamber and context during each procedure and the same tone frequency is used as the signaling stimulus. Because the subject, chamber, context, timing, and signaling stimulus are the same for both procedures, the only major difference encountered by the animal when switched between appetitive and aversive training is the consequence of the bar-press (i.e., food versus foot shock). We have used these procedures to study the involvement of the cerebellum in appetitive versus aversive learning (Steinmetz et al., 1993) and have also conducted a series of studies to examine appetitive and aversive learning in rats that are bred specifically for alcohol preference (e.g., Blankenship, Finn & Steinmetz, 1998).

Here is a brief description of our instrumental conditioning set-up: Rats are trained in standard operant boxes (30 x 24 x 21 cm) that have stainless steel sides and Plexiglas tops, fronts, and backs. Some of our boxes have hinged tops that open while others have hinged front panels that open. A variety of these boxes are commercially available (e.g., Lafayette Instruments, MED Associates, or Coulbourn Instruments). The operant training box is placed in a sound-attenuating chamber during training. One side panel of the operant box contains a response bar with a recessed food tray to one side. The floor is composed of a grid of 12 stainless steel bars through which electric shock can be delivered. The speaker for tone presentations is attached to the chamber wall about 5 cm above and behind the front panel of the operant box. A 10-W houselight is on continuously during all sessions.

We currently use a C++ runtime software/hardware system that we created (Chen & Steinmetz, 1998) to control the delivery of stimuli and record responses to a microcomputer (see Chapter 12). The delivery of stimuli occurs through interface equipment that is described in Chapter 2. Response recording is quite simple. We use the response bar as a switch to gate +5V to the A/D chip of our computer interface. Thus, whenever the rat presses the bar, +5 V is routed to the computer to indicate that the event has occurred. Our software is able to record the onset and duration of the depression of the bar, thus providing valuable information about not only the how often bar presses occurred, but when they occurred.

For appetitive training, we first must establish bar-pressing for the food reward in the animal, then switch to the signaled procedure. In their home cages, all subjects are initially exposed to the 45 mg pellets (Bio-Serve, Frenchtown, NJ) that are eventually used for reinforcement. They are then hand-shaped, using the standard method of successive approximation. Once the animals obtain 100 food pellets on a continuous reinforcement schedule within a 30 min period, they are switched to a fixed ratio partial schedule that requires four bar presses to receive a reward. After reaching a criterion of 100 reinforcements on two consecutive days, they are switched to signaled bar-press training sessions. On the signaled sessions, each rat receives a set number of trials on which a tone or white noise is presented for a period time (typically 3–8 sec for most of our experiments). A bar press made during the tone terminates the tone and activates a pellet dispenser (e.g., Coulbourne Instruments) that dispenses a food pellet into the recessed food tray. A 15 sec intertrial interval (ITI) elapses before the beginning of the next trial. After the ITI expires, a 1–8 sec pre-tone period begins. If the rat bar-presses during the pseudorandomly varied pre-tone period, the trial time clock is reset and the sequence continued until the rat withholds bar-pressing for the duration of the pre-tone period. When the pre-tone period is successfully completed, the next tone trial is immediately initiated. The pre-tone period was introduced because early attempts at appetitive training with no pre-tone period or a fixed 2 sec per-tone period showed that the rats paid little attention to the tone signal, but rather learned timing strategies to obtain food pellets. The variable 1–8 sec period prevents the adoption of these timing strategies. We find that rats can learn the appetitively-motivated response in 4–6 days of training, showing learning levels that exceed 80% learned responses.

All aversive training sessions are conducted in the same operant box and sound-attenuating chambers in which appetitive training is conducted. As in appetitive training, the signal is typically a tone or white noise stimulus. A foot shock stimulus is generated by a standard shock generator (Lafayette and Coulbourne models) located outside of the chamber. For some studies, we have passed the current through a neon grid scrambler (Model 58020, Lafayette Instruments), which was connected to the floor of the operant box via a 12-plug conductor harness that linked the grid scrambler with the 12 stainless steel bars making up the floor of the box. For other studies, we have not used a grid scrambler, but rather have implanted a small loop of stainless steel wire in the scapula region of the back and used that wire as a return path for the current (while electrifying all of the floor bars). This arrangement requires that a swivel or other device be used to route the cable to the back (see Chapter 4). We find that relatively low shock intensities can be used (e.g., 0.7 mA) and that pulsating the shock 3–4 times (e.g., 250 msec on and 500 msec off) is particularly effective in producing learning.

For aversive training, all rats receive one or two shaping sessions that consist of manual presentation of the shock that is terminated when the rat gets successively closer to pressing the response bar. This typically takes about 30 min over one or two days. Some rats learn to hold the bar down for long periods of time during shaping. To avoid this, mild "off-the-bar" shocks are given whenever the bar is depressed for more than 5 sec. Rats learn very quickly (i.e., within 5–10 min) not to hold the bar down. We use a C++ software routine for shaping. Avoidance training begins after the rat is successfully shaped to escape 100 shocks in 30 min. For signaled aversive training, a tone is presented for a period of time and a bar press during the tone presentation terminates the tone and prevents shock delivery. A bar press made after tone presentation but during shock delivery immediately terminates the shock (an escape response). Usually 100–300 trials are given daily with a 15 sec ITI. As in appetitive training, a variable 1–8 msec period

Chapter 6 211

is given before the beginning of trial (although there is typically very little bar-pressing during the ITI in the aversive procedure). Within 1-2 days of signaled training, rats show 90-100% escape responding. Avoidance learning rats reach about 50-60% by the tenth session. While this rate is somewhat lower than the appetitive learning rate, we are very pleased with it. After all, a very large literature was generated in the 1960s that claimed that rats could not learn an active avoidance response that required a lever-pressing response (Meyer, Cho & Wesemann, 1960). Our studies have demonstrated that this literature is wrong.

We have used a number of variations of these appetitive and aversive bar-pressing tasks to study a variety of behavioral and neurobiological features of instrumental learning. For example, we have introduced a delay of reinforcement feature into the appetitive task to study behavioral disinhibition in rats bred specifically for high or low alcohol preference. This is done simply by withholding reinforcement if a response is made within the first few seconds of the tone period. Others have used this type of manipulation to study the involvement of brain structures in response timing (e.g., Port, Curtis, Inoue, Briggs & Seybold, 1993). Also, we have studied the impact of shortening or lengthening the tone-food or tone-shock interval and examined when specific brain areas are engaged in the task. We have assessed the effects of cerebellar lesions on the within-subject procedures to studying when the structure is necessary for learning. These studies have shown the cerebellum is necessary for the aversive, but not the appetitive task. We have also studied fear conditioning (heart rate changes and freezing) during the aversive version of the task and shown when the freezing response may prevent learning of the active avoidance response.

OTHER INSTRUMENTAL CONDITIONING PROCEDURES

There are several other instrumental conditioning procedures that have commonly been used in studies of the neurobiology of learning and memory. These procedures include: an array of active and passive avoidance procedures that involved whole body movements; delayed alternation procedures; differential reinforcement of low rate procedures that tax response timing systems; and a variety of maze and spatial learning tasks. We briefly review some of the methodological features of these procedures here to illustrate the rich variety of behavioral paradigms that are at the disposal of the experimenter. Again, we remind the reader that this is not meant to be an exhaustive review of all instrumental conditioning procedures that are available but rather a brief look at some behavioral tasks that can be used.

Michael Gabriel and his colleagues at the University of Illinois have collected an impressive amount of data concerning the involvement of the limbic system and forebrain in a form of instrumental conditioning known as discriminative avoidance conditioning (see Gabriel & Talk, 2001). They have used rabbits as subjects in their experiments. In their task, which was adapted from a procedure described by Brogden and Culler (1936), rabbits are placed in a large activity wheel that rotates around a central axis. Because much of their work involves concomitant neural recording, the activity wheel is located within a sound- and electrical-noise-attenuating chamber. An exhaust fan and a speaker in the chamber is used to produce a masking noise (70 dB re: 20 N/m^2). The rabbits are trained to discriminate between two stimuli (typically pure tones) using an avoidance conditioning procedure. In this task, one CS (the CS+) is consistently followed by a 1.5–2.5 mA footshock US that is delivered though a grid of steel bars that actually forms the floor of the Plexiglas wheel (See Figure 6.2). The second CS (the CS-) is never followed by the footshock. Usually, 5 sec separates the CS+ onset from the US onset. On CS+ trials, the rabbit can avoid deliver of the footshock if it begins a stepping motion that causes the wheel to rotate by 2° or more. Gabriel and colleagues have reported that while only a 2° rotation is needed to avoid the foot shock US, typically the conditioned responses consist of one or more steps in the wheel such that the mean magnitude of avoidance responding for a session in a well trained rabbit is about 400° of wheel rotation. Good discriminative instrumental learning can be achieved in less than seven days using this task. In addition, this task has proven very useful for studying the neurobiology of instrumental learning (see Gabriel & Talk, 2001, for review). For example, neuronal recording can be done concomitantly with the behavioral training and unit responsiveness can be determined under a variety behavioral conditioning (e.g., correct responding to the CS+, failure to respond to the CS+, correct failure to respond to the CS-, and incorrect responding to the CS-).

Over the years, passive avoidance has been often used to assess instrumental learning that requires the inhibition of responding. Passive avoidance conditioning is extremely easy to set up and record, and this may be a primary reason why it has been popular as a behavioral paradigm. We have used a step-down version of passive avoidance to study behavioral disinhibition in rats bred specifically for alcohol preference (Steinmetz, Blankenship, Green, Smith & Finn, 2000). To accomplish this, we have taken a standard Plexiglas and stainless steel operant conditioning box and removed all response levers and food trays such that the box is completely empty. The floor is made from 12 stainless steel bars that can be electrified using a grid scrambler. Next, a small Plexiglas platform (15 x 15 x 15 cm) is placed in the box, either in one of the corners or in the middle of the operant box. Rats are placed on the platform and the time until the rat steps off of the platform onto the electrified grid is measured. One to five trials are given daily. The rats normally learn quite quickly to remain on the platform (i.e., they passively avoid stepping onto the platform). There are several ways to measure this behavior. For example, videotaping can be done and the time that elapses from placing the rat on the platform and stepping down can be determined from the videotapes. Alternatively, commercial systems are available that use motion detection or electronic resistance detection circuitry to determine when the rat steps from the platform to the floor grid. We do not think that it is a good idea to use simple observation methods for this (or any other) behavioral task. It is always a good idea to have some enduring record (e.g., videotapes) to archive the learning.

Shuttle boxes have also been employed in active and passive avoidance learning experiments. The simplest is a two-compartment shuttle box (e.g., Overmier, 1966). A two-compartment box is typically a rectangular stainless steel or Plexiglas box (50 x 100 x 70 cm) that has a wall in the center that divides the box into two equal sized rooms. The wall usually has a door in its center than can be opened (either manually or using motors and electronics). Sometimes the rooms are decorated differently (e.g., horizontal versus vertical stripes) or one side is lighted while the other side is dark. The behavioral task is quite simple. In the active avoidance task a signal might be turned on and the rat has to move to the opposite compartment to avoid a foot shock (delivered through an electrified floor grid) or other aversive stimulus, such as a loud noise. In a passive avoidance task, the subject does not move to successfully avoid the aversive stimulus. These same boxes have been used to study behavioral and neurobiological aspects of novelty (or familiarity). Rats are placed in one side of the box and the time taken to begin exploring the other side of the box is noted. Videotaping can be used to record these behaviors. Alternatively, the recording process can be automated using photobeam/detector technology to sense movement through the doorway into the second compartment. The latter movement detection method has become quite popular and is relatively straightforward. An infrared LED can be used together with an optical integrated circuit sensor to pick up movement. That is, when the infrared to detector beam is broken by the presence of the animal, a movement event is detected and can be fed to a computer or other counting device. The photobeam technology can be used in a variety of applications other than general movement detection including in a lickometer, and for pellet detection. Passive and active avoidance shuttle box systems are readily available from commercial sources (e. g., Columbus Instruments' Model PACS-30, Stoelting/Basile's Model 57530 and Model 57550, Med Associates' Model ENV-010 modular shuttle box, and San Diego Instruments' GEMINI avoidance system, which also has a freeze monitor option that uses photobeam technology).

The basic behavioral and neurobiology substrates of response timing has been studied using an instrumental procedure called differential reinforcement of low rates (DRL). We have used a version of this task to study behavioral and neurobiological aspects of response inhibition (Steinmetz et al., 2000). In our version of the task, rats are placed in the same operant boxes that we use for signaled bar-press conditioning (see above). However, in the basic DRL procedure, no signal stimulus is delivered. Rather, rats learn to suppress their bar-pressing for a period of time that is set by the experimenter to receive a food reward. Typically, rats are first shaped to bar-press for the reinforcement. Next, the DRL requirement is instituted, usually starting with small time intervals and progressing to longer time interval. For example, initially a DRL-5 schedule might be used; this requires the rat to withhold bar-pressing for a 5 sec period, after which food is available if a bar press is made within a given span of time. Rats can be easily trained to withhold bar-pressing for longer and longer period of time. For example, a DRL-30 schedule is possible, where the rat learns to withhold responding for 30 sec to receive a food reward within a 5 sec window that follows the 30 sec non-response period. Equipment-wise, this procedure was relatively easy to set up. We simply used our signaled bar-pressing software and hardware with a slight modification. The ITI for the program was set to 0 and instead of using a variable 1–8 pre-tone period, we set the pre-tone period to equal the DRL length we desired. A bar-press made during the pre-tone period reset the timers to restart the DRL interval. If the rat withheld bar-pressing during the DRL period, a 5 sec trial was delivered during which time a bar press resulted in a good reward. We simply turned the audio generator equipment off so that no tones would be delivered. Different DRL schedules were set by choosing different pre-tone period lengths.

Finally, there are a number of instrumental procedures that involve runways and mazes. These procedures have been used by the behavioral neuroscience community mainly to study the neural bases of spatial learning. While there is certainly a spatial component to these tasks, we often forget that they are instrumental in nature. Perhaps the simplest of these spatial tasks is the simple runway task. A runway task used by Capaldi, Alptekin and Birmingham (1996) to study instrumental performance serves to illustrate this class of behavioral tasks. The apparatus was quite simple; a straight gray runway (about 136 cm long and 9 cm wide, with 9 cm sides). At either end of the runway was a startbox and goalbox that were about 23 cm and 30 cm long, respectively. The startbox and goalbox were closed off by metal guillotine doors. Lifting the startbox door started a silent digital clock which was stopped when a photobeam located about 106 cm from the startbox door and 2 cm in front of a goal cup was broken by the rat. A vari-

Chapter 6

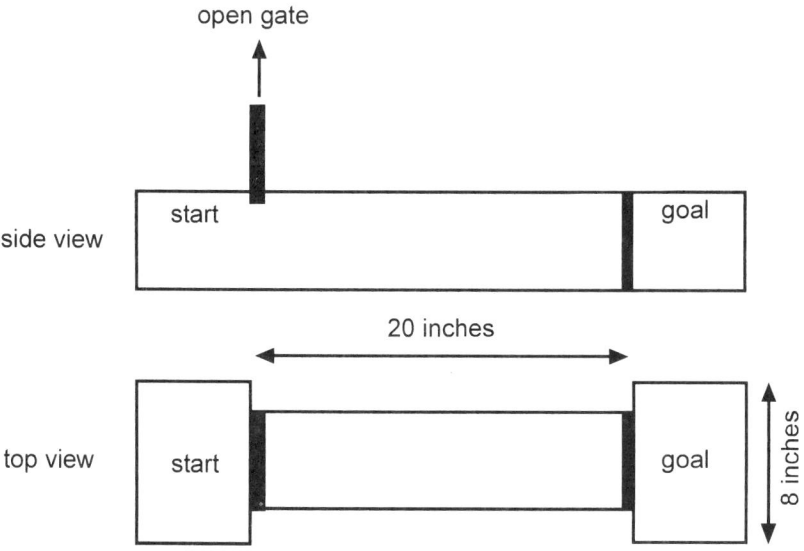

Figure 6.2: Runway or straight maze.

ety of tasks can be run using this simple device and almost all involve recording the time it takes the rat to move from the startbox to the goalbox and consume the food reward. Unlike lever pressing, the simple runway task requires little if any shaping; rats will naturally explore the runway. Figure 6.2 shows a simple runway.

Mazes have been commonly used to explore the neural bases of spatial learning and the lion's share of this research has involved assessments of hippocampal function (which, of course, has been historically thought to be involved in spatial processing). Although several mazes have been built and used over the years, we will briefly describe three types here: the simple T-maze, the radial arm maze, and the Morris water maze. All mazes share a few things in common. The subject (typically rats) may use intramaze and extramaze cues to learn and remember the correct movements to make to solve the maze. Interamaze cues refer to those cues within the maze that guide movements and these may include odors, the color of walls and floors, and the textures of the floors or walls. Extramaze cues refer to those cues outside of the maze that provide information for the subject. These cues may include the position of items in the room such as doors, windows, bookcases, and wall hangings. Normally, subjects might use both sets of cues to solve the maze.

The T-maze is a relatively simple maze that has been used for many years with rats and mice as subjects. This maze is composed of a long run-

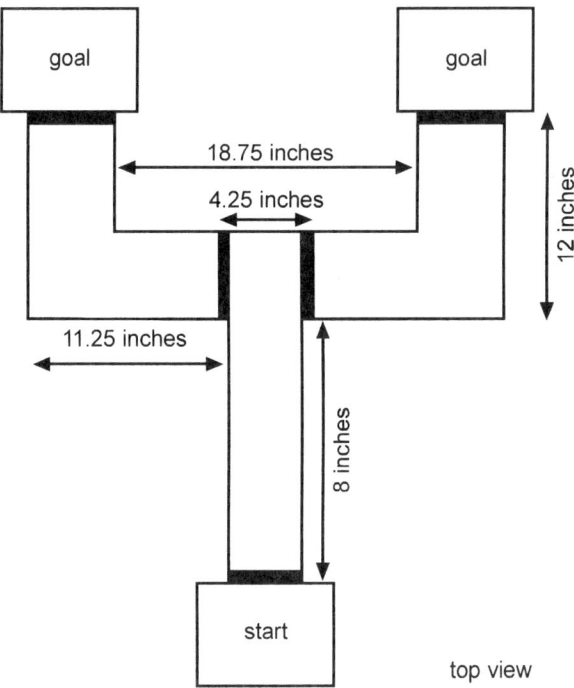

Figure 6.3: T-maze.

way with a startbox on one end and two arms that are at right angles to the runway at the other end. The Y-maze is a variation of the T-maze where the arms are at angles that are greater than 90 degrees (e.g., 120 degrees). Subjects are normally trained to run to point in the maze where the arms intersect with the runway (called the choice point) and make a decision to turn right or left based on reinforcements that are left at the end of the right or left maze arms. The decision to turn could be based on one or more different requirements set up by the experimenter, such as differences in the appearance of the end of the arm or the patterns of responding that are required (e.g., always turn right, always turn left, alternate turns, etc.). Figure 6.3 shows a simple T-maze.

The radial arm maze was popularized by Olton and a number of other researchers. Typically, the radial arm maze is composed of a central startbox with 8–16 arms that radiate from the central point. Doors usually separate the subject from the radial arms when it is first placed on the maze. Usually a subset of arms are "baited" with reinforcement and the subject's task is to find the reinforcements in the least amount time while committing the fewest errors. Depending on the experiment, errors could be defined as

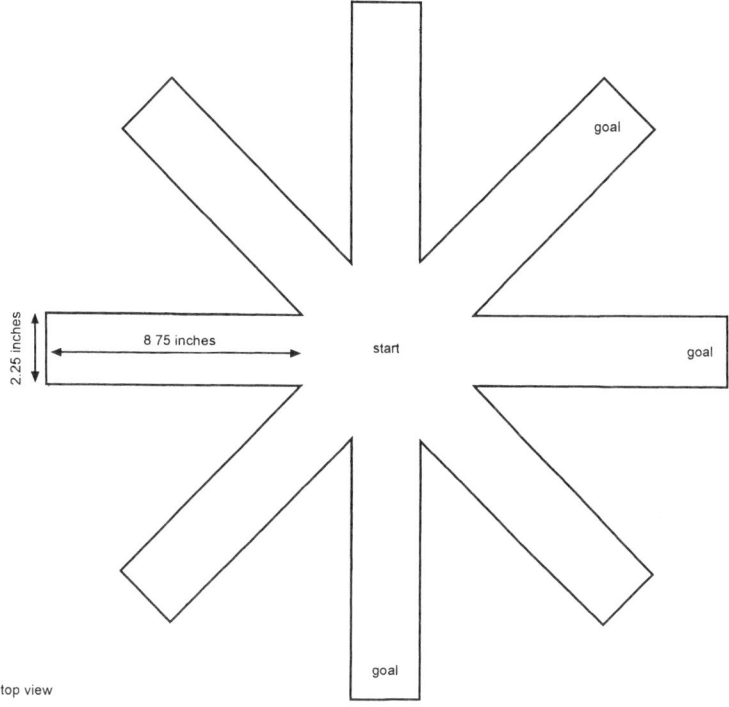

Figure 6.4: Olton 8-arm radial maze.

repeated visits to an arm that was initially baited, failure to retrieve bait from a baited arm, or visits to an arm that is never baited. Researchers such as Olton successfully used this task to study phenomena such as working and reference memory processes that are involved in running the maze (Olton, Becker & Handelman, 1979). Radial arm mazes are relatively easy to construct from either painted wood, metal or Plexiglas. See Figure 6.4 for an example of an 8-arm radial arm maze. They are also commercially available (e.g., Columbus Instruments Model 0500-1). Behavior can be monitored via videotaping or through photobeam technology that detects when the animal has entered an arm of the maze.

The Morris water maze, first described by Richard Morris (e.g., Eichenbaum, Stewart, & Morris, 1990) has become the most popular maze for studying spatial learning and memory (see Figure 6.5). In a typical experiment a large pool is filled with an opaque liquid that makes it impossible to see objects in the pool. Platforms on which rats can climb are placed in one or more locations around the maze. The task is quite simple: The rats are placed in a starting position and allowed to swim in the water until they find the platform, which allows them to escape from the water. Since the appear-

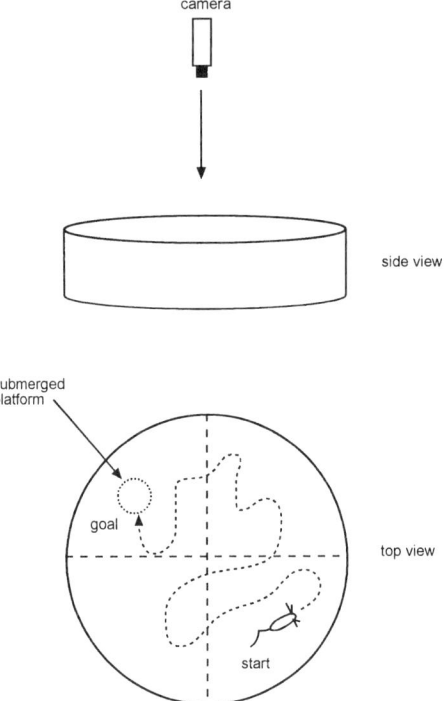

Figure 6.5: Morris water maze.

ance of the pool is relatively uniform, extramaze cues are mostly used to locate the platform (i.e., the rats form a spatial memory map based on the cues that surround the pool). To test performance variables related to the task, many experiments used a cued version of the task in which a flag or some other conspicuous cue is used to mark where the platform is located. The subjects simply swim to the cues platform to show that they have no performance deficits that may affect learning. Typically, time spent finding the platform is the dependent measure of water maze learning and memory. Videotaping is typically done to archive the movements of the animals. Also, commercially manufactured video tracking systems (e.g., Poly-track, San Diego Instruments) have proven useful for quantifying water maze performance.

Throughout the years, mazes have been used extensively to explore the neural bases of learning and memory. For example, much of Lashley's work (and his incorrect conclusion that learning and memory could not be localized) involved mazes. The radial arm maze has been used to study limbic system and frontal lobe function and the Morris water maze has been used extensively recently to study spatial learning capability in genetic mouse models. The maze has proven quite useful when coupled with brain lesion techniques and neuropharmacological methods. Given the whole body movements required for maze running, the use of the maze and other spatial tasks in neural recording studies has been somewhat limited, however. The major exception may be the elegant work by Bruce McNaughton, Phillip Best and others on the existence of place neurons in the limbic system that appear to code spatial positioning, navigation and memory (e.g., Jung, Wiener & McNaughton, 1994).

MONITORING GENERAL ACTIVITY

Perhaps one of the simplest behavioral recording situations is monitoring general activity, either in an animal's home cage or in a larger, open-field arena. General activity has been used as a behavioral index in a number of contexts. For example, behavioral pharmacologists have frequently used measures of general activity to assess general effects of drugs on locomotion and motor behaviors.

Sometimes it is desirable to obtain general activity measures of an animal over an extended period of time (e.g., for the study of circadian rhythms or to assess the long-term effects of drugs on general activity). For these experiments, the animal's home cage can be placed on a stabilimeter platform and general activity measured as movement of the home cage on the platform. Alternatively, photobeam detectors can be positioned at regular intervals around the home cage and the activity of the animal measured as interruptions of the photobeam. These data can be fed into a computer interface and the precise timing of activity can be determined over rather extended periods of time.

Some researches have used larger arenas to measure general activity. These arenas can be quite large (e.g., 1 m x 1 m) thus providing a large area for the animal to explore. A simple way to measure general activity in the arena is to subdivide the area into smaller squares by drawing intersecting lines on the floor. General activity is measured as grid crossings (i.e., the number of times the experimental subject crosses one of the grid lines). Of course, videotaping should be done to archive the activity of the animal for further analysis. Also, as with all behavioral observation experiments, the behavior under scrutiny must be operationalized. In this case, the experimenter must define ahead of time what constitutes a grid crossing (e.g., head over the line, front legs into the next quadrant, whole body across the line, etc.). The activity measurements can be automated by replacing the grid lines with photobeams that can detect when the "line" is broken.

There are a number of commercially available activity monitoring systems available. These systems tend to be relatively expensive, but for investigators with adequate funds they provide an excellent solution for activity monitoring. A system available from Columbus Instruments uses an array of infrared beams and detectors with software and a microcomputer to monitor activity in a 43 x 43 cm arena (the Opto-Varimex-3 system). Photobeam systems, such as this one, work by projecting an infrared light source onto a complementary detector. Movement is detected when the animal breaks the beam, thus interrupting input to the detector. San Diego Instruments also manufactures a relatively flexible photobeam activity monitoring system known as the Flex-Field Animal Activity System. Stoelting/Basile builds two different systems, one that uses infrared beam break technology (Model 57420) and one that uses a grid floor monitoring system (Model 57430) that can be used to measure general activity. And, an open field activity monitoring hardware and software system is available from Med Associates (Models ENV-515 and SOF-810).

Finally, it is also possible to fine tune the general activity measure by scoring individual movements that are seen in the rat. Rebec and colleagues, for example, have frequently used an observer scoring method to note individual behaviors such as grooming, chewing, licking, rearing, and head bobbing (Rebec, 1998). From videotapes (that often have unit activity recorded on the tape through the audio channel for correlating brain-behavior events) they rate a variety of behaviors at regular intervals according to duration (on a scale of 1–3) and intensity (on a scale of 0–3). This method has proven useful for correlating behavioral events with neural activity and, further, to assess the effects of drugs on the brain-behavior relationships they have established.

CONCLUDING REMARKS

A whole book or more could be dedicated to describing methods and behavioral data analysis techniques associated with fear conditioning, instrumental procedures, spatial paradigms, and general activity monitoring. While we did not intend to comprehensively review these behavioral procedures here, we hope that we have at least been able to provide some illustrations of how researchers have studied behaviors other than classical conditioning of somatic responses, especially in neurobiological studies. In writing this chapter we were struck with the commonalities in methods that actually exist and how some devices have proven very useful for a number of research applications. For example, videotaping is still a very valuable tool for recording behavioral events, and we did not even touch on the newer digital technologies that have made it possible to more automatically analyze the content of these tapes. In general, experimenters in classical conditioning are remiss when they put their animals in the chambers and do not check to observe the behavior. Also, the Skinner box is still used extensively by researchers in a number of ways, providing a rather standard experimental environment in which a variety of studies can be conducted. Motion detection devices seem to have universal applicability. In reviewing the very large literature concerning behavioral experiments we were struck by something else: The investigators who have studied behavior either alone or together with neurobiology have consistently come up with very ingenious solutions for measuring and quantifying behavior. We expect that this trend will continue for years to come.

Chapter 7

SURGICAL METHODS AND TECHNIQUES

OVERVIEW

Much of the progress in the field of behavioral neuroscience has depended on the use of animal models that involve neurobiological methods that complement the behavioral analyses that are conducted. Many of these brain-behavior experiments have involved procedures such as brain lesions, brain stimulation and neural recording that require survival surgery. A great deal of attention should be paid to this part of the experimental process because the quality of the neural and behavioral data collected after the animal has recovered from the surgery is greatly dependent on the general post-surgical health and condition of the animal. Great care must be taken to prevent infections that may be caused by the surgery, to prevent unnecessary damage to the biological system under study (which in our case is usually the brain), and to prevent undo stress before, during and after the surgery. In this chapter, we review basic procedures associated with surgery in animal subjects.

We concentrate for the most part on techniques associated with brain surgery, but much of what we cover is fully applicable to surgery involving other areas of the body as well. The precise surgery protocol will vary according to the requirements of the experiment as well as the species under study. The following is an outline for aseptic surgery we use for our rabbit and rat experiments; the protocols are, for the most part, the same for both species. This outline is the official Standard Operating Procedure (SOP) for our experiments. We require that everyone who performs experiments must be familiar with these procedures and we keep copies of the SOP with the animal log books and surgery records. We believe that it is a good idea for all investigators to create their own SOPs for surgical methods and techniques. It is our experience that standardizing the protocols for surgery provides good training experiences for students and technicians and, more importantly, increases the prospects for success while minimizing the stress to the animal.

We both have had success in training new personnel by using a general strategy of successive approximations (after all, the behavioral condi-

Surgery Record

Rabbit # _____
Study _____
Surgeon _____
Date _____
Protocol (indicate one): ___ Standard Operating Procedure (SOP), OR ___ Protocol # _____

Presurgical Evaluation:
Note Abnormalities only
___ Diarrhea ___ Dehydration ___ Ear mites ___ Other (identify _____)

Anesthesia:
Body weight: _____ kg

Pre-operative anesthesia:
___ cc Ketamine ___ cc Xylazine, OR ___ L/min Halothane, OR ___ Other (specify _____)

General anesthesia:
___ cc Ketamine ___ cc Xylazine, OR ___ L/min Halothane, OR ___ Other (specify _____)

Sugery:
Time of beginning of surgery _____
Note monitoring and complications during surgery and initial _____

Time of end of surgery _____

Recovery
Note time, observation, and initial

Time fully awake, recovered and returned to cage _____

Postoperative Recovery (date, time, observation[s], initials)
Daily observations until fully recovered; indicate animal and headstage (if applicable) condition;
indicate analgesic type (e.g., Buprenex), dose and time/date:

Fully recoverd on _____ (date and time)

Sacrifice
Date of Euthanasia _____

Figure 7.1: Example of surgical record.

tioning literature has consistently shown us that nonhuman animals are not the only ones that learn effectively by shaping strategies). Operationally, this means that the new person first watches surgery, then assists in surgery, then is allowed to perform a minor aspect of an ongoing surgery, then gradually is given more and more responsibility, until the entire surgery is performed. The hardest part for the supervisor is waiting nearby, but out of sight, when the first solo surgery is performed. The emphases in training are competence and training to handle emergency situations. In reality, the techniques associated with brain surgery are not usually difficult and can be learned by any trained technician or undergraduate.

Figure 7.1 shows a standard Surgery Record sheet that we use to

Chapter 7

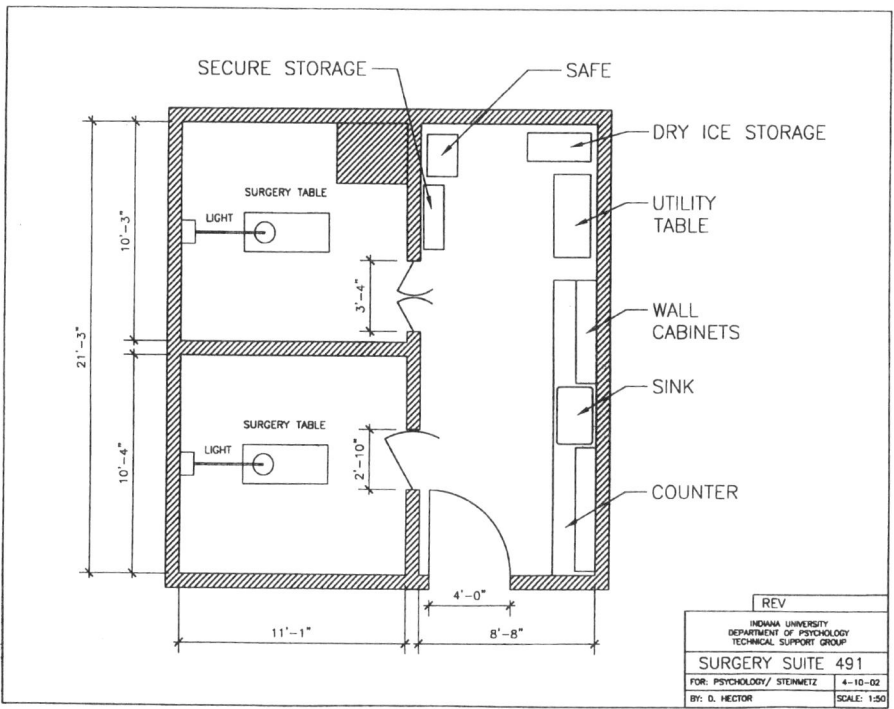

Figure 7.2: Layout for a surgical suite.

document the major concerns of survival surgery. The Surgery Record contains complete information about the animal including the unique number assigned to the animal (which is very common in laboratories), the title of the study in which the animal will be used, the name of the surgeon, and the date of the surgery. A thorough presurgical evaluation should be done and the results noted on the form. If any health problems are detected, these might be cause to delay or cancel the surgery. The weight of the animal should be recorded and details concerning anesthesia noted (type, amount given, route, etc.). The beginning and ending time of the surgery should be written down along with notes concerning any complications that arose during the procedure. The length of the recovery time should also be noted and postoperative recovery data entered (e.g., the frequency and amount of analgesics that were administered, any long term effects that have become apparent, such as motor or respiratory difficulties). Finally, if the animal is euthanized (which is common in behavioral neuroscience studies), the euthanasia date may be recorded on the Surgery Record sheet as well. The Surgery Record, as with any records like those used for monitoring drug disposition, should be made in ink and initialized at each recording. The Surgery Record is either kept in a separate log specifically kept for those records, or with the behavioral training records and data collected for each animal, depending upon the style of the experimenter.

As of the date of publication of this book, aseptic surgery is not required for all animals. While federal regulations require that aseptic surgical practices be followed for primates, dogs, cats and rabbits, many other species, such as rats, mice and pigeons do not fall under this regulation. We advise, however, that aseptic procedures be followed for all animals as the risks of infection and stress are not restricted only to the animals covered under the federal regulation. We recommend that whenever possible, a dedicated surgery suite be used for surgeries involving vertebrates. It is easy to maintain a high-quality area for surgery when a room is dedicated for this purpose. The surgery suite should contain a preparation room and a surgery room. The preparation room should have plenty of counter and storage space and also contain a scrub sink with pedals (operated by the foot, knee or elbow) that provides a means to scrub and prepare for surgery. The surgical room should have walls, floors and ceilings that are completely washable and contain no wood or cracks on surfaces where germs and bacteria can take residence. Figure 7.2 shows a layout for a functional, multi-user surgery suite. Having a good surgical facility will increase the chances for survival, which is the major goal in designing surgical procedures and practices.

GENERAL PREPARATION FOR SURGERY

An important part of the surgery actually occurs before the subject is even brought to the surgical room—it is vital that the surgery suite be cleaned and prepared before the subject is anesthetized. Literally everything that is in the surgery room should be cleaned and disinfected before the surgery takes place. Because of this requirement, it is best to minimize the number of items in the room to only those absolutely needed for surgery. After cleaning the room with an Alconox solution, we use 1 ml of Roccal disinfectant in 200 ml of tap water to further clean and disinfect the surgery area including the surgery table, walls, floors, lamps, stereotaxic equipment, and any other equipment with surfaces that can be wiped down.

It is also important that all necessary equipment and supplies are on hand before you start surgery. Table 7.1 shows a list of items we typically use in surgery (although not always every item in every surgery). In addition, we often use recording, lesion and stimulation equipment during the implantation of electrodes. These pieces of equipment may be located at a distance from the surgical table in a rack or on a cart that can be moved in and out of surgery. This equipment may include a brain recording amplifier, audio monitor, stimulator, oscilloscope, picospritzer and nitrogen gas, or a

Table 7.1: Items commonly found in the surgery room.

> Surgery Table
> Surgical Lamp
> Instrument Table
> Mayo Stand
> Equipment Rack
> Stereotaxic System
> Head holder
> Drill
> Waste Basket
> Biological waste disposal
> Plastic tub with Benz-all for cold sterilization
> Fleaker® with Benz-all with sponge forceps
> Surgical area scrub items
> > Gauze in alcohol
> > Gauze in betadine scrub or nNolvasan scrub
> > Gauze in betadine or Nolvasan
> > Sterile saline

lesion maker. Leads from the equipment are usually routed to the surgery area. The leads should be cleaned and disinfected and fronts of the equipment wiped down with disinfectant before the equipment is brought into the surgery room.

SURGICAL INSTRUMENTS

Surgery requires a number of specialized tools and instruments that must be prepared and on hand before surgery is started. Good quality surgical instruments can last for several years if treated well and cared for properly. Again, as is the case for preparing the room for surgery, making sure that all sterilized instruments that are needed for surgery are placed in the surgery room prior to preparing the animal for surgery is extremely important for assuring success of the surgery. Here is a list of instruments that we commonly use in brain surgery procedures:

Steam sterilized (Follow directions on autoclaves carefully for best results):

> microcurette
> beaker
> 2 x 2 gauze
> 4 x 4 gauze
> cotton tipped applicators

animal drape and/or towel
instrument drape (use drape from instruments)
surgery gown with towel
note paper (use inside paper from gloves)
instrument pack
    3 towel clamps (leave opened, do not lock)
    1 sharp-blunt scissors for cutting drape
    1 #3 scalpel handle
    4 hemostats (leave opened, do not lock)
    1 needle holder
    1 rat toothed forceps
    1 serrated forceps
    1 Adson forceps

Prepackaged sterilized items:

    latex gloves
    #10 scalpel blade
    bone wax
    gelfoam
    4-0 and 6-0 suture with needle
    tube of Betadine or Clinidine

Cold sterilization (1 bottle Benz-all to 4 L deionized H_2O, 15 min):

    electrodes, wires and skull screws (stainless steel)
    pencil and ruler
    drill bit
    mixing tray (e.g., Teflon sheet, Dixie cup)
    screwdriver
    wire cutter made from stainless steel
    wrench for drill
    Allen wrenches for drill and stereotaxic equipment
    bone wax
    4-0 and 6-0 suture with needle

Clean (but not sterilized):

    crib liner for animal to rest on
    surgery hat/cap, booties and mask (personal reuse)
    dental acrylic set (powder, catalyst/solvent)
    stereotaxic instrument

bars
carrier(s)
electrode holders
head leveler

The sterilization procedure may cause some difficulties for subsequent surgery. First, steam sterilization can be hard on equipment and also is somewhat time consuming. Second, while 100% stainless steel instruments can withstand frequent and repetitive sterilization, other non-stainless steel instruments (like screwdrivers, pencils, rulers, etc) do not withstand steam sterilization and also are easily ruined using cold (liquid) sterilization methods. Glass bead (heat) sterilization is an alternative: Sterilization can be achieved quite rapidly and may cause less damage than cold sterilization methods.

PREPARING THE ANIMAL FOR SURGERY

Anesthetics

The choice of anesthetics is extremely important as different species show different tolerance levels for the various anesthesias that are available for use. As with the room and the instruments, it is important to make sure that the anesthetics and delivery system are available and working properly before the animal is brought to the surgery area. This includes making sure an adequate amount of oxygen and a fully functional anesthesia delivery machine is available if gas anesthesia is being used, and that adequate numbers and sizes of syringes and needles are on hand if an injectable anesthesia is being used.

For rabbits, we have found either Halothane, Methoxyflurane, Isoflurane, or a mixture of Ketamine and Xylazine are effective anesthesia for most procedures. If Halothane alone is used it is delivered with O_2 at a rate of 0.8 - 1.2 L/min. Initial anesthesia is induced with a 5% Halothane/O_2 mixture while 1-2% Halothane/O_2 mixture is used to maintain anesthesia. For Methoxyflurane, 0.4-1.0% is given for anesthesia while 1.0-3.0% of Isoflurane is used. Halothane, Methoxyflurane and Isoflurane can also be used after anesthesia is initiated by an injection of a Ketamine/Xylazine mixture. Initially, the rabbit is given an im injection of 5-8 mg/kg Xylazine. After 10-15 min, 50-60 mg/kg is Ketamine is injected sc. Anesthesia can be maintained by Halothane, Methoxyflurane or Isoflurane using the dosages listed just above.

A mixture of Ketamine and Xylazine can be used as an anesthesia in rabbits. Ketamine Hydrochloride is a dissociative anesthetic while Xylazine is an alpha-2 adrenergic agonist. Xylazine is typically administered with Ketamine because it is a very good muscle relaxant while Ketamine is not. Some care should be exercised in using this combination, however, as the Ketamine/Xylazine combination produces decreases in cardiopulmonary function and hypothermia. Initially, the rabbit is given an im injection of 5-8 mg/kg Xylazine. After 10-15 min, 50-60 mg/kg is Ketamine is injected subcutaneously (sc). For maintenance of anesthesia, we inject a cocktail of 20 mg/ml Xylazine and 100 mg/ml Ketamine, administering 1 ml of the cocktail im every 20-70 minutes as needed (often 40 minutes).

For rats, we have typically first given an injection of Atropine (0.05 mg/kg either IP or SC) to counter decreased heart rate and increased salivation caused by anesthesia. Most often, we use a mixture of Ketamine (75-100 mg/kg, IP) and Xylazine (10 mg/kg, IP) as an anesthetic. This anesthesia lasts for about 30–40 min before another injection is necessary. For light anesthesia, a combination of Ketamine (75 mg/kg, IP) and Acepromazine (2.5 mg/kg, IP) can be used. The Acepromazine (2.5 mg/kg IM or IP) is a very effective sedative for minor procedures.

For some procedures, we have used local anesthesias (such as in putting recording EMG electrodes in the musculature surrounding the eye). Injectable Lidocaine and Bupivicaine can be placed directly into or around the site of the injury and only a very small volume is needed. A Lignocaine-prilocaine cream (e.g., EMLA cream, Astra Pharmaceutical Products, Inc., Westborough, MA) can be used for skin desensitization. Normally, the cream is applied to the skin and then covered with a bandage for about an hour before the desensitization occurs.

Preparing the Surgical Site

After a deep level of anesthesia has been induced (as verified by toe and ear pinches), the animal is prepared further for surgery. First, the eyes are lubricated with a sterile ophthalmic ointment. Next the head is shaved closely but without abrading the skin. We use an Oster electric shaver with a special blade made for animal fur for shaving the surgical area. Next, the head is thoroughly cleaned by a person wearing scrubs or a clean lab coat, nonsterile examination gloves, a head covering and a mask. To do this, the head should be wiped with an outward spiral motion, starting from center and working out using (1) alcohol (70% ETOH or 100% isopropyl), (2) Betadine scrub (which is Betadine solution that contains soap), (3) alcohol again, (4) Betadine scrub again, (5) alcohol again, and finally (6) pure Betadine (not the scrub solution). The animal is then brought into the surgery

Chapter 7

room. Normally, the head is mounted in a stereotaxic head holder, being careful not to touch the scrubbed area on the head. A final cleaning of the head is done using sponge forceps, sterile gauze, and solutions kept in the surgery room. The order of final cleaning is (1) alcohol, (2) Betadine scrub, (3) alcohol, and (4) pure Betadine.

THE SURGERY

Surgeon's Preparation

During surgery, the surgeon is primarily responsible for 1) good surgical technique, 2) knowledge of the anatomy and physiology of the organ being operated on, and 3) knowledge of the experiment. An assistant, if present, is primarily responsible for 1) assisting the surgeon, 2) monitoring sterility, 3) monitoring anesthesia level, and 4) monitoring the condition of the animal. Maintaining aseptic conditions in the surgery room requires that the surgical field be kept sterile and this requires a great deal of effort on behalf of the surgeon and the assistant. Clothing must be clean: The surgeon wears surgical scrubs, a head covering, foot coverings (booties), and a mask while the assistant wears scrubs or a lab coat, a hat, and a mask. After the surgeon puts on the hat, mask and booties, he or she scrubs for 2 minutes each on fingers, hands and forearms with an aseptic soap using plenty of lather. The surgeon keeps his or her hands above the level of the elbows after scrubbing. The assistant opens the outer wrapping that contains a sterilized surgery gown and a drying towel. The surgeon dries his or her hands on the enclosed towel—one corner of towel per hand and per arm, always wiping towards the elbow. The surgeon then unwraps the gown and puts it on without touching the outer surfaces. Sometimes the assistant can be very helpful in pulling on sleeves, being careful not to touch the outside of the gown (which will be in the surgical field). The assistant next opens the outer wrapping to sterilized gloves and the surgeon puts on the gloves without touching the outer surfaces. From this point on, the surgeon must be careful to touch only sterile fields. Any nonsterile items that must be used are handled only through sterile gauze pads. In preparation for surgery, the surgeon places wrappings on the instrument tray, Mayo stand, etc. The assistant opens the outer wrapping for instruments, gauze, beaker, blade, and any other wrapped instruments. The assistant hands over cold sterilized instruments by using sponge forceps and the surgeon places instruments on the tray. The surgeon places a drape over the animal and secures it with towel clamps. A hole is cut into the drape over the surgical site for an incision.

Basic Surgical Procedures

Throughout the surgery, the assistant opens instruments or supplies needed by the surgeon. The assistant also monitors the level of anesthesia and the respiration and heart rate of the animal. Special attention is given to the respiration rate when the skin is incised, when the periosteum is reflected, when bone is cut, when electrodes are lowered and when sutures are placed; big changes in respiration can signal that the animal has become light and more anesthesia should be delivered. Injections or adjustments to the flow rate of gas anesthesia are done by the assistant. While difficult, one can perform aseptic surgery without an assistant, although this requires good planning and technique. Whenever possible, though, an assistant should be used. We have found that undergraduate students make good assistants and that they can learn a great deal about anatomy and physiology by assisting during a brain implant or a brain lesion procedure.

The surgeon is ultimately responsible for creating and maintaining all surgery records including the drug log (especially important when controlled substances are being used), the animal log (indicating the number assigned to the animal and the procedure performed), the surgery log (detailing the events of surgery), and the postoperative care log (indicating the condition of the animal and the use of analgesics).

Consecutive Surgeries

Sometimes, a surgery involving a second or perhaps even a third animal may follow directly a surgery that has been completed. In this case, all instruments should be cleaned in sterile saline and placed in a cold sterilization solution for several minutes. The surgery table and stereotaxic equipment are wiped down with a disinfectant (e.g., Roccal) solution. New gowns, masks, gloves, etc., should be used for the second surgery.

Post-surgical Cleanup

A thorough cleanup of the surgical suite not only maintains peace and order in the laboratory, but also greatly extends the life of the equipment and instruments that are being used. The gas delivery machine and oxygen tank should be turned off. Throw away all used sharp supplies (e.g., scalpel blade, electrodes) in a sharps container. Empty trash or move waste basket outside of surgery room (janitors typically do not enter the surgery room).

Biological wastes should be disposed of properly. Wash all instruments in Alconox or an equivalent laboratory soap. Lubricate all instruments with joints (e.g., hemostats and scissors) in instrument milk (do not rinse). Allow the instruments to air dry since wiping them with a towel can dull sharp edges. Put unused gauze and cotton-tipped applicators in an agreed upon place so that they can be resterilized. Place the surgical gown on a coat rack for washing and resterilization. Put all other supplies and instruments away. Do NOT leave bone wax in Benz-all for an extended period of time as the bone was will be destroyed. Be sure to remove and clean drill bit. Wipe down the counter, instrument table, Mayo stand, surgery table, floor—use a disinfectant (e.g., Roccal) and clean from the highest surfaces first to the lowest surface last. Clean the face mask assembly if it was used to deliver gas anesthesia; leave the rubber mask in an unstretched condition. Clean preparation area, especially getting rid of any fur that may have accumulated (use Roccal or other disinfectant). Finally, clean and lubricate the shaver.

Postsurgical Monitoring

As we mentioned above, it is very important that the animal's health be monitored for several days after the surgery. Remain with the animal until it is out of the anesthesia and ready to be returned to its home cage. The animal should be kept warm using a heating pad or wrapped in a covering. During recovery, look for signs of consciousness, note the ease of recovery, note any signs of distress, and log any changes in the animal's condition. After the animal is initially returned to its home cage, it should be observed at least once per day and its condition noted in a postoperative recovery log. We look for the normal condition ('ok') or improvements or deteriorations in condition. Give postoperative analgesics according to animal protocol; for example, we give Buprenex 0.02 mg/kg twice a day for two days for rabbits. Other analgesics that can be used for rabbits include aspirin (100 mg/kg, PO in solution) and Butorphanol (5 mg/kg, IM) with Xylazine (4 mg/kg, IM). Butorphanol (2 mg/kg, SC), Nalbuphine (1-2 mg/kg, IM) and Meperidine (10-20 mg/kg, IM or SC) can be used as analgesics for rats. End the postoperative log only when the animal has recovered satisfactorily (is "fully recovered"). Animals with dental acrylic headstages are checked daily for healing or signs of infections. Infections are cleaned with cotton-tipped applicators using saline and/or hydrogen peroxide and dressed with an antibiotic. Experimental observations are put in the notes for the study.

Figure 7.3: Biela rabbit stereotaxic headholder, consisting of a single nose/mouth piece through which Halothane gas anesthesia can be delivered, mounted on Kopf stereotax frame.

STEREOTAXIC SURGERY TECHNIQUES

A key contributing factor for the great advancements seen in the field of behavioral neuroscience over the years has been the use of stereotaxic surgery techniques to localize brain areas. Using these methods, it is possible to more or less accurately place a tip of an electrode or end of a cannula in a specific location in the brain with a precision in the range of tenths of millimeters. Until imaging techniques become readily available for animal surgery, the stereotaxic method is the most accurate technique.

Figures 7.3 and 7.4 show typical stereotaxic apparatus that are set up for rabbit and rat surgeries, respectively. The basic stereotaxic appratus consists of a holding frame that secures the animal's head and also a set of calibrated bars on which micromanipulators can be placed. For rat surgeries, the head is held in place with ear bars that are placed in the ear canal and by clamping the snout of the animal in a standard bite-bar (Figure 7.4). Given the very acute angles of the ear canals of the rabbit, ear bars cannot be easily used to secure the rabbit head. Thus, two different head holding devices have been created for rabbits. Several labs that we know use a specialized head holder that was designed and built by Josef Biela (University of Cali-

Chapter 7

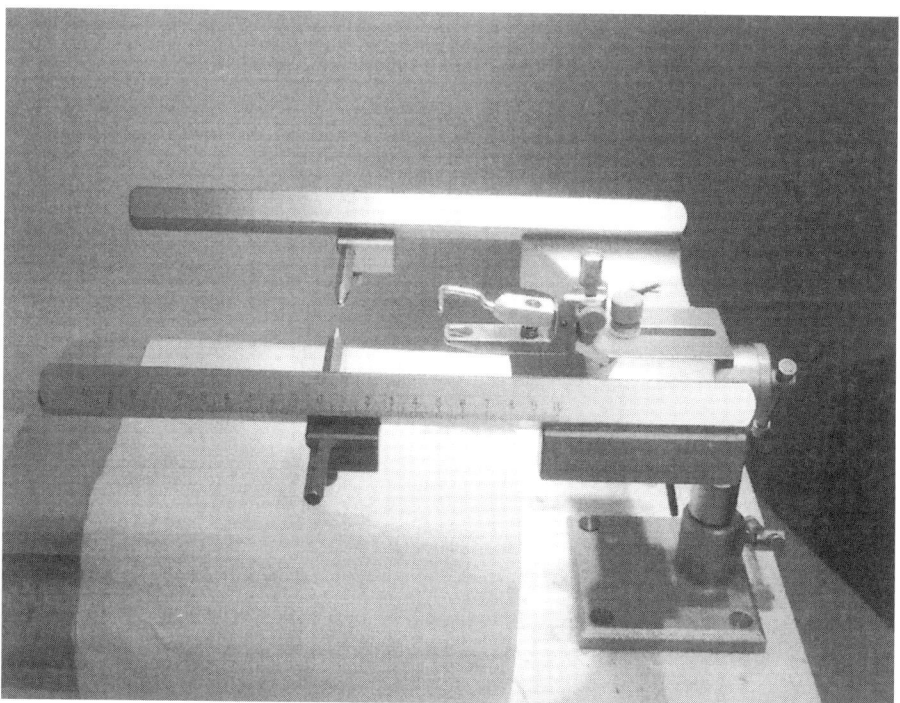

Figure 7.4: Kopf rat stereotaxic headholder, consisting of ear bars and nose/mouth piece, mounted on Knopf stereotax frame.

fornia, Irvine). This device has a metal piece that fits into the mouth of the rabbit between the top of the tongue and the roof of the mouth. The head is secured by a metal clamp that tightens over the snout. The head can be moved in three dimensions by adjusting knobs located in the x-, y- and z-planes. An alternative device is made by David Kopf Instruments and fits onto their standard stereotaxic electrode carrier frames. It is essentially a U-shaped clamp that secures the head by clamping at the left and right zygomatic bones (the temples). The head can be adjusted in the three planes by turning appropriate controls.

Microelectrodes or cannulas are secured in the micromanipulator and the micromanipulator is used to move the electrode or cannula into position. Because stereotaxic surgery involves localization in three planes, the micromanipulator must allow for movements of the electrode or cannula in three planes. Movement in the anterior-posterior plane is accomplished by sliding the micromanipulator along the calibrated bars that are part of the stereotaxic apparatus. Movements in the medial-lateral and dorsal-ventral planes are accomplished by turning knobs on the micromanipulator (see Figure 7.3). The scales on the micromanipulator and stereotaxic apparatus are Vernier. As shown in Figure 7.5, the top side of the Vernier scale is marked

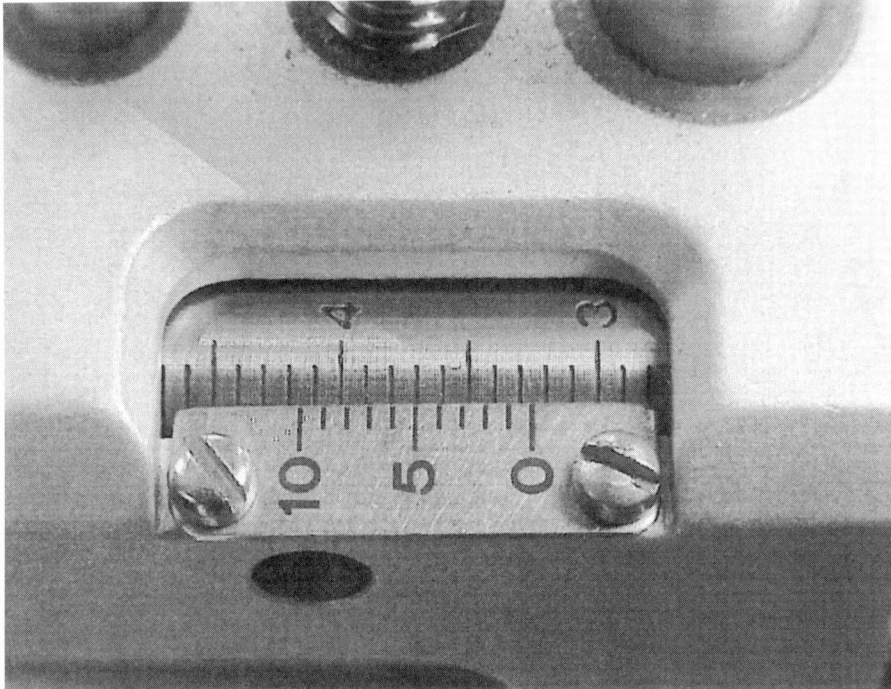

Figure 7.5: Reading the Vernier scale on the stereotax allows measurements to the tenth of a millimeter. The value here is 32.6 mm. Notice that the line for 0.6 lines up with a line on the centimeter scale, and that the lines for 0.5 and 0.7 are slightly offset and within lines on the centimeter scale.

in centimeters (3 and 4 centimeters are showing). This top scale is further divided into whole millimeters. Normally we read the scale in terms of millimeters (i.e., 30 and 40 millimeters shown). The bottom side of the scale shows tenths of a millimeter. A reading is made by first noting where the 0 (zero) on the bottom (tenths) scale lies. In Figure 7.5 the 0 (zero) lies between 32 and 33 millimeters. The value will be '32 point something.' Tenths of millimeters is read on the bottom scale by noting where any two lines best match between two top (whole millimeter) and bottom (tenths) scales. Which lines match up varies from reading to reading. In Figure 7.5, the 6 on the bottom scale lines up with a line on the top scale, yielding a reading of 6/10 of a millimeter. Thus, a reading of 32.6 would be recorded for this setting. We encourage novice surgeons to practice accurately reading the scale before attempting surgery.

The stereotaxic method involves finding a location in the brain in three dimensions by referencing an external landmark. In rabbits and rats the landmark may be a location on the skull, such as the bregma or lambda points formed by the intersection of the suture lines on the skull. For rats, the interaural line, which is determined by the location of the ear bars that

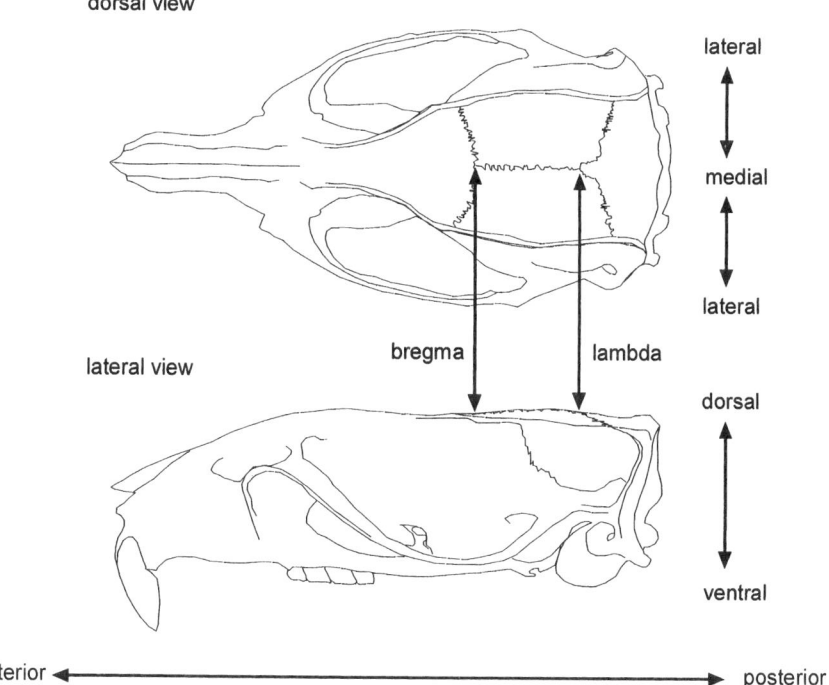

Figure 7.6: Rat skull showing the positions of bregma and lambda landmarks. Bregma is the intersection of the coronal and sagittal (longitudinal) sutures. Alternatively, bregma can be defined as the point where the sutures *would* meet if they were straight. Lambda is the intersection of the sagittal (longitudinal) and transverse sutures. Alternatively, lambda can be defined as the highest point posterior to the apex where the transverse suture meets the sagittal (longitudinal) suture. The fact that the sutures 'wander' makes the definitions of both bregma and lambda subject to one's best guess. As short-hand notations, we usually use β (beta) for bregma and λ (lamdba) for itself. Figure based on Paxinos and Watson (1982).

anchor the head in position, is often used as the reference point for finding brain regions. The basic stereotaxic technique is conceptually quite easy to grasp. The animal's head is first secured in a standard position, the coordinates for the brain area of interest are determined by consulting a standard brain atlas, and the location of the electrode tip is set by moving the electrode to the atlas site in the brain with reference to the landmark being used.

The final coordinates for locating the electrode tip are determined by consulting a standard brain atlas. Some details concerning brain atlases that are available are found in Chapter 10. The key to using brain atlases to determine brain locations is that the head is assumed to be in a standard, replicable position. That is, the brain sections used in constructing the atlas were taken from an animal that had its head positioned in a standard plane. Thus, during surgery, the head must be moved into the same standard plane that was used to construct the atlas. For the rabbit, the two standard head/brain positions that are commonly used are: (1) top of the head flat, with bregma and lambda in the same horizontal and medial/lateral planes (see bregma and lambda on the rat brain in Figure 7.6), or (2) bregma is positioned 1.5 mm above lambda in the horizontal plane, but in the same medial/lateral plane. We have typically used the latter standard head position for stereotaxic surgery.

An Example of Stereotaxic Surgery

We can use a concrete example to illustrate how stereotaxic surgery is performed. In our example, we wish to place the tip of a recording electrode into a location within the interpositus nucleus of the cerebellum in a rabbit. To begin this process, we first place the head of the animal in a standard head holding device. Preliminary phases of the brain surgery are first completed before the head is secured in the standard stereotaxic position. A small incision is made in the skin over the skull and the skin and periosteum is carefully retracted exposing the skull. Great care should be taken at this stage so that excessive bleeding from the muscle and tissue around the skull is avoided. The skull should be thoroughly cleaned and dried (using hydrogen peroxide helps) before continuing with the surgery. Next, the general area of the skull over the location in the brain where the recording electrode is to be lowered is marked and a hole drilled through the skull using a fine high-speed dental or Dremel drill. Care should be taken to drill through the skull without penetrating the dura layers. We typically make holes to accept stainless steel skull screws at two or three locations that are placed several millimeters from the implant site and then thread the screws into the holes. These screws are used as anchor points for the dental acrylic that will eventually be used to hold the electrode in place. The screws also provide a convenient place for securing a ground wire for subsequent recording.

The head is now ready to be placed into the standard stereotaxic position, where it must remain until the electrode is secured in place. For most surgeries, bregma is positioned 1.5 mm above lambda, the brain landmarks are lined up in the same medial-lateral plane, and the right and left sides of the skull are leveled. To get the head in this position, we first make sure that bregma and lambda are located in the same medial-lateral (ML) plane. For this adjustment, a large gauge needle or stainless steel cannula is placed in the micromanipulator and lowered onto bregma and its position is noted. The micromanipulator is then moved in the AP plane so the cannula can be lowered onto lambda. If the cannula touches lambda on the first try, bregma and lambda are positioned in the same ML plane. If not, the head holder is adjusted to align bregma and lambda, and the process continued until ML alignment is achieved. Next, the cannula is lowered onto bregma and the dorsal-ventral (DV) position of bregma is recorded. The cannula is next lowered onto lambda and its DV position noted. If bregma is not located 1.5 mm above lambda, the head position is then adjusted in the DV plane such that lambda is either raised or lowered. DV readings are taken at bregma and lambda until the 1.5 mm difference between bregma and lambda is achieved. Finally, we make sure that the left and right side of the skull are in balance. This is done by repeatedly lowering the cannula down onto points located 5 mm lateral from bregma on the right and left sides of the skull. When the same DV reading is obtained from both sides, the head is in the final stereotaxic plane for which the brain atlas was created and we are ready to proceed with the surgery. We note here that Kopf Instruments has made a rabbit alignment tool that greatly facilitates the head leveling operation. We have found this tool to be a very good investment.

We are now ready to position the recording electrode in the brain. The recording electrode should be mounted in the micromanipulator. Great care should be taken to not damage the tip. Indeed, depending on the tip profile, the surgeon may want to puncture the dura in the hole that has been drilled in the skull over the brain recording site using a dura hook or needle, as some electrode tips are so fragile they can break off when penetrating the dura layers. Also. It is important to make sure the electrode is very straight in the micromanipulator. In our example, we want to position the recording electrode in the left interpositus nucleus. The brain atlas that we use shows that our implant site is located 0.7 mm in front of lambda, 5.5 mm lateral to lambda, and 14.5 mm below lambda. To place the electrode in this position, we must first determine the location of the lambda landmark in three dimensions and move the electrode to the atlas location relative to lambda. We therefore lower the electrode carefully onto lambda, making sure not to damage the electrode tip, and then read the position of the electrode in the three planes. In our example, we record the following coordinates for the location of lambda: AP = 20.8, ML = 45.9, and DV = 71.3. Subtracting the atlas values from the lambda values, we thus determine that our electrode should be placed at: AP = 20.8 − 0.7 = 20.1, ML = 45.9 − 5.5 = 40.4, and DV = 71.3 − 14.5 = 56.8. Please note that the signs of these values are very important. For example, for the ML plane, depending on what side of the animal you are working, a left-side implant might require the subtraction of the atlas value from the landmark point while a right-side implant might require the addition of the atlas value. Also, the AP bar on the stereotaxic apparatus has a zero point so there are negative and positive values on the bar.

The next step involves lowering the electrode to its final location. We begin by raising the electrode off of lambda, making sure the electrode point is well clear of the skull. Next, we move the micromanipulator in the AP and ML planes using the coordinates we calculated. If we drilled the hole properly, the electrode should be centered precisely over the hole in the skull. We then slowly lower the electrode into the brain, stopping at the DV coordinate that we calculated. The electrode should now be located in the interpositus nucleus of the cerebellum in this rabbit.

Here are a few pointers for increasing the accuracy of the placement. First, take great care when leveling the head into stereotaxic position. It is a good idea to check the level from time to time during the procedure. Second, take great care in mounting the electrode in the manipulator; it should be as parallel to the manipulator as you can get it. Third, if the hole you drilled is "slightly off," re-drill it and then re-level the head. Do not simply move the electrode over slightly to accommodate the mis-drilled hole. Fourth, lower the electrode very slowly. The brain tissue is actually dimpled as the electrode is advanced through the brain and an accurate final electrode position can be found when the brain is given time to settle during and after the lowering of the electrode. Finally, do not depend solely on the landmark-based stereotaxic coordinates for determining electrode location. We typically do two things to increase the accuracy of our placements. First, we routinely record neural activity while lowering electrode. For every electrode track that is made, the activity of the brain provides a unique "signature" that can be used to find brain structures. For example, while lowering the electrode towards the interpositus nucleus you go in and out of several layers of Purkinje cells that are quite easy to identify, you encounter a thin band of white matter just dorsal to the nucleus where recordings are relatively quiet, and then you come upon a flurry of activity that is characteristic of a brain nuclei with relatively small, densely packed neurons (i.e., the interpositus nucleus). Second, we use other sets of stereotaxic coordinates when they are available. For example, we know that it is roughly 7.5 mm from the first cerebellar cortical cells we encounter after passing through the skull and dural spaces to the beginning of the interpositus nucleus. This calculation can be made and used to guide electrode placement.

Occasionally, it is necessary to use an angled approach when lowering electrodes into the brain. For example, we were interested in recording unit activity simultaneously from the interpositus nucleus and Larsell's lobule HVI of cerebellar cortex (Gould & Steinmetz, 1996). This was not easy to do because the cerebellar cortical area we wished to implant was located directly above the cerebellar nuclear site (i.e., in the same coronal and sagittal planes). To get around this problem we implanted the cortical electrode using an angled entry while implanting the interpositus nucleus electrode using the normal vertical approach. The trick in using an angled approach is to chose an implant angle that allows you to reach the implant site with the

Chapter 7

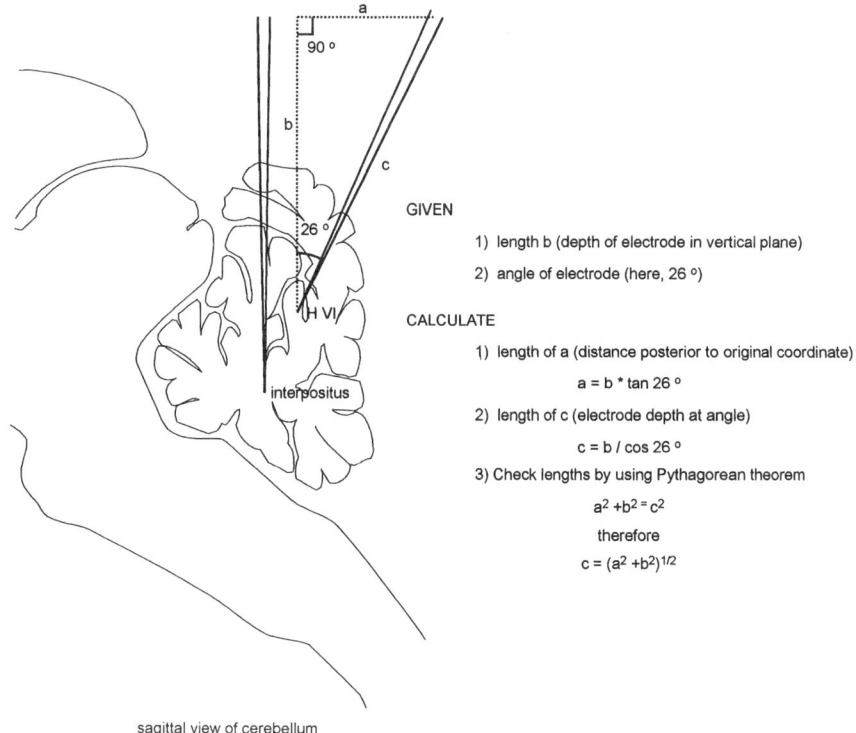

Figure 7.7: Angled electrode placement using calculated values to correct for anterior-posterior (AP) and dorsal-ventral (DV) offset.

electrode length that you have available while choosing an angle that allows you to make sure the two electrodes do not interfere with each other. For the Gould and Steinmetz (1996) study, the lobule HVI electrode was lowered at an angle of 26° relative to the coronal plane (see Figure 7.7). The depth the angled electrode is lowered is determined using the Pythagorean Theorem or a little trigonometry.

Alternatively, the corrections for anterior-posterior (AP) and dorsal-ventral (DV) for moving from the normal vertical placement to an angled approach can be empirically determined. The only read advantage of this method is that it avoids the trigonometry—often a trivial consideration since trigonometric functions are so readily available in simple calculators—but it does mean that the experimenter does not have to remember or have written down the calculations from Figure 7.7.

Lavond has used the empirical method for cold probe placements. For the following we will use the coordinates for the medial dentate nucleus of the cerebellum in the rabbit as an example. With the skull oriented so that

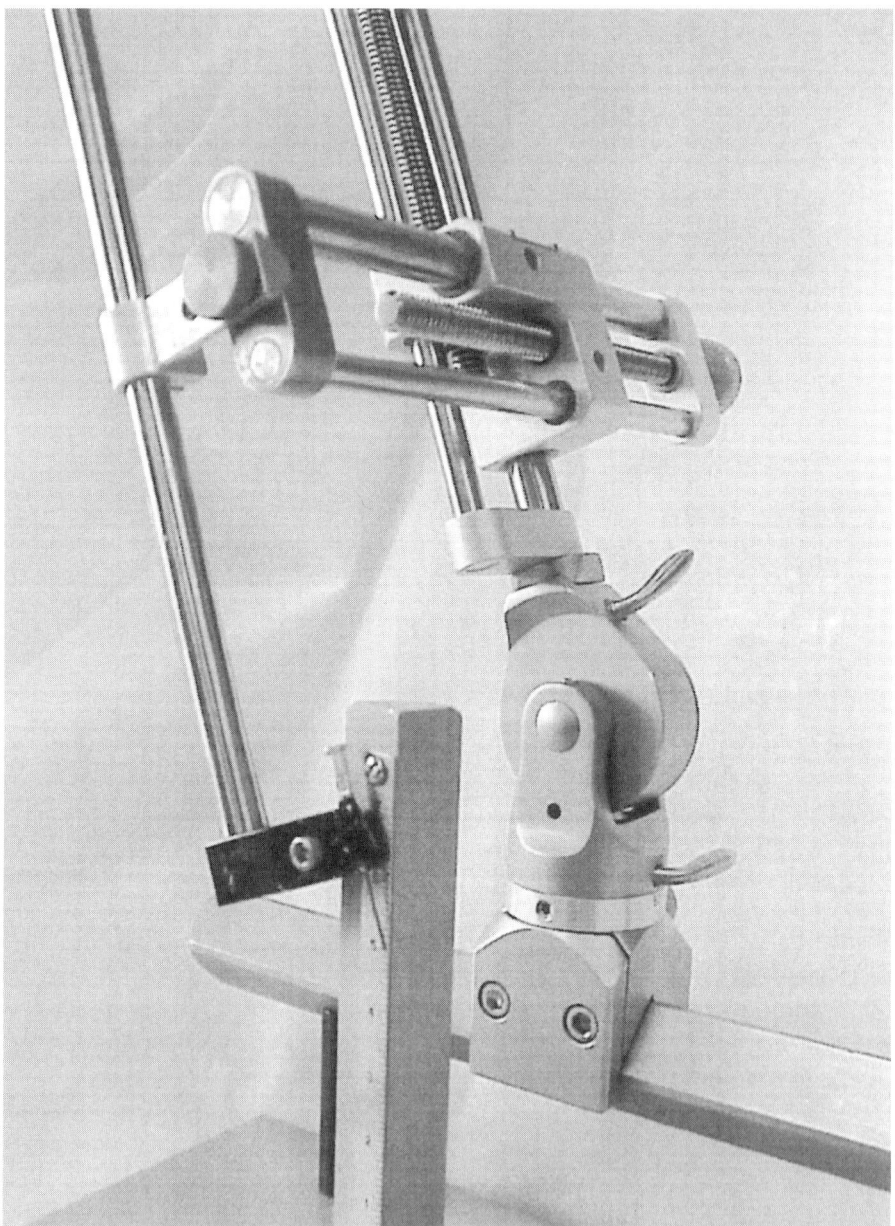

Figure 7.8: Angled electrode placement using empirically determined values to correct for anterior-posterior (AP) offset and dorsal-ventral (DV) length. The pointer (a blunted injection needle) is placed at two different positions: First, at some arbitrary point (here 50.0 mm), and second at a vertical position representing the intended vertical depth of the electrode. The AP and DV coordinates are measured at both positions. The differences between these two positions for AP and DV values represent the corrections needed for the angled placement from the original vertical placements.

lambda is 1.5 mm below bregma, the *target coordinates* are AP +0.5, ML 6.5, and DV –14.5 (anterior, lateral and ventral from lambda, respectively). The following are the steps for placing the electrode posterior to these target coordinates using a 20° (off vertical) angled approach. Figure 7.8 shows a Kopf stereotax with the electrode carrier placed at 20°. The 'flag' or 'wedge' of the electrode carrier temporarily holds a blunted injection needle in place of an electrode. The tip of the needle is placed on the vertical ruler at a position that we refer to as the 'target' (the medial dentate nucleus), in this example at 50 mm as shown in Figure 7.8—the exact location on the ruler does not matter within reason. Record the AP and DV coordinates from the stereotax. The DV target coordinate is 14.5 mm below lambda, therefore move the stereotax carrier to place the tip of the needle 14.5 mm above the initial position, to 64.5 mm on the ruler. (Note that moving up or down is somewhat arbitrary, as the difference is relative. The needle tip could be placed at 35.5, instead—14.5 mm below the original 50.0 mm—as long as the absolute difference and direction for correction are understood.) Read the new AP and DV coordinates from the stereotax.

The *difference* between the target AP and DV coordinates and the new AP and DV coordinates are the *corrections* from the vertical coordinates to the new angled coordinates. The *corrected AP* is determined by adding (or subtracting) the AP correction (a in Figure 7.7) to the target coordinates to move the electrode path posteriorly. In this example, the corrected AP is 5.3 mm behind the target AP coordinate. The *corrected DV* is simply the DV difference. In this example, the corrected DV depth is 15.4 mm from lambda.

As a check, since eyeballing the tip of the needle in the two positions is not precise, it is best to use Pythagorus' theorem to double check the measurements: To do that, know that the DV difference is the hypotenuse, the target depth (14.5 mm) is the adjacent side, and the AP difference is the opposite side (c, b and a respectively as illustrated in Figure 7.7). Note that this potential error in measurement is avoided in the previous method because the calculations are based on a single point.

To use this empirical information, now place an electrode in the carrier and position the tip of the electrode on the landmark (lambda). Add the target coordinates for AP and ML to the landmark coordinates for AP and ML. Add (or subtract) the AP correction (lambda + target AP - 5.3 mm). Drill a hole in the skull for the electrode and place the electrode to its depth using the corrected DV (-15.4 mm).

Assuming that we are satisfied with the electrode's position, it is now time to secure it in place for subsequent chronic recording sessions. We first very carefully place bone wax in the hole in the skull that surrounds the electrode, taking great care not to bump the electrode in the process. Next, we gather up any wires or leads and secure them to the plug assembly that will be used to route the recording and ground connections to the brain amplifiers (see discussion in Chapter 4). Some of this can be done before the electrode is lowered if the surgeon desires. Next, using a flat stainless steel spatula, dental acrylic is applied in layers to secure the electrode and plug assembly to the skull. The dental cement is first placed around the electrode shaft where it contacts the skull, making sure that some of it also covers at least one of the screws that were placed earlier in the skull. When these early layers of cement are thoroughly dry, the electrode can be detached from the micromanipulator, the micromanipulator set aside, and the remaining layers of dental acrylic applied over the electrode and plug assembly. Any other headstage components (such as bolts to hold an air puff nozzle) are added at this time. Care should be taken not to get cement into the holes of the plug assembly and not to get cement on the edges of the wound (as this could be a source of infection). Also, the smallest headstage possible should be built for the later comfort of the animal.

After the headstage is thoroughly dried, the wound is treated with antibiotics and sutured closed with 4-0 nylon suture. At this time, other minor procedure may be performed, such as putting suture in the nictitating membrane for monitoring behavioral performance via potentiometers or putting wire electrodes in the eyeblink musculature for EMG recordings. The animal is then removed from the stereotaxic head holder and placed in the recovery area. We typically allow a minimum of 1 week for recovery from this procedure before beginning any behavioral work.

Note that in our example, we detailed procedures for the implantation of recording electrodes in a rabbit. Very similar procedures can be used to implant electrodes into rats. Different atlases and slightly different stereotaxic equipment are typically used, but the basic procedures that are followed are very similar.

SOME NOTES ON KNOTS BEFORE LEAVING THIS CHAPTER

Since we are covering surgery in this chapter, we thought it would be a good idea to review some basic knot-tying techniques since knots are used extensively in surgery (and for other purposes, such as providing a means for attaching potentiometers to measure eyeblinks).

In Chapter 3 we discussed the use of potentiometers to measure eyeblink responses. Typically, the potentiometer is attached to the rabbit's nictitating membrane via a nylon loop that is placed into the nictitating membrane at the end of surgery when the rabbit is still anesthetized. The loop

Chapter 7

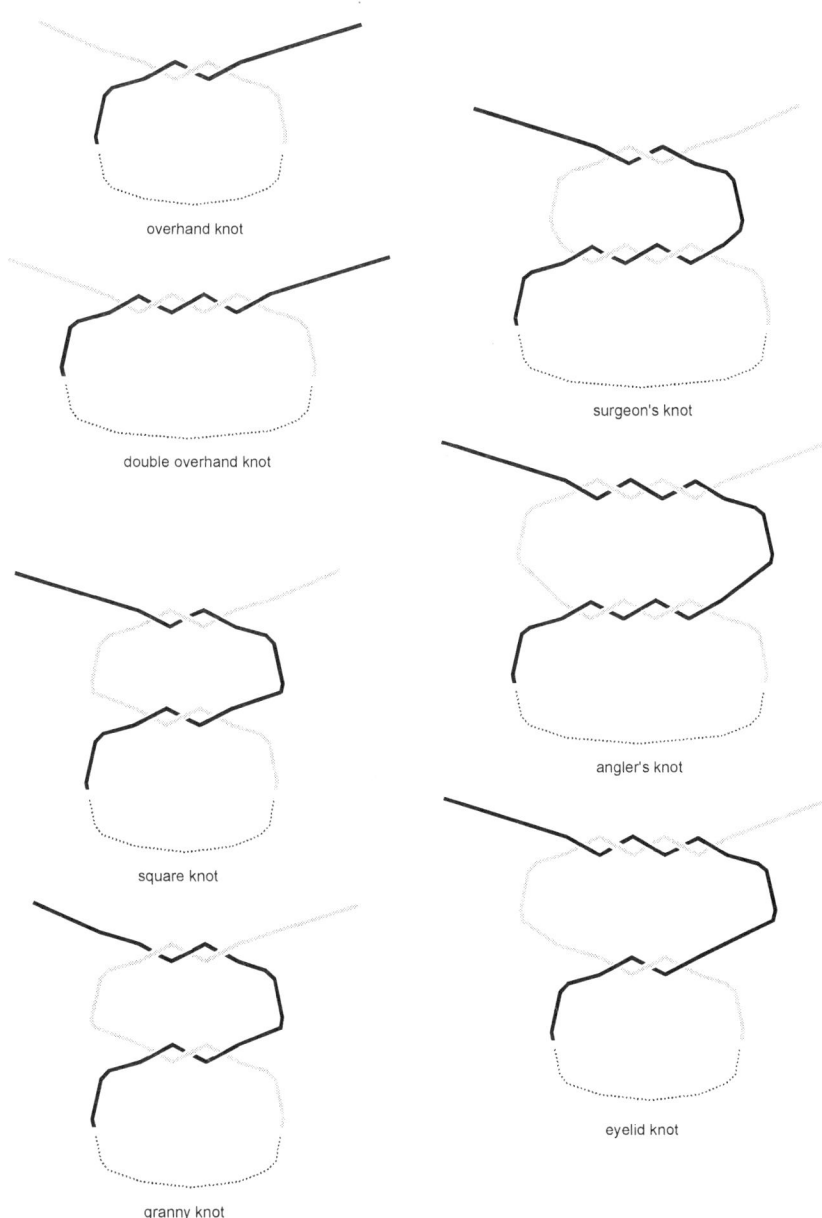

Figure 7.9: Introduction to knots. We have seen any of knots that use a double overhand knot with an overhand or double-overhand used for the eyelid suture. Knots that are asymmetric (surgeon's knot, eyelid knot) work well with slippery nylon suture.

can be placed in the nictitating membrane of an awake animal with the eye anesthetized using a local ophthalmic anesthetic. Various tools have been used to gauge the size of the loop—for example, the wooden end of a Q-tip, the shank of a surgical drill bit—but by far the most popular is the smooth shaft of an injection needle. Injection needles have the advantage that they come in a variety of standard sizes from which to choose. Ideally, the size of the loop should be as small as possible. We typically use 19 gauge needles as the largest size, and occasionally use 21 gauge needles for the smallest loops. If the loop is made too small it can be difficult to attach or remove the potentiometer linkage. One of the problems with nylon is that the knot can come undone, often either because the slippery nylon suture has not been tied tightly or because the ends of the suture have been cut too close to the knot, requiring that another loop replace the lost suture (usually a local ophthalmic anesthetic but occasionally a general anesthetic is used).

Figure 7.9 shows important knots to know. The overhand knot is the simplest knot possible. We know the overhand knot as the first half of the knot we use to tie our shoe laces. The overhand knot is simply made by forming a loop (technically a bight) and passing the end through the loop once. A double overhand knot is formed by passing the end through the loop twice. These two knots form half of the bases for tying ordinary and surgical knots. Parting from the official knot community (yes, there is one), we do not consider the overhand to really be a knot unless it is opposed to something, either against itself as when it is tightened into a stopper knot, or against another knot. If the overhand knot is not opposed but its loop is filled with an object (whether a pole or tissue) then it might hold momentarily but it invariably comes undone. The double overhand knot fares better because of its increased friction but it will still come undone. True knots are opposed to something. Ordinary and surgical knots are created by opposing a second knot against the first.

Two overhand knots, one on top of the other, can make either a square knot or a granny knot. The difference is whether both overhand knots are made the same way (for example, left over right for the first knot, then left over right again for the second knot) resulting in a granny knot, or whether the second overhand knot is tied oppositely (for example, left over right for the first knot, then right over left for the second knot) for a square knot. Because nylon suture is slippery, both square and granny knots can be pulled into a slip knot which will undo the knot. If one of these knots is tied, this is one of the reasons why an eyelid loop will come undone. Technically, the granny knot can be pulled into a slip knot formed by two complementary half hitches (a clove hitch on the standing part), and a square knot can be pulled into a slip knot formed by two opposing half hitches (a girth hitch on the standing part). By using the double overhand on either the first or sec-

Chapter 7 247

ond knot, or for both knots, these problems can be avoided because the double overhand has greater friction on the nylon suture and its topology cannot be pulled into a slip knot.

The surgeon's or ligature knot is formed by first tying a double overhand knot (left over right over right) and then tying an opposing overhand knot on top (right over left) to secure it (Ashley, 1944; Van Way and Buerk, 1986). The first double overhand temporarily helps to hold the knot from slipping. We have noticed an added feature, which is that the double (or multiple) knot can be used to span great distances (for example, when suturing skin) without puckering the tissue with the final knot. The knot that some books call a surgeon's knot is made with two double overhand knots (for example, left over right over right for one knot then right over left over left for the second knot; see Bigon and Regazzoni, 1981, and Jacobson, 1999). We like to call this an angler's knot because the principle of multiple wraps is normally used on fishing tackle. As we have noted, multiples of wraps are made by passing the end through the loop many times. Multiple passes increase the friction, making the knot relatively less likely to slip or break. Any knot actually weakens the breaking strength of the line--the one rare exception in the catalog of knots is the friction caused by the multiple wraps and reduction of stress caused by limiting the bend of the line that is seen in the Bimini twist fishing knot (Budworth, 1997). The disadvantage of the angler's knot for the use on the eyelid loop is that it is bulky and can irritate the eye.

The final knot in Figure 7.9 we call the eyelid knot. This is the knot that Don Wiesz taught Lavond for the eyelid suture. Like the surgeon's knot it is asymmetric in that the first is an overhand knot and the second is a double overhand knot, only the order of the knots differs between a surgeon's knot and the eyelid knot. In reality, probably any of the knots with a double overhand would work, although we would prefer one of the two asymmetric knots. We believe that asymmetry is an important, unrecognized feature for the integrity of both the surgeon's and eyelid knots. Remembering how to tie the eyelid knot is easy: The first knot is one pass through the loop (an overhand) pulled taut (not tight) and the second knot is two passes through the loop (a double overhand) pulled tight. Lavond has never had an eyelid knot come undone when the second knot was pulled tight and the ends were cut to the right length. The right length is long enough to allow for normal slippage but short enough to prevent irritation of the ends rubbing against the cornea--about 0.5 mm long, measured with the top edge of the scissors.

CONCLUDING REMARKS

As we pointed out earlier in this chapter, we believe that almost anyone can be trained to do stereotaxic surgery. It is our experience that most successful surgeries are due to at least three things: Good knowledge of the anatomy and physiology of the subject one is using, adequate planning and preparation for surgery, and exercising patience. We recommend that before you conduct surgery that you thoroughly study the organism with which you are working. The Life Science Laboratory Animal Series published by CRC is an excellent source of information (including *The Laboratory Rat* by P. E. Sharp and M. C. La Regina (1998) and *The Laboratory Rabbit* by M. A. Suckow and F. A. Douglas (1997)). We cannot stress enough the importance of being prepared for surgery. Everything should be laid out and ready to go before the animal is anesthetized. Also, the surgeon should be prepared for emergencies and any contingency that can come up during the procedure. Finally, many mistakes are made in this process when the surgeon is rushed or does not take enough time to carefully complete all steps of the surgery. We advise that plenty of time be allowed for the surgery and that the surgeon does not feel rushed, but rather exercises patience. Surgery may be the most critical step in the experiment. If this step is successfully carried out, fewer animals are used in the research and the quality of data is high.

Chapter 8

LESION TECHNIQUES FOR BEHAVIORAL EXPERIMENTS

OVERVIEW

Historically, the lesion method has been one of the most popular techniques used for the study of brain-behavior relationships. The basic theory of lesioning is conceptually very simple: The function of a brain area is inferred by observing the deficits that are produced when the area is removed from the brain. This method of establishing brain-behavior relationships has been used in humans; damage to a brain area through injury or degenerative disease, although not intentionally placed, can still be used to assess brain function. Indeed, much of the very early neurological literature was established by observing the behavioral performance of brain damaged individuals (e.g., Broca, 1861; Hughlings-Jackson, 1931).

Brain lesions have commonly been used to study brain-behavior relationships in animal models as well. In these experiments, researchers attempt to place lesions in restricted brain regions using a variety of techniques and then infer function of the brain areas by observing changes in behavior. This has been a powerful technique for initially establishing the involvement of a brain structure or region in behavior. However, this technique is not without its share of interpretative problems.

First, a lesion effect suggests that the area is involved in the behavior under observation, but does not prove that the brain area is the originating source for brain activity associated with that behavior. For example, if we lesion the abducens nucleus of a rat or rabbit after it has undergone classical eyeblink conditioning, we will see a loss of the eyeblink conditioned response. Does this mean that the critical cellular plasticity responsible for generating the learned response resides in the abducens nucleus? The answer is no; the lesion effect only shows that the structure is located somewhere in the neural circuitry that is responsible for CR generation and performance. In fact, in our example we know that the abducens nucleus contains motoneurons that control musculature that are involved in generating the response and that it is not a site of critical plasticity associated with the learning and memory of the response.

Second, even when a positive lesion effect is noted, depending on the lesion method used, there is a chance that the brain structure or area is not even involved in the behavior under observation. Every brain scientist knows that the brain is made up of clusters of neuronal cell bodies and axons from those cell bodies that course in bundles around the brain. Some lesion methods indiscriminantly destroy cell bodies and axons, thus a researcher cannot be certain if the lesion is affecting cell bodies or axons that are coursing in or near the cell bodies. A lesion that affects the fibers *en passant* in essence cuts off communication between two areas of the brain. Hence, the observed change in behavior could be due to this loss of connectivity and not due to loss of cell bodies in a given brain area.

Over the years, some improvements in the lesion technique have addressed the two problems outlined above. First, many researchers have coupled neural recording techniques with lesions to further define the involvement of a given brain area in a behavioral function. For example, recording in structures afferent and efferent to the brain area of interest might reveal whether or not the brain area is acting as something more than a passive conduit for critical activity arising elsewhere in the brain. Second, newer chemical lesion techniques have eliminated the problem of lesioning fibers *en passant* and still newer reversible lesion (or inactivation) methods have allowed well contolled lesions to be temporarily placed while behavioral function is assessed.

In this chapter, we review the nuts and bolts of the lesion technique. We cover the permanent lesion techniques that include aspiration, electrolytic, and chemical lesions as well as temporary inactivation methods (cooling probe and chemical). We believe that while none of these lesion techniques can provide all of the data necessary to evaluate the function of a brain structure or system, all of these lesion techniques can be successfully used to further advance our understanding of brain-behavior relationships.

ABLATION AND ASPIRATION LESIONS

Perhaps the oldest of the lesion techniques are the simple scoop and transection lesion methods, sometimes referred to as ablation lesion techniques. For the scoop lesion, the skull is removed over the area to be lesioned and the region of the brain is destroyed by carefully removing tissue with a small curette or other instrument. Obviously, the scoop technique is limited to lesions that involve the surface of the brain. Indeed, these have been used extensively in the past to study cortical function. Sometimes, the purpose of the lesion is to cut fiber tracts while not removing brain tissue or to separate a brain region from others areas of the brain. For these transec-

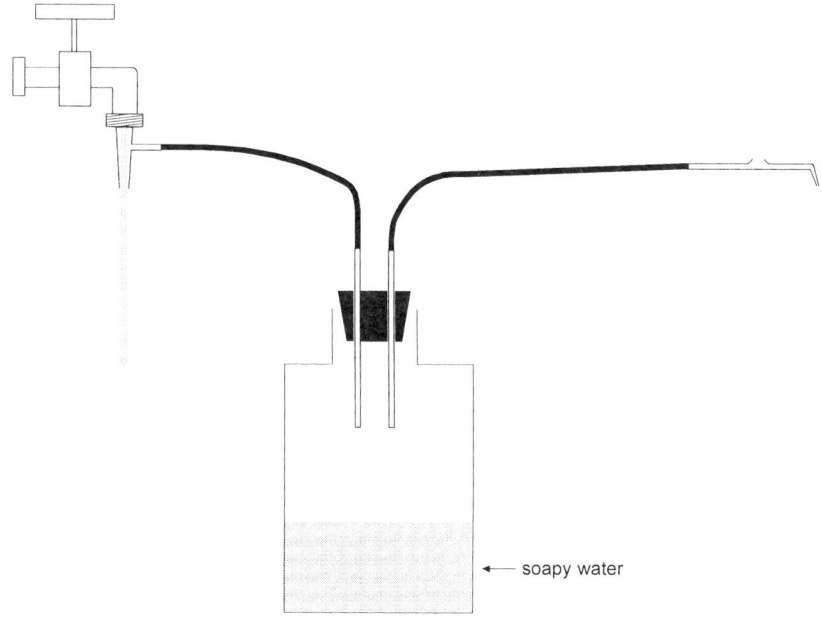

Figure 8.1: Aspirator created from a simple vacuum using water pressure.

tion lesions, a variety of specialized surgical microknives have been created, some of which allow fiber pathways that lie deep in the brain to be cut. For example, a specialized knife used to isolate the hypothalamus from the rest of the brain was described by Ellison (1972). While effective as lesion techniques, these ablation methods are relatively hard to control and when not properly done can cause excessive bleeding, among other problems.

For removing large amounts of tissue the method of aspiration is favored. This technique involves a vacuum source, a collection bottle, tubing and the pipette for aspiration. The latter is typically referred to colloquially as 'the aspirator'; however, actually the aspirator should refer to the whole system.

The vacuum source must be able to apply 2–10 lbs of suction. Specialized surgical equipment for suction is ideal. Alternatively, vacuum pumps can be secured from photographic, electronic and hobbiest supplies. Using the principle that a high pressure stream can be used to create a vacuum, aspiration can be achieved using a stream of water from a faucet as illustrated in Figure 8.1. Here, the flow of water past a small opening in the fixture creates a vacuum, pulling anything (air, in this case) with a lower pressure into the stream. For most of our aspiration experiments, we have used either a surgical aspirator or more recently suction from a vacuum line that was installed in our building. Our building vacuum line supplies about 60 lbs of suction. The advantage of using the specialized surgical equipment is that it will have a vacuum gauge, often with a regulator for adjusting the pressure. The other methods usually require educated guesses, which is not difficult—just hook the system up and test the vacuum with an aspirator by sucking up some water. In the case of the vacuum from the building, the vacuum can be regulated by adjusting the valve (i.e., to something less than the 60 lbs available), by adding small escape valves (essentially holes in the tubing that lessen the vacuum at the aspirator tip by providing alternative paths for air to enter the vacuum), or by adding a finger hole on the aspirator, described below and shown in the figure.

A sealed bottle such as an Erlenmeyer flask with a rubber stopper, for example, acts as a buffer between the vacuum source and the removed tissue. The stopper has two holes, one going to the vacuum source and the second going to the aspirator device. The bottle prevents tissue from entering into the vacuum system, conveniently collecting the tissue so that it can be disposed. We like to add soap to the water so that the Erlenmeyer flask is easier to clean after the surgery.

The tubing used for the aspirator is either made from the soft yellow rubber medical tubing or from clear plastic tubing that can be bought in the plumbing section of any hardware store. The pipette or aspirator can be made from syringes with large (e.g., 20 gauge) needles, or from glass. Glass pipettes are available that work well for this purpose. We have also used glass tubing, drawing and shaping it under the flame of a Bunsen burner; we use a Bunsen burner with the glass pipettes to shape them also. We usually add three modifications to the glass tubing or pipettes. First, we add a downward bend near the tip of the aspirator that helps maneuverability. Second, we 'fire' the very end of the tip, meaning we melt it just enough to give it a smooth surface and added strength without significantly closing the opening. The thin glass pipettes tend not to be very useful if they are closed up. Third, we melt an area of the tubing or pipette while blowing into the big end (usually with a piece of plastic tubing as an extension so our heads do not get too close to the flame) and plugging the small working end against our finger. As the glass melts the pressure causes a glass bubble to form and eventually burst. We fire the edges of the hole. This added feature is a finger hole, used to regulate the pressure while aspirating. Completely closing this hole with a finger directs the greatest amount of vacuum to the working end of the aspirator. Completely opening this hole by removing the finger entirely diverts the vacuum to the large hole, away from the aspirator tip. By varying the amount of the finger opening, the surgeon can gain some amount of control over the vacuum.

Besides the usual surgical concerns, similar to the ablation methods, aspiration has its own special problems with homeostasis, the control of blood loss. With large cerebral cortical and cerebellar cortex lesions we have been able to control bleeding with gelfoam left in place. As if those are not radical enough surgeries, others have used aspiration to remove the hippocampus or perform high decerebrations (removal of all tissue about the midbrain). Both of these procedures have been described as causing significant bleeding. We have little to no experience with these surgeries. We do know that it is said that cold sterile saline helps to control the bleeding until coagulation occurs and that chemical means of promoting coagulation, such as thrombin, are important to stem bleeding. Given the extensive nature of these lesions, 'tying off bleeders,' using hemostats to crush severed vessels, or using heat cauterizing equipment are not particularly useful techniques for controlling the bleeding from aspiration of the brain.

ELECTROLYTIC LESIONS

Electrolytic lesions have been frequently used in brain-behavior experiments. Small lesions (e.g., 100 μA, 10 sec) through stainless steel metal electrodes used for neural recordings can help to identify the location of the recording in later histology (see Chapter 10). Larger lesions (e.g., 2 mA, 2-1/2 minutes, see for example Clark et al., 1984) can be used to assess brain-behavior relationships and likewise can be identified in later histology. To create these lesions a constant current source for the electrolytic lesions is very valuable for localizing the placement of recording/stimulating electrodes (small lesions) and for assessing the functioning of a structure (larger lesions).

Theory Behind Electrolytic Lesions

Electrolytic lesions create a lesion by passing current. Ideally, this current is maintained at a known value throughout the lesion. However, as the lesion progresses, tissue damage in the animal along with changes in the properties of the electrode increase the overall resistance in the lesion circuit. It can be seen in the familiar relationship described by Ohm's Law ($V = IR$, with V being the same voltage source) that as the resistance increases, the amount of current decreases proportionately. Because it is the charge transfer, or current delivered over a period of time, that accounts for making the lesion, the major design feature for making a device to create a lesion is to have a well-regulated, constant current source.

There are two ways to achieve this property of constant currency,

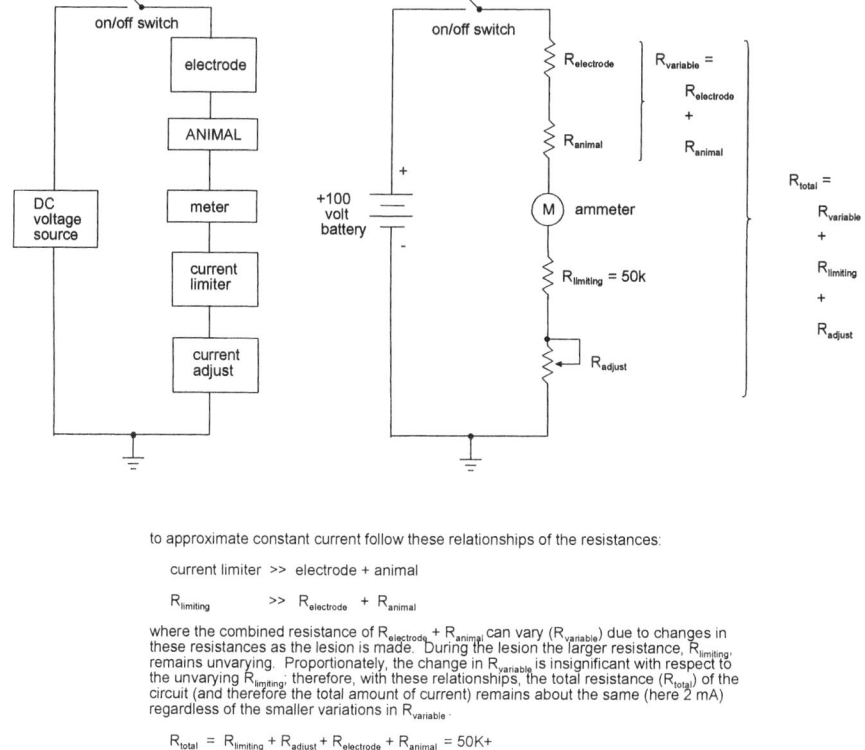

Figure 8.2: Fundamental method for creating a constant current lesion.

both of which are incorporated into the lesion maker presented here. The first method (Figure 8.2) is to use a large current limiting resistor (current limiter, $R_{limiting}$) in series with a large voltage source (DC voltage source, +100 volt battery) and the combined resistance ($R_{variable}$) of the electrode ($R_{electrode}$) and the animal (R_{animal}). As noted in Figure 8.2, the total resistance (R_{total}) is greater than 50 kΩ, making the current I of this circuit 2 mA (0.002 A). This is the maximum current achievable for this circuit. By adding a variable resistor R_{adjust} to increase the total current R_{total} we modify the lesion current to a smaller amount. Typically 100 µA of current is used for marking lesions. Most of the resistance in this circuit is accounted for by the limiting resistor $R_{limiting}$, so that small changes of resistance of the electrode and animal ($R_{variable}$) have little effect on the overall current, I. You can imagine that using a larger voltage source allows for an even larger value for $R_{limiting}$

Chapter 8

so that any small changes of resistance of the electrode and animal ($R_{variable}$) have even less of an effect on the overall current.

The second method for achieving constant currency is to use an active element that automatically changes its resistance to compensate for increased resistance of the electrode and animal ($R_{variable}$). The active element in the following circuits is a transistor, configured as a constant current source.

The Thompson Lesion Maker

The lesion maker described here is based on a device used in the Richard Thompson laboratory many years ago. Exactly who originally designed and built the first prototype of this lesion maker is not clear to us. Being curious as to how it worked, Lavond once wrote down all the components and wires, but could never straighten out the wiring into a diagram that made sense (yet the device worked perfectly). At the time, Lavond knew little about using transistors as constant current devices so having the completed diagram would not have made much sense to him anyway.

Before describing the device further, we thought we would relate a story about the dangers of taking home-built lab equipment through an airport. In the mid 1980s, a few years and many lesions after the lesion maker was built, Dick Thompson sent Steinmetz with the lesion maker to Steve Berry's laboratory in Miami, Ohio, to conduct some experiments that combined the expertise of the two laboratories. The trip from the San Francisco airport to the Cincinnati airport was uneventful, but on the return trip from Cincinnati to San Francisco, airport security and another group of law enforcement officers (possibly the FBI, it was hard to tell but they wore dark clothes and sunglasses) detained and interviewed Steinmetz about the suspicious device that was in his carry-on bag. It seemed not to matter that he had an explanation for the device (telling them it was a lesion maker seemed to make matters worse). Also, it seemed not to matter that the lesion maker had neither a timer nor explosives nor a blasting cap, let alone anything combustible (it still "looked suspicious"). Steinmetz was given a choice: either have the device confiscated or let the security personnel cut several wires inside the lesion maker to render it inoperable. Because we needed the lesion maker back in California, wires were cut. However, this proved not to be a good choice since the lesion maker never worked after that. Even though we were in possession of a spaghetti-like diagram for the device, we could not figure out which wires had been cut or where the cut ones went, although we could ascertain that more than one was severed. Thompson ended up buying a Grass lesion maker (Model D.C. LM5A) for hundreds of

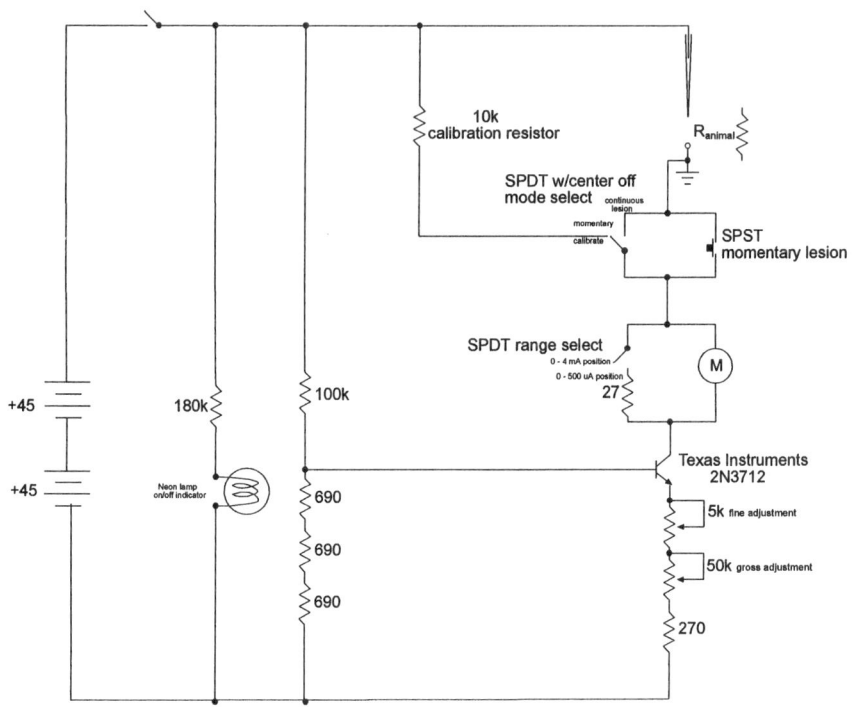

Figure 8.3: Thompson lesion maker: Active method used to create constant current lesion.

dollars to replace the one "vandalized" by the eager Cincinnati airport security force.

A few years ago Lavond ran across his notes and he worked out the original wiring, guided by knowledge about transistor circuits that he acquired in the years since originally examining the lesion maker. It seems like it is always easier to build something from scratch than it is to figure out and fix a box full of wires. Claire Cartford, working in the Lavond laboratory, found a replacement for the critical transistor stage, which turned out to be nothing more than a general purpose transistor with a working circuit that could handle 300 volts.

Figure 8.3 is the original Thompson lesion maker using an analog current meter, a part which is not that easy to find nowadays. Figure 8.4 is the same circuit modified to use a volt meter with an LCD digital display. Both are battery powered devices which is an attractive safety feature. In Figure 8.3 the original circuit uses two 45 volt batteries in series for 90 volts total. In Figure 8.4, we use ten 9 volt batteries to achieve the same 90 volts total. Two 45 volt batteries would work just as well. The point here is that the number of batteries is not important, but rather the total voltage is the

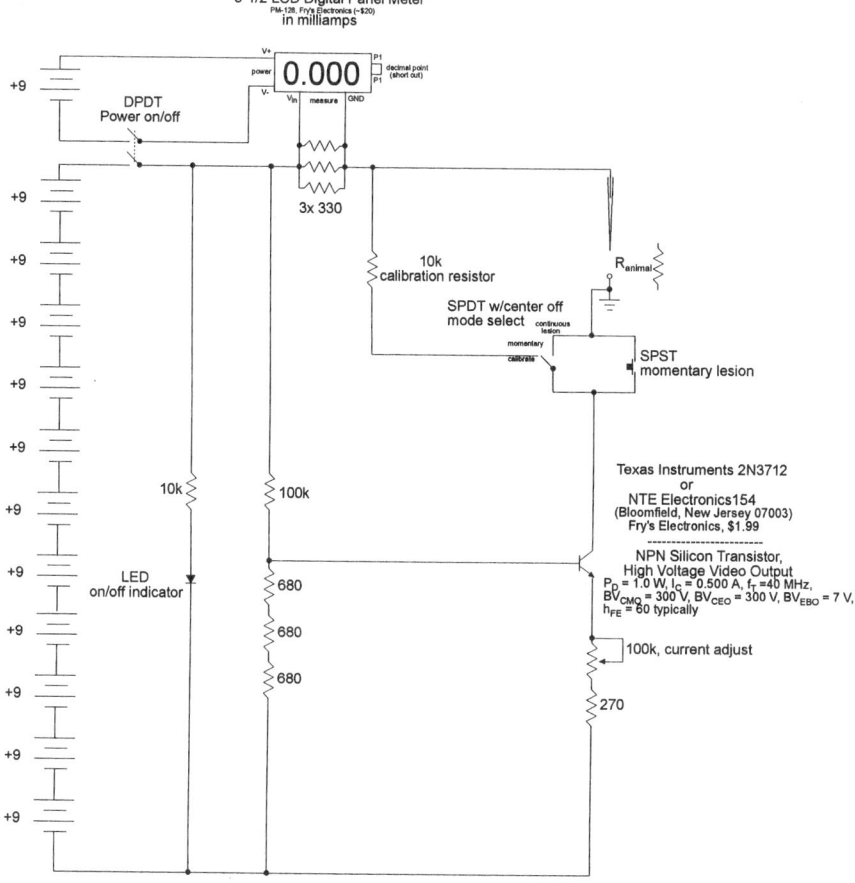

Figure 8.4: Modification of Thompson constant current lesion maker using a digital display.

objective. In fact, up to 300 volts could be used in this circuit. The 9 volt battery for the volt meter with the LCD display in Figure 8.4 must be a separate power supply.

A selector switch (SPDT with center off) allows for setting the current (calibrate position), a center position where a button press passes current through the electrode, and a third position for continuous lesion current. Note that the lesion is made by passing anodal direct current (anodal DC) with the electrode attached to the plus (anode) pole and the ground (a skull screw, for example) attached to the negative (cathode) pole of the battery stack.

Some Design Improvements

Functionally, the Grass Model LM5A that Thompson bought as a replacement for the damaged home-built lesion maker is similar to the circuits given here. However, as is usually the case, the researcher-built equipment costs considerably less; the most expensive parts being the meter and the batteries. The circuits in Figures 8.3 and 8.4 could be improved by further modification. A simple timing circuit, perhaps using the CMOS 4047 timing chip that was seen in the controller designs (described in Chapter 12), which uses either a fixed timing value or a selection of times, could be easily added to effect a relay (to handle the higher voltages and current) to pass current in the circuit.

A second improvement would be to add an integrated on/off lesion switch. Like the rise/fall switch that we used to gradually turn on and off a tone signal, an integrated on/off lesion switch for the lesion maker would help overcome what we colloquially call "electrode failure" during lesioning. With very large lesion currents delivered over long periods of time, the meter may show the current fluctuations and may completely fail by passing zero current. Often, by turning the current off and waiting a minute or so the electrode will recover and the lesion can be continued. Sometimes there are additional failures and waiting does not improve the condition. Electrode failure is due, in part, to a gas bubble formed around the electrode tip in the process of hydrolysis (the positive tip attracts negatively charged oxygen). Time seems to allow this gas bubble to dissipate. The second major reason for electrode failure is due to loss of the metal from electrode. The metal deposit in the brain is used in histological identification of the electrode tip (see below and Chapter 10). With enough of the tip removed the resistance of the remaining electrode is too high to be compensated for by the voltage of the lesion maker (a higher voltage might overcome this problem until there is no appreciable metal remaining). Electrode failure is made worse by suddenly passing a large current. This transition can be enough to cause failure of the insulation along the electrode. One way to get around this problem is to turn on the current at a low value and gradually increase (say, over 5-10 seconds) the current to the full lesion value, which is what we do in practice. Another way to prevent sudden onset is to add an integrator to the on/off lesion switch to effect this transition by putting a resistor in series with the switch and putting a capacitor across the resistor. The time of this transition is given by taking the inverted value of the capacitor, times the resistor, times twice the value of pi (that is, Period = $1/(2\pi RC)$).

Lesion Electrodes

We typically make electrodes from size 00 stainless steel insect pins (Ward's Biology) by coating them with 6-8 coats of baked Epoxylite varnish. For single unit recordings the insect pins are usually etched or ground down to a smaller diameter and then coated. The tips of coated single unit electrodes are exposed by scraping (20-40 µm) or by blowing off the insulation from the tips with a high voltage (referred to as "zapping" the electrodes) under a microscope (see Chapter 4 for more details). For stimulation electrodes the insect pins are used without making them smaller in diameter. The tips of the electrodes are exposed by scraping away 250-500 µm (0.25 - 0.5 mm) of insulation visualized under a microscope. For lesion electrodes the insect pins are also used without making them smaller in diameter. The lesion electrodes are created by exposing 500 µm (0.5 mm) of insulation from the tip under a microscope.

Locating Electrolytic Lesions

Once the brain has been cut on a microtome and the sections mounted onto chrom-alum subbed slides and dried, the electrolytic lesions can be localized by exposing the sections to the Prussian blue reaction, assuming, of course that the electrode had iron in it (as stainless steel does). This Prussian blue solution consists of two parts mixed immediately before use: Part A consists of 100 g potassium ferrocyanide per 1000 ml distilled water; Part B consists of 200 ml HCl (reagent 37%) and 800 ml distilled water. To use, mix two parts of A for every one part of B. Place mounted slides in this solution for 3 minutes then rinse in distilled water. Additionally, the sections should be rinsed in 1% sodium acetate to neutralize the acid in the Prussian blue solution, followed by another rinse in distilled water. Metal deposited by the electrolytic lesion reacts with the Prussian blue solution to yield a blue or blue-green stain identifying the tip of the electrode. We then counter-stain with any of the water based nucleic acid stains (e.g., cresyl violet, neutral red, thionin). This technique is most useful for finding small marking lesions (e.g., 100 µA, 10 sec). Larger electrolytic lesions (e.g., 1 mA, 30 sec) are obvious. It should be noted that the Prussian blue solution, even with only a 3 minute exposure, appreciably shrinks the brain tissue. Chapter 10 provides a more in-depth discussion of common histological methods.

Electrolytic versus Radio Frequency Lesion Techniques

Before leaving the topic of electrolytic lesions, we should say a few words about the radio frequency (RF) lesion technique. The RF lesion method destroys tissue by thermocoagulation, which is achieved by passing radio frequency current through the electrode. A lesion is actually produced via heat that is generated at the electrode tip. This technique became popular in the 1960s when some investigators argued that electrolytic lesions caused hemorrhaging and the deposition of metal that could lead to a "chronic irritative foci" that, in turn, could confound the results of the lesion (e.g., Reynolds, 1965). Over the years, this idea was contested by a variety of researchers (e.g, DiCara et al., 1974) and the electrolytic lesion method has continued to be more popular than the RF lesion method. We suspect that this is mainly due to the more widespread availability of electrolytic lesion makers. RF lesions require a device that can generate radio frequency current. While commercially available, compared to electrolytic lesions makers these devices are relatively hard to build from scratch.

PERMANENT CHEMICAL LESIONS

Over the past 20 years or so, a trend toward using chemical lesion methods instead of ablation, aspiration and electrolytic lesion techniques has risen. The major advantage of using this technique is that chemical lesions can spare the fiber *en passant* when delivered thus eliminating, for the most part, difficulties in interpreting whether or not the lesions affected the cell bodies that were targeted for the lesion. Three major chemicals have been used: kainic acid, ibotenic acid, and 6-hydroxydopamine (6-OHDA).

Kainic acid and ibotenic acid are both excitotoxic heterocyclic amino acids that are analogues of the neurotransmitter glutamate. These compounds strongly excite neurons in a variety of locations throughout the nervous system. It is assumed that kainic acid and ibotenic acid cause a state of continuous depolarization and increased plasma membrane permeability when applied to neurons. The constant excitotoxic depolarlization is thought to produce a continuous effort by the neuron to restore its ionic balance. The cell literally excites itself to death (i.e., dies when its energy stores are exhausted due to the continual excitation). Because kainic acid and ibotenic acid act primarily at the postsynaptic glutamate receptors, these compounds mainly destroy neurons in the injected area and not fibers *en passant*. When injected, 6-OHDA selectively destroys the axons of norephinephrine and dopamine neurons at the site of injection, which causes anterograde and retrograde degeneration of the affected neuron. Its total effect is dependent on a number of factors including the amount injected and the precise location of the injection site. This method has been used extensively in behavioral pharmacology experiments aimed at delineating noradrenergic and dopaminergic function. It is a good idea to measure norepinephrine or dopamine levels in the brain after the 6-OHDA lesion to determine the extent of the lesion.

The basic methods involved in delivering permanent chemical lesions are relatively straightforward. Typically during a surgery, the tip of a cannula is stereotaxically positioned into the site where the lesion is to be made. We have made cannulas from 22-gauge, thin-walled, stainless steel tubing. The tubing is cut to the desired length and the end is then sanded and buffed on a small grinding wheel. To prevent the cannula from clogging, we insert a stylus (fashioned from a stainless steel rod) into the cannula before it is lowered into the brain. The stylus is left in place at all times except when infusions are taking place. We cement the cannula in place using the same techniques described in Chapter 7 for anchoring recording electrodes. During infusions, the stylus is removed and a 26-gauge needle that is the same length as the cannula is inserted into the cannula. The non-injection end of the 26-gauge needle is fitted with a Teflon tube into which the needle of a Hamilton syringe is placed. The Hamilton syringe contains the chemical to be used to make the brain lesion. While some laboratories inject the kainic or ibotenic acid all at once in a single push of the syringe plunger, we believe that it is best if the substance is infused over a longer period using a syringe pump as this technique minimizes tissue damaged caused by injecting volumes of solutions into the brain.

The amount of kainic or ibotenic acid that is delivered depends largely on the size of the injection that is desired. For example, we have used kainic acid to lesion the interpositus nucleus of the cerebellum in rabbits (Katz and Steinmetz, 1997). For this study, we used 0.2 µl of 25 nmoles of kainic acid per 1 µl of saline and infused the chemical over an 8 min period in an attempt to lesion an area that was approximately 1 mm^3 in size. In another study, we used ibotenic acid to lesion the interpositus and dentate nuclear complex in the rat (Steinmetz, Logue & Miller, 1993). For this study, we used three discrete 0.2 µl injections of ibotenic acid (10 µg/µl) on each side of the brain. To increase the accuracy of the cannula placement, we have cemented a multiple-unit size recording electrode onto the cannula and recorded from the electrode while lowering the cannula during surgery (and also during the injection). Assessing the extent of the chemical lesion can be difficult as unlike electrolytic lesions, no discrete marking is typically left behind by the chemical lesion. Most researchers assess the extent of the chemical lesion by injecting radio-labeled chemical products at the end of the experiment or by counting neurons microscopically at the injection site. We will discuss these techniques further in Chapter 10.

TEMPORARY INACTIVATION/REVERSIBLE LESION TECHNIQUES

With the various advances that have occurred in the brain sciences, the lesion technique still remains the single most important method for studying the relationship between brain and behavior. In addition to the drawbacks we pointed out above, another one of the major criticisms of the lesion technique is that testing is always performed on brain damaged subjects. While this property might be valuable for generalizations to patients with brain injury, understanding brain organization of a "normal" individual is more difficult to justify with destructive lesions. One way to solve this problem is to use multiple techniques like recordings and stimulation. Another way to solve this problem is to introduce a temporary lesion for the duration of a phase of training, for example during acquisition, and then remove the effects of the lesion during later testing. Some examples are in order.

Temporary lesions can be made after an animal has learned a task to assess whether or not the structure being lesioned is involved in performance of the learned response. For example, a temporary lesion of the accessory abducens nucleus delivered after the animal has acquired the classically conditioned eyeblink response would produce an abolition of the eyeblink CR (because the abducens nucleus contains motor neurons needed for generating the learned response). If no inactivation is performed on subsequent sessions, due to the temporary nature of the reversible lesion, the eyeblink CR would eventually reappear. A more powerful use of the reversible lesion is to directly assess the critical involvement of a brain structure in the acquisition process. For example, several sessions of classical eyeblink conditioning can be given both with and without inactivation and the rate of conditioning compared with animals given all of the same experience, but with no inactivation. If no conditioned responses are seen during the initial acquisition phase, one can conclude that the inactivated brain area is a part of the neural circuitry involved in the generation of the CR. More important, however, is the relative rate of conditioning on later sessions when the brain area is no longer inactivated. If the rates of conditioning during this phase are identical for the inactivated and control animals, we can assume that the reversible lesion prevented important plasticity processes from occurring and that the inactivated area may contain neurons that encode the learning. If the rate of learning is faster in the previously inactivated animals, relative to controls, one can assume that critical plasticity processes occurred in sites other than the area that was inactivated during the initial phase of training. The former effect has been found when the interpositus nucleus of the cerebellum was inactivated during training thus providing strong evidence that basic plasticity processes important for learning the eyeblink CR occurred in this structure (Clark and Lavond, 1993; Clark, Zhang and Lavond, 1992).

The reversible lesion technique can be achieved by pharmacological agents, as in neurotransmitter or ion channel blockers (e.g., Muscimol, TTX and Lidocaine), and by the method of cooling. We will present some information on cooling first, then briefly talk about the chemical inactivation methods that are available.

Overview of the Cooling Technique

Cooling has been used for studying brain-behavior relations for some time now (Skinner and Lindsley, 1967, 1968). Brooks (1983) is a classic review. Recent understanding of the effects of cooling can be found in Janssen (1992). It should first be noted that we are talking about cooling and not freezing of the tissue. Freezing would permanently damage the tissue, and cryoprobes have been designed and used for this purpose, such as in destroying tumors. For the purpose of reversible lesions, the tissue is cooled below the basal metabolic rate necessary for synaptic conduction, which recovers to normal functioning when body temperatures are allowed to return to normal. We favor using the term 'cooling probe' rather than 'cold probe' to further distinguish the methods.

Thompson was first interested in applying reversible lesions to the problem of identifying the locus of learning in classical conditioning. A visiting professor, Merle Prim of Western Washington University, tried to achieve cooling using a probe with circulating alcohol in Thompson's laboratory; this method was not totally successful. Later, Thompson's graduate student Paul Chapman was encouraged to use a method of cooling described by Zhang, Ni and Harper (1986), a method that involves the use of Freon. We subsequently adapted the method of Zhang et al. in our own studies with a great deal of success (e.g., Clark and Lavond, 1993; Clark, Zhang and Lavond, 1992). We describe here the probes and control circuitry used in this procedure. The cooling probe is an integrated system that includes the source for a cooling agent, a valve to control the release of the cooling agent into flexible tubing, the cooling probe which may include a temperature measuring device and a heater for the shaft (thus confining cooling to the tip of the probe) and an exhaust system that may include a vacuum source.

The Cooling Probe

The cooling probe works by circulating a cooling agent through metal tubing that is placed near the tissue of interest. A common misunderstanding is that the cooling agent is not directly applied to the tissue, but rather heat is transferred indirectly through contact with the metal tubing that contains the cooling agent. In our experiments, the cooling probe is normally placed about 1.5 mm from the target, relying on the spread of cooling to affect the tissue. The cooling profiles have been measured in the original article by Zhang et al., in our own experiments by ourselves (Andrew Zhang, Bob Clark) and by Yuan Wang in Kathleen Chambers' laboratory at the University of Southern California. The probe is not placed directly into the target as its physical size would destroy the locus, which we have observed when cooling probes are inserted directly into the interpositus nucleus of the cerebellum. Several years ago, Laura Mamounas from the Thompson laboratory made a similar observation with placements of cannula in the interpositus for pharmacology experiments being conducted in collaboration with Jack Barchas and colleagues.

The original cooling probe of Zhang et al. used Freon-12 as the cooling agent. The cooling agent has evolved over the years, out of environmental concerns, to Freon-22 and now HFC-134a. Our method is identical for each, therefore we will refer to the cooling agent generically as 'the refrigerant.'

Unlike circulating alcohol which is cool throughout its travel, the refrigerant cools by expansion from a constricted space into a larger space. The refrigerant is delivered through small flexible plastic tubing to 30-gauge stainless steel tubing, which in turn opens up into a larger space confined by 19-gauge stainless steel tubing. The expansion of the refrigerant at the interface between the 30- and 19-gauge tubing is the action that causes absorption of heat. The expanded and cool refrigerant circulates inside the 19-gauge tubing from which it is exhausted. The control of the amount of refrigerant injected and the rate of its evacuation from the cooling probe affect how effectively the probe cools. Associated components control this flow, measure the temperature and heat the shaft of the cooling probe.

Figure 8.5 shows the simplified cooling probe that we use in most of our experiments. This probe is reduced to the bare essentials needed for cooling experiments: concentric 30-, 23- and 19-gauge tubings soldered together. The end of the 19-gauge shaft is soldered shut and plated with gold to prevent tissue reaction. Tapani Korhonen plates with rhodium. The refrigerant intake is the 30 gauge tubing, the refrigerant outtake is a 19-gauge extension that is soldered as a Y-joint to the 19-gauge shaft. It takes about 2 hours to construct this probe. This probe does not have a thermocouple for

Chapter 8

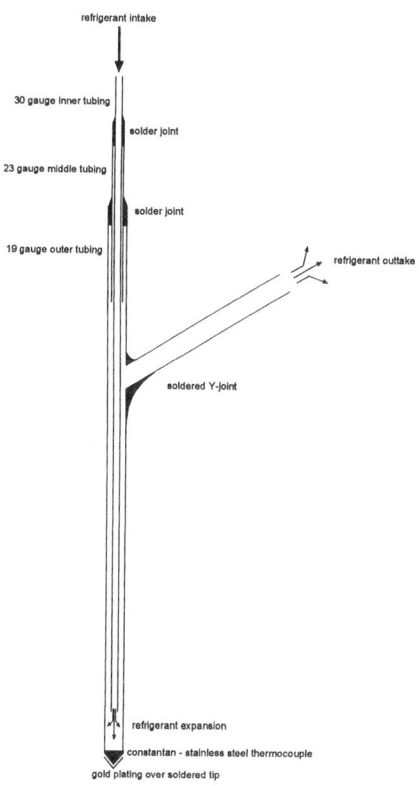

Figure 8.5: Simple cooling probe with soldered Y-joint.

measuring the temperature at the tip, or a coil to heat the shaft. For many experiments these extras are not necessary because the temperature profiles have been measured; we have past experience with effective rates of cooling; one can see the amount of frost that gathers on the exhaust tubing and this is related to the effectiveness of cooling (this is an imperfect indication, however, because it appears to be related to environmental humidity); and, most importantly, we can use the effect of cooling on behavior as an index. A probe with a thermocouple and heating coil takes us about 8 hours to construct.

The details for construction of the simplified cooling probe are as follows. The first task is to create the Y-joint between two pieces of 19-gauge stainless steel tubing. The first piece of 19 gauge stainless steel tubing is typically cut to 30 mm in length as the 'shaft' of the cooling probe. One side of the 19-gauge tubing is ground flat, nearly but not through to the lumen, 20 mm from what will be the end of the cooling probe. A second piece of 19-gauge tubing is cut about 1 cm long as the exhaust 'extension,' with one end ground to a flat (90°) angle and the other end ground to about a 30–45° angle. The end with the 30–45° angle will be soldered to the flattened area on the shaft to form the Y-joint. In preparation for joining the two 19-gauge pieces, both the shaft and the extension are 'tinned' with solder. This means that a small amount of solder is placed on the shaft around (but not on) the flattened surface and placed around the outside of the extension near the 30–45° angle. The two pieces are then soldered together. Previously tinning the two pieces helps to get a strong solder joint. The soldering is structural in the sense that it should resist a reasonable amount of pressure against it, and functional in the sense that there should not be any leaks from the lumen in the final product. The lumens are made patent (jointed together) by drilling the extension through any solder that gets into its lumen and drilling through the final bit of steel into the shaft. The drill is a high speed drill bit purchased from a hobby shop or from Small Parts, Inc. We have used an electric drill but caution is needed to prevent too much heat from being generated by the friction of the bit with the inside of the tubing. The heat created by the friction may be enough to unsolder the two pieces. To prevent unsoldering when using an electric drill we have used water to cool the pieces. To avoid the heat problem altogether, we now use a pin vice to hand drill the extension rather than an electric drill. We drill repeatedly through both lumens to remove any burrs.

    A few words about soldering. First, soldering to stainless steel requires using special stainless steel flux. Ordinary solder flux or rosin core solder will not allow solder to adhere to the stainless steel. Stainless steel flux can be found (with some trouble) in plumbing supply stores or plumbing areas of hardware stores. It can also be ordered from Small Parts, Inc. Second, as a strategy for soldering it is best to first tin each piece separately and then join the two pieces together. The joint is stronger because it avoids the dreaded 'cold solder joint.' A cold solder joint is one where the solder has not actually adhered to the metal, but merely appears to be holding. Physically exerting pressure against a cold solder joint will easily loosen the joint. A good solder joint can be recognized by the shiny and flowing appearances of the solder. Third, it is important to limit the amount of solder that gets into the lumen of the shaft (which is why the area is flattened without cutting through to the lumen) and the extension (tinning the outside of the extension helps but does not prevent solder from flowing into the lumen). Any solder that gets into the lumens must be drilled out. Drilling with an electric drill causes enough heat from the friction to undo the soldered Y-joint. Drilling by hand is a good alternative that has the added benefit that bits are not broken as often. Fourth, we tape over the edge of a table the extension with the 30–45° angle resting against the flattened place on the shaft. Any other suitable way to hold the pieces together for soldering would be helpful (e.g., a 'third hand' device with alligator clips). If possible it is helpful to position the two 19-gauge pieces so that gravity directs the flow of solder into a helpful direction (i.e., away from the lumens).

Creating the Y-joint is the hardest part in the construction of the cooling probe. Tapani Korhonen's strategy for joining the shaft and extension is much faster; if only we could duplicate his skill. He drills the shaft completely through to the lumen. He then inserts an insulated wire through the extension and into the shaft, and solders the two together. The insulated wire is then withdrawn, leaving the joined lumens. We have tried the same idea, but we either melt the insulation which adheres to the lumen or the angle of the wire is too acute such that in either case we invariably cannot remove the wire. Teflon coated wire may help.

Figure 8.6 shows an alternative way to create the Y-joint without soldering. Graduate student Ami Makena of Kathleen Chambers' laboratory at the University of Southern California worked with Small Parts, Inc. to design a manifold for joining the inner shaft, outer shaft and extension by a pressure fit. The manifold is a piece of round stock bar that has been drilled to accept these three pieces and cut off. This provides a particular advantage for Chambers' cold probes because of their small size for rats. The probes are used in rats so the whole cooling probe has been downsized so that the outer shaft is 22-gauge tubing. This 22-gauge tubing size makes it hard to drill out excess solder. Although designed for rats, the figure shows dimensions for a comparable design for rabbits.

The end of the shaft is closed with solder and ground to a rounded or pointed shape. The difficulty is that the solder can be drawn up into the lumen for a long distance. This presents a problem because drilling it out may not be practical (too much heat unsolders the Y-joint or the bits break off). We usually solve this problem by tinning around the outside edge of the shaft, then placing a small amount of solder on the iron and allowing this amount to plug the shaft. It also helps to place the shaft down, so that the solder must rise up into the shaft. About 1 mm of solder at the tip is good. If there is more solder then the excess can be drilled from the inside by hand or the outside can be ground down until about 1 mm of tip remains. Alternatively, the end of the shaft can be crushed, soldered, and then shaped by grinding. Tapani Korhonen solves the problem of closing the end of the shaft by placing a small piece of 23-gauge tubing in the end, soldering and grinding it.

Now that the Y-joint has been created, next create the inner shaft. The inner shaft consists of a 35 mm long piece 30 gauge stainless steel tubing soldered to a 1 cm piece of 23-gauge tubing near one end. The 23-gauge tubing is just a spacer that helps in joining the 30- and 19-gauge shafts to-

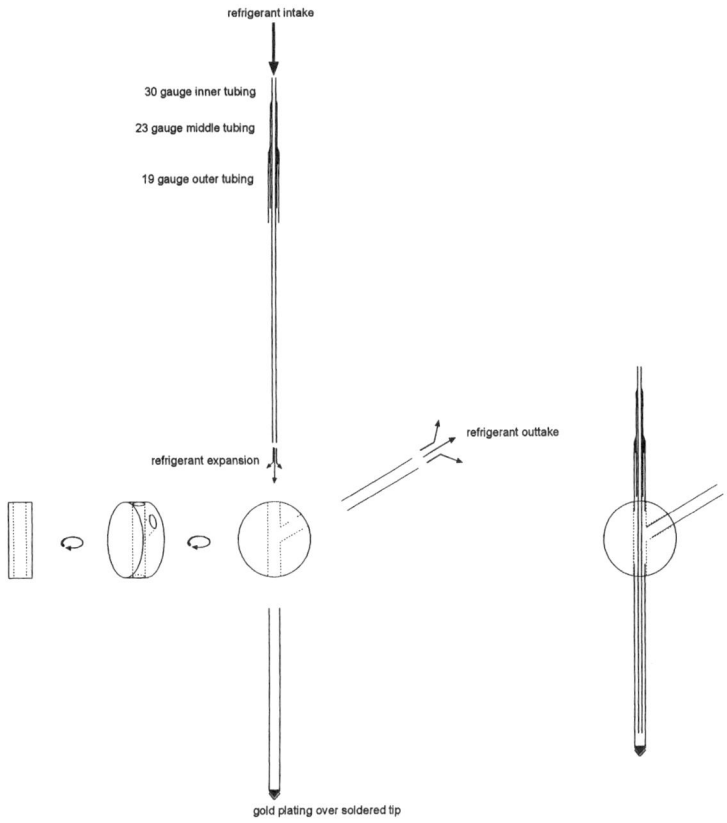

Figure 8.6: Makena's simple cooling probe with manifold Y-joint made by Small Parts, Inc.

gether. The placement of the 23-gauge tubing is not critical except to make sure that it does not block the Y-joint. The 23-gauge tubing and its solder is also used to join the plastic tubing that brings the refrigerant to the 30-gauge intake. The only tricky part about the inner shaft is to make certain that its ends are patent. We test this by using a syringe filled with water and attaching it with plastic tubing to the 30/23-gauge joint and checking the flow. A better test is to hook the inner shaft up to the refrigerant system and observe a 2-3 inch symmetrical jet of refrigerant from the end of the inner shaft—this is where the expansion of the refrigerant has its cooling effect in the final cooling probe. Clogged or poor flow can be solved by acid etching the ends of the 30-gauge tubing or by using a small tool to ream the inside edges of the tubing (Korhonen's method).

Once the inner shaft has been constructed it is placed inside the outer shaft and soldered in place. The distance from the end of the inner

Chapter 8

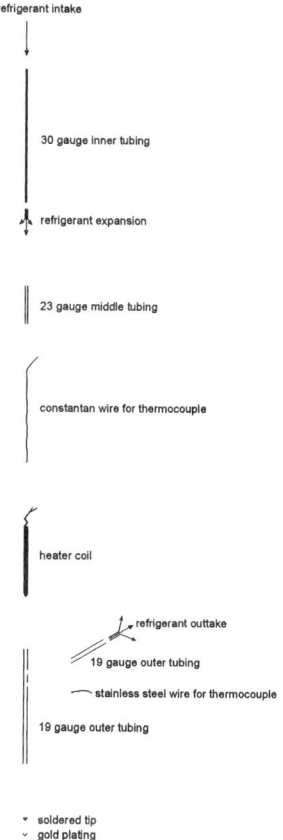

Figure 8.7: Exploded view of the components for a full version of a cooling probe.

shaft to the end of the shaft is 2 mm. This is the area where the refrigerant expands from the 30-gauge inner shaft within the 19-gauge outer shaft. The cooling probe is checked by using a syringe to push water from the 19-gauge extension backwards through to the 30-gauge. A stream of water several inches long is a good indication that the lumen are patent.

The components for the full cooling probe and their relationships are illustrated in Figure 8.7. The full version includes an internally placed constantan wire to construct a thermocouple and a coil of wire used to make an internal heater for the shaft. The thermocouple is made by soldering the constantan wire to the inside of the tip of the 19-gauge stainless steel shaft. The difference in the metals of the wire and shaft create a voltage which is proportional to the temperature. This difference can be used by itself to measure the temperature, or used in a circuit described by Zhang et al. (1986) to measure the difference in temperatures at the cooling probe and at

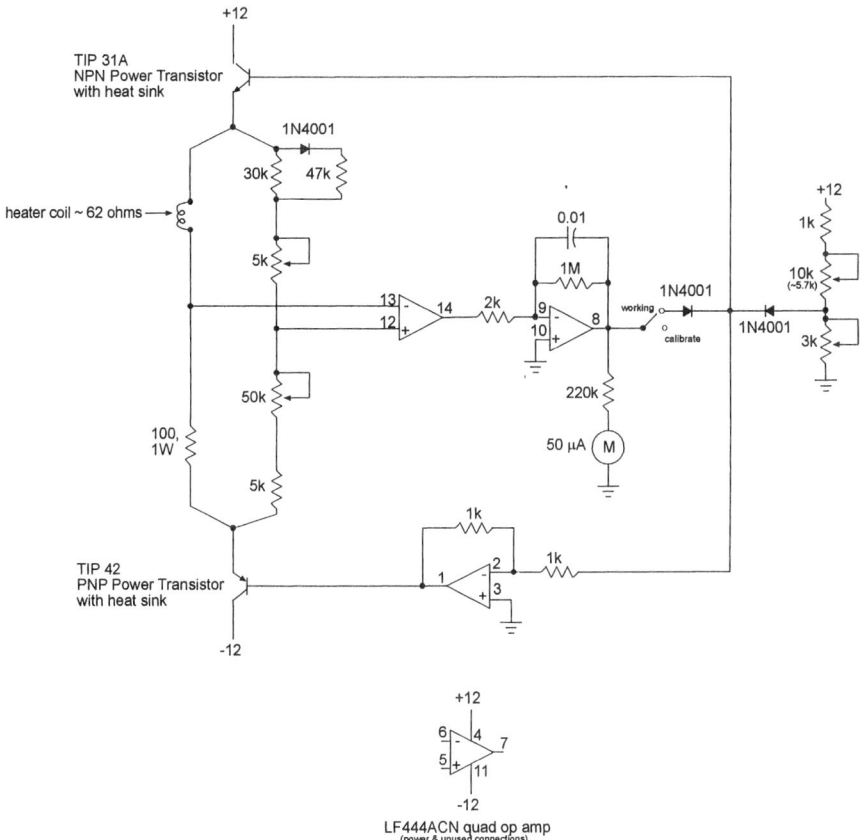

Figure 8.8: Clark's modification of circuitry to control cooling probe's internal heating coil.

a distant site on the animal (see their article for that circuit). In either case the thermocouple needs to be calibrated by placing it in hot and cold baths monitored with mercury or electronic thermometers. We use boiling water and ice cold water for the extremes.

The heater coil is used to keep the shaft at body temperature while the tip of the cold probe gets cold. It is made by taking about 11 inches of Teflon coated stainless steel wire, doubling it over, and taking the bight (the doubled middle) and winding it around a 27-gauge wire or tubing until the windings are about 15 mm in length. This coil is then placed inside the 19-gauge outer shaft, and the constantan wire and inner shafts are placed inside the coil. The heater circuit described by Zhang et al. (1986) measures the resistance of the heater coil (see their article for that circuit), which for us is about 60 Ω at room temperature. When the temperature falls, the resistance of the coil drops. The circuit compensations for this change in resistance by placing a greater amount of current through the coil, which heats it up, much like the coils one sees inside a toaster. With the increased current and subsequent increased heat, the coil's resistance is maintained at its original normal temperature, thus heating the shaft. As a result, the heater circuit is self-regulating to maintain body temperature. The heating circuit and the modifications by Bob Clark (see Figure 8.8) actually work quite well, but we are distrustful of placing elements that literally glow red like a toaster inside of an animal.

The full version of the cooling probe is illustrated in Figure 8.9. It takes us about 8 hours to construct the full version of the cooling probe. The greater difficulty in construction is related to the delicacy of the constantan wire we use to create the thermocouple and the insulation on the heater coil. The delicacy of the constantan wire is due to our inheritance of a large coil of very thin constantan wire, which breaks easily from Paul Chapman. We could get thicker wire and save ourselves some trouble. The problem with the coil is that it must be electrically insulated from the cooling probe, and the insulation is often compromised in turning the coils or with inserting the coil inside the outer shaft, or inserting the thermocouple wire and inner shaft inside the coil. A simple check with a resistance meter shows whether there is a short between the coil and the cooling probe. With rat-sized cooling probes the internal heating coil becomes impractical.

The difficulty in construction that favors making the simple probe over the full cooling probe are not critical, however, for controls for cooling along the shaft are often readily available in the experimental data. Take cooling the interpositus as an example. First, despite years of experience, multiple coordinates and recordings for placements, some cooling probes end up too lateral. Cooling probes 1.5 mm away from the interpositus in the dentate nucleus are effective in abolishing learned, classically conditioned responses. Cooling probes 2.5 mm away from the interpositus (1 mm more lateral) have no effect on classical conditioning. Second, cooling probes that are intentionally placed 1 mm higher, thus still cooling the tissue above the interpositus but not cooling the interpositus itself, also have no effect on classical conditioning. Third, cooling probes that are placed at different angles of approach to the interpositus have the same effect on classical conditioning although their shafts are cooling different areas. The effects on behavior of these different controls are much more meaningful than controlling the temperature of the shaft through heating.

Controlling Cooling

The cooling probe temperature is a function of many variables, re-

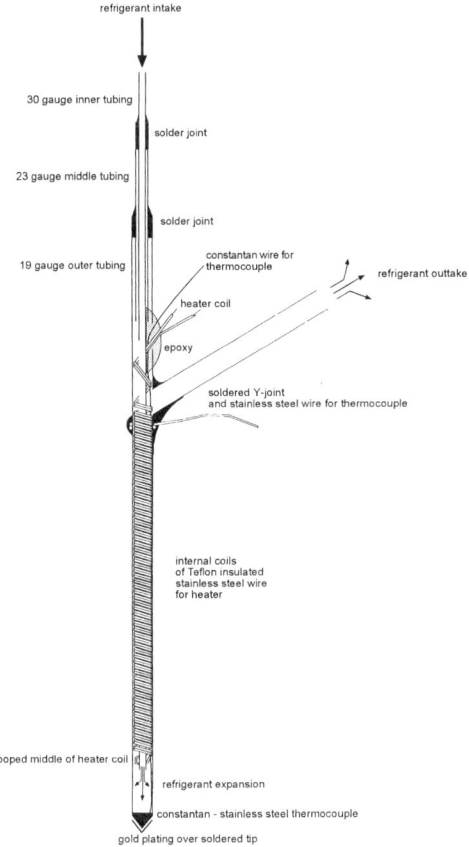

Figure 8.9: Assembled full version of a cooling probe with soldered Y-joint and internal heating coil.

ducing primarily the rate of flow of the refrigerant. Flow rate is a function of the refrigerant pressure and the smallest constriction in the path. For our purposes, the pressure in the refrigerant source does not seem to be exceptionally critical, as we have used small cans of Freon as originally described in the Zhang et al. (1986) article and now use 30 lb. tanks of HFC134a. The major difference between the two sources is that the larger tanks leak much less than the small cans. For us it was more critical to have a vacuum source on the outlet of the extension. However, Tapani Korhonen was able to create a reliable cooling system without a vacuum. That may be related to the different fuel injector (solenoid) that he used initially. We are guessing that modern fuel injectors have different size openings than the Volkswagan solenoid Zhang et al. and we use.

The solenoid is the major controlling element that determines the

Chapter 8

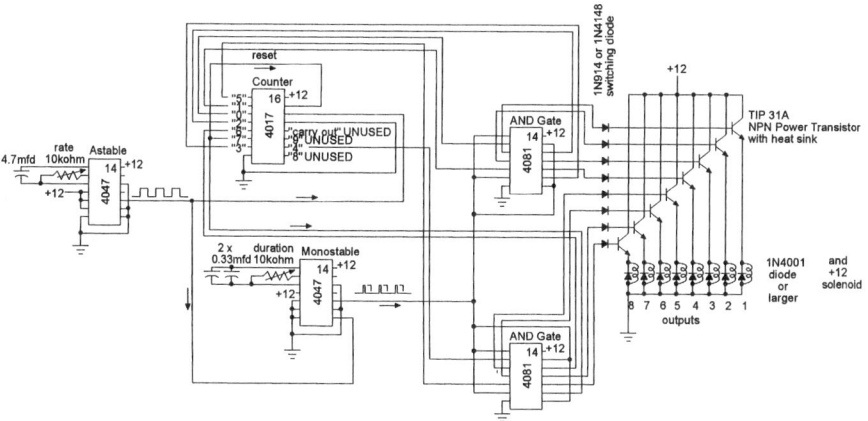

Figure 8.10: Multiplexed controller for multiple cold probes.

rate of cooling. Figure 8.10 shows our CMOS version of the Zhang et al. cooling circuit. An astable (4047) creates a time base that is both counted (4017) and generates short monostable pulse (a second 4047). The astable is used to select one of eight solenoids for cooling experiments. This could be used to control single cooling probes in eight different animals, or eight cooling probes in one animal (e.g., in the hippocampus bilaterally), or something in between. The genius of the original Zhang et al. circuit, incorporated here, is the multiplexed output so that only one solenoid is activated at a time, thus reducing the current requirements of the power supply. The monostable is used to determine the 'on' time for the solenoids. The count and monostable are logically ANDed (4081) so that only one solenoid is on at a time for a set duration. Figure 8.11 shows a second version of the basic circuit, where up to eight solenoids are controlled, but using two different time bases with separate monostable 'on' times for each half of the solenoids. The advantage of this circuit is that cooling can be individually adjusted for two different animals (or locations). Alternatively, Tapani Korhonen dispensed with the fuel injector/solenoid altogether. Korhonen simply regulates the rate of HFC134a flow by pinching the tubing with a standard screw clamp. In essence, Korhonen's is an analog solution and the Zhang et al. design is a digital (all or none) solution to the problem of regulating flow. The low-tech Korhonen solution is inexpensive, easily adjusted and very reliable from session to session.

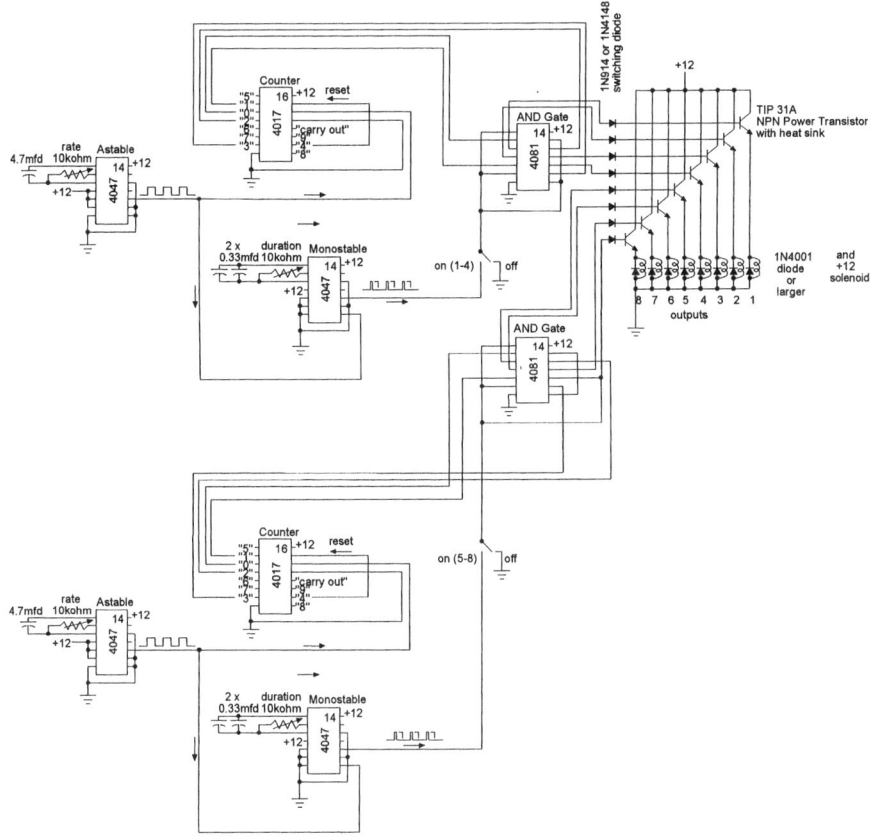

Figure 8.11: Dual multiplexed controller for multiple cold probes.

CHEMICAL INACTIVATION METHODS

Using surgical implant and delivery methods identical to those described above for permanent chemical lesions, it is possible to use pharmacological methods to temporarily and reversibly lesion selected brain areas. Three popular inactivation agents include Lidocaine, TTX, and Muscimol.

Lidocaine is a local anesthetic that is commonly used for minor surgical procedures. It can also be infused into a brain area to produce a temporary lesion. Lidocaine (or Lignocaine) is an amide local anesthetic that like other local anesthetics, slows down the depolarization of the nerve cell membrane. Its effect is based on the interaction with a specific receptor site in the sodium channel. It can have a stimulating or a sedative effect on the central nervous system. Lidocaine's effect is not specific to cell bodies or axons, so this agent is not useful when it is desirable to selectively inactivate specific components of the neuron. We have used Lidocaine successfully in a number of studies. For example, in an attempt to establish the output pathway from the cerebellum for the eyeblink CR, we infused Lidocaine in the interpositus nucleus while recording neuronal activity from the red nucleus in some rabbits, and also infused Lidocaine in the red nucleus while recording activity from the interpositus nucleus in other rabbits (Chapman et al., 1990). In this study, we infused 3 μl of 4% Lidocaine through an indwelling cannula using an injection pump set at a constant rate of 1.1μl/min. In this study and others we have conducted, our behavioral evidence suggests that the effects of the Lidocaine infusion start to subside about 5 min after infusion is stopped. How did we determine this? The answer is: With a behavioral assay. Lidocaine infusion into either the red nucleus or the interpositus nucleus produces an abolition of eyeblink CRs. We typically see the reemergence of some CRs about 5 min after Lidocaine infusion is stopped.

Tetrodotoxin (TTX) is a naturally occurring toxin that comes from the pufferfish. It can also be used as a very effective inactivation agent. TTX is an especially potent neurotoxin and it works by specifically blocking voltage-gated sodium channels on the surface of neuronal membranes. The TTX molecule consist of a positively charged guanidinium group and a pyrimidine ring with additional fused ring systems. The TTX-Na^+ Channel binding site is very strong. The TTX molecule mimics the hydrated sodium cation, it enters the mouth of the Na^+-channel peptide complex, binds to a peptide glutamate side group, among others, and then further strengthens its hold when the peptide changes confirmation in the second half of the binding event. Following the complex conformational changes, TTX is further electrostatically attached to the opening of the Na+ gate channel. As a testimomy to its potency, it is interesting to note that a single milligram or less of TTX is enough to kill an adult human. TTX has been used in a variety of in vivo and in vitro experiments designed to study the electrical properties of neurons. It has also been infused in behavioral experiments to produce a relatively long-term, yet reversible lesion. For example, Thompson and colleagues infused TTX into the superior cerebellar peduncle as a means of interrupting output from the interpositus nucleus during acquisition and performance of the classically conditioned eyeblink response (see Krupa and Thompson, 1995, for details).

Finally, Muscimol has become a popular pharmacological agent for producing a temporary inactivation of neuronal activity. Muscimol is a $GABA_A$ agonist, which means that it binds at $GABA_A$ receptors resulting in the generation of Inhibitory Post-Synaptic Potentials (IPSPs) in those cells affected by the compound. The net effect is that the cells are prevented from becoming excited, hence effectively inactivating the target population of neurons. We have used Muscimol to check whether or not our cannulas were placed correctly for the infusion of other substances (e.g., Chen and Steinmetz, 2000). For example, we have studied the involvement of cerebellar NMDA receptors in eyeblink conditioning by infusion AP5, a potential NMDA receptor blocker into the cerebellum before and after conditioning. At the end of the experiment, we infused 2 µl of 400 ng/µl Muscimol through the same cannula that we previously used to infuse the AP5 and observed the behavior of the rabbit. If the eyeblink CRs disappeared in the rabbits after Muscimol infusion, we assumed that the cannula were aimed at the location in the cerebellum that was critical for conditioning. This technique was also used by Krupa, Thompson and Thompson (1997; Krupa and Thompson, 1993) to establish that critical learning-related plasticity occurred in the cerebellum. Using an experimental design that was very similar to the cooling probe experiments that Lavond and colleagues conducted (described above), the effects of the Muscimol infusion in the cerebellum on acquisition and performance of the eyeblink CR was assessed. Similar to cooling, no savings effects were noted during training sessions that followed the Muscimol infusion (i.e., when training was conducted under the Muscimol inactivation, it appears that cellular processes associated with learning were not engaged). These data add to the growing body of evidence that demonstrates the cerebellum as a locus of important plasticity processes associated with eyeblink conditioning.

CONCLUDING REMARKS

Historically, the lesion method has been used extensively to study brain-behavior relationships. The method has not been without its critics. Over the years, people have pointed out the paradoxical nature of the method. That is, it seems odd to study function by creating dysfunction. Others have been quick to point out that after the lesion has been performed, the subject under observation is no longer 'normal.' Still others have raised the issue of the localization of the effect—has the lesion affected neurons at the target site or has the lesion affected fibers *en passant*? Even given these valid criticisms, we believe that the lesion method, when properly conducted, is an extremely valuable tool and the recent improvements in methodology have addressed some of the problems that have been raised. For example, the use of chemical lesion techniques have addressed the specificity of the lesion location issue. Also, the development of reversible lesions or inactivation methods such as cooling and Muscimol infusion have addressed issues of function versus dysfunction. It seems likely that brain lesion methods will continue to be used effectively for years to come to study brain-behavior relationships.

Chapter 9

BRAIN STIMULATION TECHNIQUES

OVERVIEW

To this point, we have covered two basic neurobiological techniques that have proven useful for the study of brain-behavior relationships; neuronal recording and brain lesions. In this chapter, we discuss methods associated with brain stimulation, another research technique that has proven useful for advancing our understanding of brain function. Brain stimulation is easy to conceptualize: Electrical current is used to activate excitable brain tissue, either cell bodies or axons, in essence providing a means to induce brain activity that is under the control of the experimenter. One way to consider this is that the brain stimulation is used as a substitute for activity that is either generated in, or projected to, neurons that are located at the point of stimulation. An example will illustrate the use of this technique.

We used electrical brain stimulation in our efforts to define the basic brain circuitry involved in classical eyeblink conditioning (Rosen, Steinmetz & Thompson, 1989; Steinmetz, Lavond & Thompson, 1985, 1989; Steinmetz, Rosen, Chapman, Lavond & Thompson, 1986; Steinmetz, Rosen, Woodruff-Pak, Lavond & Thompson, 1986). For example, brain stimulation techniques were used to study the neural pathways along which the CS was projected to the cerebellum during eyeblink conditioning. Given the huge number of neurons in the pontine nuclei that send afferent input to the cerebellum, we suspected that some of these neurons may be involved in projecting the CS to the cerebellum. Using a click stimulus (to mimic an acoustic CS) we systematically recorded from several pontine nuclear areas (i.e., we "mapped" the pontine nuclei) and identified several areas that were responsive to auditory stimuli. Many of these responsive neurons were located in lateral areas of the nuclei. Next, we placed stimulating electrodes into the areas where acoustic-related responses were recorded and substituted electrical stimulation of these areas for the tone CS normally used during training. We produced rapid acquisition of the eyeblink CR with this method, thus demonstrating that the activation of the pontine neurons by the electrical stimulation could *substitute* for the peripheral tone that was normally used. This study provided early evidence in support of the idea that neurons in the pontine nuclei provided relay points into the cerebellum for acoustic stimuli used in conditioning.

Like all techniques, electrical stimulation of the brain must be done properly to provide meaningful experimental results. Unfortunately, very few researchers take into consideration all of the factors that are important in designing and executing brain stimulation experiments, factors such as stimulus intensity, effective spread of current, and choice of electrodes. Indeed, some researchers seem to have the attitude that given a stimulator, electrode, and a subject, anyone can conduct "foolproof" brain stimulation experiments with little experience or background. As evidence of this, one of us was at a grant review panel meeting where a study using brain stimulation was being considered for funding. During the meeting, two of the panel members noted that even though the principal investigator had no previous experience with using brain stimulation, that was not considered to be a weakness because "anyone could do brain stimulation." We do not agree with this position but rather believe that brain stimulation, when done carefully, can be a powerful tool in the "bag of tricks" available for brain scientists. In this chapter, we discuss several things that should be considered when using this technique.

EXCITING NEURONS

Perhaps the easiest thing to forget when using brain stimulation is that the fundamental goal of the technique is to electrically excite individual neurons. In essence, when using this technique, we are not trying to excite a specific brain region or area, but rather are trying to excite several individual elements (i.e., neurons) that make up a brain area. This could involve the electrical activation of either cell bodies or axons or both.

Basic Fundamentals of Brain Stimulation

We can ask ourselves several basic questions concerning brain stimulation. First, what do we need to electrically stimulate the brain? We need a current source (usually a commercially available stimulator), two electrical poles (usually electrodes) that constitute a point of delivery and point of return for the current, and a living brain (or equally viable part of the nervous system). Second, how does electrical stimulation excite neurons? All students of brain science know that neurons at rest have a more negative charge inside of the cell relative to outside of the cell (ranging from -50 to -90 mV, depending on location). Assuming extracellular stimulation techniques (we will not consider intracellular stimulation techniques

Chapter 9

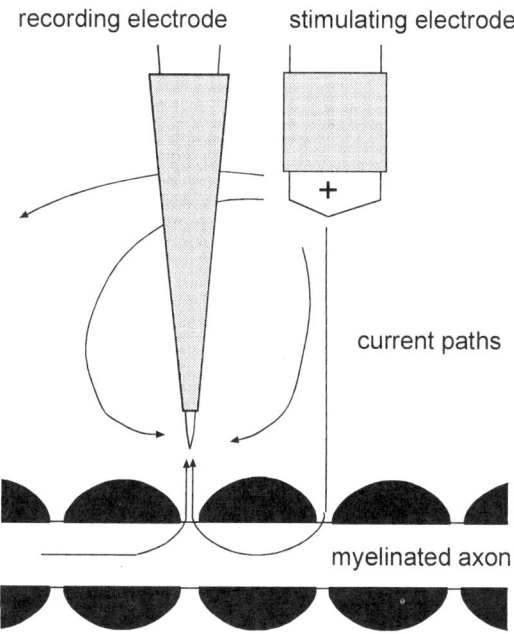

Figure 9.1: Current flow from the stimulating electrode to surrounding tissue, the recording electrode and an axon.

in this chapter) electrical stimulation works by reducing the resting potentials (i.e., depolarizing the cell). More specifically, a negative (cathodal) stimulation pulse delivered outside of the cell produces this depolarization and this change in potential leads to an action potential. Third, how does the cathodal stimulation pulse actually depolarize the membrane? The answer to this question is that the current that is delivered activates voltage-sensitive (excitable) ion channels that depolarize the membrane. More specifically, the current opens many Na+ channels and the entry of these positive ions make the cell more positive (less negative) than their resting state (see Yeomans, 1990, for a more complete explanation of this mechanism). Finally, how is the electrical circuit completed to produce current flow during brain stimulation? For typical extracellular stimulation experiments, this is shown in Figure 9.1. To effectively activate neural tissue, current must flow out of the neuron through the resistance of the membrane then complete a circuit by entering into the neuron at some distant point. It is the entry of current at the distant point on the neuron that actually produces the depolarization of the neuron (this process is referred to as anodal excitation).

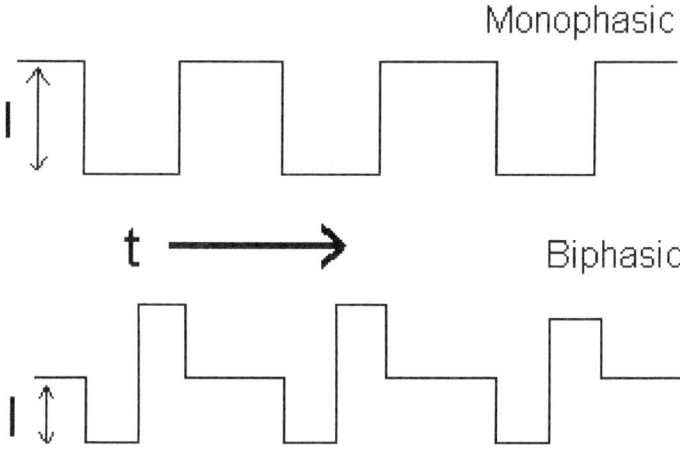

Figure 9.2: Monophasic and biphasic stimulation. Biphasic pulses mitigate metal electrode polarization during stimulation and glass electrodes clogging during iontophoretic injections.

Choosing a Current Source for Brain Stimulation

Most current sources for brain stimulation (commonly called "stimulators") generate square wave pulses because this type of waveform is most effective at activating neurons (see below for more details about the stimulus used in brain stimulation). Weighing the cost of a commercial stimulator against the complexities of building one, we do not advise that the researcher try to build their own stimulators. Many different kinds of stimulators are commercially available from companies such as Grass Instruments and WPI, and these devices generate precise, highly regulated waveforms that are easy to manipulate.

Models of stimulators vary somewhat in their features but there are some commonalities. The stimulator generally provide two different types of output: monophasic, biphasic or DC (see Figure 9.2). Monophasic output is a square-wave pulse of one polarity (either plus or minus) while biphasic output equalizes current flow in either direction. DC (direct current) outputs are not common on stimulators made nowadays, due to the potential hazards of inadvertently delivering DC (lesion) current to nervous system sites. As depicted in Figure 9.3, three different output modes are also typically avail-

Chapter 9

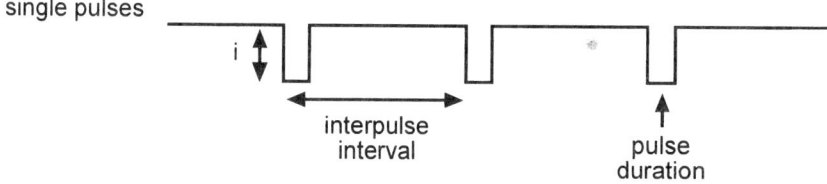

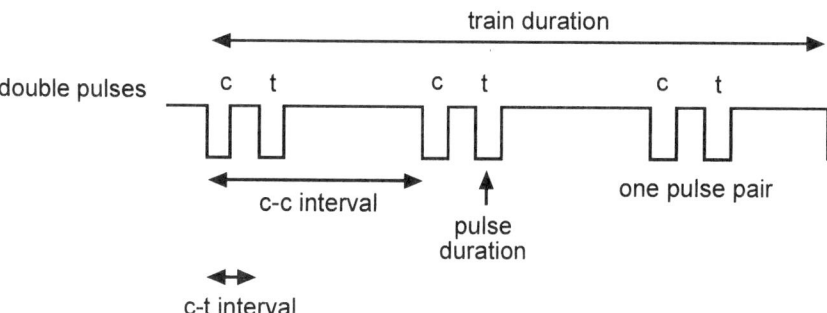

Figure 9.3: Stimulation parameters include current (i) intensity, duration and frequency. Temporal parameters of stimulation include single, twin (double) and train patterns. In double stimulation, the first pulse is the conditioning (c) pulse and the second pulse is the test (t) pulse. Both c-t and c-c intervals can be varied. Trains are repeated patterns of stimulation.

able on stimulators: single (one stimulus pulse is delivered); twin (two successive pulses are delivered with the inter-pulse interval set by the delay function of the stimulator); and trains (a series of current pulses delivered with the duration and frequency set by function knobs on the stimulator). Generally, the frequency, duration, and voltage of the stimulator output are set with separate controls. The ranges of these stimulus parameters vary somewhat from stimulator to stimulator. Some stimulators have controls for setting output delays (relative to a pulse used to trigger the stimulus). The delay function is helpful for positioning the response on output devices, such as oscilloscopes, computers, and polygraphs. All stimulators have polarity switches for setting the polarity of the square-wave output. Many stimulators can be triggered by other devices using "sync in" or "triggered output" settings. These are circuits that look for an input voltage to trigger the stimulator instead of a manual control. This mode is particularly useful for remote control of the stimulator via a computer interface. Finally, most stimulators have switches that allow them to be operated in continuous or single pulse mode.

Two other devices are sometimes used together with brain stimula-

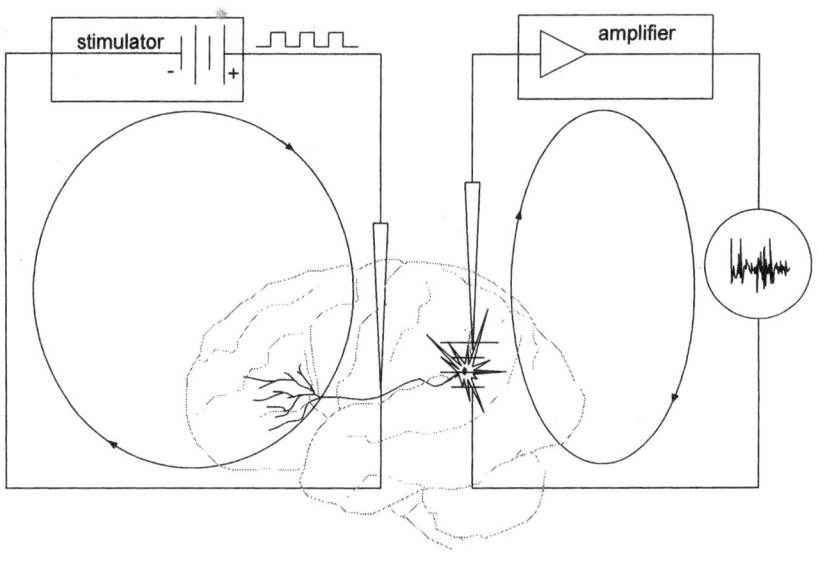

Figure 9.4: Stimulation artifact in the recording is controlled by confining the flow of current in the respective paths. Here the stimulator is shown as one battery source and the neuron as a second battery source for two separate circuits. Isolation is typically achieved by controlling the stimulator through an optical (optoisolator) or inductive (transformer) element. These and other concepts for stimulation and recording are expanded further in Figure 9.7.

tors; constant current units and stimulus isolation units. First, the property of having 'constant current' is essential for brain stimulation experiments. It is very important to think of brain stimulation as delivering current and not voltage. We excite neurons by passing current, and not voltage, through them. Hence, when reporting parameters used in stimulation experiments, it is fairly useless to report stimulation voltages and much more informative to report stimulation currents. Furthermore, to interpret results of stimulation experiments, it is important that stimulation current be held constant during an experiment and this requires special circuitry which may be a separate piece of equipment (e.g., when used with Grass Instruments stimulators) or built into the stimulator (e.g., the WPI stimulators). Over the course of the experiment, the total resistance in the stimulation circuit may vary due to changes of resistance in the subject or changes in resistance due to alterations of the electrode tip. Constant current circuits monitor the resistance in the circuit and introduce voltage (using basic Ohm's law calculations) to maintain a steady current during stimulation (for the concept of constant currency, see Figure 8.2 in the chapter on lesion techniques). Second, experiments that involve simultaneous stimulation and recording benefit from using stimulus isolation. Some researchers might find stimulus isolation units (SIUs) useful. Often, the voltage used for stimulating is much larger than the voltage response recorded from the preparation and this may produce a large stimulus artifact in the record that could obscure all or part of the neuronal response being recorded. The SIU, when inserted between the stimulator and the preparation, greatly reduces the stimulus artifact by isolating the stimulating current from the rest of the system (see Figure 9.4). Artifact is typically removed using optical (shown) or inductive isolation.

Parameters of Electrical Brain Stimulation

As we pointed out above, it is well known that the most effective waveform for extracellularly activating neurons is a sharp rise time, rectangular pulse of negative current (i.e., cathodal stimulation). For this arrangement, the stimulating electrode nearest the tissue is negative (cathodal) relative to the positive (anodal) return path electrode that is typically placed a distance from the targeted neurons. Great care should be taken not to deliver anodal stimulation as neural tissue can be damaged by this technique when metal ions are emitted from the electrode (see Chapter 8 for information about electrolytic lesions). Remember, the effective electrodes for extracellular stimulation and lesions are opposites: cathodal stimulation and anodal lesion. For at least two reasons, relatively short-duration (0.1 – 0.2 msec) pulses are better than longer duration (greater than 0.4 ms) pulses for brain stimulation. First, short duration pulses reduce the amount of charge needed to excite the neuron and this minimizes the potential for damage. Second, when the pulse is shorter than the absolute refractory period of the neuron, it is difficult to excite the neuron twice.

Frequency of the brain stimulation is a very important parameter for experiments. As a general rule, the frequency of brain stimulation that is used should be within a range that the neurons in the brain region are capable of firing. Theoretically, one could argue that neurons are capable of firing 500 to 1000 times per second (based on a 1–2 msec firing duration). This is not realistic, however, as maximum firing rates recorded in areas around the brain rarely exceed 100 spikes per second. Perhaps a better rule of thumb for choosing stimulation frequency is one cited by Yeomans in his 1990 book, *Principles of Brain Stimulation* (an excellent book for information on brain stimulation, by the way). According to Yeomans, ". . . the frequency of stimulation should be in a range for which all neurons can be expected to fire one action potential per pulse. . . " (pg. 26). Practically, this means a frequency in the range of 1 to 40 Hz is a good starting point.

Current and train duration are other parameters that can be altered during brain stimulation. Most often, increasing the stimulation current increases the area of stimulation and thus activates more neurons. One must be very careful about setting the current parameter, however, because the larger the stimulation area, the more likely it is that neurons outside the target area will be stimulated. Indeed, if these areas inhibit the neurons targeted for stimulation (an arrangement that is not rare in the nervous system), increasing the stimulation current will actually decrease the behavior under observation. Increasing or decreasing the duration of the stimulation train could alter the effectiveness of the stimulation, especially if through some recurrent process or inherent property more and more of the neuronal population is recruited as the train duration increases. Researchers should also be aware that sometimes neurons respond to both onset of a stimulation train as well as offset of the stimulation train.

BRAIN STIMULATION ELECTRODES

"It takes two" is a saying that is very appropriate for brain stimulation electrodes—it takes two electrodes to stimulate brain tissue, one electrode that delivers current from the stimulator to the preparation and a second electrode that provides a return path for the current back from the preparation to the stimulator. With that said, there are basically two arrangements for placing the two electrodes; *monopolar* stimulation and *bipolar* stimulation. For monopolar stimulation a single electrode is placed in the brain and the return electrode is placed on the surface of the brain, on the skull, or sometimes on the surface of the body (such as attached to an ear lobe). Usually a large, flat surface is used for the return electrode (such as a stainless steel skull screw). For bipolar stimulation, the two electrodes are positioned in the brain (and usually somewhat close together). Which method is preferred? Most experts agree that unipolar stimulation is preferred for several reasons: (1) Perhaps most important is the fact that the field of stimulation is not well understood for bipolar stimulation. Consider the situation where two electrodes are placed in the same nuclei and positioned less than 1 mm apart. When a pulse of stimulation is delivered, one electrode is the cathode and one is the anode thus simultaneous cathodal and anodal stimulation is delivered to the same structure. Things can get very complicated under these conditions when biphasic, instead of monophasic, pulses are used. (2) Bipolar electrodes double the amount of damaged tissue in an area (although concentric electrodes may alleviate this problem; see below). (3) Monopolar electrode configurations that include the return current electrode on the brain eliminate the possibility of stimulating brain tissue through the return electrode. Plus, because it is typically designated as the anode and not located within brain tissue, there is little chance for tissue damage with a monopolar setup. (4) In general, the current density spread is much better understood for monopolar electrode setups where the current density around the cathodal electrode dissipates somewhat uniformly as you move away from the electrode. Why then does anyone ever use bipolar stimulation? There is one advantage of this arrangement—when recording in brain tissue simultaneously with stimulation, stimulus artifacts are greatly reduced with bipolar stimulation because the return electrode is near the source electrode.

For brain stimulation experiments we have conducted, we have found that using epoxy-insulated 00 stainless steel insect pins produce the best results (see Chapters 4 and 8 on recording and lesioning for the preparation of these electrodes). For most experiments, electrodes with impedance tips ranging from 200 Ω to 1 MΩ seem to work best. Practically, we have used insect pins with exposed tips of 100 to 250 µm. Tips of these impedances can be obtained by scraping the insulated electrode with a surgical scalpel while looking through a dissection microscope. For unipolar stimulation, the current delivery electrode is lowered into the desired brain region while a stainless steel screw placed in the skull acts as a current return electrode. We have also used two insulated and scraped insect pins place inside the brain for bipolar stimulation experiments. These electrodes can either be glued together, with one exposed tip slightly above the other or placed side by side with 0.5 – 1.0 mm separating the tips (depending on the brain area to be stimulated). Alternatively, concentric bipolar stimulating electrodes are available from suppliers such as FHC, Inc and A-M Systems. These electrode contain a center current delivery electrode that is surrounded by a second current return electrode (like an electrode within a tube). They come in a variety of sizes and impedances with various tip profiles. These electrodes work very well and, depending on their diameter, may cause only minimal tissue damage.

SPATIAL EFFECTS OF BRAIN STIMULATION

An important, yet frequently overlooked, consideration for electrical stimulation experiments in the relative spread of the current during stimulation (i.e., the spatial effects of brain stimulation). In essence, the number of neurons that are stimulated by the electrical current is determined by the spatial effects of brain stimulation. Yeomans (1990) identifies three critical spatial steps on which a stimulus-response relationship depends: (a) placing the electrode such that current is delivered from the electrode through the brain tissue in a manner that electrically excites neurons, (b) reaching a local current density that will excite neurons in the desired population, and (c) exciting enough neurons in the population to produce the response that is desired.

For the most part, the extent of the field of excitation produced by electrical brain stimulation is determined by the current-distance relationship of the neurons that are at the tip of the electrode. The current-distance function can be obtained by systematically lowering a stimulating electrode past a neuron, delivering a pulse of current, and then noting the current required to elicit a single action potential at several points along the microelectrode track. Systematic studies have demonstrated that for myelinated axons, the lowest thresholds for eliciting responses are at nodes of Ranvier (e.g., Tasaki, 1959), while for cell bodies the lowest thresholds for eliciting responses are at the initial segment (e.g., Gustafsson & Jankowska, 1976). Hence, minimum stimulation currents can be used when the stimulating electrode tips are placed close to nodes of Ranvier (for axons) or near initial segments (for cell bodies).

Yeomans (1990) has identified two "second-order effects" of electrical brain stimulation: *block* and *electrode orientation*. Block refers to a stimulation-produced disruption of action potentials. A block can be caused by a variety of things including the collision of action potentials due to high rates of stimulation, strong inward current relative to outward current, and hyperpolarization produced by a stimulation cathode (i.e., outward current flow at a cathodal can cause inward current flow at a distance point which can, in turn, block the conduction of action potentials). The orientation of the electrodes can also determine the relative effectiveness of electrical stimulation. For example, bipolar stimulating electrodes can be placed either longitudinally or transversely with respect to the axons or cell bodies being stimulated (see Figure 9.5). Current thresholds are much lower when electrodes are placed longitudinally because current flows with ease from the anode through the neuron toward the cathode. Threshold current is higher for transversely oriented electrodes because the transverse current is similar along the length of the neuron, hence current cannot exit the neuron effectively.

We encourage readers interested in a more in-depth discussion of the intricacies of brain stimulation to consult Yeomans' discussion of the "ideal bundle model" for brain stimulation (Yeomans, 1990, pp. 47-53). A few highlights are worth mentioning here, however. In most behavioral experiments that involve brain stimulation, the major question asked is: How many neurons must be excited in a population to produce the behavior in which I am interested? The answer to this question is relatively complicated and depends on several factors. First, the density and distribution of neurons is important. In general, the number of axons or cell bodies that is stimulated by a single pulse of stimulation increases linearly with current (assuming a rela-

Chapter 9   287

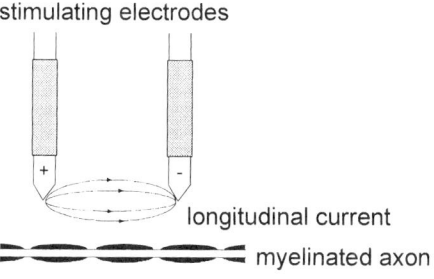

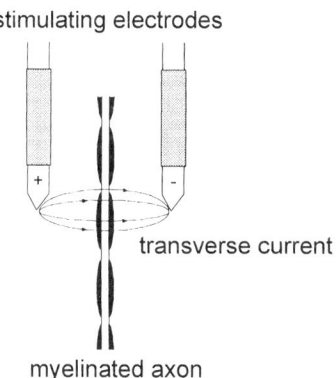

Figure 9.5: Current flow through axons oriented longitudinally or transversely to bipolar stimulating electrodes. Stimulation is more effective (i.e., less injected current evokes a neural response) with a longitudinal orientation.

tively uniform distribution of axons and cell bodies). Second, the size and shape of the tip of the stimulating electrode affects excitation. For a spherical tip, there is a direct relationship between stimulus current and the number of neuronal elements that are activated. Things are more complicated when the electrode tip is not spherical (as is the case when sharpened electrodes are used). Third, the location of the electrode tip within the neuronal population determines, in part, the effectiveness of the stimulation. Simply put, when a stimulating electrode is not placed in the direct center of the stimulation field the number of neuronal elements that are stimulated is dependent on the area of intersection of two circles that can be constructed—a circle that represents the axonal or neuronal population and a circle that can be drawn around the electrode. Generally, the more that these two circles overlap, the more neuronal elements are activated (see Figure 9.6). Finally, choice of stimulation frequency is very important. The number of action potentials produced by the brain stimulation is highly dependent on stimulus frequency. On one hand, relatively low frequencies may fail to recruit

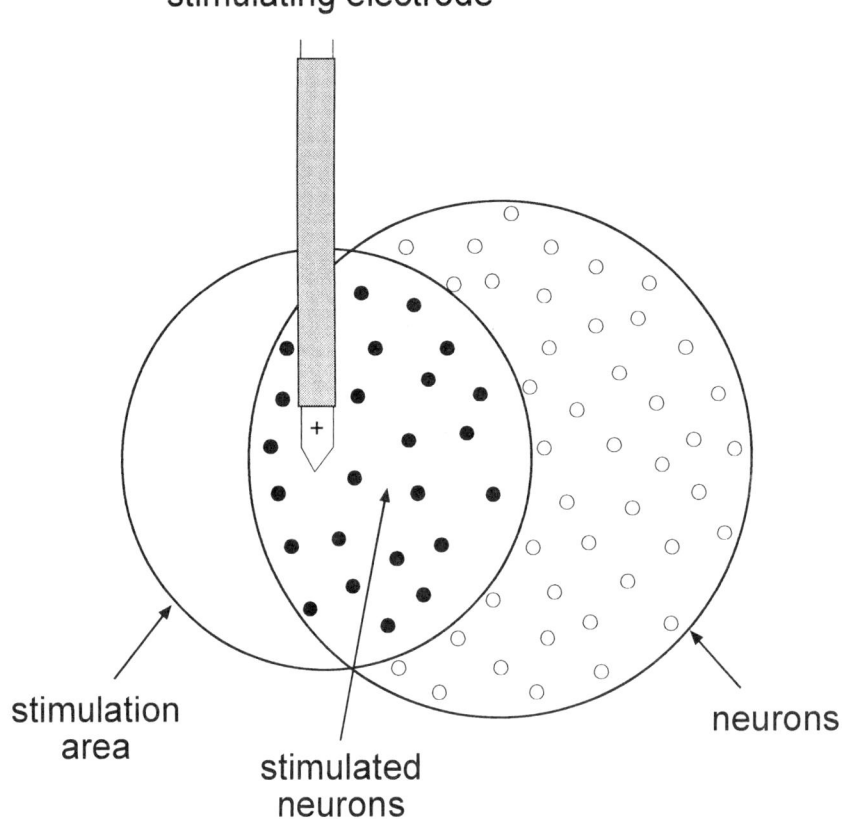

Figure 9.6: Overlap of the stimulation current and the neuronal population is one of the factors that determines the number of neurons that are affected by the stimulation. Other factors are neuron density, the size and shape of the stimulating electrode tip, and stimulation frequency.

enough neurons to produce the desired behavioral effect. On the other hand, relatively high frequencies may reduce stimulation effectiveness because due to inherent properties or population effects (e.g., nearby inhibitory influences) the neurons are incapable of tracking the stimulation (i.e., firing at the frequency being used in the experiment).

SOME APPLICATIONS OF BRAIN STIMULATION

Over the years, we have successfully used brain stimulation techniques in a variety of situations. As a graduate student, Steinmetz used peripheral nerve stimulation to study spinal cord plasticity processes in two model systems of learning and memory (e.g., Beggs, Steinmetz, Romano & Patterson, 1983; Steinmetz & Patterson, 1985). Steinmetz and Lavond have used brain stimulation to study the neuronal pathways that project the CS and US to the cerebellum during the classical conditioning of skeletal muscle responses (e.g., Steinmetz et al., 1986). Also, we have used brain stimulation together with brain recording to study brain structures and systems involved in classical eyeblink conditioning (e.g., Gould, Sears & Steinmetz, 1993). As examples of how brain stimulation techniques can be used, we briefly describe these experiments here.

The flexor reflex response in the mammalian spinal cord can be classically conditioned by delivering paired presentations of electrical stimulation to a peripheral nerve as a CS and electrical stimulation to the ankle skin as a US (e.g., Beggs et al., 1983). The CS is a relatively low intensity train of electrical stimulation delivered to a cutaneous nerve (such as the superficial peroneal nerve) while the US is a very intense electrical stimulation of the skin that overlies the ankle. The animals in these acute experiments are first anesthetized, then spinalized. Conditioning is measured as either an increase in the CS-evoked EMG response recorded from the flexor muscle in the leg or an increase in the CS-evoked whole nerve response recorded from a motor nerve that innervates the flexor muscles. For this preparation, sensory nerves (for CS delivery) and motor nerves (for CR recording) are dissected from the hind leg and positioned over bipolar hook electrodes made from platinum-iridium wire. The nerves are typically cut and the distal ends of the nerves crushed and secured to the bipolar electrodes by suturing them to the distal electrode of the pair. Using similar methods, Steinmetz conducted a series of experiments that examined a nonassociative form of spinal cord plasticity known as peripherally induced spinal fixation (e.g., Steinmetz & Patterson, 1985).

As discussed briefly at the beginning of this chapter, we have also effectively used brain stimulation techniques to delineate the pathways that carry the CS and US to the cerebellum during classical conditioning of skeletal muscle responses. The logic of these experiments was quite straightforward. We reasoned that if you substituted electrical stimulation delivered to a brain area suspected to be the part of a pathway important for transmission of the CS or the US, that conditioning would occur if the stimulation was paired with the US or the CS, respectively. Using this method, we discovered that stimulation of the basilar pontine nuclei served as a very effective CS, thus suggesting that this structure was a critical relay for CS information to the cerebellum (Steinmetz et al., 1986). We also substituted electrical stimulation of the inferior olivary complex for the peripheral air puff normally used during eyeblink conditioning (Mauk et al., 1986). First, we discovered that dependent on the precise location of the stimulating electrode within the dorsal accessory portion of the inferior olive, a variety of different behavioral responses could be elicited by the stimulation (e.g., eyeblinks, head-turns, limb movements, etc.). Second, pairing a tone CS with the IO-stimulation US produced robust conditioning thus demonstrating that the IO stimulation could effectively substitute for the peripheral stimulation. We concluded this series of experiments by successfully obtaining classical conditioning by pairing a pontine-stimulation-CS with an inferior olive-stimulation-US (Steinmetz et al., 1989). Most often, chronically implanted bipolar stainless steel stimulating electrodes were used to deliver the electrical stimulation in these experiments.

We have also used electrical stimulation to explore patterns of convergence of CS and US pathways in the cerebellum (Gould et al., 1993). In these experiments, we placed stimulating electrodes in either the pontine nuclei or the inferior olive (or both) and systematically lowered recording electrodes through the cerebellar cortex or the deep cerebellar nuclei. At each recording site, we delivered single pulses of stimulation at various stimulus intensities and looked for activation of a population of neurons in the cerebellum, thus establishing connectivity patterns and putative locations of CS-US convergence. The elicitation of field potentials by electrically stimulating inputs to the population of neurons from which recordings are made is a classic means for studying input-output or stimulus-response relationships. Indeed, this method has been extensively used by researchers interested in LTP and LTD and other cellular plasticity processes (e.g., Bliss & Lomo, 1973).

REMOVING STIMULATION AND ENVIRONMENTAL ARTIFACTS FROM NEURAL RECORDINGS

Good neural recordings can be impaired by a number of factors that create artifacts or noise (nonneural signals) in the record. One source of artifact is the stimulation we introduce into the neural tissue. As discussed below, there are a number of ways to deal with stimulation artifact, including consideration of the recording amplifier's supply voltage, using bipolar stimulation, implementing stimulus isolation (optical or inductive), placing the orientation of the electrodes with respect to the recording electrode, and using a sample-and-hold circuit. A second source is environmental artifact that is mostly caused by 60 Hz signals that are transmitted through the air by wall-powered equipment (especially lights), but also may be higher frequency noise that originate from computers and their monitors. Turning off the source, putting the source at a distance, and shielding the recording from the source are the major techniques to handle environmental sources. Shielding, in turn, can introduce another problem, the 'ground loop,' which causes high frequency noise on top of the neural signals, which can be eliminated with careful attention to grounding. Finally, power supplies for the equipment may introduce high frequency noise. This is particularly true of switching power supplies. The solutions are to use a linear power supply or, much better, to use batteries as the power supply for critical, low amplitude signals. We briefly consider these issues next.

Stimulation Artifact and Filtering

Stimulation can wreak havoc when simultaneously trying to take neural recordings. The stimulation signal can be far larger than the neural signals, hiding neural activity from view. Further, subsequent filtering that is meant for conditioning the neural recordings can smooth out and extend the duration of the stimulation artifact. As a result, the short duration stimulation artifact can be lengthened so much in time that neural responses may be masked. One strategy is to use as little filtering as possible, especially early in the amplification process. That is, amplification without filtering near the recording electrode will increase the size of the neural signal relative to the stimulus artifact (if the stimulation artifact is saturated then it cannot get any larger; see saturation, below). Once the neural signal is of an adequate size then filtering can be applied in later stages of amplification.

Differential Recording

One important technique to eliminate stimulation artifact is to use differential recordings between two electrodes and ground, where a common signal on the two nearby electrodes is cancelled out leaving only the signals that are unique to each recording electrode. The common signal could come from stimulation or from environmental noise (60 Hz signals transmitted by fluorescent lights, for example, are environmental artifacts). It should be noted that almost any appliance (light, motor, computer) that plugs into the wall can act as a transmitter source of 60 Hz electrical noise signals (artifacts) through the air. The heart rate amplifier that we previously described uses differential recording to cancel out common 60 Hz (and other) noise, leaving just the heart's activity in the record (see Figure 3.10). Differential recording is potentially applicable for eliminating common sources of noise such as 60 Hz or stimulation artifact from neural recordings.

Saturation

Differential recording is more effective for eliminating environmental noise, however, mainly because the size of the stimulation artifact may be too large for the active elements (op amps) used for differential amplifiers. Saturation occurs when the input signal to the amplifier is greater than the power supply for the amplifier. A common consequence for semiconductors that are powered by ±15 volts (or less) is that the amplifier may temporarily turn off; worse, the op amp may be permanently damaged. Differential amplifiers made from devices that have higher voltage supplies, for example, older technology high-voltage vacuum tubes or newer technology high-voltage semiconductors, may be a solution for handling these large voltage stimulation artifacts.

Orientation

The orientation of the two recording electrodes to the stimulating electrode or bipolar stimulating electrode may also influence differential recordings. Ideally, the two recording electrodes should sit on the same isopotential line for current spread. If the electrodes are on different current lines then one electrode will 'see' more current than the other. As a result, the stimulation signal is not fully 'common' (not the same) in both electrodes, therefore the differential amplifier cannot cancel them out. One painstaking way around this problem is to selectively amplify (or attenuate) one of the signals to match the signal on the other differential electrode, thus cancelling out the now common signal. As this requires individual adjustment and specialized circuitry it is not a common strategy for neural recordings.

Shielding

Proper shielding is important for preventing environmental artifacts from obliterating neural signals. The major source of environmental artifact is 60 Hz noise. Solid metal or metal mesh screens that are grounded will intercept transmitted signals when placed between the source of noise and the recording electrode/initial amplification. Lavond's running chambers are metal acoustic chambers from Industrial Acoustics that we have grounded. In stubborn or critical places (acute surgeries or slice experiment laboratories) we have constructed Faraday cages (metal screen that is grounded) that contain the experimental recording and preparation. We have seen others use sheets of aluminum foil that are grounded to intercept transmitted noise from 60 Hz sources.

Keeping it at a Distance

Distance itself is an effective deterrent for many sources of environmental noise: The further away an offending source of noise, the better. Remotely located light sources are good. A relatively new source of transmitted noise comes from the widespread use of computers and computer monitors to control experiments and collect data. Keeping these sources away from the recording and preparation is often an effective and simple solution.

Turning it Off

Nothing could be more effective or cheaper than simply turning off the transmitter. Turning off fluorescent lights, in particular, often results in major improvement in the recorded neural signal. The general rule is, whenever there is 60 Hz noise, turn lights and equipment off systematically to see when the noise goes away (i.e., what caused the noise). Note that 60 Hz noise is easy to identify on an oscilloscope by observing the standing wave when the oscilloscope is set for 'normal' or 'line' horizontal sweeps. This setting triggers a sweep across the screen in synchrony with the 60 Hz line frequency coming from the power company. Battery powered flashlights are used to illuminate notes and stereotax Vernier scales without introducing this noise.

Acting Like an Antenna

Critical pieces of metal such as the stereotax or electrode carrier can act as antennas to pick up smaller 60 Hz sources and convey them to the recording environment. The experimenter himself can also act as an antenna when being near or touching the apparatus. Requiring rubber gloves during aseptic surgeries helps eliminate the latter by insulating the person from the metal, but not the former. Whether a metal antenna or a human antenna, grounding eliminates the problem.

Ground Loops

The tendency is to ground absolutely everything to eliminate 60 Hz noise. Unfortunately, this can introduce another source of noise, the dreaded 'ground loop.' A ground loop exists when a voltage source has two or more paths, of unequal lengths, to return to ground. The critical feature here is 'unequal lengths.' The same signal traveling over two or more unequal lengths means that it takes an unequal amount of time for the same signal to reach the same ground point. In essence, different ground points could exist at different voltage levels. The different arrival times result in a 'ringing' in the recording. In general, ground loops thus created can be identified by high-frequency noise in the recording, and are easily distinguished from 60 Hz noise signals. Ground loops are exacerbated by the tendency in an experimental situation to ground multiple locations to more than one ground point, for example, to ground the animal to the recording amplifier, to ground the electrode shields to the stereotax, to ground the stereotax to the oscilloscope, and to ground the Faraday cage to the equipment rack holding all the electronics. The best strategy and simplest solution is, first, to begin with as little grounding as necessary, second, to only ground those things that need to be grounded, and third, only then to ground everything to just one point, usually to the equipment rack that is directly grounded through the wall socket.

Stimulus Isolation

Earlier in this chapter we touched briefly on the issue of preventing stimulation artifacts in neural recordings by using stimulus isolation (see Figure 9.4). By far this is the most important consideration for this source of noise. By defining and maintaining the stimulation and recording circuits separately, much of the stimulation artifact problem can be avoided. Stimulus isolation is typically achieved through optical or inductive means, where the stimulator modulates an isolated, battery powered stimulation source. An optoisolator or LED emitter/detector pair are used for optical isolation. An isolation transformer, usually where the input and output coils have the same impedance, works by inductance of the current from the input coil to the output windings.

Sample-and-hold

Neural amplifiers exist that turn off or hold the neural signal during

Chapter 9

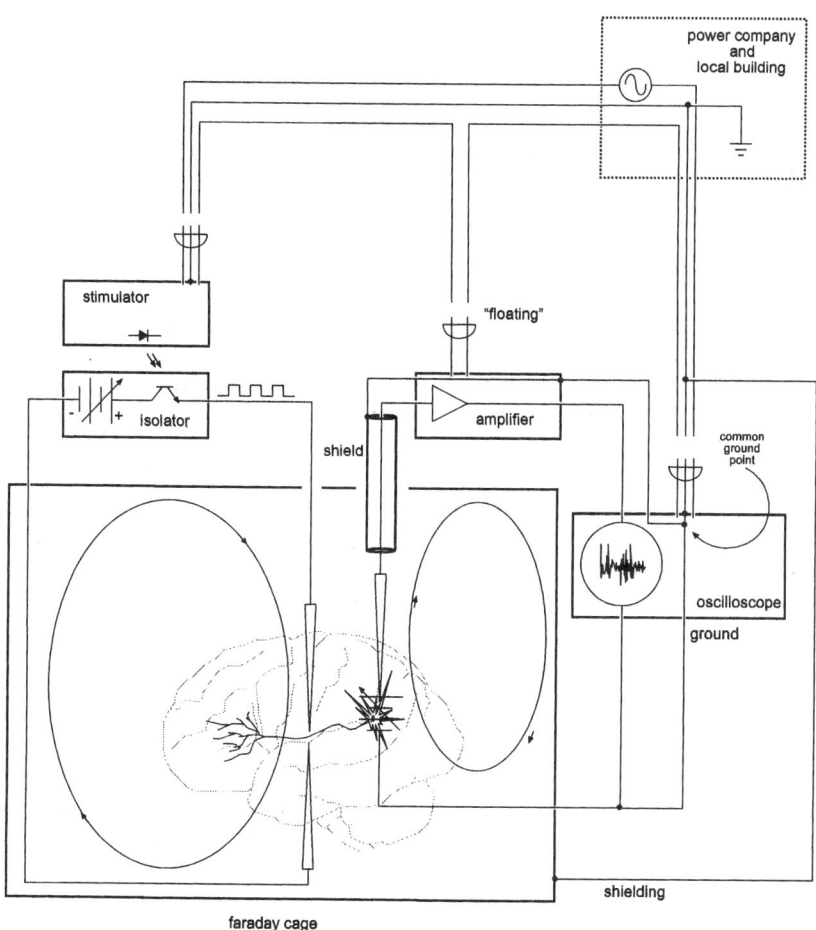

Figure 9.7: Features of stimulus isolation, recording isolation, avoidance of ground loops through a single ground point, and shielding.

stimulation. Since the stimulation pulse is usually of short duration, very little neural activity is lost. The great advantage of this technique, however, is that the stimulus artifact is not distorted and lengthened by saturating the recording amplifier or by coming through the filters in the recording system. Some specialized op amps incorporate this sample-and-hold feature, or sample-and-hold can be constructed using a circuit similar to the one we used for the eyelid response in Figure 3.8.

Power Supply Noise

The power supply for the equipment may add noise to the recording. This is particularly true for switching power supplies which depend on very fast oscillations to charge capacitors to achieve the desired voltage. This added noise counters the advantages of switching power supplies being relatively cheap and lightweight. A linear power supply, heavier and larger because of the transformer and capacitors, may be a solution. Batteries are ideal in critical applications such as powering the voltage-follower op amps that may be directly attached to the recording electrode.

Bring it All Together

Figure 9.8 shows a composite of much of this discussion. The most important considerations from above are included. The figure shows separate stimulation and recording circuits made possible by using an optical interface (optoisolator) in the stimulator circuit. Stimulus isolation can also be achieved through induction (an isolation transformer) instead. Bipolar stimulation electrodes help to confine stimulation to a small part of the nervous system. The figure also illustrates the importance of establishing a single ground pathway to avoid unwanted ground loops. In this figure, the single ground comes through the oscilloscope, and the grounds for all subsequent elements for the recording and shields are through this point. Since the recording amplifier and oscilloscope are both powered through three-prong plugs (hot, neutral and ground) to the wall, ground loops can only be eliminated by 'floating' the recording amplier. That is, the recording amplifier is *not* grounded through the wall. The A-M Systems Model 1800 recording amplifier, for example, has a strap that can be disconnected to float this amplifier. If this feature is not provided, then care must be taken if 'cheaters' are used to convert a three-prong plug into a two-prong plug without the ground wire. Also shown are two shields, one for the recording electrode and wiring, and the other for the environment as a whole (Faraday cage).

CONCLUSIONS

If done correctly, the use of electrical stimulation of the brain to study basic brain-behavior relationships is a very powerful tool. As we discussed above, the major criticism of the technique is that the relative size and specificity of brain stimulation is often hard to define and largely unknown. This is the same type of criticism that is often raised about brain lesion techniques. However, several steps can be taken to improve the use of this technique: 1) Experimenters should become familiar with the relative arrangement of cell bodies and fibers in the area being stimulated. Well-placed electrodes, combined with knowledge of the organization of the brain area to be stimulated, may help control the extent of the stimulation and thus minimize the chances of stimulating non-target tissue. 2) Take great care in the choice of stimulation parameters. For example, high frequency and high intensity stimulation should be avoided. 3) Be sensitive to the differences between stimulating with bipolar versus monopolar electrode arrangements. The spread of current is dependent to a large extent on the choice that is made. 4) Whenever possible, vary the current in a systematic fashion to create input-output functions that describe the effect of stimulation on the behavior you are observing. 5) Whenever possible, use neuronal recording in conjunction with brain stimulation either near the site of recording or in a target neuronal population that is to be activated by the stimulation. The recordings may verify the effectiveness of the stimulation. 6) Last, similar to lesion techniques (and most of not all other experimental methods), brain stimulation alone cannot provide conclusive evidence for the role of a brain area in generation and maintenance of the behavior under observation. Brain stimulation is but one of a variety of tools that is available to the behavioral neuroscientist for experiments designed to study brain-behavior relationships.

Chapter 10

HISTOLOGICAL METHODS

INTRODUCTION

A very important part of research involving the study of brain-behavior relationships is verifying the effectiveness or extent of the brain manipulation that was performed. When brain recording or brain stimulation has been done, the precise location of the tip of a recording or stimulating electrode must be determined. When brain lesion or brain infusion has been used, it is important to establish the extent of the lesion or the spread of the compound that was infused into the brain. Researchers use histological methods to establish electrode position or lesion extent. We include a wide range of techniques in our definition of histological methods including the perfusion and removal of the brain from the animal as well as the entire process of cutting the brain, mounting it onto slides, staining the tissue, examining the slides under a microscope, and recording the results. Basic histological technique has been described in a number of places (e.g., Clark and Clark, 1971; Coolidge and Howard, 1979; Hart, 1969; LaBossiere and Glickstein, 1976; Lauber, 1970; Oakley and Schafer, 1978; Presnell and Schreibman, 1997; Sheehan and Hrapchak, 1980; Skinner, 1971; Wolf, 1971) and interested readers are encouraged to look at these sources for more details than we can provide in this relatively brief chapter. Also, we have not attempted to cover the variety of histological methods that are associated with neural tract tracing and many other neuroanatomical techniques that are available to investigators (e.g., HRP tracing, Golgi methods, and Ca^{++}-sensitive dyes, to name a few). A number of good books and articles are available on these techniques and readers should consult them for details concerning these procedures.

PERFUSION: FIXING THE BRAIN FOR SECTIONING

Perfusion refers to the process of embalming tissue in preparation for histological analysis. This process begins by overdosing the animal with a general anesthetic until it is brain dead. The blood of the animal is then replaced with physiological saline and then the saline is replaced with formalin that effectively fixes the tissue. The tissue of interest (e.g., in our case, the brain) is then removed for sectioning and staining.

Before continuing this presentation of perfusion techniques, we believe that it is important to step back for a moment and consider the issues and implications that arise at the time a perfusion is to be done. The inescapable fact here is that in order to carry out this phase of research, the animal has to be killed so that the brain can be removed to verify the brain manipulation that was used earlier. Without a doubt, for most researchers, the perfusion is the most unpleasant part of animal research. The fact is that an animal is being killed (i.e., "euthanized," "sacrificed" or "sac'd"). The research must be important enough to justify the death of an animal. The researcher must be comfortable with his or her judgment thereby accepting the idea that ultimate benefits of the research that lead to the euthanization of the animal outweighs the cost of the animal's life. The researcher must accept responsibility for his or her actions: There is no excuse for killing an animal merely at the request of someone else. Individuals who are not comfortable with this aspect of research should not participate, should never be forced to participate, and should be encouraged to pursue other interests.

There is also another unpleasant side to the perfusion process: The blood associated with perfusions can be an unpleasant thought and unsighlty vision. For most, this aspect of a perfusion is relatively easy to overcome with experience. The other issue, the killing of an animal, is not; neither author has gotten "used to" performing perfusions and this fact has provided us with a strong motivation to make the perfusion as quick and painless to the animal as possible. Indeed, if one judges the research valuable and elects to do animal research in which the deaths of animals are an end result, then the perfusion must be done as quickly and painlessly as possible for both the animal and the researcher. For the animal a quick and painless death requires skilled administration of an anesthetic overdose, skilled judgment of the cessation of life, and a skilled perfusion technique.

Novice experimenters should never be allowed to perform a perfusion by merely reading instructions, such as those that follow in this chapter. Similar to surgical instructions, perfusion instruction should take place by using the behavior modification technique called "successive approximations," which means that the novitiate first watches an experienced person perform a perfusion and then participates in limited steps in a perfusion with an experienced person performing the majority of the perfusion. Next, the person performs most of a perfusion then undertakes the entire perfusion with the experienced person nearby. Then, and only then, is the individual ready to perform a solo perfusion. The exact number of steps may be fewer or greater depending upon the skill of the novitiate. For someone who has never seen a perfusion the technical process should be described so that he or she knows what to expect (i.e., the person should know what will happen before it actually happens). The novice should not be made to feel embarrassed if he or she has to leave the room. However, he or she must not abandon the animal—someone (an experienced person) must finish the job. The experienced person should also routinely review the perfusion procedure before each perfusion so that all the necessary equipment and solutions are available.

Overview of the Perfusion Process

The procedure described here is a basic, general plan for perfusion. Variations and details are described later. The animal is brought into the perfusion area and overdosed with an appropriate, approved euthanizing agent (e.g., pentobarbital, Euthasol, Euthanasia-5). When there is cessation of breathing, the animal is placed face up on a perfusion rack. Another check is made of breathing and the toes may be pinched hard to see the absence of reflexes. The skin is then incised along the sternum from the top to the bottom of the rib cage. The rib cage is opened with a pair of scissors and the ribs are spread with retractors. The heart is located and the pericardium clipped with a pair of surgical scissors and then removed. The left ventricle, right auricle/atrium, the septum and apex, and the aorta are identified. The heart is held in order to place the perfusion needle into the left ventricle (it may then be placed up into the aorta). The right auricle/atrium is cut with scissors. Saline is fed into the circulatory system until the blood is washed out (e.g., the liver, eyes, lips and/or tongue are pale). The saline is replaced with formalin, which is the tissue fixative. Formalin also reacts with muscle and causes contractions and limb movement. To a person seeing a perfusion for the first time, the formalin-induced muscle contractions can be quite alarming. However, the animal is not being revived by the formalin: The contractions of the muscle to formalin infusion are due to chemical reactions between the formalin and muscle, and not due to a nerve action. The formalin-induced movements can be quite deceiving; for example, Lavond once had a veterinary technician suggest to him at this point that the animal needed more anesthetic. The animal had been long dead at this point. Eventually, the opposing contractions of extensor and flexor muscles, along with the preserving properties of formalin in the muscle, cause the animal to become rigid. The head is then removed from the body to facilitate extraction of the brain, which is accomplished by systematically removing the skull that surrounds the brain. The extracted brain is placed into a bottle with formalin and marked for identification.

Perfusion protocol

Presented here is the general protocol for a perfusion. To facilitate the presentation, we have organized the procedure around the order of tools that are used:

1) **Initial Check.** Before beginning the process, the experimenter should check to see that all the tools and chemicals needed for the perfusion are ready to go. Once the procedure has begun, the quality of the perfused tissue will depend to a great extent on the smooth execution of the steps involved. Before beginning, the experimenter should flush the perfusion line with saline to insure that no formalin was left in the line from the last perfusion.

2) **Euthanasia.** Euthanasia is achieved by overdosing the animal with a lethal dose of a general anesthetic. We have used sodium pentobarbitol (50-65 mg/ml; Abbott Laboratories), Euthasol or Euthanasia-5 solution, a pentobarbital-based compound (Henry Schein, Inc.). For rats administer 1 cc or more pentobarbital intraperitoneally (ip). Cessation of breathing should occur in 2-10 minutes. Administer 0.5 cc more if needed. For rabbits, administer 4 cc pentobarbital intravenously (iv) into the marginal vein of an ear. Cessation of breathing occurs within seconds. Administer 1.0 cc more if needed. If Euthasol or Euthanasia-5 solution is used, dosages can be cut by about 50–75%. Indications of death include cessation of breathing (although local diaphram movement may be observed), paleness of the eyes and other vascularized structures (e.g., lips), and no reaction to "hard" toe pinch (use a hemostat—if the animal reacts to a pinch then you should not open its chest up with a scalpel blade).

3) **Lesion maker.** Normally, at this point, we mark the location of recording electrodes by passing a small amount of current through each electrode (positive pole) referenced to ground (negative pole). See Chapter 8 for a discussion of lesion techniques. A value of 100 µA for 20 sec makes a mark that can be easily seen with a Prussian Blue reaction (see staining protocol below) of 0.5 to 0.8 mm, assuming of course that you are using stainless steel electrodes. A duration of 10 sec may be appropriate for more precise localization although the mark will be harder to see. A current value should be chosen that is appropriate for your purposes. The lesion may be made after the perfusion rather than before (see below). The advantage of placing the marking lesion at this point in the procedure is that the brain will not have shrunk and pulled away from the electrode. In practice, however, we have not found this to be a problem with lesions made after the perfusion. The disadvantage to making the lesion at this point directly after the

Chapter 10   303

overdose is, of course, that you are in a hurry to get into the heart before the it stops its intrinsic beat and blood coagulation begins.

4) **Perfusion Rack.** The animal is placed on a perfusion rack, face up, and over a sink that is located in a fume hood. Turn on the water and the hood light and fan. Because formalin is used in the perfusion process, *all perfusions should take place in an approved fume hood.* The hood sash can be up during early stages of the perfusion process when the incision into the thorax is made and perfusion needles are placed and the heart is cut, but the hood sash should be down when formalin is being infused into the preparation.

5) **Glove(s).** Wear a glove on the hand closest to the animal's head, or one on each hand. Disposable exam gloves or reuseable formalin-resistant rubber gloves can be used. This will prevent formalin from making contact with the skin of the experimenter.

6) **Scalpel.** Identify the top and bottom extents of the sternum and cut the skin with a #10 scalpel blade. Use enough force to cut through the skin and muscles, but be careful not to cut through the rib cage, where you can damage the heart. If you inadvertently cut into the heart at this point you must continue with the following steps and act fast but cautiously. Identify what you have cut, determine if it is necessary to ligate the wound, and identify the parts of the heart critical for continuing the perfusion. A normal perfusion is still possible. In the worst case, the aorta may be cut or torn. If this is the case, perfusion can still be accomplished by dissecting the common carotid artery of the neck. With skill there is no need to panic because the perfusion can still be saved. Experience is required because the common carotid and vagus nerve are covered by a tube-like structure. The trick is to place the perfusion needle into the carotid, not its covering.

Be aware that there many be some local muscle reactions when cutting the skin. Be certain to distinguish between these local reactions and more global reactions that would require additional anesthetic. The anesthetic can be directly injected into the heart at this point by directing a needle between the ribs near the sternum and towards the heart. Aspiration of the syringe should produce lots of blood. Inject any drawn blood and anesthetic back into the heart. In rats it may be easier to inject it into the liver.

7) **Scissors, large.** Cut open the rib cage up the middle (through the sternum) from top to bottom. In practice, the scissors usually cut the ribs immediately lateral to the sternum. Use one hand to press the ribs together from the sides so that the sternum is slightly lifting up. This stabilizes the thorax and places the sternum as far from the internal organs (especially the heart)

as possible. We use a pair of large sharp-blunt scissors with curved blades. These are dull, so others do not like to use them because it takes some effort. We place the sharp tip a small distance (about 0.5–1 cm) just under the xiphoid process (the abdominal end of the sternum). Always move the scissors a short distance in, lift up (to get away from internal organs), and cut. Repeat until the ribs are cut to the top of the rib cage (to the clavicular insertion). Do not cut into the neck which would interrupt the flow of the perfusion solutions from the heart to the head.

For rats you should additionally use the scissors to cut the diaphram laterally. Cut as close to the ribs as possible in order to avoid cutting other organs and the vessels that go through the diaphram (the descending aorta) on top of the spinal column.

8) **Retractors**. For rabbits you can use retractors to open the chest. Be certain the retractors are closed then insert them into the chest. Spread the ribs with the retractors. Next, identify the heart. The heart should be beating because it has an intrinsic rhythm generator. (The heart can be removed, placed into a saline solution, and will continue to beat.) Ideally, it is hoped that the perfusion process will proceed rapidly enough that the heart is still beating at this point in the procedure. The beating heart can be used to help pump the perfusion fluids through the body.

As an aside, we believe that before doing your first perfusion, you should practice holding, opening, and closing the retractors because the action of the retractors seems opposite to handling a pair of scissors. The thumb goes into one loop of the retractor. The middle finger goes into the other loop. The index finger (i.e., the one between the thumb and middle finger) is used to unlock the ratcheting mechanism. If you do not hold the retractors this way you may accidentally spread its arms and then be unable to unlock and close them without taking your fingers out of the loops. For rats retractors may be cumbersome. By previously cutting the diaphram you may not need the retractors at all.

9) **Sponge forceps**. Use the closed sponge forceps, or a finger, to dissect away the adhesions of the heart. The heart is attached only at the anterior (or rostral, i.e., nose-ward direction) where the aorta and vena cava attach. The heart can be gently held with fingers or forceps (do not clamp the forceps). Surrounding the heart is a sac (the pericardium), which contains lubricating fluid. This sac should be removed so that the parts of the heart can be identified. In some instances, the sac may have been removed when the ribs were cut with the scissors. Use a small pair of scissors to carefully cut the sac, or use forceps (sponge or rat toothed) to grab the sac and tear it away. Do not waste time removing all of the sac; remove just enough to expose the heart.

Chapter 10									305

Identify these parts of the heart: left ventricle, right atrium (the auricle is the earlike protrusion of the atrium), the aorta, the septum, and the apex.

10) **Perfusion needle**. For rabbits, clamp the heart with the sponge forceps at its apex where the septum is obvious. This is the strongest part of the heart. For rats the forceps will tear the heart apart so use your gloved fingers or nontraumatic forceps to hold the beating heart. You may cut the left ventricle first with scissors, scalpel, or another needle of the same size as the perfusion needle. This prevents a "plug" of heart tissue from clogging the perfusion needle or from being purged during perfusion and lodging itself somewhere in the body like an embolism. Do not cut an excessively large hole. Notice, however, that a small hole will seal up and very little blood will be lost; that is, you do not have to worry about spurting blood. Insert the perfusion needle into the left ventricle. The perfusion needle is previously blunted (e.g., a 19-gauge blunt needle; a 14 gauge needle with a round tip and side holes; a plastic tubing). Do not lodge the tip of the needle against the inside of the ventricle; do not push the needle completely through the heart; and do not push the needle into the left atrium (the wrong direction for the flow of blood). Some masking tape wrapped around the perfusion needle can be used to prevent the needle from going in too far. Alternatively, the needle can be threaded up into the aorta (where it can be seen through the translucent aorta and felt with your fingers). Gently pull the heart towards the hindlimbs to keep the heart and aorta from collapsing.

The more daring researcher may wish to clamp off the descending aorta with forceps. The advantage is that the perfusion will be of the head and upper body but not of the lower body; therefore, less solution will be used and more perfusion pressure will be applied to the brain. The disadvantages are the time it takes to find the right place to clamp, and the possibility of tearing the descending aorta. For standard formalin perfusions we normally do not clamp the descending aorta; for perfusions for anatomical tracing experiments where special perfusates are used we normally do.

11) **The Perfusion rig**. A system of tubing must be used to deliver the saline (and then formalin) from containers to the needle that is inserted into the heart or ascending aorta. Basically any airtight tubing system can be used. The general idea here is that saline and formalin are placed in large glass beakers or plastic containers and then tubing is used to transport the saline and formalin from the beakers to the needle. The system is gravity-fed, meaning that the beakers or containers containing the solutions are located above the preparation. The pressure of the system depends upon the height of the solutions above the animal. A good height to begin with is about one meter. A very simple system is to place the solutions in closed carboys that

have valved spigots near their bases. Latex hosing can be fitted onto the spigots and connected, via a Y-connector to a single length of hosing, which can be fixed to the needle. Once the spigot is opened, either the saline or formalin runs through the hose to the preparation. A hemostat or hose clamp placed on the final common hose can serve as a final "valve" to control flow of the perfusate into the animal.

Alternatively, we have used a dual IV infusion rig for perfusions (VWR Scientific Products). This rig consists of two filtered intake stages that can be placed in beakers or containers of saline and formalin. Tubing attached to the intake ends carries the solutions to a Y-connector that ends in a needle-compatible outlet (thus making it very easy to attach the rig to the perfusion needle). Two clear siphon bulbs, placed in the two lines of tubing between the solutions and the Y-connector, allow for the initiation of flow in the gravity-based system. These bulbs also serve another important purpose—the flow of solution can be monitored in the bulb during the procedure. Hose clamps are also part of the rig; two on the lines come from the beakers and one on the line goes from the Y-connector to the needle. These clamps provide excellent means for controlling flow in the system.

12) **Scissors**. Assuming the needle is in position and you are ready to infuse the saline, the right atrium is next cut. The auricular ("ear") part of the atrium is the protruding flap and is easiest to cut. The solutions introduced into the body through the ventricle must escape somewhere, and this hole in the atrium is the place. The blood and perfusion solutions will escape through this hole, fill the thoracic cavity, and overspill into the sink. Use running tap water to flush these solutions.

13) **Saline**. Next, open the valve to the saline. The purpose of saline is to wash out the blood so that it does not coagulate and block capillaries. In practice, it is normal procedure to use saline, but the pressure behind formalin is usually enough to ensure a reasonably good perfusion, even if saline is not used. For rats use about 50 ml of saline. For rabbits use about 100 ml of saline. Look for saline running out of the atrium of the heart, and/or for pale liver, tongue, lips, and eyes. [*Recipe for Saline:* 0.9 g NaCl per 100 ml distilled H_2O].

As we mentioned above, pressure is created by gravity feed of the solutions to the level of the animal. The bottles containing the perfusion solutions should be about 1 m or higher when using a 16-gauge perfusion needle or larger. Higher distances are used with smaller needles (remember, larger gauges are smaller needles). The flow rate can be regulated by varying height of the solutions or varying the openings of valves that are placed in the line. Obviously, it is easier to use valves.

There are two alternatives to using a gravity feed system. Pressure may also be created by using a pump. We typically use a pump when doing special perfusions for anatomical tracing experiments. Alternatively, saline and formalin can be placed in large (e.g., 50cc) syringes and injected through the perfusion needle into the animal. Large syringes work particularly well if the perfusion must be made through the common carotid artery in the neck. Although this method could be done routinely, it requires some skill and we usually reserve this method for desperate situations when the heart is torn or the carotid is penetrated by the perfusiton needle.

14) **The circulatory system**. Some basic knowledge of the circulatory system will help you understand how perfusions work. The left ventricle is larger than the right ventricle because it pumps blood to the majority of the body, while the smaller right ventricle pumps blood to the lungs. In perfusions, the lungs are bypassed. Solutions are introduced into the left ventricle, which connects by valves to the aorta. The aorta, in turn, distributes blood to thorax, abdomen and hind limbs (descending aorta) and to the thorax, upper limbs, neck and head (ascending aorta). Arteries reduce to arterioles and then to capillaries. Capillaries gather into venules and then to veins to return blood to the jugular vein, to the vena cava and then to the right atrium of the heart. The right atrium of the heart is cut in perfusions. Perfusions, therefore, bypass the lungs (right atrium to right ventricle to pulmonary artery to lungs to pulmonary vein to left atrium to left ventricle).

15) **Hood**. Turn on the hood to evacuate any formalin fumes.

16) **Formalin**. After a sufficient amount of saline has been circulated through the system, infusion of formalin is begun. How much is enough saline in the system? This can be determined by examining the outflow from the right atrium. When the outflow is reasonably clear (i.e, free from blood), formalin infusion can begin. In our experience, this requires up to 750 ml of saline to be infused in the rabbit (and at 50-100 ml, minimum considerably less in rats). To begin the infusion of formalin, close the valve to saline and open the valve to formalin. In a moment the animal will start to move. As we covered above, this movement is a muscle reaction to the formalin; the animal has been long since dead at this point. The more movement, the better, as one is certain that the solutions are going to all parts of the body. Clear fluid (formalin) coming from the nose indicates that the vessels in the sinuses or lungs have ruptured due to the amount of pressure. Some people think this is a good sign, and others think it is a bad sign. For a good perfusion, look for stiffness of the joints (especially of the jaw and neck) and for the cessation of movements. For rats use about 100-300 ml of formalin. For

rabbits use about 200-750 ml of formalin, depending on the histological procedure that follows. For poorer perfusions (i.e., ones in which there was not a lot of formalin-induced muscular reactions) perfuse by time (5-20 minutes) or volume (400-1000 ml).

17) **A recipe for formalin**. There is often a lot of confusion about what is meant by 10% formalin. This is a traditional histological measurement that bears little resemblance to modern standardized measurement. The rationale is as follows. Formaldehyde itself is a gas. Water saturated with formaldehyde gas is called formalin and contains about 37.5% formaldehyde. A 10% formalin solution takes one part of formalin (37.5% formaldehyde gas in water) and adds nine parts of water. The 1:9 ratio makes it a 10% solution of the original stock. The original stock, of course, was 37.5% formaldehyde. The final 10% formalin solution, therefore, is really about 4% formaldehyde. This is why, when making formaldehyde from paraformaldehyde, you make up a 4% solution. (Unlike the 37.5% formaldehyde solution called formalin; a 4% paraformaldehyde solution will contain no added methyl alcohol—about 12% added as a preservative. One should be aware that certain histological procedures, like HRP reactions, may be compromised by the alcohol). *Recipe for a 10% formalin solution*: Add 1 part formalin (37.5% formaldehyde) to 9 parts of distilled water.

Since formalin causes muscles to contract, and the heart is a muscle, the hole in which the perfusion needle is placed will begin to enlarge as the formalin causes the heart muscle to contract. You may have to take the sponge forceps off of the apex of the heart and clamp the ventricle against the perfusion needle to stop leakage of formalin out of the heart through this hole.

18) **Sucrose**. In addition to a formalin perfusion, or instead of a pure formalin perfusion as described above, you can perfuse with 10% sucrose-formalin. The advantages of sucrose are: it slightly shrinks and hardens the brain (which helps in extraction), and it prevents cells from bursting when freezing the tissue for histology (cryoprotection). *Recipe for a sucrose-formalin solution*: Add 10 g sucrose to each 100 ml of 10% (described above).

19) **Soak and clean up**. Turn off all the perfusion solutions. Remove the sponge forceps, retractors, and perfusion needle and place them in soapy water with the scissors to remove the blood. Flush out the perfusion line so that someone who perfuses after you does not accidentally begin with formalin rather than saline.

20) **Scalpel, scissors or shears**. At this point, the head is separated from the body so that the brain can be extracted. This is done by cutting all the soft tissue down to the vertebral column. At this point either cut the spinal column (scissors or shears) or twist the head. Cut away excess soft skin and muscles to expose the cranium. It should be noted that this step can be skipped: The head does not have to be separated from the body before brain extraction, but it facilitates brain removal by taking advantage of the natural hole in the skull, the foramen magnum, where the spinal cord enters the cranium. Be especially careful with a scalpel so that you do not cut yourself. Always keep in mind the thought that if you should slip you want the knife to go away from you.

21) **Lesion maker**. You can mark the location of recording electrodes by passing a small amount of current through each electrode (positive pole) referenced to ground (negative pole). See Chapter 8 for details. As we pointed out above, a value of 100 µA for 20 seconds makes a mark seen with Prussian Blue reaction (see staining protocol) of 0.5 to 0.8 mm. Choose current and timing values appropriate for your purposes. The major advantage to waiting until this point in the procedure to place the making lesion is that you are not as hurried as you might be right after the overdose is given.

22) **Vice grips**. You can remove dental acrylic headstages by gripping the plastic with a pair of vice grips and applying a steady outward force. Be careful not to twist the vice grips while pulling off the headstage as a twisting action may cause the electrodes being held in the brain to damage the perfused tissue. Never use rongeurs (next) to remove the headstage as this inappropriate use will damage them.

23) **Rongeurs**. Bone rongeurs are used to chip away the vertebrae and cranium from the brain and spinal cord. Start at the vertebrae, work up to the foramen magnum, and continue to cranium. To use the rongeurs, slip the lower jaw a short distance (1–2 mm) under the bone (gently depressing the nervous tissue with the rounded bottom of the tool) and grip a small piece of bone between the jaws. Keep in mind that, should you slip, you want the rongeurs to slip away from the brain and away from your self. You may have to apply a great deal of pressure on the rongeurs to cut the bone, so you can inflict a great deal of damage to the brain or your hand if you slip. To cut with the rongeurs squeeze the bone between the jaws and rotate the rongeurs away from the specimen at the same time. Continue this movement a little at a time until brain is fully extracted. Normally, we take about 3–5 minutes to extract a brain because we are being careful to perserve as much of the brain as possible. Persons who extract living brain tissue for LTP ex-

periments, for example, can remove the hippocampus in seconds without regard to the condition of the remaining tissue. Rongeurs (also called bone chippers) are sometimes referred to as "pliers" by dentists. These are *not* pliers. These are expensive ($80–120 or more) instruments. Do not clamp down the rongeurs on themselves (i.e., with nothing in between them). Do not grab dental acrylic or stainless steel skull screws with them. And, do not try to cut wire with them.

24) **Postfix**. Place the extracted brain into a jar with formalin or 10% sucrose-formalin. A poorly perfused brain or one perfused at a lower sucrose concentration will float. The common histological practice (and possibly mythology) is to use a brain that has sunk as an indication that a post fixed brain is ready to be cut. Label the jar for later identification. A 30% sucrose-formalin solution is sometimes used as a postfix. This shrinks the tissue even more, distorting the ventricles (making them appear larger), but affording even greater cryoprotection.

25) **Disposal**. Experimenters should consult with their animal care staffs for procedures to follow to dispose of the animal's body and other remains from the perfusion process.

26) **Cleanup**. Use soap, water and a brush to clean, then dry and put away all tools. Clean the perfusion rack and the sink. Use soap and water to clean the perfusion sink; clean out any fur/hair that remains in the drain. Clean the inside walls of the hood. Check that the perfusion lines are cleared with saline. Be sure the perfusion lines are all closed. Turn off the hood. Put away the sodium pentobarbitol. Rinse out the syringe and infusion set.

27) **Records**. Like all other phases of experimentation, it is very important that good records are kept concerning the perfusion process. Log the amount of sodium pentobarbitol (or other agent) used. Enter the euthanization date into the animal logbook and make sure that the animal care staff is informed that the animal has been euthanized.

BRAIN EMBEDDING AND SECTIONING TECHNIQUES

After the brain is extracted from the cranium in the perfusion procedure, the brain is stored in a 10% formalin solution indefinitely until brain sectioning and histology procedures are performed. Some researchers prefer to store the brain in a sucrose-formalin solution (e.g., 10 or 30% sucross in 10% formalin). The sucrose imparts the property of "cryoprotection," meaning that the cells do not readily burst when the tissue is frozen for cutting. It also shrinks the tissue, particularly with the higher sucrose concentrations. The folklore is that the brain is ready to be cut when it sinks in the postfix solution. Postfixing in the perfusate (10% formalin alone) may result in immediate sinking yet better histology would be had if one waited a few days or a week. Postfixes with sucrose cause immediate floating and therefore sinking is more obvious when the brain is supposedly ready. Whether sinking is related to the tissue becoming impregnated with sucrose (which is the idea) or to sucrose diffusing into the ventricles (which is unimportant for the quality of the tissue) or to what relative degrees of both, is uncertain.

The brains may be cut on a microtome or cryostat directly from the postfix. Sometimes the brain tissue breaks up easily when sliced (e.g., the hippocampus or the folia of the cerebellum). To aid in holding these areas of the brain together during sectioning, the brain may be embedded in a block of material before being sliced. We normally use an embedding medium made of gelatin and albumin. To embed a brain, place the brain in a cardboard or plastic container, reheat a small portion of an embedding gelatin solution, and pour this matrix around the brain in the container. Insect pins can be used to keep the brain from floating. Throw away unused, heated gelatin; do not reuse. Place the gelatin-surrounded brain in formalin fumes until hardened. A covered staining dish with approximately 1 cm of formalin in the bottom can provide a container for this stage. If desired, the block can then be placed into the perfusate solution indefinitely. (It is wise, however, to remove the insect pins as these will rust over time.) This matrix plus brain can then be frozen and cut on a microtome. The sections retrieved from the blade are like small pages that can easily be mounted onto slides and dried. By cutting off the upper right corner of the embedding gelatin, for example, it is easy to maintain the proper orientation of the sections on the slides. Good gelatin will not stain with cresyl violet. Bad gelatin (old, reheated) has bacteria in it and stains darkly with cresyl violet. Here is the recipe for the embedding gelatin mixture that we normally use:

Embedding gelatin:

6 g gelatin (175 bloom)
60 g albumin (chicken egg, Sigma)
400 ml distilled H_2O

Dissolve gelatin in 100 ml of the distilled H_2O at 70°C then cool to room temperature. Separately, stir albumin in remainder of distilled H_2O (300 ml) at room temperature for 10-15 minutes. When both are cool and well mixed, combine the two solutions. Filter out lumps with cheese cloth. Store in sealed containers in refrigerator. The shelf life at 4° C for the mixture is about two weeks.

As a practical matter, most of the time we do not use a sucrose postfix, opting instead to wait a few days or at least a week before cutting the brain on a sliding microtome or on a cryostat (which is basically a microtome mounted inside a freezer). Regardless if a sliding microtome or cryostat is used to section the brain tissue, the ultimate goal is to generate thin brain slices (generally 20–50 μm thick) that can be mounted onto glass slides for further histological processing. For generating brain slices on a sliding microtome, the brain (either embedded or not) is placed on a metal stage and deep frozen by using a freezing unit that is connected to the stage or by surrounding the blocked tissue with dry ice for a period of time. Prior to freezing the brain onto the slicing stage, the embedding block or the brain itself can be notched on either its right or left side to allow proper orientation of the slice onto the glass slide later in the process, as mentioned above. The block of tissue is secured to the freezing stage using either distilled water or a tissue-mounting compound (e.g., O.C.T. Compound, Sakura Finteck, Inc.). After the brain is thoroughly frozen, the corners of the freezing stage can be raised or lowered to make sure the brain is placed symmetrically on the stage for slicing. A well-sharpened blade should be used in the microtome and it should be slid over the top of the tissue with a smooth, even stroke. If cut properly the slice of brain tissue should adhere to the blade at the point where the blade contacts the brain block. Slicing on a cryostat device is not much different from slicing on a sliding microtome. On most cryostats, the brain is fixed in position and the brain sections taken by rotating a blade mounted on a drum that is turned by a crank located on the outside of the freezing chamber. When a sliding microtome is used, the slices are wiped from the microtome blade with a small paintbrush and transferred to wells filled with 20% alcohol where they remain until the sections are mounted onto glass slides. When a cryostat is used, the slices are immediately transferred to the glass slides after they are cut.

It is important that the glass slides be coated with a sticky substrate before brain sections are mounted on them. If plain glass slides are used, the brain sections tend to shrivel and fall off easily during the staining process. We call the process of coating the slides with a sticky substance "subbing" the slides. Most often the glass slides are 'subbed' (coated) with chrome-alum. You should either buy cleaned slides or clean your own in a solution of hydrochloric acid (1 part HCl to 2 parts distilled H_2O). Rinse the slides in distilled H_2O then dip them into the chrome-alum subbing solution. Throw away unused subbing solution. Make sure the slides are dried, which can be done is an oven at low temperature overnight. Drying without an oven works for less critical histology. What is important is to dry or bake in a dust-free environment. Here is a recipe for the chrome-alum solution:

Chapter 10

Chrome-alum solution

1.75 g gelatin (300 bloom, Sigma porcine, type II)
0.175 g chromium potassium sulfate (chrome-alum)
350 ml distilled H_2O

Heat the distilled H_2O at a low temperature. Add gelatin and dissolve completely. Dissolve chromium sulfate. Cool. Clean the glassware used immediately, otherwise it will be sticky.

Needless to say, the slides must be subbed well before sections are mounted onto them. The brain sections are mounted onto the chrome-alum subbed slides by floating them in distilled H_2O or 20% ethyl alcohol or a mounting solution made with gelatin. The latter helps if there is difficulty getting the sections to adhere to the chrome-alum coated slides. If the brain or embedding medium was notched before sectioning, the right side of the brain can be distinguished from the left side of the brain by finding the notch and mounting the sections on the slides such that the notched side is oriented on the proper side of the slide. The slides are laid flat for drying in the air or placed on a slide dryer. Again, if the slides must be very clean then dry them in a dust-free environment. Here is a recipe for the mounting solution:

Mounting solution

0.5 g gelatin dissolved in 290 ml warm distilled H_2O
210 ml 95% ethyl alcohol

Dissolve the gelatin completely in distilled H_2O heated to 55°C. Cool the gelatin to the same temperature as the alcohol, and add the alcohol to the gelatin solution (the gelatin solidifies/clouds if added to alcohol). When thoroughly dried the slides can be stained.

BRAIN STAINING METHODS

Most often, histological stains are used to identify regions of cell bodies and fibers in the brain sections that have been mounted on the glass slides. The most popular stains highlight nucleic acids, particularly those associated with the rough endoplasmic reticulum and the nucleus, which together mark cell bodies, particularly those cells involved in secretions, like the neuron (neurotransmitters). Common stains in this category, all staining the same thing, include eosin-hematoxylin, cresyl violet, thionin, neutral red, and Safranin O. Because the cell body stains are charged stains we refer to them as water-based, as opposed to lipid-based stains used for myelin like iron hematoxylin (Weil-Weigert). The common cell body stains are basically the same, differing mainly in their color and stability. We use cresyl violet for most of our studies, often with the option for counterstaining for the metal left by a marking lesion (Prussian blue). We have had some luck (and some failure) using Weil-Weigert for fibers (Wolf, 1971); when it works it is very impressive, otherwise we end up with tissue that is light brown all over. We have had better experience with the Kluver-Barrera (Wolf, 1971) combined cell body and fiber stain, which can be used separately for staining just fibers or cells. We include these three staining protocols in the following. A recent article by Simmons and Swanson (1993) describes the thionin Nissl stain with a protocol. Their advice on staining is important for Nissl stains in general. The general protocols for three different staining techniques follow. For all of these techniques, careful and well-timed steps are critical for generating a successful final product.

Cresyl violet stain for cell bodies with optional Prussian blue for electrolytic marking lesion

The following protocol stains neural tissue mounted onto slides for normal cells bodies (rough endoplasmic reticulum) and optionally counterstains for iron deposited by electrolytic lesion through a stainless steel electrode (e.g., 100 µA for 10 seconds). Larger electrolytic lesions can also be identified, but Prussian blue is usually overkill because those lesions are fairly obvious.

1. 100% ethyl alcohol — 2–5 min
2. Histoclear — 2–5 min
3. 100% ethyl alcohol — 2–5 min
4. 70% ethyl alcohol — 2–5 min
5. 50% ethyl alcohol — 2–5 min
6. Distilled H_2O rinse — 2–5 min
7. Prussian blue (optional) — 1 min

Comment: Metal deposited by electrolytic lesion will stain blue/green.

Recipe:
20 g potassium ferr*o*cyanide (note spelling; this is *not* Weil-Weigert)
+ 20 ml of 37% HCl

Chapter 10 315

+ 280 ml distilled H₂O

Comment: Can be made as two stock solutions but do not mix ferrocyanide and water until use.
Also, remember to add acid to water, not the other way around.

7a.	Distilled H₂O rinse	1 min
7b.	1% sodium acetate neutralizer	1 min

   Recipe:
   3 g NaOAc
   + 300 ml distilled H₂O

   7c. Distilled H₂O rinse   1 min

8. Cresyl violet   2–5 min

   Comment: A good stain at this stage is dark all over the tissue. Steps 9–16 will differentiate and take out stain from the background. It is a good idea to knock the staining tray to dislodge bubbles that may be caught between slides, thus preventing the
   stain from reaching the tissue.

   Recipe:
   1 g cresyl violet (certified) + 10 ml acetic acid (glacial)
   990 ml distilled H₂O.
   Heat at 60°C for 2-3 hours, cool overnight, reheat and filter.

9. distilled H₂O rinse   1 min
10. distilled H₂O rinse   1 min
11. 50% ethyl alcohol   2–5 min

   Comment: Note that steps 11 through 15 say 2–5 minutes each. Some people actually do this. What we actually do is spend a total of about 5 minutes for steps 11 through 15, taking about one minute at each step.

12. 70% ethyl alcohol   2–5 min
13. 95% ethyl alcohol   2–5 min
14. 95% ethyl alcohol & 2-3 drops acetic acid   2–5 min

   Comment: Watch this step very carefully. Once the stain starts fading in the alcohol-acid, it goes fast.

15. 100% ethyl alcohol   2–5 min
16. Butanol   5 min

Comments: A considerable amount of background 'redness' may still be in the tissue when this step is begun. However, butanol does a very good job of 'getting the red out' and differentiating cell bodies from background. The sections can stay in butanol longer than 5 minutes.

17. Histoclear   minimum of 5 min

Comments: The sections can stay in Histoclear almost indefinitely. This is not a good practice, however, only because it occupies the staining tray.

18. Coverslip the slide

Modified Weil-Weigert fiber stain

The following protocol stains neural tissue mounted onto slides for normal myelinated fibers.

1. Distilled H_2O rinse   1 min
2. Acetone   10–20 min

Comment: We usually leave sections in for 20 minutes while we are simultaneously staining alternate sections for cresyl violet.

3. Distilled H_2O rinse   1 min
4. Iron hematoxylin at 50–60°C   60 min

Comments: Some say that aged hematoxylin (e.g., months or years) works best, however, we have had good stains from freshly made hematoxylin. A good stain will give you sections that look black all over (differentiation, below, will remove excess stain).

Recipe:
    2 g hematoxylin in 20 ml 100% ethyl alcohol

+ 180 ml distilled H2O
+ 8 g ferric ammonium sulfate in 200 ml distilled H_2O

5. Distilled H_2O rinse   1 min
6. Differentiation   N min

Comment: Differentiation is the process by which excess background staining is removed to reveal the stained elements of interest (in this case, myelinated fibers). Sections start to differentiate between 1 and 10 minutes after placed in this solution. When the sections start to differentiate, the process occurs very rapidly. Therefore, you have to be vigilant to be certain that not too much stain is taken out. A good guide is the cerebral cortex will look light brown and the corpus callosum will look black. Also look for detail in the fibers; for example, the medial lemniscus in the thalamus will show fine blackened detail. Individual sections will stain differently, so use the majority of sections as a guide.

Recipe:
12.5 g potassium ferr_i_cyanide (note spelling; this is *not* Prussian blue)
+ 12.5 g sodium borate
+ 500 ml distilled H_2O

7. 70% ethyl alcohol   2 min
8. 95% ethyl alcohol   2 min
9. 100% ethyl alcohol   2 min
10. Xylene or histoclear   2–5 min
11. Coverslip the slides

Modified Kluver-Barrera stain

The following protocol stains neural tissue mounted onto slides for both normal cell bodies and normal myelinated fibers.

1. Distilled H_2O rinse   1 min
2. Luxol blue at 50-60°C   2–24 hr

Recipe:
3 g Luxol blue

+ 0.3 ml acetic acid
+ 300 ml 100% ethyl alcohol

3. Distilled H_2O rinse	1 min
4. Fiber differentiation	15 sec

Recipe:
5 ml 3% ammonium hydroxide
+ 25 ml distilled H_2O
+ 270. ml 100% ethyl alcohol

5. Distilled H_2O rinse	1 min
6. Safranin O or neutral red	2–10 min

Recipe:
3 g safranin O or neutral red
+ 300 ml distilled H_2O

7. Distilled H_2O rinse	1 min
8. Cell differentiation	N min

Recipe:
3 ml 100% acetic acid
+ 297 ml 100% ethyl alcohol

Comment: Differentiate to your preference

9. Distilled H_2O rinse	1 min
10. 75% ethyl alcohol	2 min
11. 95% ethyl alcohol	2 min
12. 100% ethyl alcohol	2 min
13. Xylene or histoclear	2 min
14. Coverslip the slide	

INSPECTING THE PROCESSED TISSUE

After gluing a coverslip over the section (e.g., Permount for normal histology, DPX for low fluorescence) and allowing it to dry well (the edges dry in about one day, the center dries completely in a few days or a week), the slides are ready for inspection. For electrode localization and identification of nuclei and tracts, a light microscope is ideal. For whole views of the histological section we have used a number of methods; a stereomicroscope, a light projection microscope which puts an image onto a page (where it can be traced), a drawing tube that attaches to a microscope that puts a false image onto a page (where it can be traced), photography with a 35 mm camera fitted with a macrolens (the photographs can be scanned into a computer), a 2 x 2 slide scanner used to scan histological slides (this is how we got our images for our atlas), a CCD camera used to capture the image to a computer, or a digital camera (we have used a Sony Mavica which creates files on a disk that can be used by a computer). We use Adobe Photoshop™ to edit the images (rotate the image, 'repair' torn or missing tissue caused by inexperience or accident, remove background, improve the contrast, etc.). Outlines of the sections and of nuclei and tracts are exported as Adobe Illustrator™ files, and imported into the Macromedia Freehand™ drawing program where labels, scales, dashes and stroke widths are added. Our brainstem atlas, both the histology and the drawings, are included in the CD; it serves as an example of the end-product of the histological methods described here. We do not alter original findings. We might alter noncritical tissue like an atlas used for illustration. The motivation and rewards happen when other researchers replicate the findings, which can only happen with accurate and honest representations of the data.

ASSESSING THE EXTENT OF A BRAIN LESION

In the past, it was relatively difficult and time consuming to quantify the extent of a brain lesion. One method that was commonly used was to use a projection microscope to project an image of a brain section onto a sheet of paper, then tracing the extent of the brain lesion looking for missing tissue and signs of gliosis. One strategy for assessing the lesion size was to cut out the lesion from the piece of paper and weigh the piece of paper to get a relative idea of the quantity of missing tissue. Nowadays, thanks to computers, it is much easier to more precisely determine the size of a lesion. One method we have used is to use digital photography to take pictures of several sections of brain tissue that contain the structure to be lesioned. Using a number of available programs (e.g., Macromedia Freehand™) the outline of the structure of interest can be traced for all of the sections that were photographed and a three-dimensional representation of the structure can be created. The volume of the structure can be then calculated. Next, the extent of the lesion in the same brain structures can be outlined, a three-dimensional view of the lesion created, and the relative volume of missing tissue in the structure can be determined.

ASSESSING THE EXTENT OF AN INJECTION MADE INTO THE BRAIN

More and more researchers are using brain infusion techniques to study brain function, such as the infusion of the GABA competitive agonist, Muscimol, into the brain to temporarily inactivate a brain area. There are techniques available to assess the relative size of the infusions that have been made. The simplest method is to inject India ink dye (or cresyl violet stain) in place of the substance normally injected. This ink yields a permanent stain that is a first approximation of the extent of the injection. Of course, this dye does not mimic the molecular size and natural diffusion of the original chemical. For this, a radioactively labeled version of the chemical of interest is normally used, usually a tritiated version (e.g., H^+-Muscimol). Some commonly used chemicals have readily available tritiated versions. For a price, less common material can often be manufactured at the researcher's request. The tritiated chemical is normally injected at the same rate and concentration as the unlabeled chemical that was used in the original experiment. Often, time after injection is also a factor, being a determinant for the time of the perfusion. Obviously, the longer one waits, the greater the diffusion. One wants to retrieve a sense of the diffusion that mimics the experimental condition. The spread of the chemical is indirectly determined by assessing the resultant labeling of film or emulsion that has been exposed to the radioactivity. The individual university/institution and federal or state regulatory entities will require special training, records and handling for using and disposing of radioactively labeled chemicals and animals.

While radiolabeling is often required by grant and article reviewers, the method is much less satisfactory than often recognized or acknowledged for at least three reasons. First, the size of the grains in the film or emulsion determines the resolution. In general, film is less precise than emulsion that has been placed directly on the histological sections, but film is much more conveniently used. One can use much higher power on a microscopic inspection of a section than on a piece of film. Second, both film and emulsion are susceptible to "background labeling," either coming from the tissue itself, from the slides (hence the reason for cleaning slides and then drying them in a dust-free environment), partial exposures of the emulsion or film (for ^{14}C they are sometimes stored in lead to attenuate exposure), or any environmental source of contamination. The problem this background labeling causes comes when trying to determine the exact edge of the labeled chemical—background causes uncertainty about that edge so it is "fuzzy" and requires a best guess. Third, even if the extent of labeling can be determined, this does not mean that an effective concentration exists throughout. Knowing the 'effective spread' is not the same thing as knowing the 'physical spread.' Practically speaking, everyone adopts a fairly arbitrary criterion for the definition of the effective edge of the diffusion.

The reality is that radiolabeling is not much better than injecting a dye. A much better experimental solution is to slightly vary the location of injections using the same amount of injection. By plotting effective and ineffective placements one can 'hone in' on the critical locus. An added benefit is that stereotaxic surgery is not a precise methodology, so that there will be natural variations of placements from one animal to the next. This means that special attempts and 'control groups' to vary the location of the injection cannula often are not necessary—the placements will occur naturally as normal consequences of the variability of the experiment.

BRAIN ATLASES

At the end of this section (which actually concludes this chapter) we have supplied a list of brain atlases that have proven useful for stereotaxically implanting recording and stimulating electrodes, locating sites for brain lesions and infusions and for generally identifying brain structures and areas. See Chapter 7 for a discussion of using the brain atlases during stereotaxic surgery. One point of emphasis should be raised, however, at this point. Almost every atlas made is unique from the perspective the relative planes of section that appear in each atlas can be rather specific (e.g., level skull versus angled skull and zero point defined relative to a skull landmark versus an ear-bar position) and that the sizes of animals used for the various atlases can be quite different. The researcher should consider at least these two factors when selecting an atlas to use. We have included in the list some helpful human atlases as well.

ATLASES

Berman, A.L. and Jones, E.G. (1982). The Thalamus and Basal Telencephalon of the Cat: A Cytoarchitectonic Atlas with Stereotaxis Coordinates. Madison, Wisconsin, University of Wisconsin Press.

Brodal, A. (1940). The cerebellum of the rabbit: a topographical atlas of the folia as revealed in transverse sections. *Journal of Comparative Neurology*, 72, 63-81.

Cooley, R.K. and Vanderwolf, C.H. (1979). The Sheep Brain: A Basic Guide. London, Ontario, A.J. Kirby Co.

Dearmond, S.J., Fusco, M.M. and Dewey, M.M. (1989). *Structure of the Human Brain: A Photographic Atlas*. New York, Oxford University Press.

Girgis, M. and Shih-Chang, W. (1981). *A New Stereotaxic Atlas of the Rabbit Brain*. St. Louis, Warren H. Green Inc.

Haines, D. (1991). *Neuroanatomy: An Atlas of Structures, Sections and Systems*. Baltimore, Maryland, Williams & Wilkins.

Hendelman, W.J. and Morrissey, J-P. (1988). *A Student's Atlas of Neuroanatomy*, 2nd edition. Ottawa, Canada, Univeristy of Ottawa Press.

Mai, J.K. (1997). *Atlas of the Human Brain*. New York, Academic Press.

Martin, J.H. (1989). *Neuroanatomy: Text and Atlas*. New York, Elsevier.

McBride, R.L. and Klemm, W.R. (1968). Stereotaxic atlas of rabbit brain, based on the rapid method of photography of frozen, unstained sections. *Communications in Behavioral Biology*, Part A, *2*, 179-215.

Meesen, H. and Olszewski, J. (1949). *A Cytoarchitectonic Atlas of the Rhombencephalon of the Rabbit*. New York, Karger.

Nieuwenhuys, R., Voogd, J. and van Huijzen, C. (1988). *The Human Central Nervous System: A Synopsis and Atlas*. New York, Springer-Verlag.

Olszewski, J. and Baxter, D. (1954). *Cytoarchitecture of the Human Brain Stem*. New York, Karger.

Paxinos, G. and Watson, C. (1982). *The Rat Brain in Stereotaxic Coordinates*. New York, Academic Press.

Schaltenbrand, G. and Wahren, W. (1977). *Atlas for Stereotaxy of the Human Brain*. Stuttgart, Georg Thieme Publishers.

Swanson, L.W. (1992). *Brain Maps: Structure of the Rat Brain*. New York, Elsevier.

Chapter 11

CONTROLLING CLASSICAL CONDITIONING AND OTHER BEHAVIORAL NEUROSCIENCE EXPERIMENTS

INTRODUCTION

The microcomputer is arguably the most powerful tool available to behavioral neuroscience researchers. As we have pointed out in a number of places in an earlier chapter, the microcomputer has become indispensable for research. Microcomputers are used to deliver stimuli, acquire data, and perform data reduction and analyses, including statistical analyses used for hypothesis testing. Databases are stored on microcomputers and they are even used to write the papers that report and interpret the results that are obtained in the study. In this chapter, for the most part, we concentrate on the use of microcomputers to control experiments. We begin with a brief history of the equipment that has been used to control classical conditioning experiments. We also give a brief history of our experience with the development of controllers and computers for classical conditioning as they evolved in the Thompson laboratory and later in our individual laboratories at the University of Southern California and Indiana University. We thank Richard Thompson for some corrections to this history. We next provide some details about simple controllers that were used prior to the advent of microcomputers and then present, in detail, the computer systems we currently use in our laboratories at the University of Southern California (Lavond) and at Indiana University (Steinmetz). We conclude with a brief presentation of some other systems that are available.

A BRIEF HISTORY OF CLASSICAL CONDITIONING CONTROLLERS

A complete review of classical conditioning and the technology to support it is well beyond the scope of this handbook (although we did present some background information on classical conditioning in Chapter 1).

Here, we wish to acknowledge only the bare essentials with three names, Ivan Pavlov, Isadore Gormezano, and Richard Thompson. Ivan Pavlov discovered and characterized autonomic classical conditioning in the early 1900s. His work became widely known with the 1927 translation, by Anrep, of his work. Pavlov's experiments supported a cortical localization for learning and memory, although recovery of function was the norm. The discovery of somatic classical conditioning by the American, Twitmeyer (who predated Pavlov), has had little if any impact on psychology. Beginning in the early 1960s, Isadore Gormezano promoted the rabbit nictitating membrane paradigm for studying classical conditioning behavior. Gormezano moved from measuring nictitating membrane responses with a ruler on polygraph paper to a sophisticated online system controlling trials and analyzing behavior using a personal computer, a system that was called the FIRST system (Scandrett & Gormezano, 1980). Richard Thompson combined the rabbit paradigm with collection of neural unit activity in the early 1970s. Thompson was the first to automate classical conditioning online for both behavior and neural activity, using a Digital Equipment Corporation PDP-12 computer. Thompson's work localized somatic eyelid classical conditioning to the cerebellum and he and his colleagues continue to actively use this paradigm to study brain-behavior relations for learning and memory.

Between the time of Pavlov and the time of Gormezano the technology for classical conditioning was primitive but effective. Figure 1.2 in the introductory chapter is an early photograph given to Thompson by Ernest Hilgard that illustrates the conditioning equipment used in the 1930s. The young Hilgard is being prepared as a subject for an eyeblink conditioning experiment. The experimenter is seen adjusting the automated wooden face-slapper, used to deliver the US. Even in those days psychology was an empirical science, Thompson likes to say. The onlooking experimenters are Clark Hull, postdoctoral student Helen Peak, and Walter Shipley.

By the time Gormezano began conditioning rabbits, the technology had progressed to relays and solid state equipment, with the eyeblink data recorded on polygraph paper. Polygraph recordings were used as a backup through the middle of the 1980s, even in Thompson's laboratory. The permanence of polygraph tracings was trusted more than fragile computer records. But, whereas the recordings remained primitive, technological improvements were being made in the controllers. During this time the UC Irvine ROM controller and Solomon's KIM-1 microcomputer began running Thompson's experiments.

Thompson's interest in classical conditioning began in graduate school at the University of Wisconsin from his major professor W.J. Brogden. Brogden did classic work in animal conditioning. Brodgen himself had trained with W. Horsley Gantt. Gantt in turn had trained with Pavlov. (Like geneologies, intellectual family trees can often show relatedness within short degrees.) Another of Thompson's professors at Wisconsin was David Grant. Grant had trained with Ernest Hilgard and was a major figure in human eyeblink conditioning. Gormezano trained with Grant. Indeed, in the mid-1950s Thompson and Gormezano were fellow graduate students at Wisconsin. Twenty years later, Thompson, who was very impressed with Gormezano's development of the rabbit eyeblink-nictitating membrane conditioned response preparation beginning in the early 1960s, adopted the paradigm to study the neural bases of learning and memory. Michael Patterson, a Ph.D. student of Gormezano at the University of Iowa, became a postdoctoral fellow of Thompson's in the early 1970s. Patterson introduced the rabbit classical conditioning paradigm to the Thompson laboratory at this time. Gormezano later took a sabbatical as a visiting professor with Thompson in the early 1970s.

Until the early 1970s, classical conditioning data was collected on polygraph paper and analyzed by hand. The Thompson laboratory ran its experiments with relays and with Coulbourne or Grason-Stadler solid state modules. Around 1971, Thompson's laboratory began training rabbits for classical nictitating membrane/eyelid conditioning using a Digital Equipment Corporation PDP-12 computer. This computer cost about $100,000 with hard disk and accessories, and had the capability to simultaneously run four rabbits using programs written by Richard Roemer and colleagues (Roemer, Cegavske, Patterson & Thompson, 1975). Thompson believes this may be the first online system for classical conditioning, where the stimuli were delivered and the data collected and analyzed at the same time. However, the PDP-12 system was notorious for crashing several times a day, causing a loss of data. Because of this basic unreliability of the PDP-12, the system was converted to an offline analysis system. Data for behavior and neural activity were collected onto a 4-channel reel-to-reel tape, and the tapes were played back to the PDP-12 computer for analysis. After many years of service the PDP-12 was retired but continued to serve a very useful purpose: Many of us appreciated that fact that the PDP-12 continued to give off heat therefore functioning as a relatively expensive room heater. The PDP-12 was eventually sold as scrap parts for $500.

Since the PDP-12 was not necessary for running the experiments, Thompson had the shop of the Department of Psychobiology, University of California, Irvine develop a ROM controller (which is described below). This consisted of a rack-mounted unit with rotary thumb switches for selecting interstimulus intervals and durations of tones and air puffs. In addition, a programmed ROM module for trials fit into a plug, allowing for normal delay conditioning, trace conditioning, or unpaired conditioning. At this time, data were still collected to a 4-channel tape recorder, and analyzed by the PDP-12.

Even greater flexibility for controlling experiments was afforded by Paul Solomon's KIM-1 (6502) microprocessor based system, which interfaced with the various TTL equipment used to present tones and air puffs (Solomon & Babcock, 1979; Solomon, Weisz, Clark, Hall & Babcock, 1983). By playing a tape recorder into the KIM, code could be loaded with different training parameters. The fundamental timing algorithm was based on the execution cycles for 6052 instructions multiplied by the KIM-1 computer clock rate. [This published code is not entirely correct. See below for an explanation.] Data were still collected to a 4-channel tape recorder, and analyzed by the PDP-12.

In the summer of 1980, all of these different systems were still in use in the Thompson laboratory. Data collection was centered on 4-channel tape recordings, which were played back to a new PDP-11 computer that had just been programmed for analyzing data. Apparently, it took a considerable amount of time and effort for Thompson's laboratory to convert the PDP-12 code into PDP-11 code, most likely because the machines were somewhat incompatible at the machine code level.

At some point, Lavond designed and built two small 6502 based computers and interfaces. The software was based on the timing algorithm used by Solomon and colleagues, although the published timing code was not exactly correct (this was confirmed by checking the code on a running KIM-1). Lavond's classical conditioning programs were on ROM, rather than tape as for the KIM-1, and different types of training sessions could be selected by setting switches. Furthermore, minor changes in the training parameters could be altered using front switches that were located on the front panel of the controllers. Still, the data was collected to the 4 channel tapes.

Steinmetz, a graduate student of Michael Patterson, joined the Thompson laboratory as a postdoctoral fellow in 1983. Steinmetz introduced personal computers into Thompson's laboratory. The 6502 units soon became obsolete when Steinmetz introduced the Apple II computer and Gormezano's First programs to the Thompson laboratory. The system developed by Gormezano and colleagues used an Apple II personal computer and a proprietary programming language called First. The First language is described by Scandrett and Gormezano (1980). The name First is a play on the Forth language name, Forth being a "fourth generation" language then in existence, developed to control telescopes. Because record sizes limited names to five letters in early computer systems, "fourth" became truncated to "forth". Thus, Forth came first, First came after Forth, and First came out first by outperforming Forth (at least in terms of floating point).

To get away from the relatively peculiar First language and the requirement for a specialized computer board with a floating point processor to run the program written in the language, Lavond and Steinmetz designed alternative interfaces and wrote the software for Apple II computers and then converted the software for use on the more popular IBM-PC computers (Lavond & Steinmetz, 1989). We used the Forth language because it allowed both machine and high level programming in a small RAM space, because books and vendors of Forth were fairly easy to obtain at the time, and because First was related to Forth and could therefore serve as a model.

Steinmetz had programmed classical conditioning in First and other languages such as Basic and Fortran. Lavond had experience with Thompson's PDP-11 analysis programs and the numerous controllers in the laboratory and with programming in Basic, Forth, Fortran and several machine codes (Z80, 6502, 6800, 8088). In addition, several articles were particularly helpful in aspects of our implementation of classical conditioning. The single most important of these was an article written by Thompson, Berger, Cegavske, Patterson, Roemer, Teyler and Young (1976) for its descriptions of timing parameters and data analysis with z-scores. Also of interest were papers by Berger, Latham and Thompson (1980) and Berger, Rinaldi, Weisz and Thompson (1983) describing unit analysis techniques and cross-correlation methods. We also described data collection and analysis with z-scores and cross-correlations in our paper on the effects of cerebellar lesions on nictitating membrane and eyelid EMG behavioral measures (Lavond, Logan, Sohn, Garner and Kanzawa, 1990).

Currently Forth has declined in popularity and, perhaps, is now only of interest to those in robotics or in studying archaic computer languages. Books or magazines can no longer be readily found related to Forth. Memory and computer speed limitations are no longer as important considerations as when 64K to 256K of RAM were all that was available in computers. Popular programming languages like C++ incorporate many of the powerful programming features of Forth, such as combined interpreter and compiler functions, and integration of C++ programming with assembly language programming. The popular object oriented programming, whereby a data record can incorporate many different data types, is a subset of Forth's powerful extensibility—the ability to create new programming structures. For example, although Forth originally did not come with a case command (in case of 1 do this, in case of 2 do that, etc.) that programming structure can be created in Forth, which Lavond has done in more recent updates. Lavond still uses the programs written in Forth although he has programmed in C++, whereas Steinmetz has moved on to programs written in C++. We describe each of these systems later in this chapter in somewhat greater detail. We are hoping that these descriptions will help the reader see and understand how hardware and software are used to control behavioral neuroscience experiments. First, however, we provide some information here on some simple classical conditioning controllers that predated the microcomputer-based systems.

SIMPLE CLASSICAL CONDITIONING CONTROLLERS

Here, we consider three designs for hardware that will control simple classical conditioning experiments. With some amount of effort the three designs could be modified to give more sophisticated features of classical conditioning. They could not, however, compete with the control and analysis of experiments that would be offered by computer programs running on a microcomputer. While these three designs could actually be used for simple classical conditioning, and would have been the core of classical conditioning experiments before personal computers became readily available, we summarize these devices here to illustrate three approaches that have been used in the past to determine the timing of events. On a more practical level, portions of these designs could be used to effect other timing-dependent devices, for example to make a stimulator.

The three controllers described here have the following features for classical conditioning in common: The intertrial interval is 29 seconds and the trial is 1 second long (30 seconds from CS to CS). Trial presentation is continuous, meaning that none of the examples will stop after a certain count of trials. All the trials are paired CS-US trials. The CS lasts for 350 msec. The US lasts for 100 msec and terminates with the CS. Thus, the relationship of the CS and US describes simple delay classical conditioning. The interstimulus interval, therefore, is 250 msec. In addition to these parameters, the 250 msec baseline that precedes the CS also has an output. The entire trial duration additionally has its own output. The result for all three controllers is that there are six output signals that are available: trial duration, baseline, CS, US, interstimulus interval, and intertrial interval. Note that the CS type and US type are not specified; the types are determined by additional hardware that is attached to those signal lines, for example, CS to tone and US to air puff.

Analog Controller

Figure 11.1 shows a controller made from six timers, one for each of the six signal lines mentioned above. Many timer chips could be used for this application, but we prefer to use the CMOS 4047 (equivalent to the 90438) individual timers for clarity, ease of use and flexible timing options. Our second choice would be the CMOS 4528 dual timer, but the venerable

Chapter 11

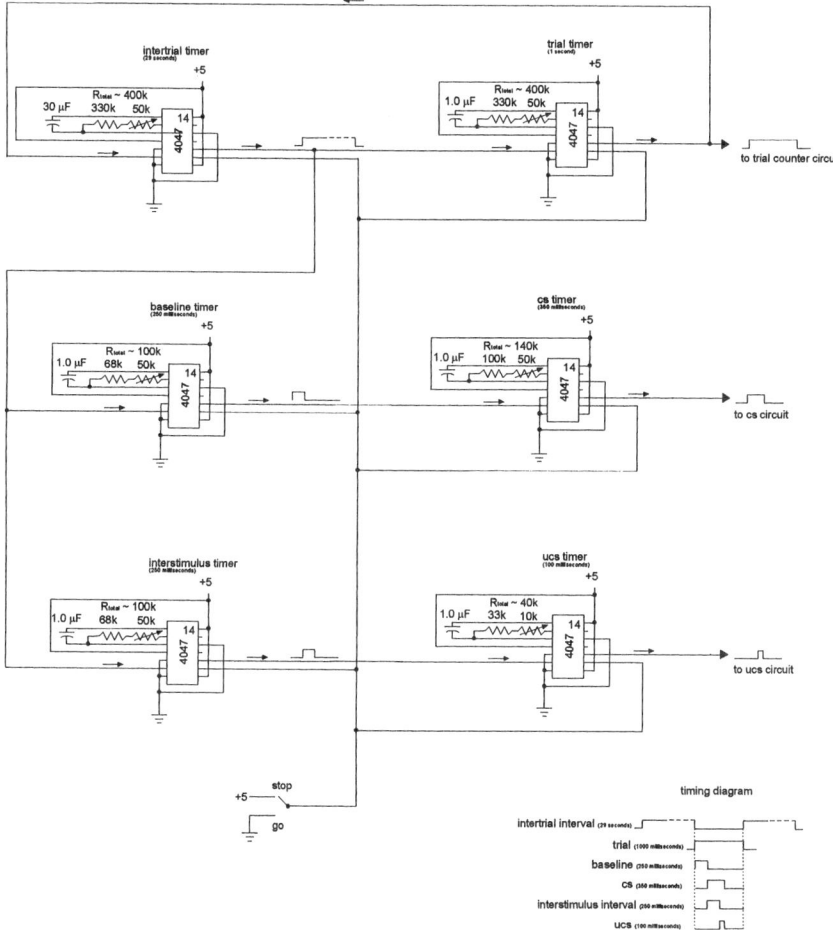

Figure 11.1: Analog controller.

555 or 556 dual timer are also possibilities. The timing is straightforward. With the switch set to "go" the intertrial interval timer is activated; choosing "stop" resets all of the timers. At the end of the intertrial timing two timers are triggered, the trial duration timer and the baseline timer. The end of the trial duration timer triggers the intertrial interval. The end of the baseline timer triggers both the CS timer and the interstimulus interval timer. The end of the interstimulus interval timer triggers the US timer.

For the most part, the internal configurations of the six 4047 timers are the same, the major exception being the choice of timing resistors and capacitors. The values shown here will yield the specified timing. The timing is calibrated by adjusting potentiometers that are associated with the timers and by monitoring the results on an oscilloscope. For this reason we refer to this controller as an "analog controller." Measuring these relatively long time intervals on a typical analog oscilloscope means that there is a degree of imprecision in the settings. Greater precision can be obtained on a digital storage oscilloscope with a scrolling feature. The other two controller designs achieve a greater degree of precision by counting time intervals. The advantage of the analog controller is that simply tuning the poteniometers allows for quick and easy adjustment for different experiments.

A similar kind of analog controller can be achieved with software programming of the 8253 programmable interval timer. Each 8253 semiconductor chip has three independent timers so it would take two 8253 chips to achieve the same design. One of the great advantages of the 8253, however, is that the precise time can be set within the limits of the computer's clock. A second advantage is that the computer can be doing other things while the timers are counting, for example, collecting information about the animal's behavior or neural unit activity, analyzing data, displaying the data on the monitor, monitoring conditions for decisions about more sophisticated designs (for example, if the animal makes a CR then do not give a puff of air to the eye, turning a classical conditioning situation into an operant one).

Decoded Bin Controller

Figure 11.2 shows a second strategy for creating a controller. As with the first design, the end of a trial activates a 4047 as the intertrial timer. The 4047, in turn, turns on a gate for a second 4047 set up as an astable. This strategy differs from the analog timer. The astable creates a series of output pulses set at a frequency of 10 Hz that is fed into a 4017 decade counter. The decade counter's outputs count every 100 msec interval, for a total of nine useable intervals (100, 200, 300 ... 900 msec). The tenth interval (000 msec) is active during the intertrial intervals (the stop condition). Using this strategy, classical conditioning events can occur at 100 msec intervals, which does not match the original design parameters for conditioning. For that, we need to measure 50 msec time intervals. This precision is achieved in the current circuit by inverting the clock output and combining counted outputs through logical AND gates. With this decoding strategy we can achieve intervals every 50 msec (50, 100, 150, 200, 250, 300 ... 950).

These decoded times are used to set the beginning or end of an

Chapter 11

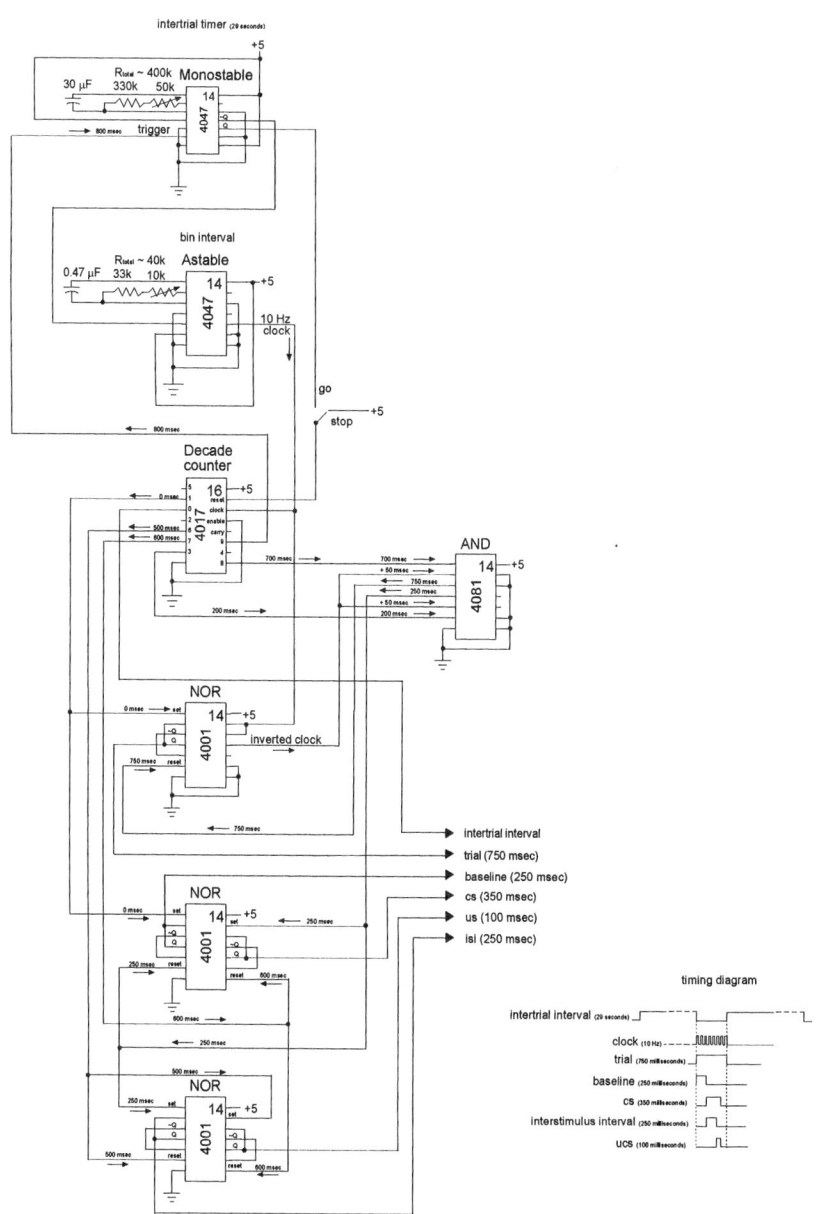

Figure 11.2: Decoded bin controller.

event. This is achieved by using NOR gates configured as simple set-reset flip-flops. The first decoded time sets or resets the flip flop whose output is the signal the beginning or end of an event, respectively. For example, one decoded time sets a flip-flop to turn on the CS while a second decoded time is used to reset that same flip-flop and turn off the CS.

There are two advantages to this design over the analog timer design. First, the 10 Hz clock can be more precisely set than the 4047 outputs in the first design. This can be achieved with either an oscilloscope or a frequency counter. Second, changes in the timing parameters can be easily selected, without further timing calibration, by choosing the appropriate decoded outputs for setting and resetting the flip-flop.

There are disadvantages to this design as well. The decoding of the 50 msec intervals is marginally complicated. Worse, chosen events must occur at 50 msec intervals. This problem can be overcome by increasing the astable rate, for example, to 10 msec intervals (100 Hz) which will yield 5 msec intervals in this design (using the inversion of the clock signal). Increasing the rate, however, means that either we are limited to very short intervals because there is only one decade counter (trials could only be 90 msec long), or that we need to cascade additional 4017 decade counters to extend the counting to achieve longer intervals. This latter strategy (using the carry out of a 4017 decade counter as the input to the next 4017 in the series) increases the burden for decoding, requiring additional AND gates.

This controller design introduces the concept of a "bin" and is referred to here as a "decoded bin controller." A bin is simply a unit of time. In Figure 11.2, the bin width is decoded as a 50 msec interval. One way to have a computer act similarly is to set one of the timers on an 8253 programmable interval timer (it has three timers) to give a signal every at every bin interval. Once the appropriate number of bins have been counted, the CS can be turned on. While waiting for the next bin, the computer could be collecting, analyzing and displaying data. At the next bin, the counter can be checked and the appropriate act carried out. When the count reaches the appropriate time, the CS and the US can be turned off, and then the current data can be collected, analyzed and displayed. In other words, the bin width becomes the interval when decisions are made about intratrial events and data.

Bin Controller

Figure 11.3 shows another bin controller, this one requiring no decoding because the time base signal is twice as fast (20 Hz) and a second counter has been added to increase the counting range. The two counters

Chapter 11

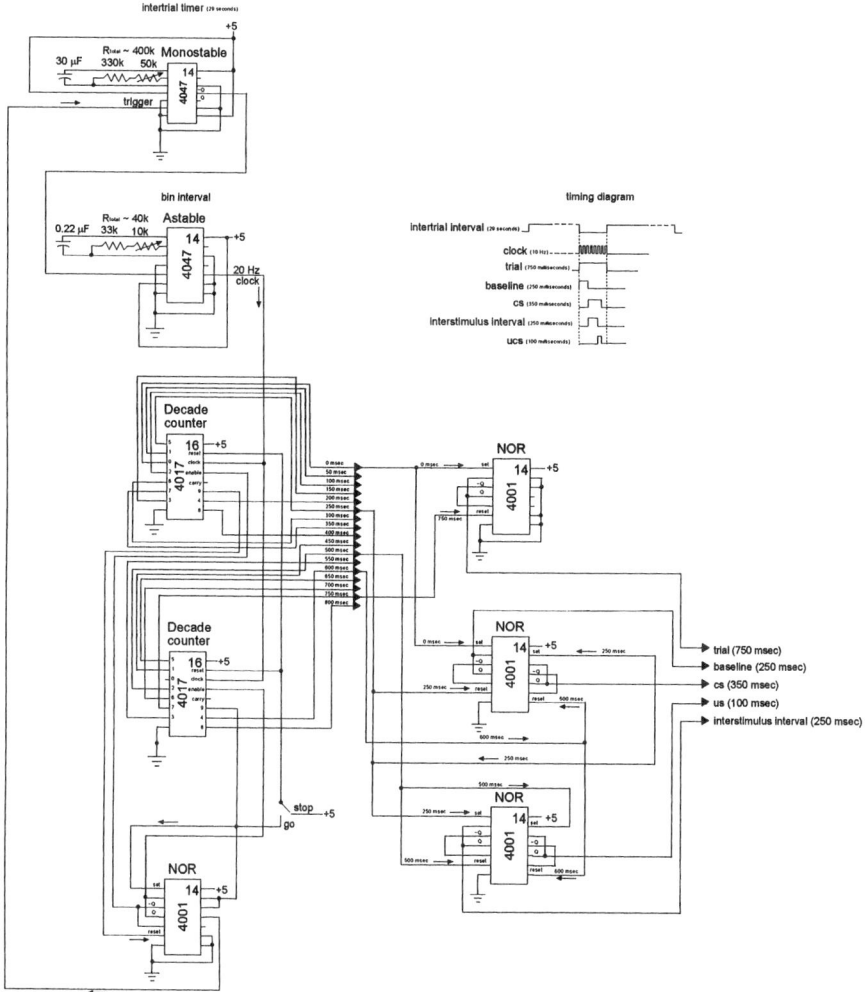

Figure 11.3: Bin controller.

have a different relationship than the multiple counters suggested in the decoded bin controller as a modification of Figure 11.2. The suggestion there was to have the carry out of the first 4017 counter act as the input to the second 4017 counter, which is the usual way of hooking these up (a cascaded design). That design creates a "units" counter (the first 4017) and a "tens" counter (the second 4017). But the design here is a little different. The two 4017 counters are configured like a "ring counter," so that the first 4017 counter handles the first 9 counts and by using a NOR flip-flop the second 4017 counter handles the remaining 8 counts, for a total of 17 counts. As a

Table 11.1: Equivalent software configurations for the bin controller.

bin number	time (msec)	input	comment configuration (set / reset)
0.	0	0001 / 0000	turn on trial duration
1.	50.	0000 / 0000	
2.	100.	0000 / 0000	
3.	150.	0000 / 0000	
4.	200.	0000 / 0000	
5.	250.	0010 / 0000	turn on CS
6.	300.	0000 / 0000	
7.	350.	0000 / 0000	
8.	400.	0000 / 0000	
9.	450.	0000 / 0000	
10.	500.	0100 / 0000	turn on US
11.	550.	0000 / 0000	
12.	600.	0000 / 0110	turn off CS and US
13.	650.	0000 / 0000	
14.	700.	0000 / 0000	
15.	750.	0000 / 0001	turn off trial duration
16.	800.	0000 / 0000	

result, the outputs of the two counters automatically decode 50, 100, 150, 200, 250 ... 800 msec. No AND gates are needed for decoding half bin intervals. As with the decoded bin controller, NOR gates are configured as set-reset flip-flops to control the onset and offset of stimulus events.

The great advantage of this design is that the selection of bin intervals is more readily apparent. Further, this design is easily extended to faster clock rates by additional 4017 counters (and NOR gates). In essence, because there are no AND gates, no decisions have to be made by the hardware. A similar design can be created in software by creating a table that contains a history of the outputs as bins are counted. The following table is the software equivalent of the bin controller:

In this table, each bin is related to its associated time since the onset of the trial. For each, at the count of the fifth bin the time is 250 msec because each bin is 50 msec in duration (20 Hz). The output configuration refers to the inputs of four different set-reset flip-flops, with the first four zeroes (0000) referring to the set operation and the second set of four zeroes (0000) referring to the reset operation. Two different computer outputs, one for setting and one for resetting the flip-flops, are needed. For both the set and reset output configurations, the first flip-flop is not being used (the leftmost digit is always set to zero in the table), the second flip-flop is related to control of the US, the third flip flop is related to control of the CS, and the fourth flip-flop (the rightmost digit) is related to control of the trial duration. At the fifth bin, for example, the CS is turned on (indicated by a 1 for "on" in the corresponding position). At the 12th bin both the CS and the US are turned off (indicated by the 1s in the corresponding positions).

ROM Controller

Figure 11.4 shows an even simpler scheme using a table of output values that avoids any decision making at the time of the trial and avoids using flip-flops. For this example we adapted a scheme we used previously for a cardiotach device. The one drawback is that the ROM controller requires prior programming of the ROM's values.

Here the clock output of an astable (20 Hz equals 50 msec per bin) is the input to a CMOS 4040 binary counter. The outputs of the 4040 counter then act as address lines to a 2716 ROM. The result is that successive addresses of the ROM are addressed and their data are successively placed on the data output lines of the 2716 ROM. The ROM must have been previously programmed with a correct table of values for the events during a trial (see Table 11.2). Here, the lowest bit is used to code the trial duration, then the next digit code successively code for the baseline, CS, US and interstimulus interval. The meaning of each digit is entirely dependent upon the equipment attached to each output—if a tone is hooked up to the output for an air puff then only a tone will happen. At every new bin the data is simply placed onto the output data lines of the ROM, where a 1 must always be present whenever an event must occur or, importantly, continue to occur. The latter is the reason why flip-flops are not necessary. Table 11.2 shows the successive data for classical conditioning:

We know of two antecedant systems that used a table scheme. The first was a controller, built for Thompson by personnel from the University of California, Irvine, that used one ROM to code the intertrial intervals and a second ROM to code the trial types and events. The second system was the KIM-1 microcomputer software for classical conditioning by Solomon and colleagues (Solomon & Babcock, 1979; Solomon, Weisz, Clark, Hall, & Babcock, 1983), which used different tables for different conditions (e.g., tone alone, trace trials, etc). The advantage of the KIM-1 microcomputer was that the data tables were stored on cassette tape and played to the KIM-1, which stored the tables in RAM. Finally, between using the KIM-1 and Apple II computers we designed and built our own 6502 microprocessor computer with programmed ROM based upon the programming concept of Solomon and colleagues.

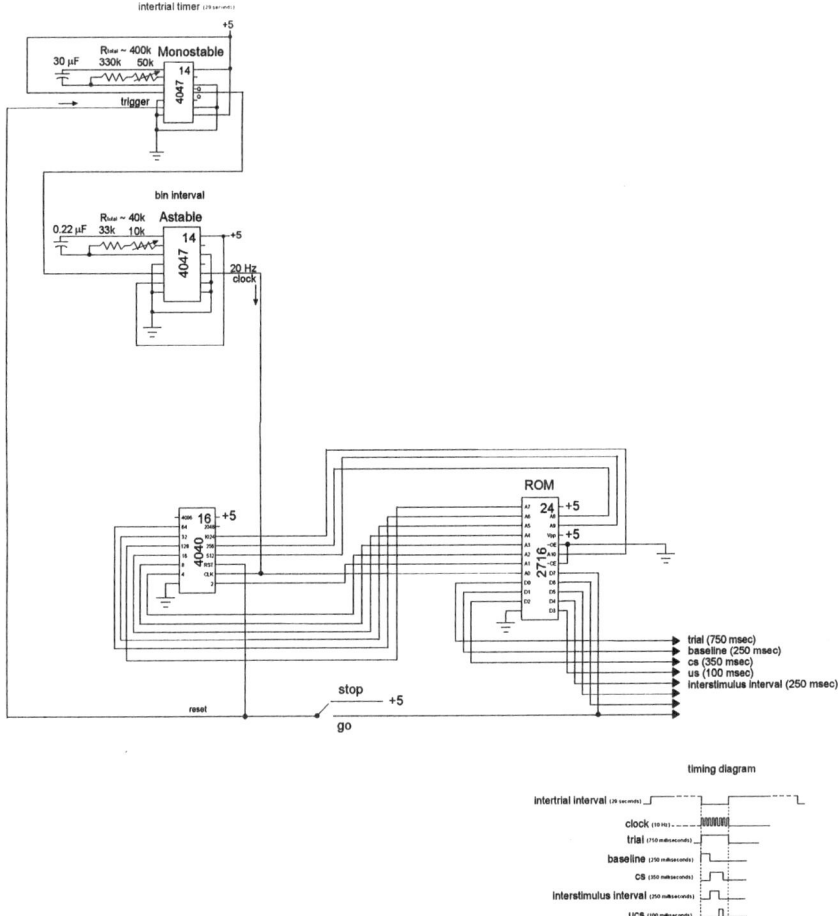

Figure 11.4: ROM controller.

Summary of Early Controllers

The controllers described here are practical in the sense that they can be used to effect classical conditioning. They have some important limitations as already indicated in that they do not automatically shut down after giving a proscribed number of trials and every trial generated is a paired CS-US trial (i.e., with no CS- or US-alone presentations possible). With greater effort these features could be attained with additional hardware. The greatest drawback of these systems by far is that these controllers are only good for presenting stimuli with preset timing relationships. They do not collect and analyze data. Years ago, these controllers would have been used in con-

Chapter 11

Table 11.2: Output data for ROM controller.

bin number	time (msec)	output configuration	comment
0.	0.	00000011	turn on trial & baseline
1.	50.	00000011	keep trial on
2.	100.	00000011	keep trial on
3.	150.	00000011	keep trial on
4.	200.	00000011	keep trial on
5.	250.	00010101	turn on CS & isi, & keep trial on
6.	300.	00010101	keep CS, isi & trial on
7.	350.	00010101	keep CS, isi & trial on
8.	400.	00010101	keep CS, isi & trial on
9.	450.	00010101	keep CS, isi & trial on
10.	500.	00001101	turn on US, keep on CS & keep trial on
11.	550.	00001101	keep US, CS & trial on
12.	600.	00000001	turn off CS & US, and keep trial on
13.	650.	00000001	keep trial on
14.	700.	00000001	keep trial on
15.	750.	00000000	turn off trial
16.	800.	10000000	reset & begin intratrail time

junction with polygraph records of the behavior, which would have to be hand scored, or with tape recordings to be played back to a computer for analysis, as with Thompson's Digital Equipment Corporation PDP-12 computer and later the Digital Equipment Corporation PDP-11 computer (we have been told that the even numbered computers were commercial models, the odd numbered ones were development computers, which supposedly explains why their numbers are not chronological). These controllers would not likely be used today because they would have the same problem of handling the data. With the availability of personal computers, first with the Apple II and later with the IBM-PC and its clones, programs were written to control the stimuli and trials, to collect the data, and to analyze and display the data online. The significance of the development and availability of these computers in every day life as well as in research should not be underestimated.

COMPUTER-BASED CONTROLLER SYSTEMS

The Forth System

We present here a relatively detailed summary of the Forth-based system that we have used to run classical conditioning experiments. While it is unlikely that the average reader will adopt this system for use in his or her laboratory, we believe that providing some detailed information about this system serves at least one important purpose: The reader can gain some excellent insights into the general design of hardware and software solutions for controlling behavioral neuroscience experiments.

IBM/Clone Interface

We have previously described an interface card for the IBM PC/XT and clones (Lavond & Steinmetz, 1989). Here, we describe an updated version of that same circuit. We colloquially refer to this circuit as the "Forth (interface) board" in comparison to the "First board" designed by Gormezano and Scandrett (1980). There is a fundamental difference, however. The First computer language *required* the board because it contained a necessary floating point processor (i.e., the First language only worked on computers that had the board with the floating point processor). In contrast, the interface described here is not needed to run the Forth language or to analyze stored data that was previously collected using Forth or any other language. It is really a misnomer to call the interface the "Forth board" because the interface has nothing to do with the Forth language. Rather, the Runtime programs, written mostly in Forth but with key assembly language passages, needs the interface board to deliver the CS and US and to collect the behavioral and neural data. It should be noted that any programming language could access the interface board described here.

The interface board plugs into any standard edge connector of an IBM-type motherboard. It does not use the extended edge connector. The standard edge connector provides power, address, control, and data lines to and from the interface board. The pins on the standard edge connector are labeled 1 through 31 on each of sides A and B. Table 11.3 shows the standard edge connector with the contact definitions. These contacts are indicated by the labels on the figure for the interface card. For example, contact A31 is address line 0 (A0) which connects to the A/D converter (ADC0808) on pin 25, two 8253-5 Programmable Interval Timers (use the -5 version) on pin 19, and two 8255 Programmable Peripheral Interfaces on pin 9.

Figure 11.5 shows the circuitry of our interface card. The interface

Chapter 11

Table 11.3: IBM-PC edge connector contacts.

B side	function	A side	function
(rear of the computer in this direction, top view)			
B1	Ground	A1	I/O Ch Ck
B2	DRV reset	A2	Data line 7
B3	+5 volts	A3	Data line 6
B4	IRQ2	A4	Data line 5
B5	-5 volts	A5	Data line 4
B6	DRQ2	A6	Data line 3
B7	-12 volts	A7	Data line 2
B8	reserved	A8	Data line 1
B9	+12 volts	A9	Data line 0
B10	Ground	A10	I/O CH RDY
B11	not MEMW	A11	AEN
B12	not MEMR	A12	Address line 19
B13	not IOW	A13	Address line 18
B14	not IOR	A14	Address line 17
B15	not DACK3	A15	Address line 16
B16	DRQ3	A16	Address line 15
B17	not DACK1	A17	Address line 14
B18	DRQ1	A18	Address line 13
B19	not DACK0	A19	Address line 12
B20	Clock	A20	Address line 11
B21	IRQ7	A21	Address line 10
B22	IRQ6	A22	Address line 9
B23	IRQ5	A23	Address line 8
B24	IRQ4	A24	Address line 7
B25	IRQ3	A25	Address line 6
B26	not DACK2	A26	Address line 5
B27	T/C	A27	Address line 4
B28	ALE	A28	Address line 3
B29	+5 volts	A29	Address line 2
B30	OSC	A30	Address line 1
B31	Ground	A31	Address line 0
(front of the computer in this direction, top view)			

has 4 available A/D converters (8 are available with a change in addressing), 6 independent timers/clocks, and 48 digital input or output signals. The A/D converters are used to collect data about the position of the eyelid or leg flexion. One of the timers/clocks is used to time the intratrial binwidth (bin), four timers/clocks are used to count neural unit activity, and one timer/clock is unused. Eight of the digital signals are used to control classical conditioning (tone, light, unused/reserved, air, shock, tape, synchronization pulses, pulse out) through external signals to the appropriate equipment, one digital input signal is used as a trigger for off-line data collection from tape and a

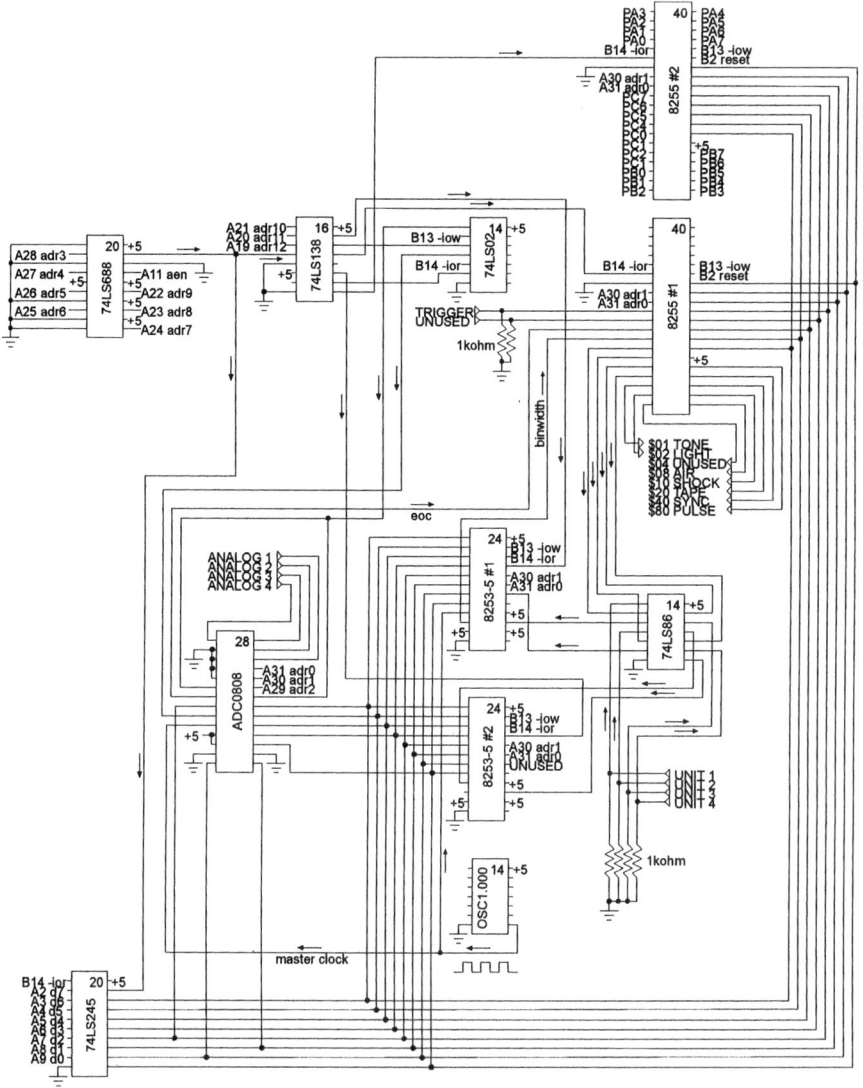

Figure 11.5: Forth interface card.

second digital input signal is unused/reserved, four output signals are used internally on the interface board to control unit collection, one digital input signal reads the end-of-conversion (eoc) from the A/D converter and another one reads the binwidth. In all, a total of 12 digital signals are accounted for, leaving 36 unused on the figure. Of these 36 remaining digital signals, the audio.scr programs (described in CD) use 8 outputs to select audio attenuation and 5 outputs to select the tone frequency, leaving 23 unused digital signals. With the possible exception of the hardware on the interface card that is associated with internal board connections, the actual functions of the four A/D converters, two timers (the one being used for binwidth and an unused one), and 46 out of 48 of the digital signals, all depend upon programming and the equipment hooked up to the signals.

The circuitry of the interface card can be broken down into signals associated with the computer, internal interface connections, and external signals. The signals associated with the computer include data lines (D0-D7), address lines (A0-A12), control lines (AEN, not IOW, not IOR, reset), and power (+5 and ground) found on the IBM-type edge connector. The data lines are buffered through the 74LS245 bus transceiver, with the direction of data flow controlled by not IOR. The address lines are decoded by the 74LS688 digital comparator and the 74LS138 address decoder for a base address of $0390. So far, this address space has not seemed to interfere with upgraded computers through the 80486 versions although it was designed for the original 8088 computers. Most of the large scale integration chips (LSI: 8253 and 8255) and the 74LS245 bus driver are compatible with the read/write signals on the edge connector. To use the A/D converter required deriving read/write signals along with address decoding. The internal signals between chips on the interface include, for example, the 8255 #1 which reads the binwidth for the intratrial timing events counted by 8253 #1, reads the end-of-conversion signal for the ADC0808 A/D converter, and sets the unit counters on 8253 #1 and #2 to zero at the beginning of each bin. The external signals include digital output signals to effect tones and air puffs, for example, digital input signals for the unit counters, and analog input signals for reading the position of the eyelid. In addition, there is one completely unused timer/counter on 8253 #2, eight completely unused digital signals on 8255 #1 (port A), and 24 unused digital signals on 8255#2 (ports A, B and C in the hardware literature; assigned as ports D, E and F in the audio.scr software). Not shown is that the second 8255 (#2) is used with the audio.scr software to control the selection of tone frequencies (5 output signals on port D) and attenuation (8 output signals on port E).

Figure 11.5 differs from the originally described interface by extending its capabilities for the audio programs (adding a second 8255 and writing the audio.scr programs), two additional unit counters (adding a second 8253-5), and by replacing the timing signal for the A/D converter with the 1 MHz oscillator (OSC1.000) instead of deriving it from the computer's oscillator (edge connector pin B30 is now unused). The latter change made the A/D conversion consistent and independent of the speed of the computer the interface was being run on. Not shown is that small disc capacitors (e.g., .001) should be placed as directly as possible onto the power supply pins of chips that change rapidly, the OSC 1.000 and the 74LS245 being prime example.

Table 11.4: Memory map of the hardware addresses on the interface card.

Address	Software Label	Chip, Function	Program Function
$0390	**time0**	8253 #1, Timer/counter #0	intratrial binwidth (mode 3)
$0391	**time1**	8253 #1, Timer/counter #1	unit counter #1 (mode 0)
$0392	**time2**	8253 #1, Timer/counter #2	unit counter #2 (mode 0)
$0393	**cw8253#1**	8253 #1, Control register	
$0790	**'0808wr**	ADC0808, A/D converter #0	write & start A/D channel #0
$1791	**'0808wr** +1	ADC0808, A/D converter #1	write & start A/D channel #1
$1792	**'0808wr** +2	ADC0808, A/D converter #2	write & start A/D channel #2
$1793	**'0808wr** +3	ADC0808, A/D converter #3	write & start A/D channel #3
$1794	**'0808wr** +4	ADC0808, A/D converter #4	write & start A/D channel #4
$1795	**'0808wr** +5	ADC0808, A/D converter #5	write & start A/D channel #5
$1796	**'0808wr** +6	ADC0808, A/D converter #6	write & start A/D channel #6
$1797	**'0808wr** +7	ADC0808, A/D converter #7	write & start A/D channel #7
$0B90	**porta**	8255 #1, Digital I/O	not used
$0B91	**portb**	8255 #1, Digital I/O	classical conditioning outputs bit 0 = tone bit 1 = light bit 2 = unused bit 3 = airpuff bit 4 = shock bit 5 = tape recorder control bit 6 = CS & US sync pulses bit 7 = pulse out
$0B92	**portc**	8255 #1, Digital I/O	interface control signals bit 0 = unit counter #1 bit 1 = unit counter #2 bit 2 = unit counter #3 bit 3 = unit counter #4 bit 4 = read binwidth timer bit 5 = read A/D eoc bit 6 = read unused ext input

Chapter 11 343

Table 11.4 (continued)

			bit 7 = read ext trigger input
$0B93	**cw8255 #1**	8255 #1, Control register	
$0F90		address decoded but not used (could be D/A converter, for example)	
$1390	**time3**	8253 #2, Timer/counter #0,	unit counter #3 (mode 0)
$1391	**time4**	8253 #2, Timer/counter #1	unit counter #4 (mode 0)
$1392	**time5**	8253 #2, Timer/counter #2	unused (mode 0)
$1393	**cw8253#2**	8253 #2, Control register	
$1790	**'0808rd**	ADC0808, A/D converter #0,	read A/D channel #0
$1791	**'0808rd** +1	ADC0808, A/D converter #1	read A/D channel #1
$1792	**'0808rd** +2	ADC0808, A/D converter #2	read A/D channel #2
$1793	**'0808rd** +3	ADC0808, A/D converter #3	read A/D channel #3
$1794	**'0808rd** +4	ADC0808, A/D converter #4	read A/D channel #4
$1795	**'0808rd** +5	ADC0808, A/D converter #5	read A/D channel #5
$1796	**'0808rd** +6	ADC0808, A/D converter #6	read A/D channel #6
$1797	**'0808rd** +7	ADC0808, A/D converter #7	read A/D channel #7
$1B90	**portd**	8255 #2, Digital I/O	audio frequency selection $00 = 1 KHz (not a mistake) $01 = 0.5 KHz (not a mistake) $02 = 2 KHz $03 = 3 KHz $04 = 4 KHz $05 = 5 KHz $06 = 6 KHz $07 = 7 KHz $08 = 8 KHz $09 = 9 KHz $10 = 10 KHz $11 = 11 KHz $12 = 12 KHz $13 = 13 KHz $14 = 14 KHz $15 = 15 KHz $16 = 16 KHz $17 = 17 KHz $18 = 18 KHz $19 = 19 KHz $1A = 20 KHz
$1B91	**porte**	8255 #2, Digital I/O	audio dB attenuation selection bit 0 = 0.5 dB bit 1 = 1 dB bit 2 = 2 dB bit 3 = 4 dB

Table 11.4 (continued)

			bit 4 = 8 dB
			bit 5 = 16 dB
			bit 6 = 32 dB
			bit 7 = 64 dB
$1B92	**portf**	8255 #2, Digital I/O	not used

All the parts cost about $100 for this interface, and it takes about eight hours to construct using wire wrapping technique. Commercially available interfaces with address decoding and buffering of control and data lines are also available, to which LSI chips could be added. Commercially available interfaces with the LSI chips are also available.

The hardware interface is nothing without the appropriate software programs. Although these programs are described in our original publication (Lavond and Steinmetz, 1989) their simple assembly language code bears repeating along with equivalent Forth code. Forth converts (compiles) the assembly language into native machine language which should be used for fast operations like controlling stimuli, timing and data collection. Forth or any other more familiar, higher level language can be used to write the initial software for testing the operation of the interface. We routinely use Forth level programming to debug problems with the interface (chips and sometimes particular pins on a chip go bad) or to test ideas before committing them to assembly language.

Table 11.4 shows the memory map of the addresses for the hardware components on the interface card. The addresses are necessary for programming and using the devices (e.g., $0B91). The labels associated with the addresses are used for convenience of identification in the software (e.g., **portb**). The individual chips and their functions at the given address are identified (e.g., 8255 #1, Digital I/O). Their use in the program (e.g., classical conditioning outputs) are identified, maybe with additional information about how they are used (e.g., bit0 of portb controls the tone). In the remainder of this chapter we will give specific programming examples for each of the main components.

Digital input/output (8255)

The two 8255s are currently used to provide 8 external interface outputs to control stimuli and events for classical conditioning (**portb**); 5 external interface outputs to control tone frequency with the audio.scr program (**portd**); 8 external interface outputs to control tone attenuation with the audio.scr program (**porte**); 4 internal interface outputs to control the four unit counters (the lower half of **portc**); 2 internal interface inputs to read the bin-width and end-of-conversion (eoc) for the A/D conversion (two inputs on the upper half of **portc**); and 2 external interface inputs to read the external trigger for off line data analysis and an unused external input (the other two inputs on the upper half of **portc**).

To use the three 8-bit ports of the 8255s requires programming their control registers to configure the ports as inputs and/or outputs. For classical conditioning, the first 8255 (8255 #1) has the 8 bits of **porta** configured as inputs, the 8 bits of **portb** configured as outputs, the lower 4 bits of **portc** configured as outputs, and the upper 4 bits of **portc** configured as inputs. The second 8255 (8255 #2) is similarly configured for its three 8-bit ports, labeled **portd**, **porte** and **portf** respectively.

To set the control register of 8255 #1 the commands are

MOV AX, $98	; $98 is the code for configuration from
	; the Intel data sheets for the 8255
	; The dollar sign ($) forces the following
	; number to be interpreted as a
	; hexidecimal (base 16) number.
MOV DX,cw8255#1	; cw8255#1 is the address $0B93
OUT DX,AL	; writes to the control register.

In Forth the same commands to set the control register would be

hex	; interpret all subsequent numbers as
	; hexidecimal (base 16)
98 cw8255#1 !	; write configuration to the control
	; register
decimal	; interpret all subsequent numbers as
	; base 10

The commands **hex** and **decimal** are made explicit here for good programming. In general, we explicitly invoke **hex** before any number or series of numbers that are base 16, commonly done when using assembly coding in Forth or accessing interface equipment, and return to base 10 (**decimal**) as a matter of practice.

To read **portb** use the following assembly code or the respective Forth code.

MOV DX,portb	; portb is the address $0B91
IN AL,DX	; read the port into the lower byte of

portb pc@   ; register A and
                                 ; read portb

To turn on the tone, write to **portb** use the following assembly code or the respective Forth code.

MOV AX,$01   ; tone is controlled by the state of the
                                 ; first bit of portb
MOV DX,portb   ; set the destination
OUT DX,AL   ; write $01 to the port

and the corresponding Forth code is

hex
01 portb pc!   ; write $01 to the port
decimal

To turn off all the equipment attached to **portb**, which could be a part of the initialization process to ensure that nothing is turned on accidentally, the assembly code is

MOV AX,$00
MOV DX,portb
OUT DX,AL

and the corresponding Forth code is

0 portb pc!

Although it is not particularly good documentation, **hex** and **decimal** were not invoked here because zero in any base is still zero.

The forgoing operations can be made into Forth subroutines, called **words**, by creating definitions (**reset, y** and **n** in this example) bracketed by a colon (:) at the start of each definition and a semicolon (;) at the end of the definition. The word **reset** is defined in the Runtime software for initializing the LSI chips on the interface board. The definitions for **y** and **n** are handy for debugging the hardware when control over the tone is helpful:

hex
: y (--) 1 portb pc! ;
: n (--) 0 portb pc! ;
decimal

Once these are defined, pressing **y\<return>** turns on the tone, pressing **n\<return>** turns off the tone (and everything else on **portb**). The definitions are compiled into the Forth dictionary (usable subroutines) and these can be used repeatedly as long as Forth is active.

Note that when invoking the command **y\<return>** that not only is the tone turned on, but that all other stimuli (light, air, shock, etc.) associated with **portb** are turned off. The same is true for the assembly code above. To turn on both the tone (at bit $01) and air puff (at bit $08) at the same time requires setting the corresponding bits and writing once ($09) to the port. In assembly language the code is

MOV AX,$09
MOV DX,portb
OUT DX,AL

and the Forth code is

hex
09 portb pc!
decimal

All other bits will be set to zero, turning off everything else attached to **portb**. Because we may not know the states of any other equipment hooked up to the port, we can turn on the tone and leave everything else the same by first reading the port, adding the bit to turn on the tone, and writing it back to the port. The assembly language code for turning on the tone without changing anything else is

MOV BX,$01	; we want to turn on the tone
MOV DX,portb	
IN AL,DX	; read the current pattern on the port
OR AL,BL	; combine the tone to the current pattern
	; to yield the combined pattern
SUB AH,AH	; get rid of the upper byte of the A
	; register (housekeeping)
OUT DX,AL	; write the combined pattern back to the
	; port

After reading the current pattern on **portb**, the OR operation allows the tone bit to be set. This new pattern is then written back to **portb**. The equivalent Forth command is

```
hex
01 portb pc@ or portb pc!
decimal
```

Here we have placed several Forth commands on the same line: read the port, OR the tone pattern to the current pattern on the port, write the new pattern back to the port. For documentation purposes the same Forth commands can be written on separate lines, making the operation easier to see, along with added comments (in parentheses) to help us keep track of the operations in

```
hex
01 ( tone_bit)
portb pc@ ( current_pattern)
( tone_bit current_pattern) or ( new_pattern)
( new_pattern) portb pc!
decimal
```

Here our comments are overkill just for emphasis. Notice also that Forth works like Hewlett-Packard calculators in that the two numbers are given, followed by the OR operation (current_pattern 01 or) which is characteristic of Forth. The Forth code can be made into a Forth definition (subroutine) like the following

```
hex
: turnon ( bit_number --)
   portb pc@ or portb pc! ;
decimal
```

Now any bit of **portb** can be turned on very easily. For example, to turn on the tone use **01 turnon<return>**, to turn on the air puff use **08 turnon<return>**, etc. The Forth definition for **turnon** is a more general case that allows us to easily turn on anything hooked up to **portb**.

The complementary operation, to turn off the tone without affecting anything else that may or may not be on **portb**, is accomplished with the following assembly and Forth codes.

```
MOV BX,$01              ; we want to turn off the tone
MOV DX,portb
IN AL,DX                ; read the current pattern on the port
PUSH AX                 ; save a copy of the current pattern on
```

Chapter 11

```
                        ; the stack
AND AX,BX               ; identify whether the tone is currently
                        ; on or not
NOT AL                  ; get the complement in preparation to
                        ; turn the tone off
POP BX                  ; move the copy of the current pattern
                        ; into the B register
AND AL,BL               ; zero the tone bit but leave all other bits
                        ; the same, yielding the new pattern
SUB AH,AH               ; get rid of the upper byte of the A
                        ; register (housekeeping)
OUT DX,AL               ; write the new pattern back to the port
```

and in Forth the code is

hex
01 not portb pc@ and portb pc!
decimal

The Forth code can be turned into a definition (a Forth subroutine), here called **turnoff**, for the general case to selectively turn off anything hooked up to **portb**.

hex
: turnoff (bit_number --)
  not portb pc@ and portb pc! ;
decimal

The command **01 turnoff\<return>** will selectively turn off the tone without affecting anything else that may or may not be turned on **portb**. Similarly, the command **08 turnoff\<return>** will selectively turn off the air puff without affecting anything else that may or may not be turned on **portb**. The routines **'pset** and **'preset** in the Runtime program achieve the same generality in assembly language as with the definitions of **turnon** and **turnoff** in Forth.

With this arsenal of commands, stimuli for classical conditioning can be turned on or off at will. The assembly language coding is used when speed is a concern, as it is during a trial. Recall that each trial is made up of a number of data points (**pts**), each data point consisting of a fixed interval of time known as the binwidth (**bin**) that is controlled by **time0** of 8253#1. The output of **time0** is a square wave which is read by bit 4 ($10) on the upper half of **portc** of the 8255#1. The software reads the square wave transitions on this bit to determine when a binwidth has occurred. If the binwidth is set for 4 milliseconds (**4000 bin !** in Forth code) then the computer has 4 milliseconds to set the output pattern on **portb** for the stimuli, as well as to read and store the values on the A/D converters and the unit counters. Practically speaking, we have found that the old 4 MHz IBM clones can do all this with 2 millisecond time bins. The Forth code looks compact, and Forth is generally as fast language, but its threaded interpretation takes too long with the older computers. With faster computers, the Forth code might substitute for the assembly code. A language like C++, which might be set to compile into optimized native (machine) code, could be fast enough to directly run classical conditioning experiments.

Analog to digital conversion (ADC0808)

The analog to digital converter is fairly straightforward to use with a few caveats in mind. Writing anything to one of the A/D converters starts the conversion process. The assembly language code for starting the conversion of the first A/D channel (channel 0 at '0808wr) is

```
MOV AX, $00           ; any value will do
MOV DX,'0808wr        ; '0808wr is the address $0790
OUT DX,AL             ; start the conversion process
```

In Forth the equivalent command is

```
hex
00 '0808wr pc!
decimal
```

For the second A/D channel use the next address (**'0808wr+1**, $0791), etc. As currently configured, 4 of the 8 A/D converters are available for collecting analog data. When the A/D conversion is completed, the A/D converter sends out an end-of-conversion signal (eoc). This eoc signal is detected by 8255#1 on bit 5 of **portc**, indicating that a valid conversion has taken place. If a second conversion is started by writing to any of the other A/D addresses on the ADC0808 converter, then the first conversion is abandoned and the second conversion started. Care must be taken to wait for the end-of-conversion signal, then read the value, then start another conversion. The assembly and Forth codes to wait for the eoc signal are

MOV DX,portc

```
loop1   IN AL,DX
        AND AL,$20      ; look for changes on bit 5 only
        JE loop1        ; continue looping until bit 5 is logic 0
loop2   IN AL,DX
        AND AL,$20      ; look for changes on bit 5 only
        JNE loop2       ; continue looping until bit 5 is logic 1
```

and the Forth equivalent

```
hex
: eoc ( --)
  begin ( loop until bit 5 is logic 0)
        portc pc@
        20 not and
  until
  begin ( loop until bit 5 is logic 1)
        portc pc@
        20 and
  until ;
decimal
```

Here the Forth code must be a definition (subroutine), here called **eoc**. The reason for this is that the looping structure **begin-again** can only be compiled, it cannot be interpreted (where the commands can be simply typed directly from the keyboard for execution) in the original version of Forth we used. In more recent versions of Forth looping structures may be allowed in the interpreter. After starting the conversion, typing **eoc<return>** will wait for a valid A/D. Both the assembly and Forth language here uses code to create a software 'debounced' switch, meaning that the loops are exited only after a negative-to-positive transition of the signal coming from **eoc** onto bit 5 of **portc**. Once a valid A/D conversion has been made, the A/D converter can be read with the following assembly and Forth codes.

MOV DX,'0808rd
IN AL,DX

and in Forth

hex
'0808rd pc@
decimal

The rate of conversion of the ADC0808 is a function of its clock (pin 10), which comes from the OSC1.000 (1 MHz clock). This places the A/D conversion near the upper end of its range. The documentation for the ADC0808 indicates that a 640 KHz clock results in a sampling rate of 10 KHz, so our sample rate is something above 10 KHz with a 1 MHz clock. Using the Nyquist rule that any signal should be sampled at twice its frequency, this means that a 10 KHz sample rate can accurately reflect a 5 KHz signal. The fastest bin-width we have ever used successfully with this interface uses a bin-width of 2 milliseconds, or 500 Hz, which is well within the sampling rate of this interface for a single A/D channel. For 4 A/D channels, the effective sampling rate with 2 millisecond bins is 2000 Hz, which is still well within the sampling rate of the ADC0808 on this interface.

In the Runtime program, once the A/D data is read it is stored into an array for the appropriate A/D channel (**nmrv**) using a pointer (**^nmrv**) into the array. The assembly language subroutine **'datsr** collects and stores the A/D and unit count data. The array is a RAM space that has been allocated as a disk buffer by **get-buffer**. After updating the buffer, the data is then saved on disk if the appropriate switch is set with **safety off**. **Safety on** is used for debugging to prevent saving the data to disk, but it is also used when giving trials with very short intertrial intervals such as pseudorandom conditioning.

Binwidth timing (8253)

The 1 MHz oscillator (OSC1.000) is a master clock that is used to create an oscillator for timing events during the trial (bin-width) through its attachment to timer/counter **time0** (pin 9) on 8253#1. **Time0** takes the 1 MHz signal and divides it by a derived factor to create the desired bin-width. For example, to create a 1 KHz oscillator requires 1000 counts of the 1 MHz master clock. Similarly, to create a 4 millisecond bin-width requires 4000 counts of the 1 MHz master clock. Technically, any bin-width from 1 μsec (0.001 msec) to 65535 μsec (65.535 msec) could be achieved with **time0**. The value 65535 is the largest value that can be held in **time0**. The unit value of **bin** is in microseconds. The value of **bin** is directly stored into **time0**. This storage is accomplished in the assembly code for **trial**. The equivalent Forth code would be **bin @ time0 !**. Practically speaking, however, the Runtime and Summary programs assume that only whole milliseconds are used (1, 2, 3, 4 ... 65 msec) to eliminate decimal computations since older versions of Forth do not have a floating-point processor (as First does). Newer versions of Forth may include floating point, either through a math coprocessor or through software implementations. Traditionally Forth avoided

Chapter 11 353

floating point by dropping the decimal for calculations and simply formatting the printout with decimals, as we do with the Summary printouts. The advantage of avoiding floating point is that the calculation speed is substantially faster. With faster and faster computers, however, this is less and less a concern.

The following assembly and Forth code sets up time0 as a timer creating a 250 Hz (4 msec binwidth) squarewave generator.

```
MOV DX,cw8253#1      ; $36 is the code for configuring a
                     ; square wave timer (mode 3)
MOV AX,$36
OUT DX,AL
MOV DX,time0
MOV AX, $0F          ; the high byte of $0FA0 which is
                     ; decimal number 4000
OUT DX,AL
MOV AX, $A0          ; the low byte of $0FA0 which is the
                     ; decimal number 4000
OUT DX,AL
```

and in Forth the code is

hex
36 cw8253#1 pc!
0f time0 pc!
a0 time0 pc!
decimal

Now that time0 is giving out square wave pulses at the rate of 250 Hz (one pulse every 4 msec) the binwidth can be detected on bit 4 of **portc** on 8253#1 using the following assembly language or Forth code for a software debounced switch.

```
        MOV DX,portc
loop1   IN AL,DX
        AND AL,$10       ; look for changes on bit 4 only
        JE loop1         ; continue looping until bit 4 is logic 0
loop2   IN AL,DX
        AND AL,$10       ; look for changes on bit 4 only
        JNE loop2        ; continue looping until bit 4 is logic 1
```

and in Forth

```
hex
: binwidth ( -- )
  begin ( loop until bit 5 is logic 0)
        portc pc@
        10 not and
  until
  begin ( loop until bit 5 is logic 1)
        portc pc@
        10 and
  until ;
decimal
```

As with the example of **eoc** above, here the Forth code must be a definition (subroutine), here called **binwidth**. The reason for this is that the looping structure **begin-again** cannot be interpreted (it must be compiled within a definition). The definition is formatted for ease of viewing and documentation, but the entire definition could be on a single line. After starting the squarewave generator at the 250 Hz frequency, typing **binwidth<return>** will wait for the end of a 4 msec period. By repeatedly executing **binwidth** (in a loop within **trial**) successive binwidths are identified. By this polling routine, stimulus events and data collection can occur at 4 msec intervals. In the Runtime program this successive polling of binwidths is accomplished in the assembly language routine **trial**, where the current bin is kept track of with **bin#** and the routine terminates when the total number of required bins (the C register of the microprocessor is initially loaded with **pts** and the C register decrements each time through the loop; the loop is repeated until the C register reaches zero).

Unit counting (8253)

The four unit counters (**time1**, **time2**, **time3** and **time4**) need to be initialized as counters rather than as timers with assembly or Forth codes similar to the following.

```
MOV DX,cw8253#1
MOV AX,$70            ; $70 is the code for configurating a
                      ; counter (mode 0)
OUT DX,AL
```

and in Forth the code is

Chapter 11

**hex
70 cw8253#1 !
decimal**

For **time1** the control register is **cw8253#1** and the mode 0 instruction code is $70; for **time2** the control register is **cw8253#1** and the mode 0 instruction code is $B0; for **time3** the control register is **cw8253#2** and the mode 0 instruction code is $30; for **time4** the control register is **cw8253#2** and the mode 0 instruction code is $70.

To begin the count, each counter must be zeroed by writing $00 to each counter, twice (once for the high counter register, the second time for the low count register). A peculiar feature is that this count is not actually entered into the count register until one external input is applied to the counter. This has an important consequence that we will address momentarily. This external input is supplied by the corresponding output from the lower byte of **portc** of 8255#1. The code to set up **time1** with a count of zero ($0000) is

**MOV DX,time1
MOV AL,$00
OUT DX,AL
OUT DX,AL**

and the code to pulse bit 0 on and off, without affecting any other bit on **portc**, thus setting the counter to zero is

**MOV DX,portc
IN AL,DX
OR AL,$01
OUT DX,AL
AND AL,$FE
OUT DX,AL**

The first OUT using $01 sets bit 0 to logic 1, the second OUT using $FE (the complement of $01) sets bit 0 to logic 0. To accomplish both of these functions in Forth the code is

**hex
00 time1 pc!
00 time1 pc!
portc pc@**

dup
01 or portc pc!
01 not and portc pc!
decimal

Here we explicitly use the complement of $01 to reset the configuration on **portc**. For **time1** use the code $01 and its complement $FE; for **time2** use $02 and its complement $FD; for **time3** use $04 and its complement $FB; and for **time4** use $08 and its complement $F7. Both the assembly language and Forth code here is overly cautious. The upper half of **portc** is configured to be input, so writing anything to that half of **portc** will have no effect. The four bits of the lower half of **portc** are used one at a time to set the respective unit counters. The bare bones assembly and Forth code are

MOV DX,time1
MOV AL,$00
OUT DX,AL
OUT DX,AL
MOV DX,portc
MOV AL,$01
OUT DX,AL
MOV AL,$00
OUT DX,AL

and in Forth the code reduces to

hex
00 time1 pc!
00 time1 pc!
01 portc pc!
00 portc pc!
decimal

For **time1** use $01 to set the counter; for **time2** use $02; for **time3** use $04; and for **time4** use $08.

Once each counter is set to zero with the external pulse it then counts additional pulses by decrementing the counter from zero to -1, -2, -3 etc for successive counts. There are two peculiar features to deal with. The easiest is to recognize is that the complement of the count equals the actual count (i.e., not -3 becomes +3). This conversion is easily handled by the software (**NEG** is used in assembly language, **not** is used in Forth). The second peculiar feature is that the input pin for each of the counters is occupied by an internal interface connection from the lower half of **portc** of 8255#1 which is necessary to set the count, yet units need to also use this pin. This problem is solved on the interface by routing the internal connection and the pulses from a unit discriminator through the inputs of an exclusive OR (XOR, 74LS86) on the interface board. A pulse from the internal connection or, alternatively, a pulse from a neural unit will set the counter to zero, and the other will decrement the counter by one count. The order does not matter. By using an XOR if both pulses occur at exactly the same time there will be no count, but since the internal pulse is very short in duration in comparison to the 200 μsec pulse coming from the discriminator, there is virtually no chance that both pulses will turn on and off at exactly the same time. As a result, the XOR allows for accurate counting of external pulses from discriminated units while at the same time allowing the internal connections to set the counts properly.

The counters are set to zero in synchronization with the bin-width using the polling routine for transitions on bit 4 on the upper half of **portc** (8255#1). At the end of the next and subsequent bin-widths up to **pts** each unit counter is latched, read, and reset. The following assembly and Forth code latches and reads the units counted on **time1**, and converts the count from a negative to a positive number.

```
MOV DX,cw8253#1
MOV AX,$40
OUT DX,AL              ; latch count for unit 1
MOV DX,time1
IN AL,DX
MOV AH,AL
IN AL,DX
XCHG AH,AL             ; reads and assembles value of unit 1
                       ; count
NEG AX                 ; complement count to yield positive
                       ; number
```

and in Forth the code is

hex
40 cw8253#1 pc! (latches counts for unit 1)
time1 pc@ time1 pc@ combine (reads and assembles value of unit 1 count)
not (complements count to yield positive number)
decimal

To latch **time1** use **cw8253#1** and $40; for **time2** use **cw8253#1** and $80; for **time3** use **cw8253#2** and $00; and for **time4** use **cw8253#2** and $40. The counts are read in two separate operations, first reading the lower byte of the count and the second reading the upper byte of the count. These bytes are then combined into a word-length (two byte) number (**XCHG** and **combine**) that represents the negative count. The positive count is achieved by complementing (**NEG** and **not**) the original negative value. The counter is immediately reset to zero using the previous code. In the Runtime program, once the unit data is read it is then stored into an array for the unit's data (**unitv**) using a pointer (**^unit**) for indexing into the array. The assembly language subroutine **'datsr** collects and stores the A/D and unit count data. The array is a RAM space that has been allocated as a disk buffer by **get-buffer**. After updating the buffer, the data is then saved on disk if the appropriate switch is set with **safety off**. **Safety on** is used for debugging to prevent saving the data to disk, but it is also used when giving trials with very short intertrial intervals such as pseudorandom conditioning.

The trial

Assembly language routines like those described above are responsible for coordinating the stimulus events and data collection for each trial during classical conditioning. The hardware controllers we described earlier in this chapter could be used as models for writing the software. The simple controller consisting of 4047 timers, for example, could be implemented by using the six timer/counters on the 8253 PIT chips (they would not be available for unit counting). Essentially this was the company's engineer's solution for programming classical conditioning using Datawave. This solution is so limited that Dave King, then a graduate student in Thompson's laboratory, used Datawave's Discovery package to create a more satisfactory program. Another solution based on the controllers is to create software tables for the succession of stimulus events that occur during a trial as with the ROM and KIM-1 controllers. Data collection routines could be added. The solution we used is based on the look of Thompson's PDP-12 and PDP-11 computer programs (we did not actually look at the code) and Steinmetz's classical conditioning programs written in First (from which Steinmetz explained what he wanted and Lavond independently wrote the assembly code for the Apple II and later the IBM-PC). For each bin, the time (**bin#**) is compared with a variable associated with each stimulus. The variables represent the time (in number of bins) from the beginning of the trial when the tone should be turned on (**toneon**) and off (**toneoff**), those times being compared with the current time (**bin#**) in the trial. If there is a match, for example with **toneon**, then the tone stimulus is set (made a logic 1) using the assembly routines above for the 8255 digital output, similar to the Forth code for **turnon**. The actual Runtime assembly subroutine to control turning on and off the tone is **'tone-on/off**, which is similar to the following assembly code

```
         MOV AX,bin#
         CMP AX,toneon
         JNE skip1
         CALL 'SYNCPULSE
         MOV BX,$01        ; bit $01 on portb controls the tone
         CALL 'PSET        ; call subroutine to set bit on portb
         RET
  skip1  CMP AX,toneoff
         JNE skip2
         MOV BX,$01        ; bit $01 on portb controls the tone
         CALL 'PRESET      ; call subroutine to reset bit on portb
  skip2  RET
```

The actual assembly code in the Runtime looks a little different only because it is a Forth assembler, following the reverse-Polish convention of Forth and the addressing conventions of Forth. Note that since **'tone-on/off** is a subroutine (it is **CALL**ed from somewhere else in the program) that the subroutine is left with the return statement (**RET**). Also note that a synchronization pulse is given whenever the tone is turned on (**CALL 'SYNCPULSE**). Functionally the assembly code is similar to the following Forth code.

```
hex
: tone-on/off ( --)
  bin# toneon =
  if 1 pset synchpulse then
  bin# toneoff =
  if 1 preset then ;
decimal
```

Similar routines are checked for turning on and off a light stimulus (**lton, ltoff, 'light-on/off**), air puff (**airon, airoff, 'air-on/off**), shock (**shkon, shkoff, 'shock-on/off**), and pulse (**pulseon, 'pulse-on**). The pulse can be placed anywhere during the trial. It is a quick on/off pulse used to trigger an oscilloscope so that, for example, a fast oscilloscope sweep can be made in the middle of the CS period when no stimuli are changing but there might be a model of unit activity of interest. The beginning of the trial, **toneon, lton** and **shkon** are associated with synchronization pulses to mark their onsets. Because it takes the air puff command some time to turn on the solenoid and the air takes time to travel down the tubing, the air puff must be turned on before it actually arrives at the eye, so no synchronization pulse is associated with **airon**. **Shkon** is used instead to create a synchronization pulse for the air puff's arrival at the eye.

The trial consists of a loop counting from zero to the number of **pts** (the total number of bins) at the pace set by the bin-width (**bin**). This loop is found in the assembly language routine **trial**. The key features of this loop are in the following assembly language and Forth codes.

```
         MOV AX,$00
         MOV bin#,AX          ; zero bin pointer
         MOV CS,pts           ; set C register for number of bins in
                              ; loop
loop1    MOV DX,portc         ; wait for synchronization with binwidth
loop2    IN AL,DX
         AND AL,$10
         JE loop2
loop3    IN AL,DX
         AND AL,$10
         JNE loop3
         CALL [VEC1]          ; begin tone, light, air, shock, etc., data
                              ; collection subroutines
         CALL [VEC2]
         CALL [VEC3]
         CALL [VEC4]
         CALL [VEC5]
         CALL [VEC6]
         INC bin#             ; increment intratrial pointer
         LOOP loop1
```

Note that the subroutine **CALL**s are indirect, calling the routines pointed to by the vectors (**vec1**, **vec2** etc.). The Forth equivalent is

```
hex
: trial ( --)
 0 is bin#
 pts 0
 do
  i is bin#
  binwidth ( synchronization)
```

Chapter 11 361

    vec1
    vec2
    vec3
    vec4
    vec5
    vec6
 loop ;
decimal

At every bin **trial** checks to see what stimuli need to be turned on or off. Because different trials may be tone-alone trials or paired tone-air puff trials, for example, the trial type is controlled by setting vectors that are used by **trial**. For example, if the trial is a paired tone-air puff trial then the first vector (**vec1**) may point to the subroutine for checking the tone (**'tone-on/off**), the second vector (**vec2**) may point to the subroutine for checking the air puff (**'air-on/off**), the third vector (**vec3**) may point to the subroutine for checking the shock (**'shock-on/off** which is needed for the synchronization pulse for the air puff but shock is not actually hooked up to the animal), and the fourth vector (**vec4**) may point to the subroutine to collect A/D and unit data (**'datsr**). Vectors that are not used (**vec5** and **vec6**) are set to a no-operation subroutine (**'noop**). A tone-alone trial would be set up the same way except that the vectors for air puff and shock (**vec2** and **vec3** in this example) would point to **'noop**. The vectors are set up before **trial** is invoked, using Forth definitions for **tone-air trial** and **tone trial** for these two examples. The function of **tone-air** and **tone** are to set up the vectors that are used by **trial**. The word **trial** also includes other functions: preventing interrupts, adding debugging commands from the keyboard, detecting offline synchronization pulses, and turning the tape recorder on and off.

Intertrial interval

    The last function to consider is the timing of the intervals between trials. The controllers covered above were particularly limited in that they always gave the same interval between trials. Time itself can be a conditioned stimulus, teaching the subject to blink after a certain period of time. This conditioning is avoided by varying the interval between trials. The Runtime program uses Forth to pseudorandomly choose an intertrial interval (ITI) that has at least a **minimum** number of seconds between trials and can have a **range** for additional pseudorandom seconds added to the minimum number. The number is chosen with **random**, which is a software algorithm that appears to give a random sequence of numbers, but in reality will give the same sequence of numbers with a given **seed** value, and for that reason is pseudorandom rather than truly random. The timing uses the on board (motherboard) 55-msec accumulated timer (read with **@time** and converted into seconds by **elapsed**) at the beginning of the ITI (**start**). The end of the ITI is calculated from the beginning of the ITI by **ending**. The ITI display is updated for the count down to the next trial by **lap** and **.time** (print time). The intertrial interval is managed by **timing** which sets up the ITI, displays the count of the time, displays the data from the previous trial and gives the current status of the A/D converters, allows the ITI to be indefinitely delayed (**inhibitor**), allows for the execution of Forth interpretive commands (**interpreter**), and keeps track of the ITI until it has elapsed and the next trial is initiated.

Written for the original IBM PCs, the Runtime program's operations conflict with more advanced PC features like TSR (Terminate Stay Resident) programs, especially those related to the clock used for the intertrial interval. Part of the problem is that this version of the Forth language is not relocatable, meaning that it must reside in a certain RAM space, which can conflict with newer programs. On these computers either avoid loading TSR programs or boot up in DOS. In addition, the Runtime program now includes a function to reset the clock to zero (**!time**), which messes up the time of day. Since our computers are dedicated to training animals we have few problems. A newer Windows version of the Forth language from LMI can solve some of these problems, but we have no impetus to spend the time porting the code.

The C++ System

A major disadvantage of using the Forth system is that few programmers are well versed in this programming language. For this reason, the Steinmetz laboratory developed a hardware and software system for controlling learning and memory experiments that use PC-compatible computers as platforms to run control software that is written in C++ (Chen & Steinmetz, 1998). Interfacing is accomplished using standard, commercially available boards manufactured by Keithley Metrabyte, Inc. Similar to the Forth board, the Keithley boards use 8253 and 8255 chips to control timing and digital I/O functions. We have primarily used this system to generate classical conditioning experiments and to record behavioral and neural data. However, we have modified the programs to run a number of other learning paradigms including a variety of basic instrumental and operant procedures (Steinmetz, Logue & Miller, 1993). We present here an abridged version of a paper that appeared in *Behavior Research Methods, Instruments and Computers* (Chen

& Steinmetz, 1998). Interested readers are directed to that paper for more details about this system.

The system runs on the most current IBM PC-compatible computers that are available, although old 386 and 486 machines are fine for most applications. A 640 x 480 resolution color monitor is required. Given the early unreliability of the Windows operating system, we developed this system using the DOS operating system although we are currently working on a Windows compatible system. The interface board linking the computer and the peripheral hardware used for stimulus delivery and response recording is the DAS-1602 (Keithley Metrabyte, Inc., Tauton, MA). This board seems ideal for our application because it includes (1) 12-bit A/D resolution, (2) up to 16 single-ended A/D channels, (3) data transfers that can be performed by program control, interrupts, or direct memory access (DMA); (4) 32 programmable I/O lines, and (5) a Borland C++ compatible function library available for interface programming (thus doing away with the need to write a lot of low-level code). This system has allowed us to control and record classical conditioning behavioral and neural data simultaneously from four experimental chambers. The software is written in Turbo C++, an object-oriented programming environment.

The computer-human interface

Because it was our goal to build a system that was very easy to operate and that did not require extensive computer experience or expertise to use, we designed a user-oriented interface for the system. This interface system, described extensively in an earlier publication (Chen & Steinmetz, 1998), allows the user to enter and change parameters that are used during behavioral experiments (such as ITI, ISI, type of CS, type of US, trial duration, trial sequences, session structure, etc.). The user goes systematically through the dialogue boxes that appear in the user interface to enter the correct values for each session that is to be run. Alternatively, the user can load a stored set of values, then proceed directly to the running of the experiment.

Program Structure

Data structures

Parameter values and other information needed to control the delivery of stimuli and the acquisition of data are stored as lists in the C++ environment. One collection of these lists is group together as PARADIGM, which contains all the information and parameters needed to define the training paradigm including paradigm identifiers, intertrial interval parameters, information concerning the number of trials and how the trials are to be blocked, trial timing information such as trial length, interstimulus interval length, the onset and offset of all stimuli to be presented, an array of stimulus types, and a table defining the order and arrangement of trials within the session.

EXP_BOX is a another collection of lists that contains three general categories of information: (a) channel addresses, (b) system configuration, and (c) run-time variables. First, channel addresses for analog input, port and line addresses for all of the digital input and output are stored in EXP_BOX. This information is stored in a file on the hard disk named "iomap.sys" and is loaded when the program is started. Second, system configuration information also appears as part of EXP_BOX. That information is entered by the user via the User Interface that is described above. Chamber number and activity switch, names of the paradigms to be run in each chamber, a variety of information concerning data sampling (e.g., A/D gain and sampling rate and digital sampling bin size), criterion for detecting excessive A/D input prior to each trial, and display parameters are all included in the system configuration. Like the PARADIGM list, this information is stored to hard disk every time a session is initiated as the file "system.cfg." Users can also save this information in configuration files that can be loaded into the system prior to running. Third, run-time variables are stored in EXP_BOX. These variables include all of those needed to control the running of a session in each of the chambers such as an identification of the paradigm that has been selected to be run, a text-string describing session to be run, the name of the file into which raw data is to be stored, counters for the intertrial interval, blocks and trials, and calculations for data that are displayed after each trial (e.g., response amplitudes, onset latency, etc.). The operations of the collection of parameters stored in EXP_BOX are actually the core of the program and take care of all of the basic chamber operations such as the count down intertrial interval timer, trial trigger, and online displays.

Signal control and data collection

After the program is started and the user executes the RUN command, a software switch to each chamber is checked by the program to determine which of the chambers are to be used during the session. If the software switch is off, the chamber is removed from the "running list" (i.e., those chambers to be serviced during the session). Another empty list, the "finished list," is also created when the program is started. During the run, if a chamber finishes a session, it is removed from the running list and placed on the finished list. The program keeps looping until all of the chambers set up in the running list have finished and been moved to the finished list such that the running list becomes empty. At this point, the program follows the finished list to print out summary reports and initiate routines to archive data files for each chamber.

The intertrial interval is controlled by a time counter set up individually for each chamber. The counter is initialized with a pseudorandom number within a range designated by the user for each paradigm. The program reads a clock continuously in the looping process. If the clock ticks (one second more than the previous reading), the counter of each chamber is updated (decremented). If the counter reaches zero, a trial is triggered in the chamber. For the intertrial interval countdown routines, the operation of the clock is a normal C++ library function. Values for other functions, such as trial triggering and counter manipulations and display, are stored among the group of parameters that constitute EXP_BOX.

Trial control and data collection

To begin triggering a trial in a chamber, the program first resets some of the parameters on the DAS-1602 interface board using configuration information for the chamber that was entered by the user before the run. The parameters include things like sampling rate, A/D channel selection, and number of sampled points. Because the interface board is reset every time a trial is triggered, each chamber can be operated independently from the other chambers. This system is very flexible: Different paradigms with different stimuli selection and timing configurations can be operated in each of the chambers if the user wishes.

In our application, there are typically two sources of data that are collected; analog behavioral data that are entering the computer via an A/D channel and digital data, such as TTL pulses that are input via I/O ports. The analog data normally encode the position of an eyelid during conditioning while the digital data represent concomitantly recorded discriminated action potentials from an area of the brain. Because of interface hardware limitations, we transfer the A/D data with DMA mode while reading the I/O ports with software polling. The polling code is written in a manner that makes the polling as quick as possible. This enables us to read I/O input pulses as narrow as 100-200 μsec (a common output for commercial window discriminators such as those manufactured by F. Haer, Inc.). Assuming four chambers being used, there are six data buffers for each chamber, one for A/D data from the current trial, one for averaged A/D data for a block of training, two for digital data from the current trials (one for each channel) and two for summed unit activity across a block of training trials (one for each channel).

Keithley MetraByte Inc. provides a library of functions to monitor the operation of the DMA transfer. Because the data transfer is controlled precisely by a timer on the interface board the counter can be used as a time index for stimuli delivery signals and also to determine bin onsets and offsets for digital data sampling. Before the DMA data transfer begins, onsets and offsets of stimuli switched on for a given trial in a given chamber are calculated and scaled according to the units defined by sampling interval parameters selected before the run. The counter for transfer is continuously compared with the time for a candidate stimulus operation (onset or offset). Once they are equal, that operation is executed and the next candidate is selected. At the same time, the border of the digital data bin is compared with the transfer counter. Once the counter crosses the border of the bin, a new bin is set as well as a new bin border. In this way, digital data are collected into bins that are established across time (i.e., within a trial). Digital pulses are counted by detecting the rising-edge of the pulse, i.e., once a high voltage (logic '1') is read following a low voltage (logic '0'), the program counts one in the current bin. Using this scheme, it is important that the polling loop time be less than the digital pulse width so that no pulses are missed.

Data file format

During each session, the program stores the raw A/D and digital data for each chamber into a data file set up by the user prior to the run. The data file consists of two parts, a header which contains all of the information concerning the chamber configuration and paradigm parameters for that session, and a data part which is organized in a trial-by-trial format. For each trial, a block of A/D data is followed by two blocks of digital data (a block for each digital data channel). Since the A/D resolution on the DAS-1602 interface board is 12 bit, each A/D data point is two bytes, with low byte first and high byte second (e.g., IBM-PC short integer format). Each digital data point is one byte. The size reserved for the header is 1024 bytes. The size of the data part is determined by the sizes of the blocks of A/D data and digital data stored with the respective block sizes determined by the total number of trials run, the trial length, and the sampling rate.

More recently, we have used waveform-matching techniques to separate action potentials that have been recorded during the behavioral experiments we conduct. We have chosen to employ a second computer to collect the neural data, mostly because we can record multiple channels of neural data at relatively high speeds. In this configuration, one computer controls the delivery of the training trials using the C++ software and hardware system while a second computer, equipped with the Spike2 software and hardware system (CED, Ltd.) is responsible for recording behavioral and unit data. Data acquisition on the second computer is triggered by an output pulse that is generated from the first computer. While this system at first blush seems cumbersome given that two computers are required, we have found it very effective for collecting behavioral and neural data.

Other Systems

Other computer systems for controlling behavioral experiments have been described. For example, Palya and colleagues use a controller that can be programmed in either a BASIC-like or an ALGOL-like procedure specification language. This system was designed to function as a remote peripheral processor in a network with any computer acting as the network supervisor and each serial port on the host computer supporting several simultaneous experiments (Palya & Walter, 1993; Palya, Walter & Chu, 1995). Other examples of systems that have been recently designed to control the delivery of stimuli or the acquisition of data during simple conditioning experiments include an interface described by Markham (1993) for controlling external devices via the parallel port of IBM PC-compatible computers and a system designed by Penttonen, Salmi, Hamalainen and Meriluoto (1994) for controlling classical conditioning experiments. A WorkbenchMac- or PC-based system designed by McGregor (1996) is available for use in animal learning, psychopharmacology and behavioral neuroscience experiments. Also, an IBM PC compatible system for use in classical eyeblink conditioning experiments has been described by Thompson, Moyer, Akase, and Disterhoft (1994).

Chapter 12

IMPORTANT ELECTRONICS FOR CLASSICAL CONDITIONING EXPERIMENTS

INTRODUCTION

Over the years, we have used a variety of simple electronic circuits in our research. These circuits have proven most useful for interfacing between computer control circuits and the variety of pieces of equipment that are used to deliver stimuli or record responses. Most of these circuits are quite easy to build and have a number of applications that can be generalized to a variety of research situations. We present these circuits in this chapter.

POWER SUPPLIES

There are generally three types of power supplies that are commonly used: battery, linear and switching power supplies. The choice of power supply to be used depends on a number of factors including the requirements for the voltage, current, wattage, noise, price and, of course, convenience.

Battery Power Supplies

Batteries are good choices for power supplies under a number of conditions. Batteries are small, lightweight, and independent of the electrical grid (i.e., the power coming out of the wall sockets). For example, simple counting or timing circuits that use CMOS technology could use batteries. To run TTL chips from a battery supply requires a voltage regulator (a zener diode or a 7805 voltage regulator or equivalent; see below). Batteries are also a good choice when electrical noise is a problem, such as when the 60 Hz noise from the electrical grid energizing a linear power supply could be introduced into the amplification of a neural unit recording. Batteries are also a good choice when low signal levels (e.g., unit recordings) need to be amplified without introducing high frequency noise caused by a switching power supply. The Brakel or Verzeano amplifier once used by Thompson

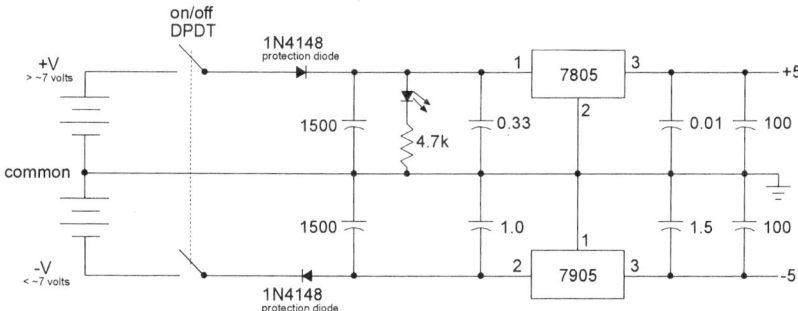

Figure 12.1: Regulating battery supply to ±5 volt.

and his colleagues is a good example (Brakel, Babb, Mahnke, & Verzeano, 1971). The Brakel or Verzeano amplifier uses pure batteries without the regulators described next. Finally, batteries are a good choice when it is important to electrically isolate the subject from the power grid when there are concerns about safety. Batteries are used in amplifiers and stimulators for the purpose of isolation and/or to avoid ground loops. Our lesion maker circuit (see Chapter 8), where we are interested in constant current output rather than constant voltage output, makes use of high voltage and stimulus isolation. On the other hand, batteries are not good choices when large currents are required or when long times of usage are required.

Figure 12.1 shows a ±5 volt power supply created by two batteries (e.g., 9 volt transitor batteries, 12-volt lantern batteries) and two voltage regulators (7805 for +5 V and 7905 for -5 V). The upper (+5 volts supply) or lower (-5 volt supply) voltages could be constructed separately as needed. This design is overkill (i.e., quite over-engineered), including two diodes to protect against accidentally putting the batteries in backwards, an on/off light (which may drain the positive battery more than the circuit using this power supply), large capacitors for smoothing the supply lines (1500 μF and 100 μF at the input and output of the regulated sections) if the supply is far from the batteries or the powered circuit, respectively (they go next to the batteries and to the powered circuit). The only really important capacitors are those immediately surrounding the regulators, the 0.33 μF and 0.01 μF capacitors on the input and output of the 7805 regulator and the 1.0 μF and 1.5 μF capacitors on the input and output of the 7905 regulator. These four capacitors should be directly soldered onto the regulator leads. The regulators require at least 2 V more on their inputs than their outputs will deliver. Thus, 5 V supplies require at least 7 V batteries. The two-volt difference basically allows the regulator to compensate for variations in voltage demands of the circuit to which it is attached, thus allowing for a constant voltage out-

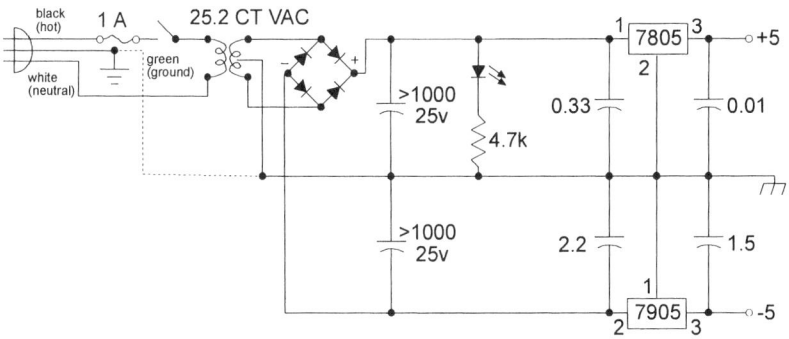

Figure 12.2: Linear ±5 volt power supply.

put. We used this supply to power our heart rate amplifier, and reduced versions for various other projects.

Linear Power Supplies

Linear power supplies are created by converting AC (alternating current) line voltages from the wall (the power grid) into battery-like steady DC (direct current) voltages. Figures 12.2 to Figure 12.7 are a compendium of linear power supply circuits that we have come across and used over the years. Linear power supplies can generally be conceived of as consisting of two parts, a section to convert AC to DC, and a section to regulate the voltage. The section to convert AC to DC involves a transformer (standard or center tap), diodes (a pair or a bridge diode) and a large capacitor. The regulating section involves a fixed (7805, 7905) or variable (317, 337) voltage regulator and its associated capacitors. It should be noted that fixed regulators are regularly available for 5, 12 and 15 V, and that circuit designs are available for higher voltages using these regulators. The fixed designs presented here use transformers appropriate for 5 V systems.

Figure 12.2 shows a ±5 volt power supply. The fuse should be on the hot (black) supply coming from the wall, which is also the smaller slit on the wall socket. The transformer has a center-tapped output, which is used for the common ground between the +5 and -5 voltages. A full wave bridge diode rectifies the AC into a positive signal, smoothed by the large filter capacitors (≥1000 µF). An LED and resistor are used as a power-on indicator. Alternatively, lights are available which could be placed on the input side of the transformer, after the on-off switch. The 7805 regulates the +5 volts and the 7905 regulates the -5 volts. Note that the regulated output is on pin 3 for both regulators, but that the input and ground pins are opposite. The capacitors should be soldered directly onto the regulator leads, taking care to protect the regulators by using a temporary heat sink between the solder point and the regulator. If the distance from the regulated outputs is far then 100 µF capacitors can be added between the outputs and the common ground at that distant location.

The power supply can be left "floating," where there is no connection of the regulated power supply side back to earth ground (green) of the wall socket. This feature is important in helping to reduce noise caused by ground loops. A ground loop is created when a signal can take more than one path to reach earth ground. If those paths are different distances or have different resistances, then that same signal reaches the ground point at different times through the different paths. This generally creates a ringing effect, observed as high frequency noise on the system. This high frequency noise is particularly a problem where high gain amplification is one of the functions of the circuit, but it can also wreak havoc on digital signals. In a system with several power supplies, they should all be tied to each other through their common grounds, just as one would have to do with separate battery supplies, but only one of those power supplies should be connected to the wall's earth ground (green). Sometimes ground loops are unknowingly created when using equipment like an oscilloscope, which, by virtue of its own three-pronged power cord, provides a second path to earth ground. In effect, as an analogy of the Heisenberg principle, the effort to measure the circuit with the oscilloscope alters the normal functioning of the circuit. The solution is to 'float' either the circuit or the oscilloscope, meaning to somehow disconnect it from the wall ground so that its ground must go through the rest of your circuitry. This can be hazardous if not done properly.

Finally, some basic safety concerns should be taken into consideration. Obviously, do not work on anything that is plugged into the wall. Be obsessive about knowing where the wire and plug are located at all times. A fuse rated for the transformer output is placed on the black (hot) lead. An on-off switch is chosen that is rated for the current rating of the transformer. A good practice is to get a DPDT (double pole, double throw) switch that exceeds the transformer ratings, and split the line between the two poles, thus halving the current on each side of the switch. Cover all contacts from the wall to the transformer with wire nuts, electrical tape, heat shrink tubing, or electrical insulating paste. The whole supply should be enclosed in a metal box that does not allow easy access (i.e., it is screwed shut). The earth ground (green) from the wall is attached to the metal chassis that houses the power supply. This connection is your safety for shorts of the power supply against the box. To be really safe, buy a commercial power supply.

Chapter 12

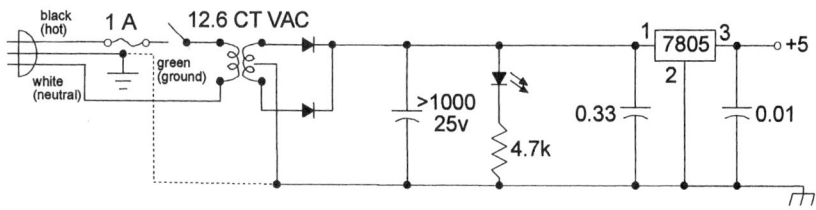

Figure 12.3: Linear +5 volt power supply using a center-tapped transformer.

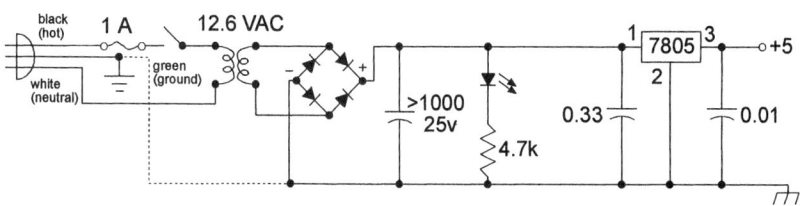

Figure 12.4: Linear +5 volt power supply using simple transformer and bridge diode.

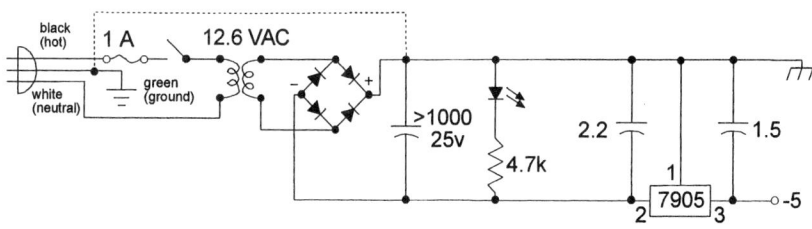

Figure 12.5: Linear -5 volt power supply using simple transformer and bridge diode.

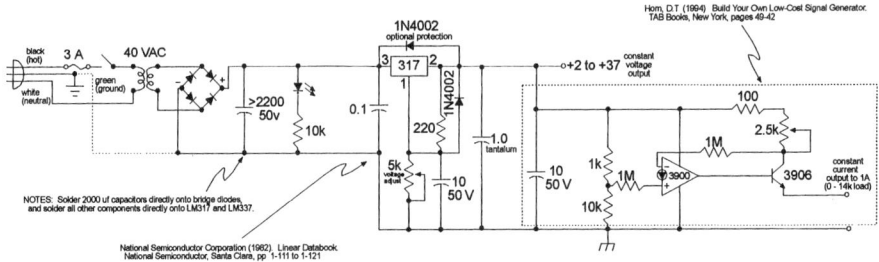

Figure 12.6: Variable +2 to +37 volt power supply with optional constant current output.

The remaining power supplies are straightforward. Figure 12.3 shows a +5 volt power supply. Everything said about the previous power supply applies for this power supply, except that instead of using full wave rectification here the AC to DC conversion begins with half wave rectification using the center tap from the transformer. Figure 12.4 shows a +5 volt power supply using full wave rectification. A center tap transformer is not used. The common ground for the +5 volts is taken from the negative side of the bridge diode. Figure 12.5 shows a -5 volt power supply with full wave rectification. Notice that the negative side from the bridge diode is the input to the 7905 voltage regulator, and that the positive side of the bridge diode is the common ground. We cannot think of an instance where we needed a -5 volt power supply by itself. When we have needed -5 volts it has always been in conjunction with a +5 volt supply.

Figures 12.6 and 12.7 use variable voltage regulators for positive (LM317) and negative (LM337) power supplies, respectively. The circuits are complicated by having variable resistors for changing the value of the regulated voltage output and protection diodes for the regulators. By using a center-tapped transformer and a larger bridge diode the two circuits can be combined yielding a variable plus/minus power supply. This is the power supply we use in the laboratory to prototype circuits that are greater than ±5 volts, but it has rarely made it into the final circuit of any project. The modification by Horn (1994) in the variable positive supply (Figure 12.6) yields a constant current output which might be useful for stimulators, lesion makers or etching purposes.

The problem with linear power supplies is that they can induce 60 Hz noise onto other parts of your system. This problem can be attenuated by shielding (a grounded metal plate or grid between the power supply and the circuit) and by distance (the further away the power supply the less transmitted signal; it falls off by the inverse square of the distance, therefore it does not take a great deal of distance to attenuate the noise). The 60 Hz noise is particularly disturbing around amplifiers used for very small signals such as neural recordings. Therefore, it is not a good idea to have a linear power

Chapter 12

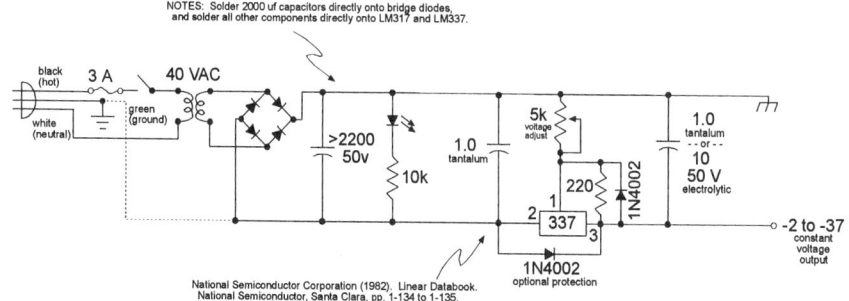

Figure 12.7: Variable -2 to -37 volt power supply.

supply inside the testing chamber. Batteries would be a better option.

Rather than constructing them, linear power supplies can be purchased. Given the component costs and costs of putting everything into a box (stand offs, grommets, switches, knobs, box, feet), and with safety concerns, it is generally cost effective to purchase rather than to build. Sometimes linear power supplies can be bought that are small enough that they can be placed inside the electronic project. For example, simple DC adaptors that plug into the wall usually consist of a small center-tapped transformer and two diodes; more sophisticated ones might have a regulator and capacitor.

Switching Power Supplies

One big disadvantage of a linear power supply is that the transformer makes it bulky and heavy. Switching power supplies avoid the transformer thus making them lightweight. They are also readily available, in greater varieties and options, than linear power supplies. Their one large disadvantage is that they generate high frequency noise, which rides on top of the supply lines. For digital equipment this high frequency noise is not generally a problem, but if the switching supply is used to power amplifiers for very small signals then this can be a problem. For neural recordings, the best choice of a power supply is a battery, followed by a linear supply, then a switching supply. For digital signals and for large voltages used by relays and solenoids (+12 and +24 V, for example), switching power supplies are the choice. We generally use a switching power supply that has ±5 and ±12 volt outputs with at least 1.0 A output for most of our interfacing requirements. The +12 V output can be used to interface with 12 V relays. The ±12 V outputs can be used to power 24 V solenoids using the optoisolator circuits (see discussion of interface circuits later in this chapter). Switching power supplies can be expensive, but the multiple output voltages that some suppliers offer make them cost effective. The switching power supply works well for much of the equipment that we make for other laboratories. Switching power supplies are, in our opinion, too complex, too readily available at reasonable prices, and not worth the effort to construct ourselves.

REALITY IN THE LABORATORY: EXAMPLES OF TRUE-LIFE PRACTICAL CONSIDERATIONS AND SOLUTIONS

Multiple power supplies are common in experiments that bring together diverse equipment. Ideally, we try to keep the number of power supplies as low as possible. For Lavond's equipment, all interface circuits for air puff solenoids, tones, white noise, rise/fall switches, lights and transistors, nictitating membrane boxes, relays, and height discriminators with oscilloscope monitors are powered by a ±15 volt supply from Haer which is down-regulated to ±5 and ±12 volts, and which surely is not an approved use in the warranty (which must be expired by now, anyway). The auditory attenuator and frequency selector are powered by a ±5 and ±12 volts switching supply. The computer interface (see Chapter 11) is powered by the IBM PC clone. The lesion maker (see Chapter 8) and calibration signal generator for checking the recording system (see Chapter 4) are both battery powered.

Interface Circuits

Interfacing two pieces of digital equipment is the problem that got us interested in electronics in the first place. Two different pieces of equipment, made by different manufacturers or made at different times using the currently available technology, may or may not be compatible with each other. For example, a computer output of 0 to +5 volts may have to operate a piece of equipment that expects the voltages of traditional relay programming equipment to go from -28 to 0 volts, respectively, such as is common for equipment made by companies such as Coulbourn Instruments or Lafayette Instruments. Today, the TTL interface is fairly standard, where zero volts equals logic 0 (off) and +5 volts equals logic 1 (on). But often in the past, the output levels and meanings of those levels were likely to be incompatible with the expected inputs from another piece of equipment. Solutions to smoothing the transition between two electrical systems are useful in themselves and useful for some design considerations for input (how to con-

Chapter 12 377

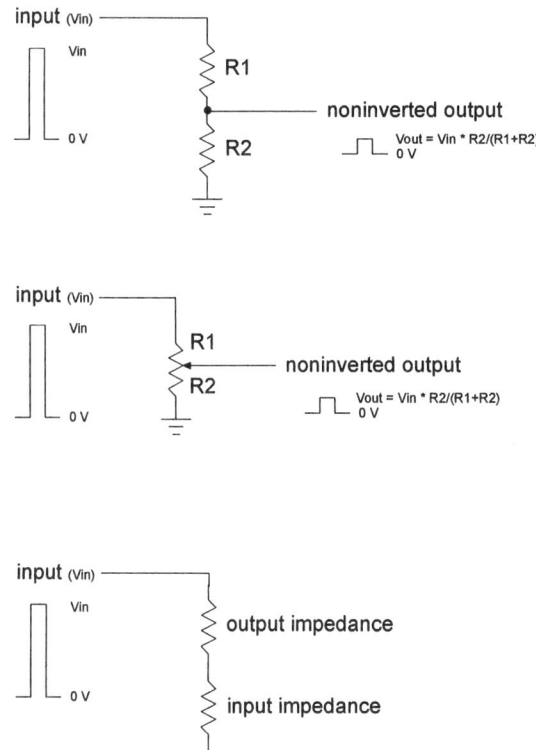

Figure 12.8: Using a voltage divider for reducing the size of the output signal.

vert a continuous wave into a digital input) and output devices (how to interface a solenoid or relay). While we are going to focus primarily on interfacing digital signals, we will point out when we get to them that the optoisolator circuits can also be used for analog signals.

Voltage Divider

The simplest interface occurs when the input voltage is larger than the required output voltage. This problem can be solved by one of two simple voltage dividers as illustrated in Figure 12.8. In the first example, R1 and R2 are fixed resistors whose values are chosen to yield a ratio of the input voltage. The output voltage (V_{out}) is equal to V_{in} times the value of R2 divided by the total resistance (R1 + R2). For example, if the input voltage (V_{in}) goes from 0 to +15 V and the desired output voltage (V_{out}) is 0 to +5 V, then R1 = 2 x R2. Choosing typical values, if R2 equals 1K ohm, then R1 equals 2K ohm. The second example in Figure 12.8 merely substitutes a variable resistor (potentiometer) but the idea of ratios is the same.

The qualifications of using a voltage divider include the following. Ideally, the size of the resistors should be as small as possible so that the maximum current is available at the output. However, small resistors mean that greater current flows through the resistors and, therefore, that the current may exceed a resistor's current ratings. A typical 1/4 watt resistor, for example, having a value of 100 ohms with 10 volts applied will generate one watt of power (where 10 V = 0.1 A times 100 Ω in the formula V = IR, and 1 W = 10 V times 0.1 A in the formula P = VI), and would burn up. This might suggest that larger resistor values, which restrict the amount of current, would be advantageous.

However, a second qualification is that the output impedance of the circuit is increased with larger resistances. That is, if the voltage divider with large output impedance is the input to another circuit, which has a small input impedance, then there may not be enough voltage to drive the second circuit. This is shown in Figure 12.8 where we can think of R1 as the output impedance and R2 as the input impedance. With a large resistor as the output impedance there is relatively little voltage drop left to cross a smaller input impedance. This sort of mismatch is most often seen with 8 ohm speakers (a really low input impedance). It can also be noticed in the distortion seen when measuring voltage drops across high impedance electrodes (output impedance) when using an oscilloscope (typically 1 MΩ, this can be a relatively small input impedance when compared to the impedance of the electrode). This impedance mismatch is a big reason not to use a voltage divider for voltage translation. One way to get around this problem, however, is to place a voltage follower operational amplifier between the voltage divider circuit and the following circuits.

We should note that we are being rather liberal here by implying that the terms resistance and impedance can be interchanged. This is not really true. Impedance is a product of resistance, capacitance and frequency of an AC signal. The impedance of an electrode is typically measured with a 1 KHz sine wave signal, and is an indication of the resistance of the exposed tip and the capacitance of the electrode's insulation, among other sources. Resistance is a small part of impedance, referring to a property in a DC circuit.

Finally, it should be noted that the shape of the signal in Figure 12.8 is a digital square wave, but in fact, any analog signal can be reduced by using a voltage divider.

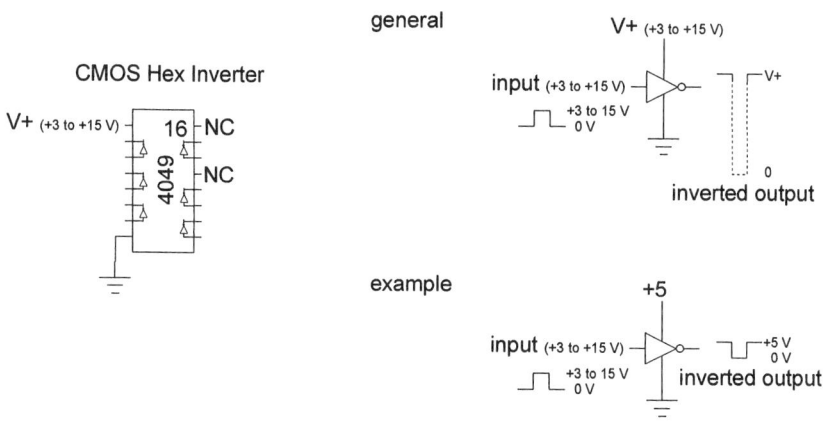

Figure 12.9: Using a TTL hex inverter to decrease the output signal.

CMOS Interface

Like the voltage divider circuit, the CMOS 4049 hex inverter chip is useful for converting a large input voltage into a small output voltage, but has the additional advantage that the output impedance of the 4049 is acceptable. The 4049 in Figure 12.9 is powered by the supply voltage you wish to see on the output. In the example, the supply voltage on pin 1 is +5 V. The input to one of the 4049's inputs (it is a device having six different inverters) is anything from +3 to +15 V. This is the single instance that violates the general rule that the input voltages cannot exceed the power supply of the chip. The output is inverted and varies from 0 V (with a logic 1 on the input) to +5 V (with a logic 0 on the input). Thus, the 4049 could be used to translate CMOS levels to TTL output levels. The inverted output is preferred because the 4049 can sink more current than it can source. A sink means that as the 4049 output goes low a device such as a relay hooked up to a +5 V supply will turn on with a logic 1 on the input of the 4049 as we will see in other circuits (like the relays in the cooling circuits). Alternatively, the 4050 is the non-inverting version.

TTL Interface

The TTL 7406 hex inverter in Figure 12.10 is useful for interfacing a +5 V TTL device with one that requires a greater voltage, for example a computer to a relay or solenoid. The air puff circuit, for example, could be

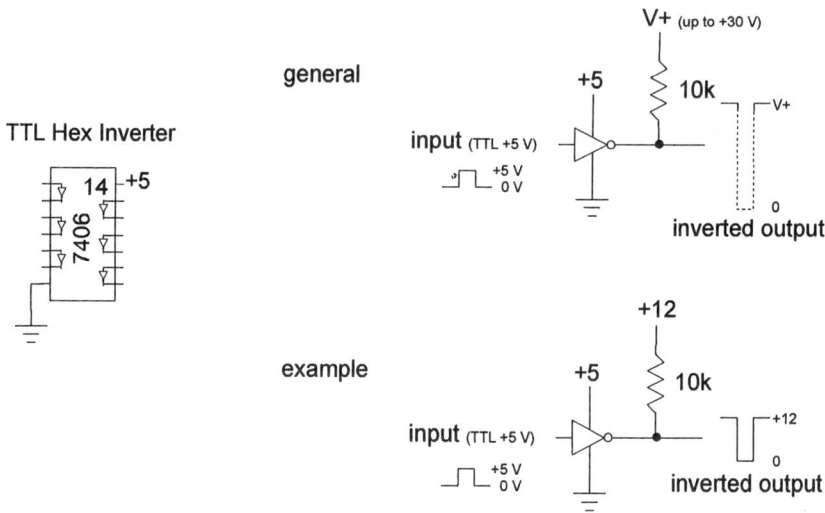

Figure 12.10: Using a TTL hex inverter to increase the output signal.

built so that a TTL level turns on a +24 V relay. Notice that, like the CMOS 4049, the TTL 7406 is an inverted signal, but that using it as a sink rather than as a source yields the familiar logic: A logic 1 on the input causes the solenoid to turn on. The 7407 is the non-inverting version.

Optoisolators

Figures 12.11 and 12.12 show several different configurations to translate TTL signals into a variety of different outputs. Figure 12.11 shows non-inverted output signals. Figure 12.12 shows the corresponding inverted output signals. Although these two figures use TTL as examples, a 10 V CMOS supply and signal could be used by substituting the 220 Ω resistor with a 470 Ω resistor.

There are several advantages to using optoisolators. One advantage is that the input and output of the optoisolator are physically and electrically isolated. The input side of the optoisolator is an LED. The light from this LED transfers the signal to a phototransistor (4N25) associated with the output side of the optoisolator. Typically there are 1200 to 2500 volts of isolation between the LED and the phototransistor, meaning that anything that goes wrong on the output side (a large voltage surge) will not damage anything (like your computer) which is hooked up to the input side of the optoisolator. That degree of safety is a big advantage. The costs for this security

Chapter 12   381

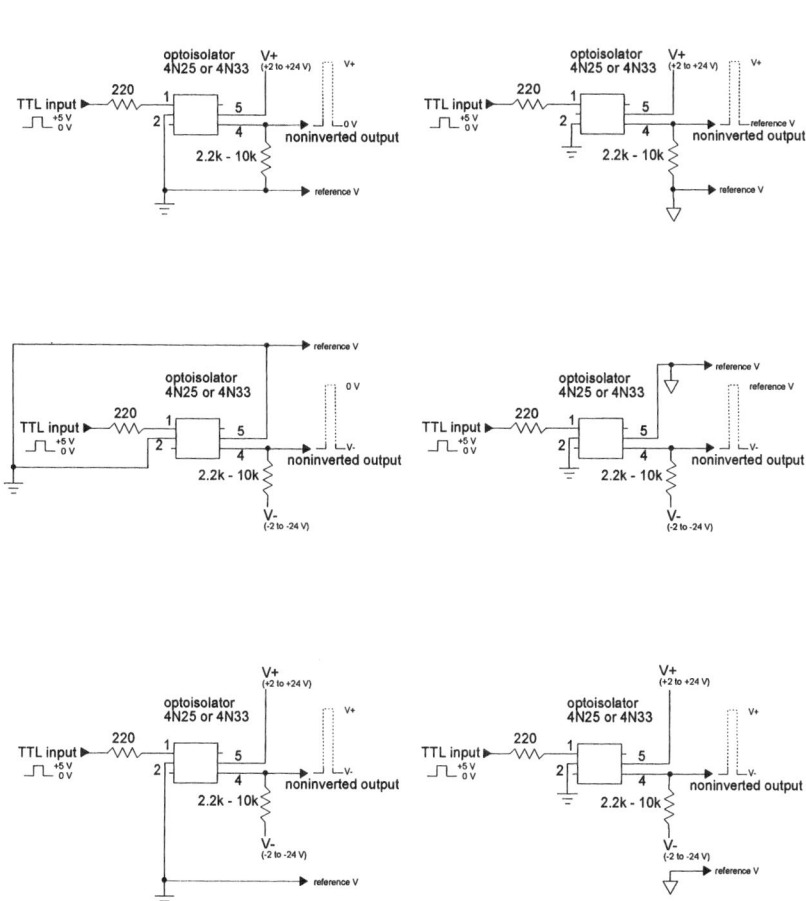

Figure 12.11: Using an optoisolator to increase or reduce a noninverted output signal.

are that optoisolators can be slow and there is some loss (as opposed to gain) of the signal from the LED to the phototransistor, which can be compensated for in part by using a photoDarlington output design (4N33).

Another great advantage of using an optoisolator is that the power supplies for the input and output sides can be different and completely isolated from each other (right hand column in Figures 12.11 and 12.12). This reduces the contamination of certain kinds of electrical noises; for example, it reduces the chances of creating a ground loop. It also means that, for example, a battery supply could be used as the output. Using this idea, the output battery side of the optoisolator could be used to deliver an electrophysiological shock stimulus to a nucleus in the brain but, because it does not share input and output sharing the same power supply   input and output with separate (isolated) power supplies

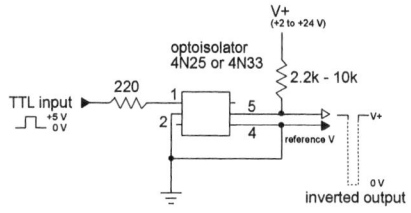
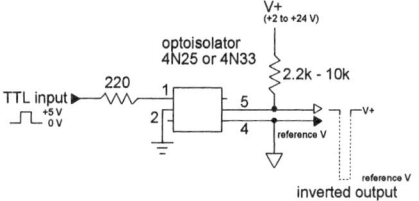

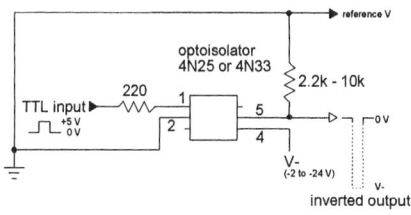
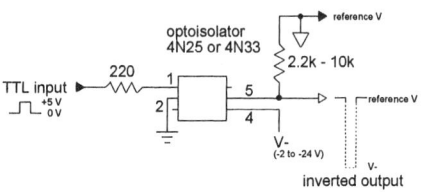

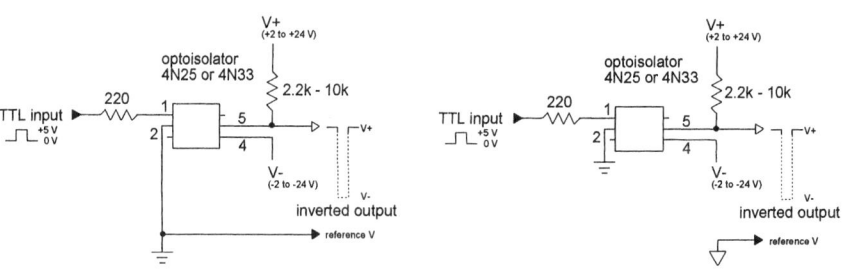

Figure 12.12: Using an optoisolator to increase or reduce an inverted output signal.

a return path with the recording apparatus, will not create as bad a shock artifact in the recording. Alternatively, the same power supply can be used for both sides (left hand column in Figures 12.11 and 12.12).

Finally, Figures 12.11 and 12.12 illustrate the versatility of the optoisolator in selecting the range of the outputs. The top figures show outputs that range from 0 to some positive voltage that is determined by the power supply on the output. The range can be from +2 to +24 V. The middle figures show outputs that range from some negative voltage (determined by the power supply on the output) to 0 V. The old relay systems used to vary their signals from -28 V to 0 V, with 0 V being logic 1. When solid state programming equipment first became available, it was made to be compatible with relay systems. Therefore, this design is useful when interfacing with old relay systems or with older solid state equipment that was made to work with the old relay systems. Today, we can choose the relay levels so this is not a usual design. The bottom figures show outputs that range from some negative to some positive voltages (determined by the power supply on the output). This design is more commonly seen when used to control AC or analog signals.

Digital Versus Analog

Although the above examples illustrate digital signals, they can also be used to transfer analog signals from one voltage range and rectification (inverted or noninverted signals). The versatility, safety through its isolation, inexpensiveness and availability of the optoisolator make it a favorite for interfacing from one system to another. In reality, although we have tested the concept, we have never seen a need to use an optoisolator circuit with an analog signal. We would consider this approach if we were recording from human subjects. The major technical solutions for this approach are first, to bias the optical elements so that they operate in the middle of their voltage ranges; second, to amplify the output from the optoisolator; third, to power the human side with batteries; and fourth, to make certain that neither side of the optoisolator is electrically connected to the other side. These are rather trivial design considerations to implement.

An Interface for Recording Data to Tape

The first online computer system developed by Thompson and colleagues involved a PDP-12 computer that crashed often enough in the middle of training that an alternative to running trials and saving data was developed. The PDP-12 would be used for offline analysis of data saved to magnetic tape in experiments using solid state controllers. At that time, four-channel, reel-to-reel tape recorders were available. The four channels were used to record behavioral and neural data with synchronization pulses for each trial (pre-CS, CS and US onset pulses) recorded on one channel, two unit recordings on each of two channels, and the fourth channel containing a mixed 'Vetterized' signal consisting of analog signals for the behavior (nictitating membrane) and the EEG of the two unit channels. The Vetter

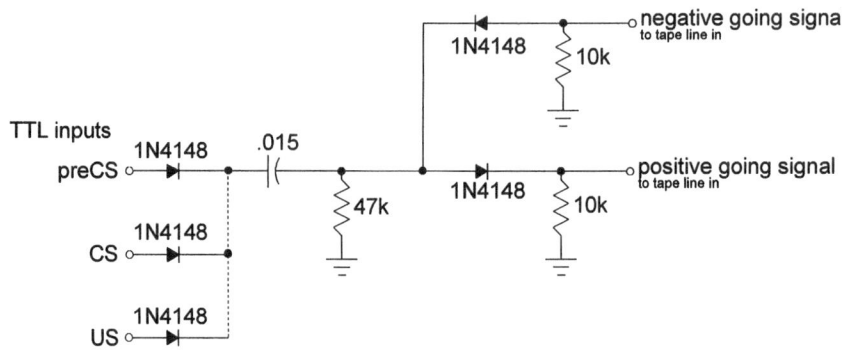

Figure 12.13: Mixing synchronization pulses for recording to tape using passive components.

company device takes a slowly changing signal and frequency modulates it into a higher frequency that can be recorded onto tape. By modulating different analog signals (nictitating membrane, two EEG) to different high frequency ranges the signals can be mixed together and recorded on the same tape channel. The signals must be separated (filtered) and demodulated to restore the original signals (or approximations thereof). The Vetter device handles the modulation, mixing, separation and demodulation.

As an aside, the availability of precision waveform generators (e.g., the 8038 used in Chapter 2) and operational amplifiers for mixing and filtering can be used to create mixed, frequency-modulated signals. The problem comes in demodulating the signals, which requires problems introduced by integration. However, voltage-to-frequency and frequency-to-voltage modulators/demodulators are available. A problem is that the modulation and demodulation actually delays the data in a frequency-dependent manner, meaning that there are different delays for slowly versus rapidly changing signals, both of which can be components of the same changing behavioral response. Thus, compensation is not simply a matter of subtracting a constant time from any signal. Fortunately, online data collection with multiple channels removes much of the need for this strategy.

Unlike the frequency modulated channel, the other channels having the unit and synchronization pulses are easily interfaced. The industry standard is that line-in and line-out signals are in the range of 1 volt peak-to-peak or less. With VU (volume units) meters on the tape recorder it is a simple matter to adjust the gain of all four channels for optimal recording. Unit recordings go directly from the amplifier to the line-in. The synchronization pulses are filtered to remove DC components then placed onto the line-in connector.

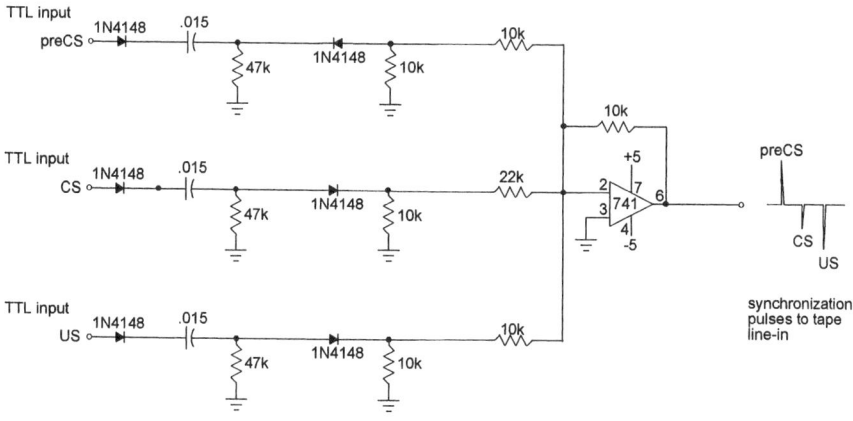

Figure 12.14: Mixing synchronization pulses for recording to tape using active components.

Figure 12.13 shows a circuit for taking a short TTL pulse from the control equipment, marking a stimulus (preCS = beginning of trial, CS onset, US onset), and converting it into an AC signal suitable for recording onto tape. The same circuit can be used to record stimulus onset markers onto a polygraph channel in parallel with heart rate, as in Chapter 3 (Heart Rate Amplifier), although this circuit is not really necessary (but it is compatible) with the capabilities of a polygraph. The first 1N4148 diode in Figure 12.13 at the input has two functions. It protects the TTL source from capacitor discharge and allows multiple TTL signals (preCS, CS, US), each with their own diode, to be mixed onto the same tape channel. The capacitor and resistor create a filter to remove DC components of the TTL signal. The values shown here work well for very short TTL pulses; longer pulses would require different values. The filtered signal has both positive and negative waves and could be placed directly onto tape or a polygraph. If only a positive going wave is desired then the second diode and 10 kΩ resistor to ground can be added. Negative going waves are achieved with the second diode's polarity reversed. The polarity might be important for later height discrimination.

A more sophisticated version for mixing synchronization pulses onto the same tape channel is seen in Figure 12.14. A separate diode-and-filter circuit for each signal (preCS, CS and US) would have to be added together with an operational amplifier to achieve the mixed signal. By manipulating the polarity and the height of the signal, the height discriminators can be used to selectively pick off each of the three signals in this example. By this means of varying the polarity and height, a computer set up for off-line analysis can detect the trial type (CS alone, paired CS-US, etc) of the corre-

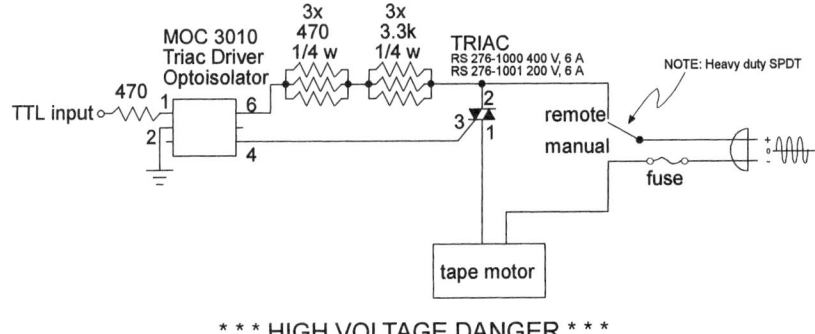

*** HIGH VOLTAGE DANGER ***

Figure 12.15: Interface for controlling the power to a mechanical tape recorder so that it only records during a classical conditioning trial and not during the intertrial interval. This saves tape and makes reviewing the data much faster.

sponding data channels for unit activity and behavior. Note that in Figure 12.14 the final polarities of the signals are inverted from those found in Figure 12.13 because of the inverting input of the operational amplifier.

Off-line data analysis has the distinct disadvantage that it takes considerable time to collect and analyze the data. The time factor can be improved, if only slightly in our opinion, by only recording the trial data to the tape, and not recording during the intertrial interval. This control saves a considerable amount of tape as the trials usually last only about 1 second out of every 30 seconds of intertrial interval. Some tape recorders have remote controls that can be interfaced for this purpose, usually by routing the remote signals through a relay. We do not know of a tape recorder that has a TTL compatible input for controlling recording. It is also impossible to find 4-channel reel-to-reel tape recorders as originally used by Thompson. The magnetic tapes, besides, deteriorate horribly in only a few years. VCRs have been adapted for data storage, having the advantage that the original data is saved and the tapes can hold a tremendous amount of data. Nevertheless, if one were still inclined, data could be saved onto stereo cassette tapes. Remote controls are usually not available for the inexpensive cassette recorders, but can be added if the cassette recorder has mechanical switches on it. Mechanical switches are the type where you physically press and hold the Play and Record buttons at the same time and they will stay in place, regardless of whether the recorder is turned on or off. This is a critical feature for the following circuit. Soft switches which only activate the record function if the recorder is turned on will not work here.

Figure 12.15 shows a circuit for controlling the AC power supply from the wall. We incorporated this design into an AC power strip as a safety feature, setting up one of the sockets for remote control of a stereo cassette tape recorder with mechanical switches. The tape recorder is plugged into the remote socket and the mechanical record and play buttons are pressed down. When a TTL signal is applied to the 3010 optoisolator triac driver in Figure 12.15, AC power is switched through the triac to the socket, thus turning on the tape recorder.

To accurately record, the tape must be started shortly before the trial begins in order for the motor to get up to speed. This lead time is accomplished in the Runtime software with the timing variable **before** which is used by the subroutine **lead** to start the tape recorder an empirically determined number of seconds before the baseline of the trial actually begins. The timing variable **after** is used by the subroutine **trail** to keep the tape recorder on for a short time after the trial has ended. It is not so important to keep the recorder on to collect data at the end of the trial as it is to give a short gap between trials on the tape recorder so that the computer has time to save the collected data to disk between trials. Normally data is collected online so the tape function is turned off with the soft switch **tape off**. To collect data to tape set the soft switch to **tape on** and save the synchronization pulses on one channel, data on the other channel. To collect the data saved to tape, play the data back to the computer with the soft switch **live** set for off-line analysis (**live off**) and feed a height-discriminated-output from the tape channel containing the synchronization pulses to the trigger input on the computer interface board (see Chapter 11). For normal online data collection and analysis set **live on**.

The quality of the tape and the tape recorder bias makes a difference in accurately recording units, for example. The dynamic response of the tape should be the highest frequency possible. Metal tapes and tape recorders with buttons to select for metal recording and playback work well. Some tape recorders supposedly will automatically select the bias for the tape being used but our experience questions that reality. Find a tape recorder with explicit buttons for the type of tape. The quality of the recording is also a function of the speed of the motor: the faster the motor, the better the recording of high frequency responses. The obsolete 4-channel tape recorders used by Thompson were fast and usually included a selection of even faster speeds. The speed of cassette tape motors is very slow and therefore not as good for storing data. The poor tape speed can be partially compensated for by using the better tapes and bias.

There is now a variety of options for storing data. Multichannel cassette recorders can be found which overcome the data limitations of the two-channel stereo cassettes. VCR recorders have already been mentioned. DAT recorders are a newer option for storing onto tape. Getting away from tape to other computer media, Minidisc recorders, writeable CDs, ZIP drives, and hard drives are also options with the added advantage that the data has been converted by A/D into numerical values. The advantage of online analysis is that the data set is reduced tremendously. The disadvantage of online analysis is that one can never recover the original data to re-analyze it in a different manner.

APPENDIX A. SUPPLIERS

The following list of suppliers are just the ones that I personally use. The list is more or less in order of use, the first source being the most used, the last the least used. In this list I note the kinds of items that I usually buy from each.

For electronic parts, Jameco Electronics is my favorite -- good prices, selection and service -- but it requires some forethought to order the parts. A reasonable alternative for electronics parts is Fry's Electronics which I would usually travel to on weekends. Fry's is a chain that started in Silicon Valley that now has numerous stores in the Los Angeles area. For immediate needs, the Radio Shack across the street is expedient for integrated circuits but it is more expensive than Jameco or Fry's. The exception is that Radio Shack is a good source for switches. The local electronics store ITC is the best place for buying standoffs for circuit boards. Most of their electronics seem to be related to communications which I have not found to be too useful.

Sources for some nonelectronic items are also listed. The main items of interest are stainless steel tubing and screws (Small Parts), solenoids for the airpuff (General Valve), stainless steel insect pins for electrodes (Ward's Biology), polarized film (Edmund Scientific), gold plating (JNT) and gold pins (Avnet). Good stainless steel solder flux is difficult to find. Small Parts sells stainless steel solder flux, but the best I have comes from a local hardware store (New York Hardware and Trading) which may not carry it any more because it is so rarely sold. Plumbing suppliers should also be a source of stainless steel flux.

The reader should be able to find local sources for other items. Plastic and plastic glues are bought at plastic specialty stores in preference to hardware stores as the former carry larger bottles of the two part 5-minute epoxy.

What is not in this list are common behavioral neuroscience vendors. For example, not listed are A-M Systems (amplifiers), Frederick Haer (discriminators), Lafayette (animal testing), David Kopf and Narishige (stereotaxes), Fine Science Tools and George Tiemann (surgical instruments), etc. We find these companies at the Society for Neuroscience annual conventions and we assume the reader is familiar with these and similar companies already.

Name:	Jameco Electronics
Type:	mail order
Location:	1355 Shoreway Road, Belmont, CA 94002-4100
Phone:	1-800-831-4242
Web:	www.jameco.com
Items:	Integrated circuits, transistors, diodes, resistors, capacitors, power supplies, crystals, sockets, IBM-PC/clone circuit boards/cards

Name:	Fry's Electronics
Type:	chain of electronic retail stores
Location:	Manhattan Beach, Burbank, Woodland Hills, etc.
Items:	Integrated circuits, transistors, diodes, resistors, capacitors, sockets

Name:	Radio Shack
Type:	chain of electronic retail stores
Location:	local
Web:	www.radioshack.com
Items:	integrated circuits, diodes, resistors, capacitors, switches, wire, small copper-patterned circuit boards

Name:	Small Parts
Type:	mail order
Location:	13980 N.W. 58th Court, P.O. Box 4650, Miami Lakes, FL 33014-0650
Phone:	1-800-220-4242
Web:	www.smallparts.com
Items:	Small stainless steel hypodermic tubing, wire, screws

Name:	Edmund Scientific
Type:	mail order
Location:	101 East Cloucester Pike, Barrington, NJ 08007-1380
Phone:	1-800-728-6999
Web:	www.scientificsonline.com
Items:	Plastic polarized film

Appendix A

Name: Ward's Biology
Type: mail order
Location: P.O. Box 92912, Rochester, NY 14692-9012
Phone: 1-800-962-2660
Web: www.wardsci.com
Items: Stainless steel insect pins

Name: ITC
Type: local electronics chain
Location: 2772 West Olympic Blvd, Los Angeles, CA 90006
Phone: 213-388-0621
Items: Standoffs for circuit boards

Name: General Valve
Type: mail order
Location: 202 Fairfield Road, P.O. Box 1333, Fairfield, NJ 07006
Phone: 201-575-4844
Items: Solenoids and fittings

Name: JNT Manufacturing
Type: mail order
Location: Mead's Lane RR2, Box 870, Stormville, NY 12582
Phone: 1-800-552-4568
Items: Gold plating kit and supplies

Name: Avnet Electronics Marketing
Type: mail order
Location: 1301 West Geneva Drive, Tempe, AZ 85282
Phone: 1-800-320-7499
Items: Cannon CTA gold pins (male 031-9540-000, female 030-9542-001)

Name: Epoxylite Corp.
Type: mail order
Location: 9400 Toledo Way, Irvine, CA 92713
Phone: 1-800-424-9300

Items:	Epoxylite baking varnish we use for insulating electrodes (6001-m varnish, 6001-s thinner)
Name:	California Fine Wire
Type:	mail order
Location:	360 S. 4th Street, Grover City, CA 93433
Phone:	1-800-489-5144
Web:	www.calfinewire.com
Items:	Spools of stainless steel and constantan wire
Name:	VWR
Type:	mail order
Phone:	1-800-932-5000
Web:	www.vwrsp.com
Items:	Large corks used for dipping electrodes into insulation (Cat. no. 23420-560 for 10 regular corks)
Name:	New York Hardware Trading Co.
Type:	local hardware store
Location:	410 W. 8th Street, Los Angeles, CA 90014
Phone:	213-614-0023
Items:	Stainless steel solder flux
Name:	South Bay Plastics
Type:	local plastics store
Location:	20820 S. Normandie Avenue, Torrance, CA 90502
Items:	Plexiglas, plastic adhesives, large bottles of two-part 5 minute epoxy

APPENDIX B. RABBIT ATLAS

When the Thompson laboratory found that the cerebellum was critically important for classical eye blink conditioning in the early 1980s there were no stereotaxic rabbit atlases available showing the cerebellum. Dave McCormick made an atlas that has since been photocopied and used by generations of students in the Thompson laboratory. If we remember correctly, the original McCormick and Thompson atlas was never published because the journal wanted photographs of the histology to accompany the drawings. Many years later, sometime in the 1990s, when we were exploring the brainstem more systematically, we used McCormick's original histology as the basis to create an atlas that used all the histological sections McCormick had collected. Our intent was twofold: First, to extend their atlas using the available material rather than to start from scratch; and, second, to have machine-readable images and drawings that we could use to overlay data accumulated from recording or lesion experiments. Since our atlas is based on McCormick's histology, by definition our atlas is not complete for the entire extent of the rabbit brain. This atlas extends posteriorly from the diencephalic–mesencephalic junction, i.e., from the posterior commissure to area postrema. For this atlas, we scanned each of the sections at every half millimeter as identified by McCormick. We electronically removed gelatin and imperfections from the background of the images, and then outlined the external and internal features, using the drawing features of Adobe Photoshop. These lines were then imported into Macromedia's Freehand, where the drawings were realized by defining in the lines, scaling the images, and adding text. The drawings are scaled to the original histology, but the vertical (DV) rulers are not anchored to any landmark such as bregma or lambda. Philosophically, we were not concerned that the original images were not cut perfectly symmetrical—that's the reality of typical histology, and has the added advantage that intermediate sections are represented on the right/left halves. Both our original images and the drawings are available on CD. Perhaps a complete atlas with new material will be a future project.

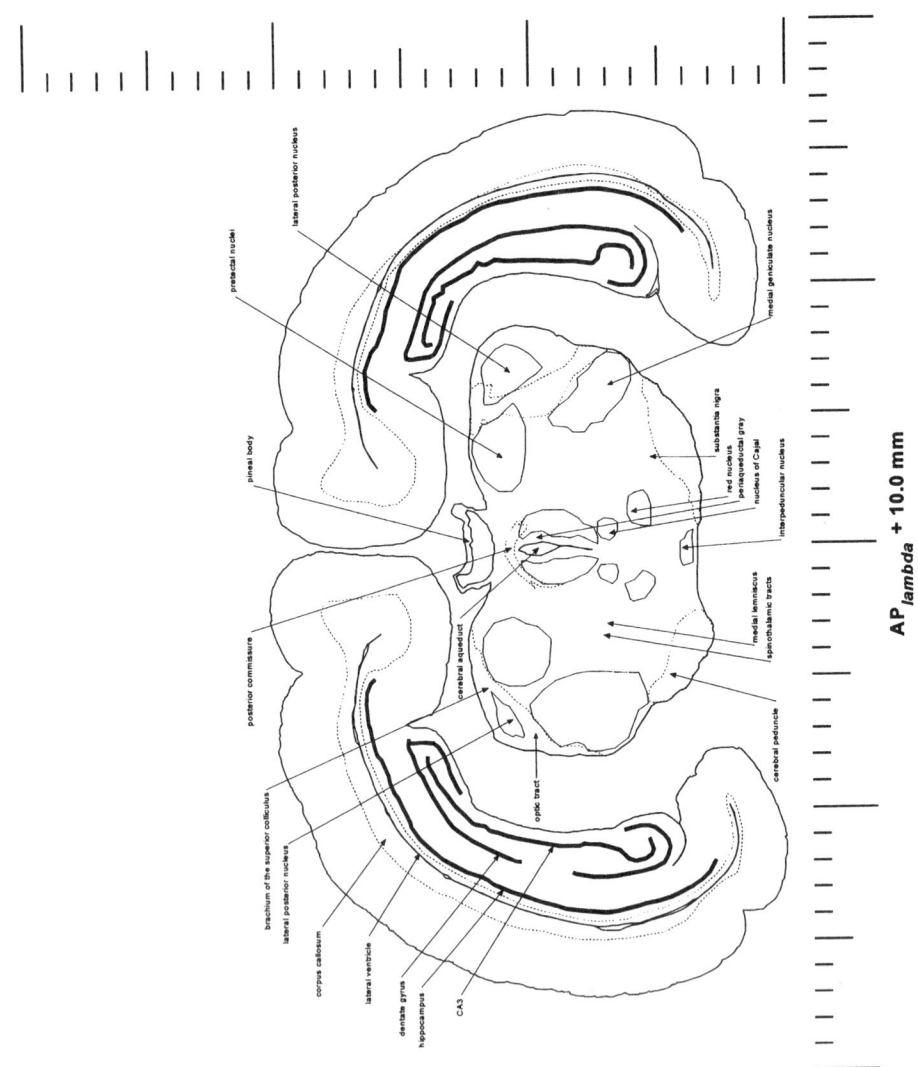

Appendix B

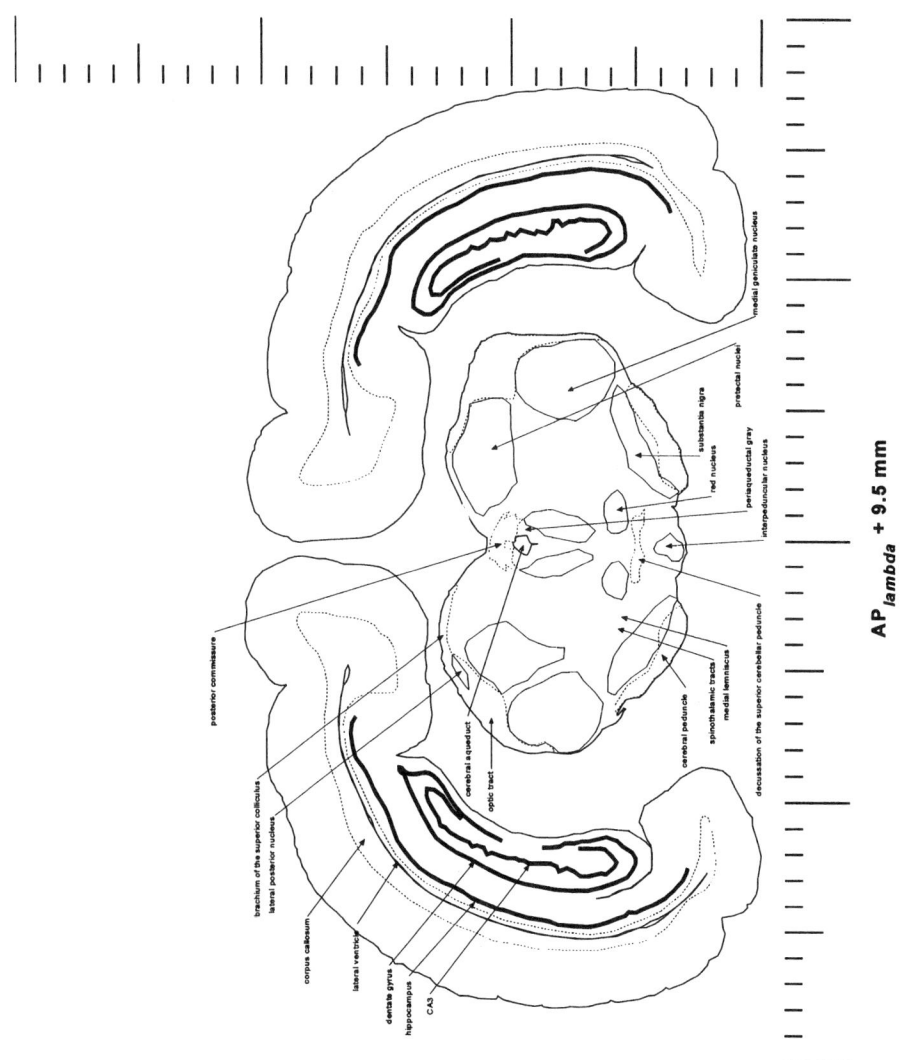

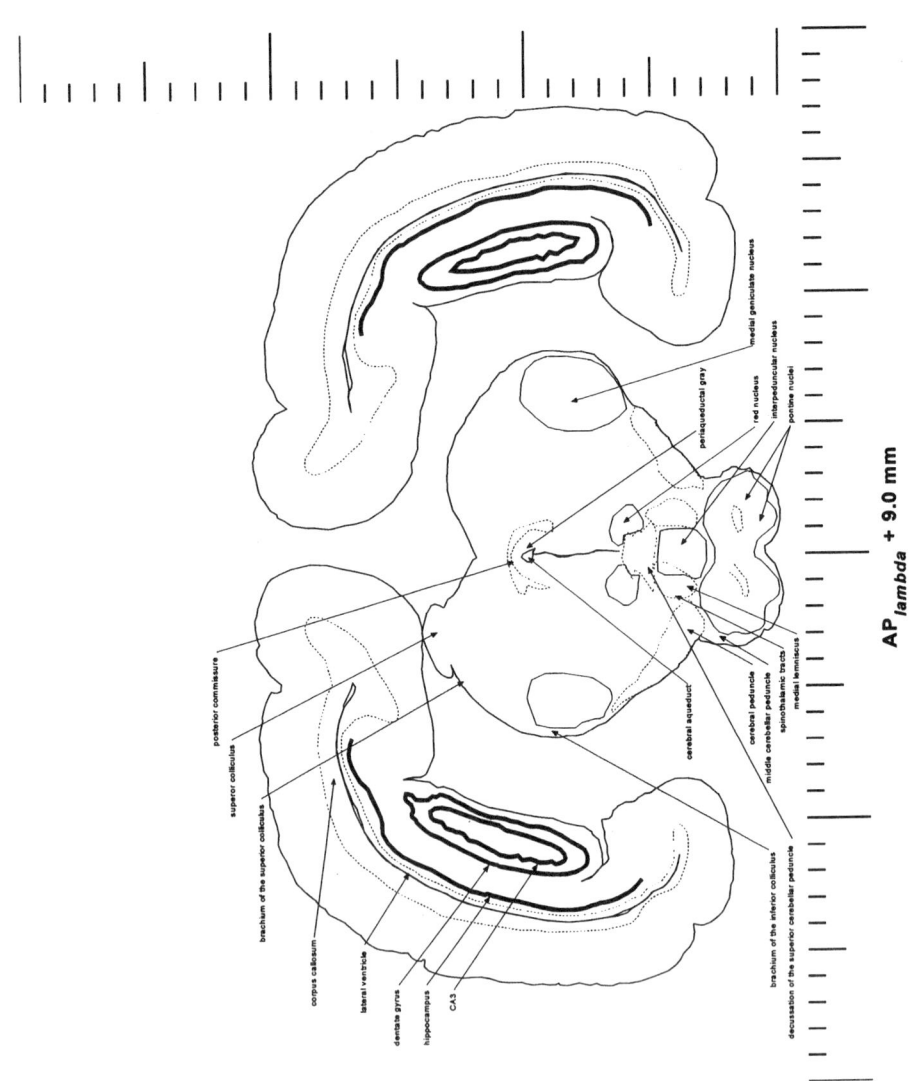

Appendix B

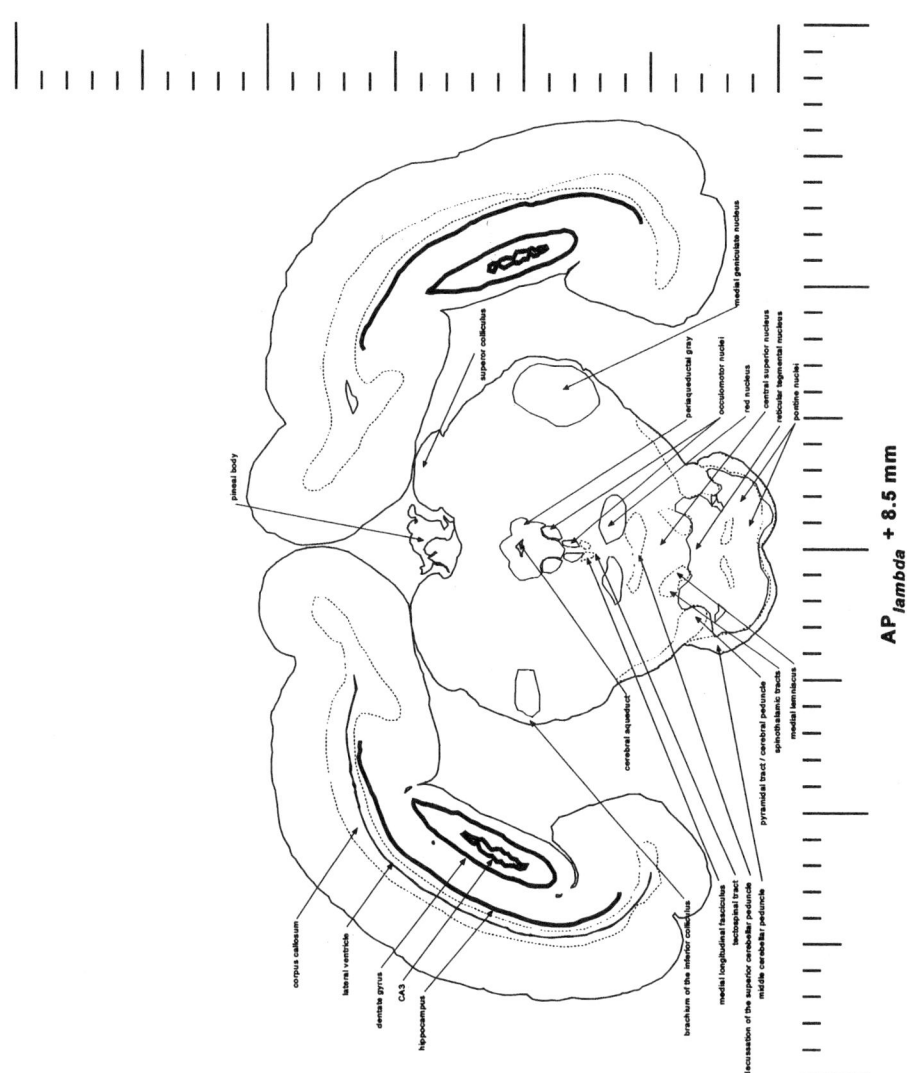

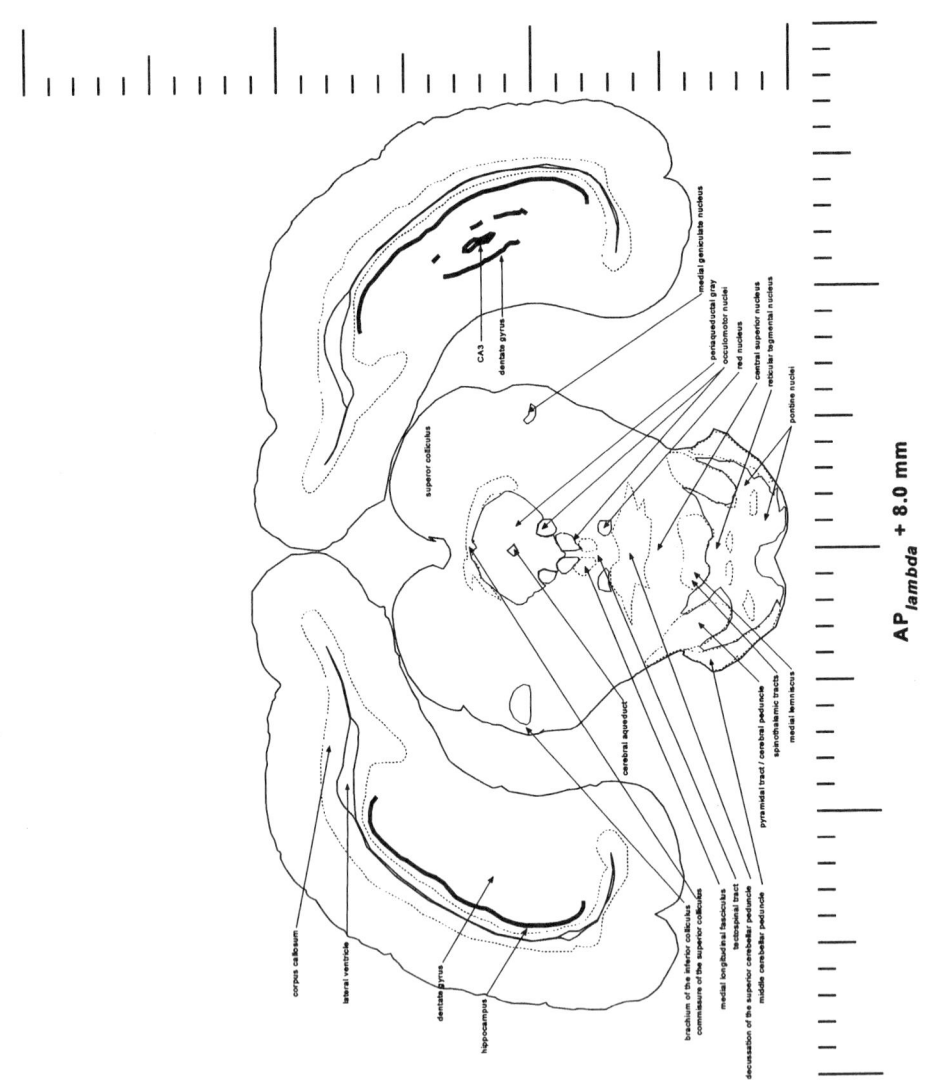

Appendix B

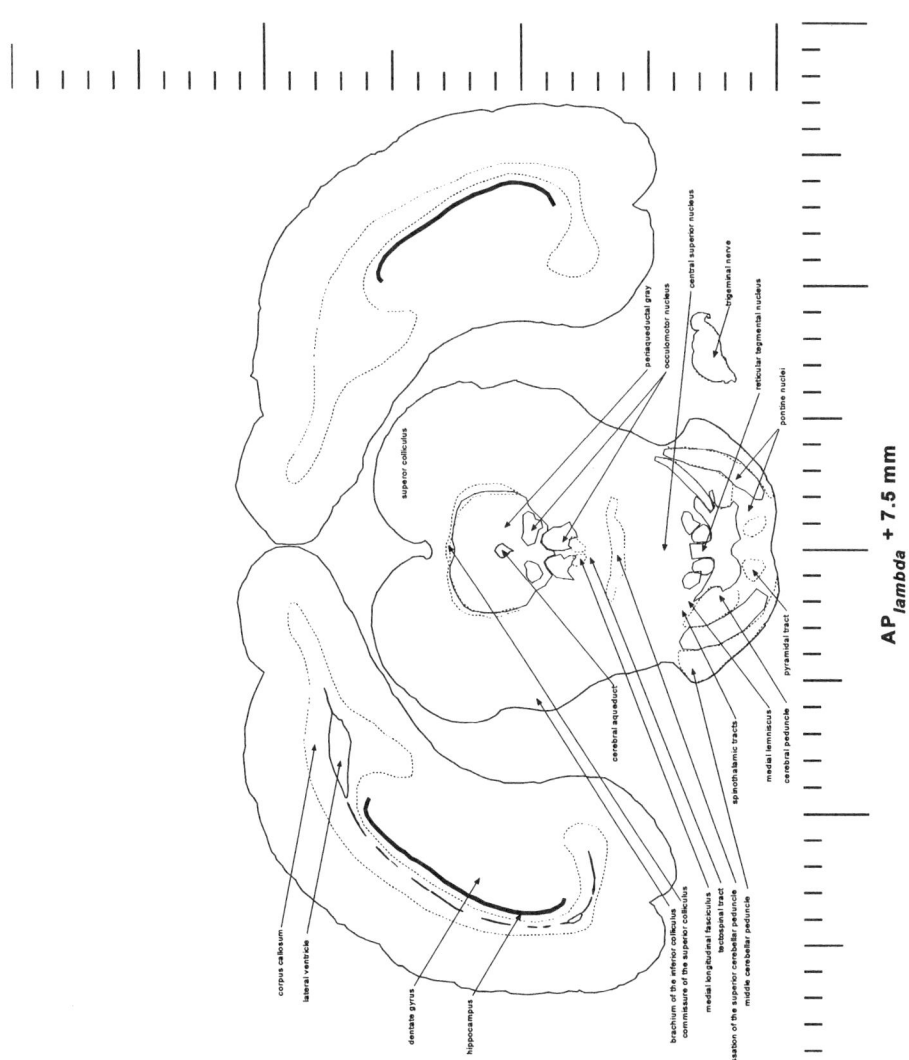

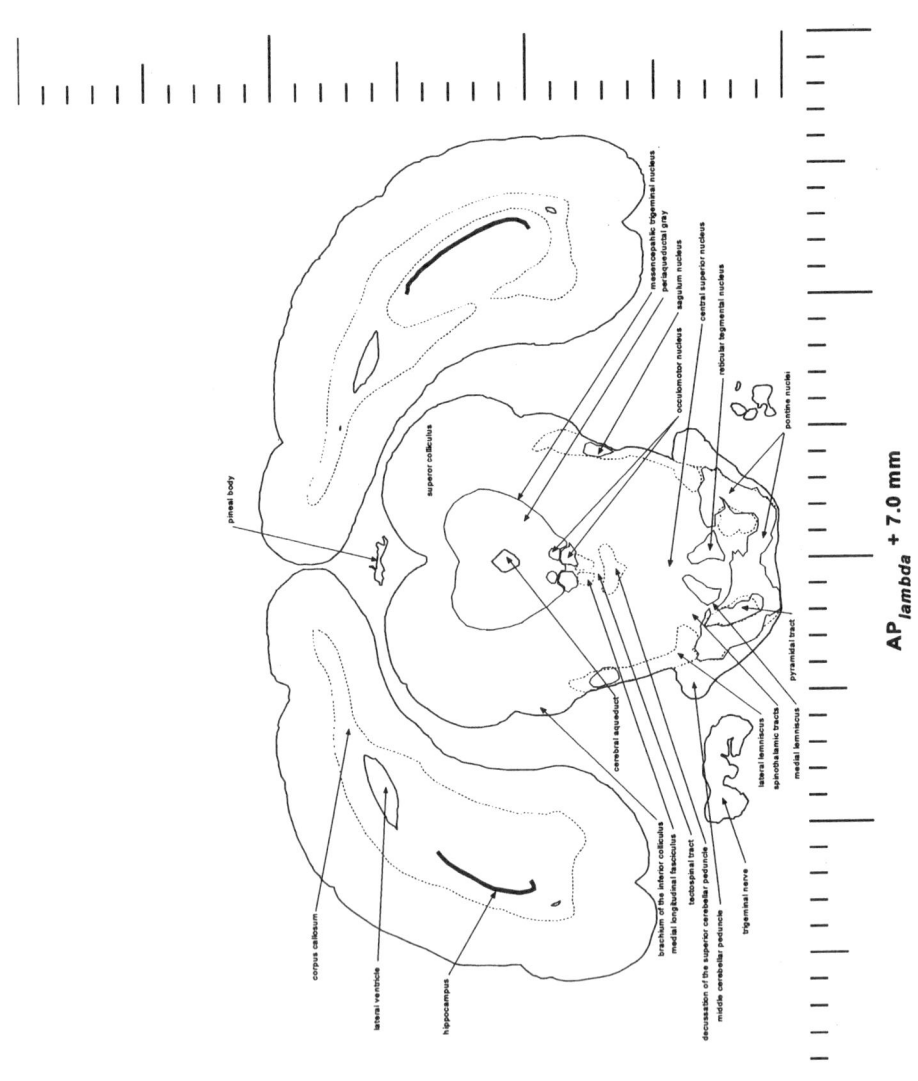

Appendix B

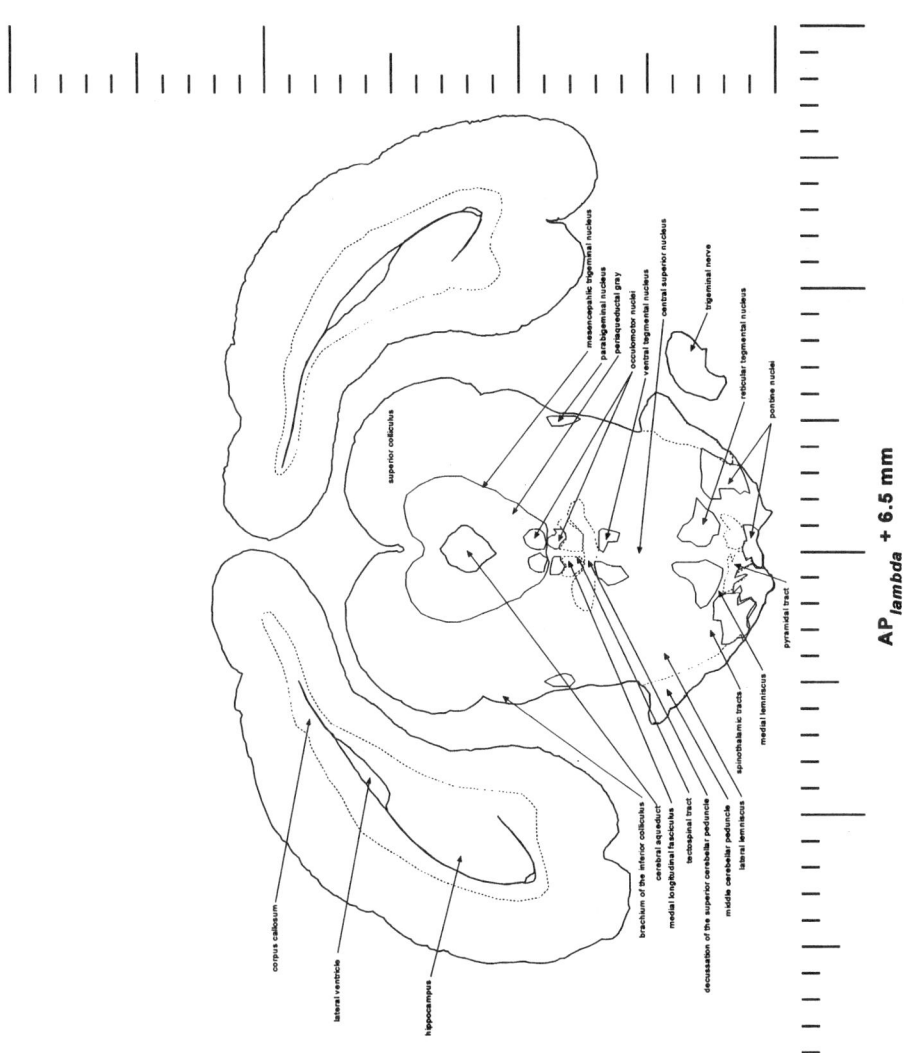

AP$_{lambda}$ + 6.5 mm

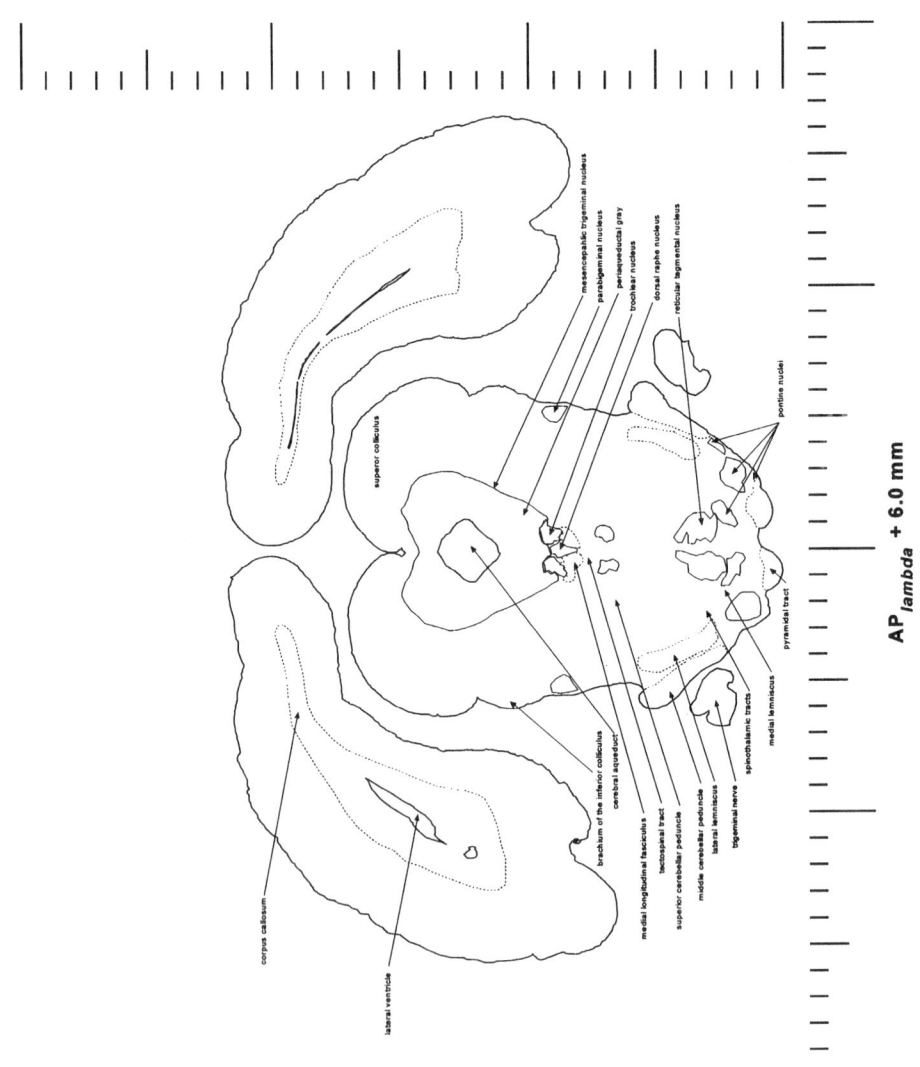

Appendix B

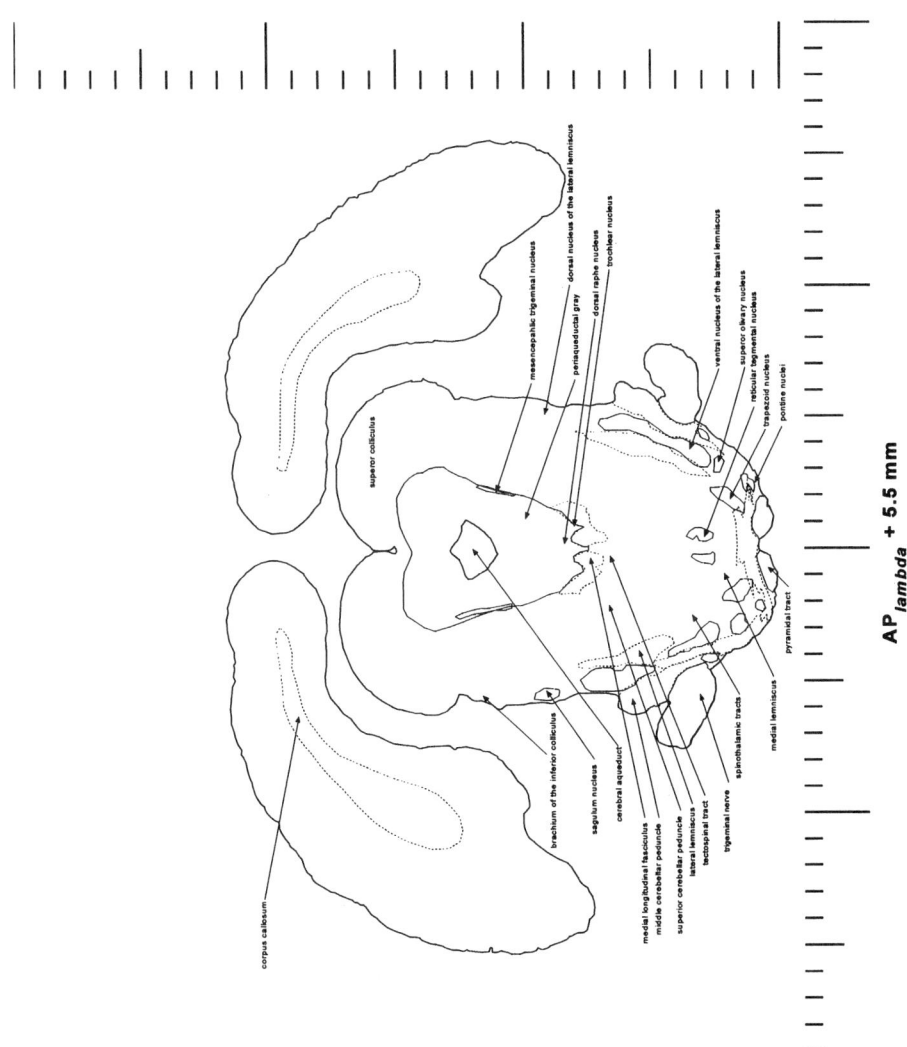

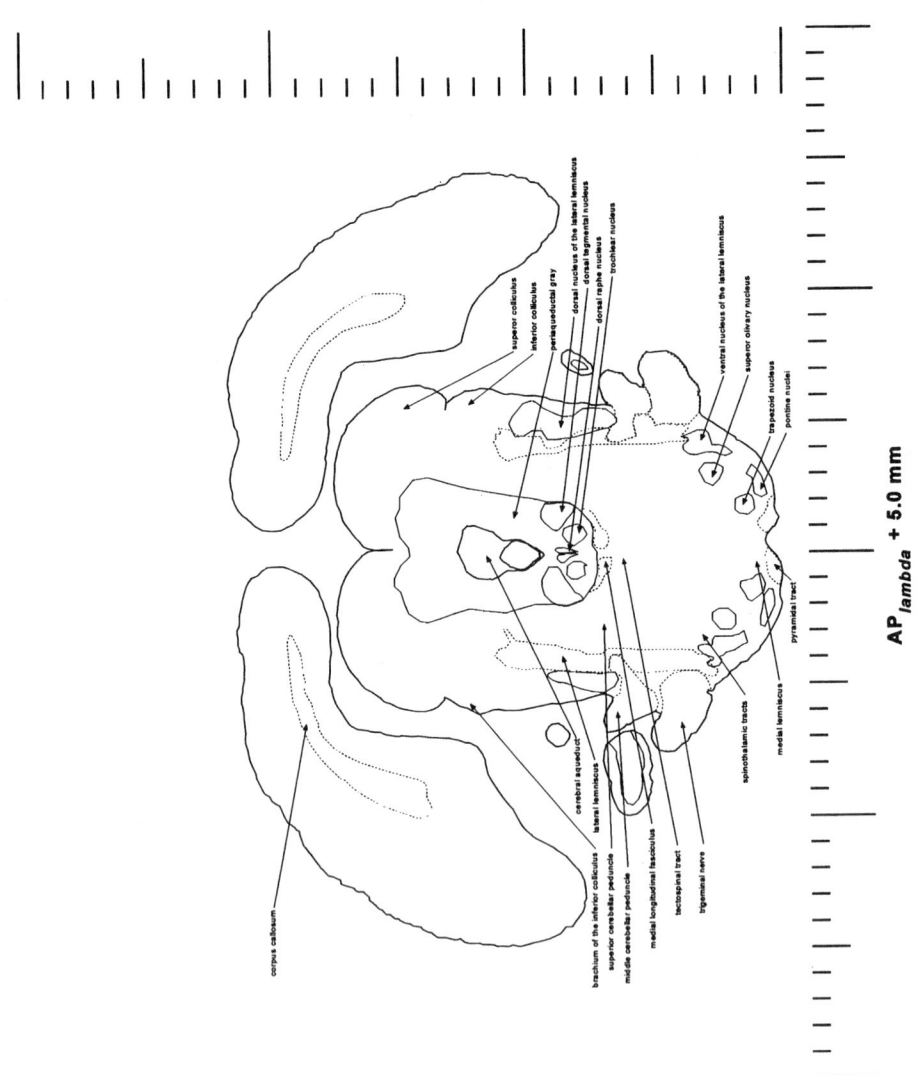

Appendix B

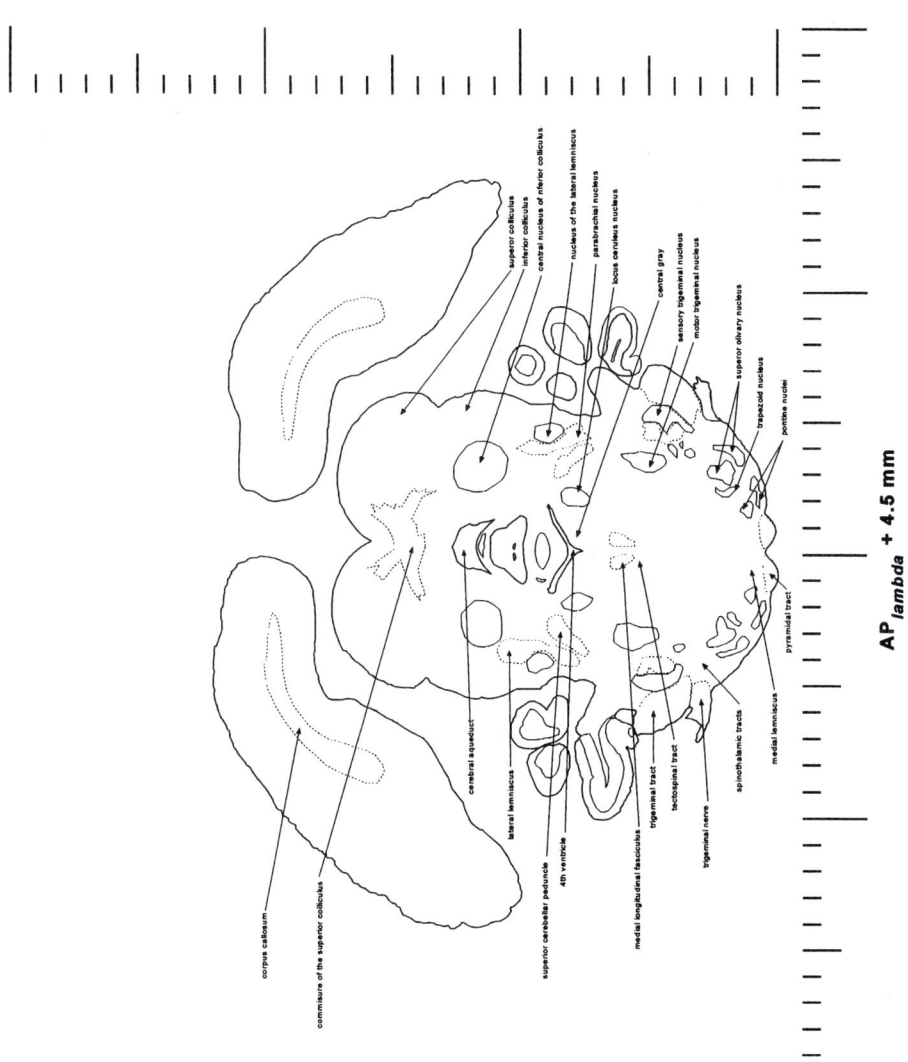

AP_{lambda} + 4.5 mm

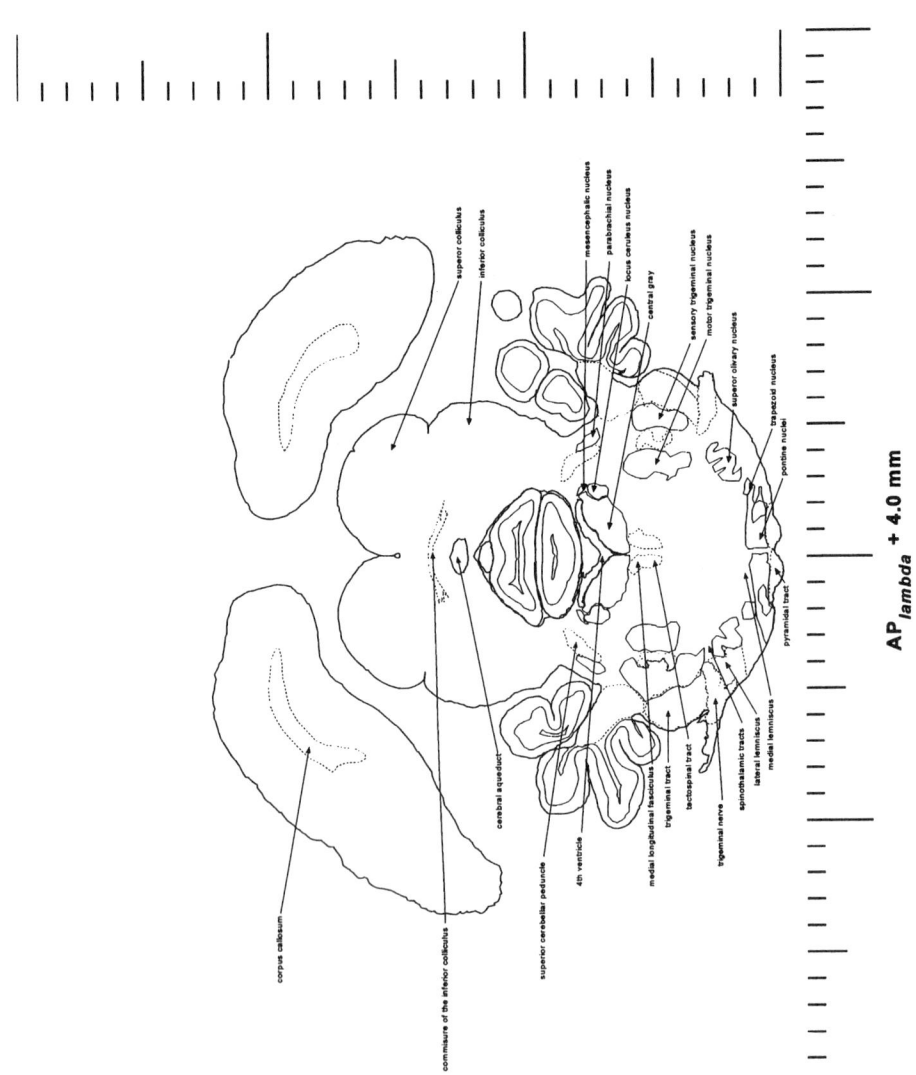

Appendix B

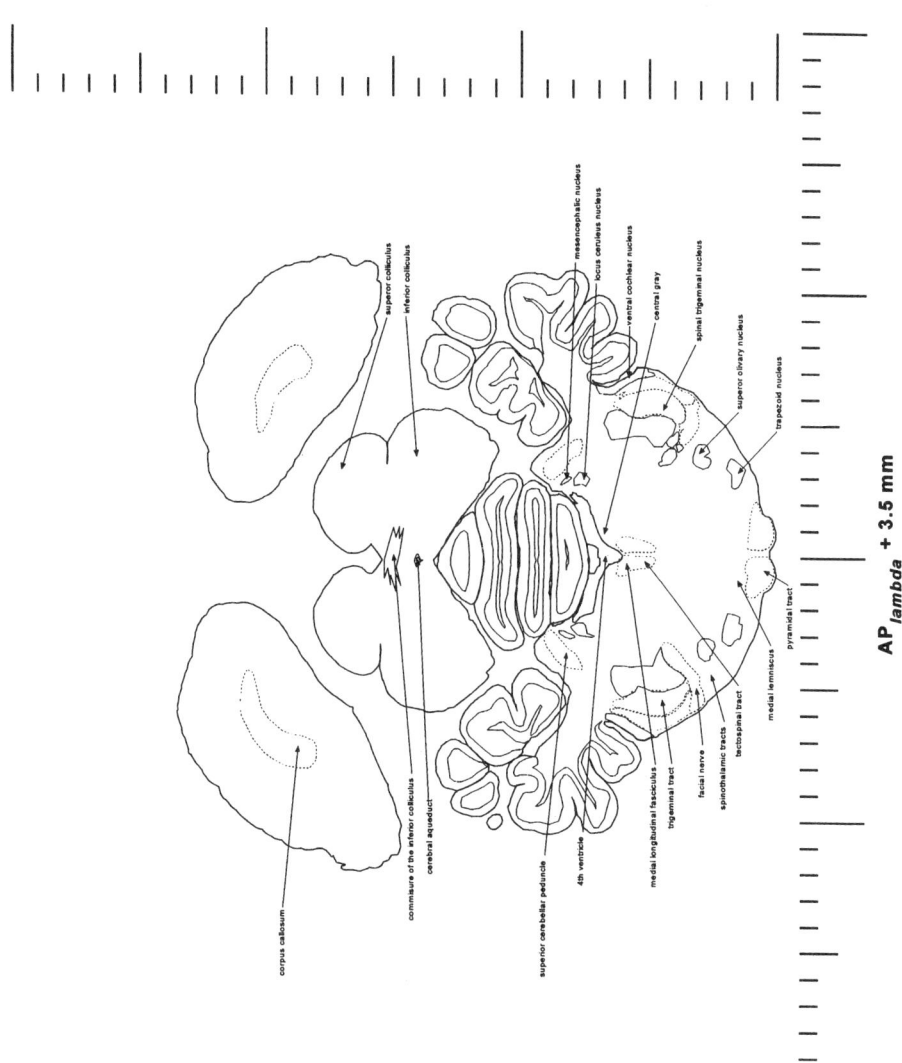

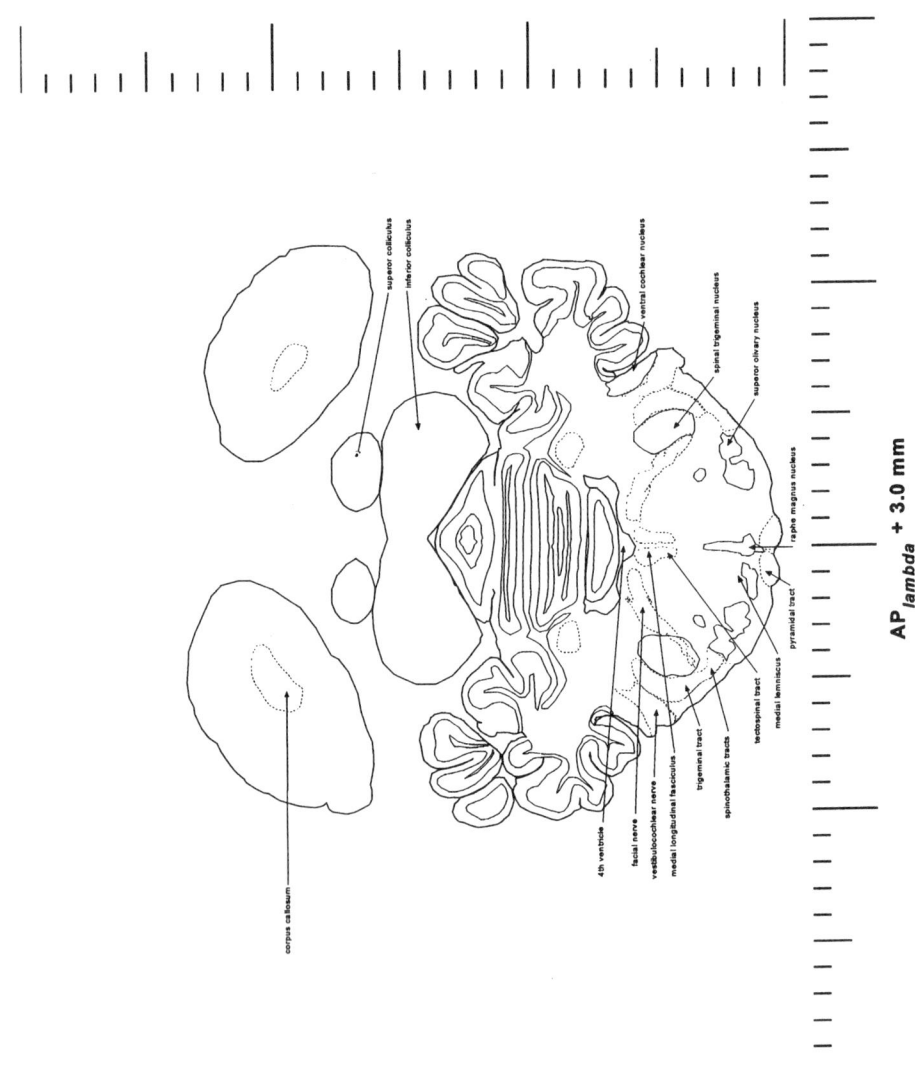

Appendix B

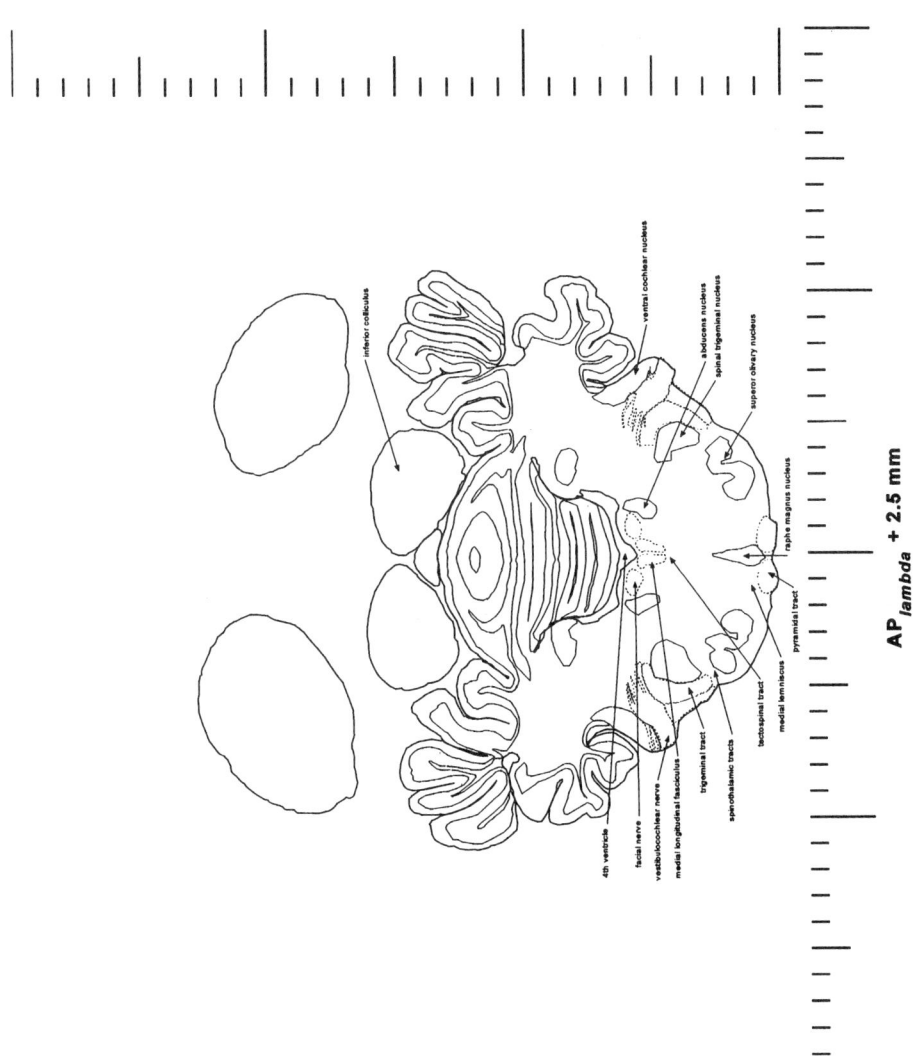

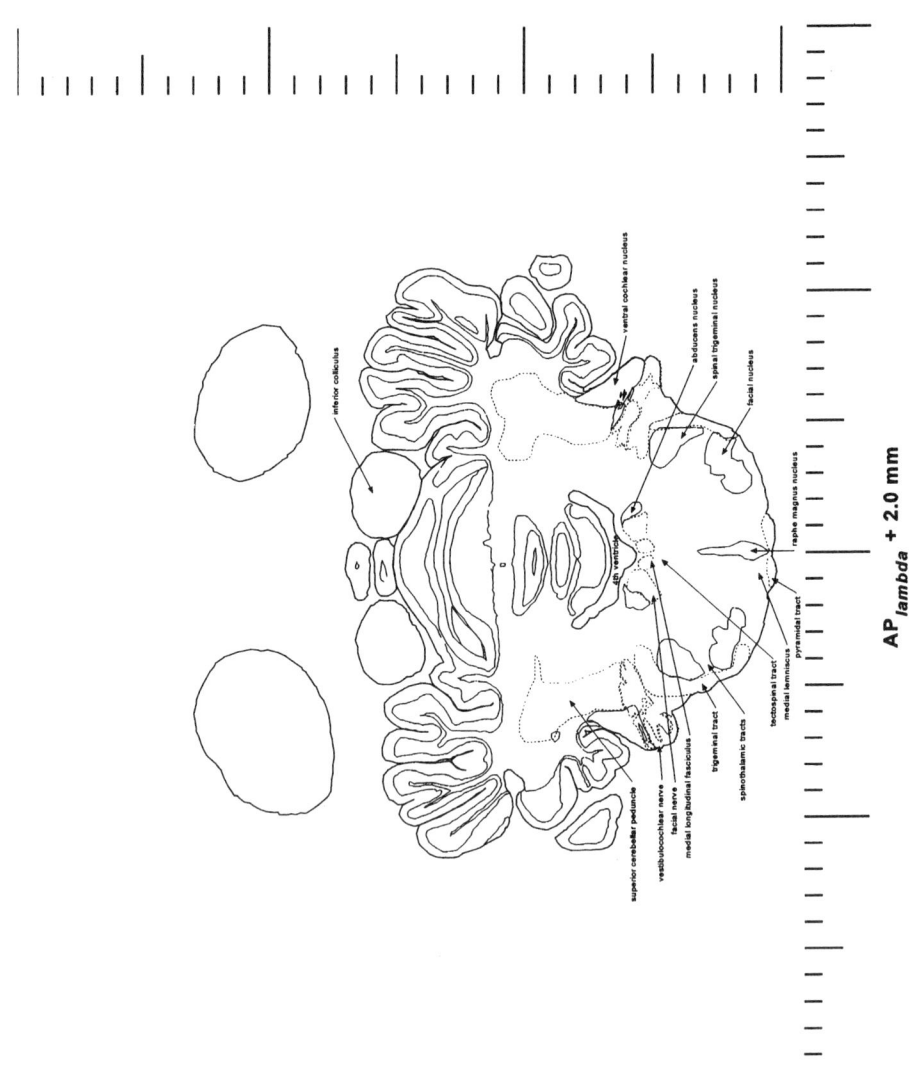

Appendix B

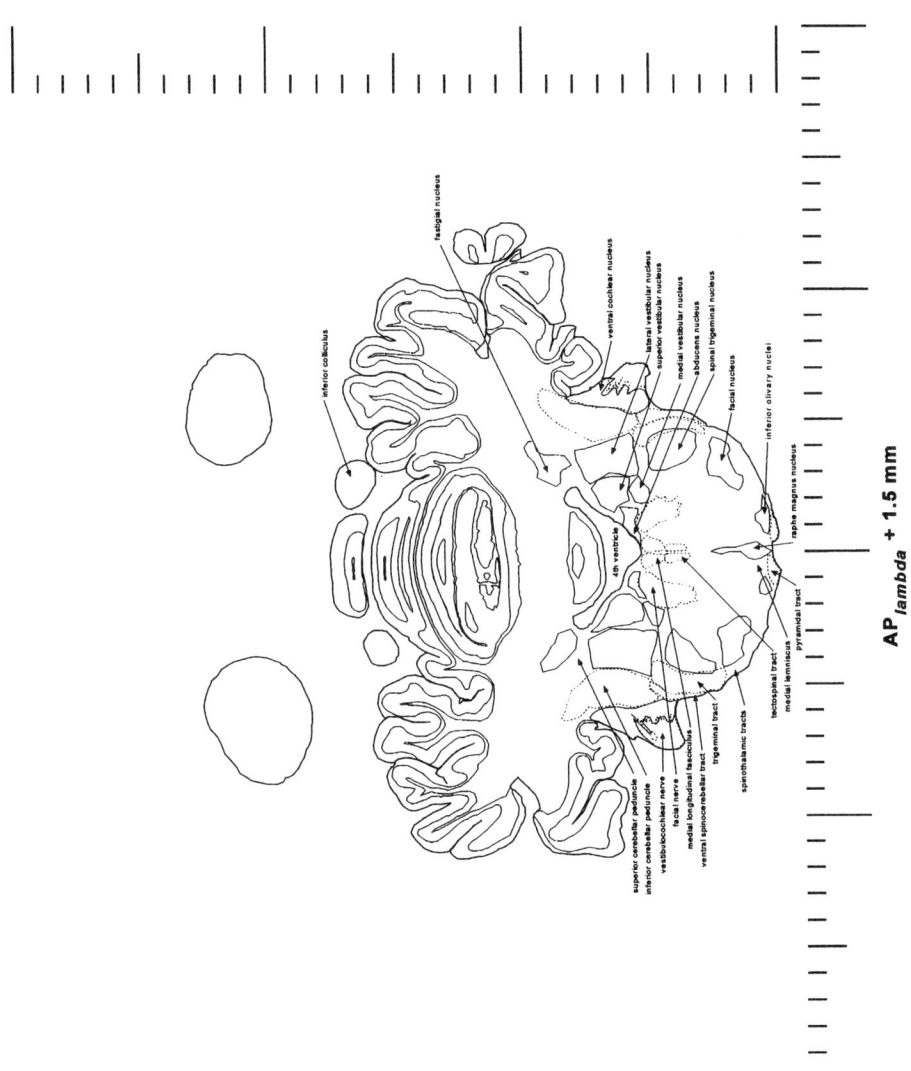

AP_{lambda} + 1.5 mm

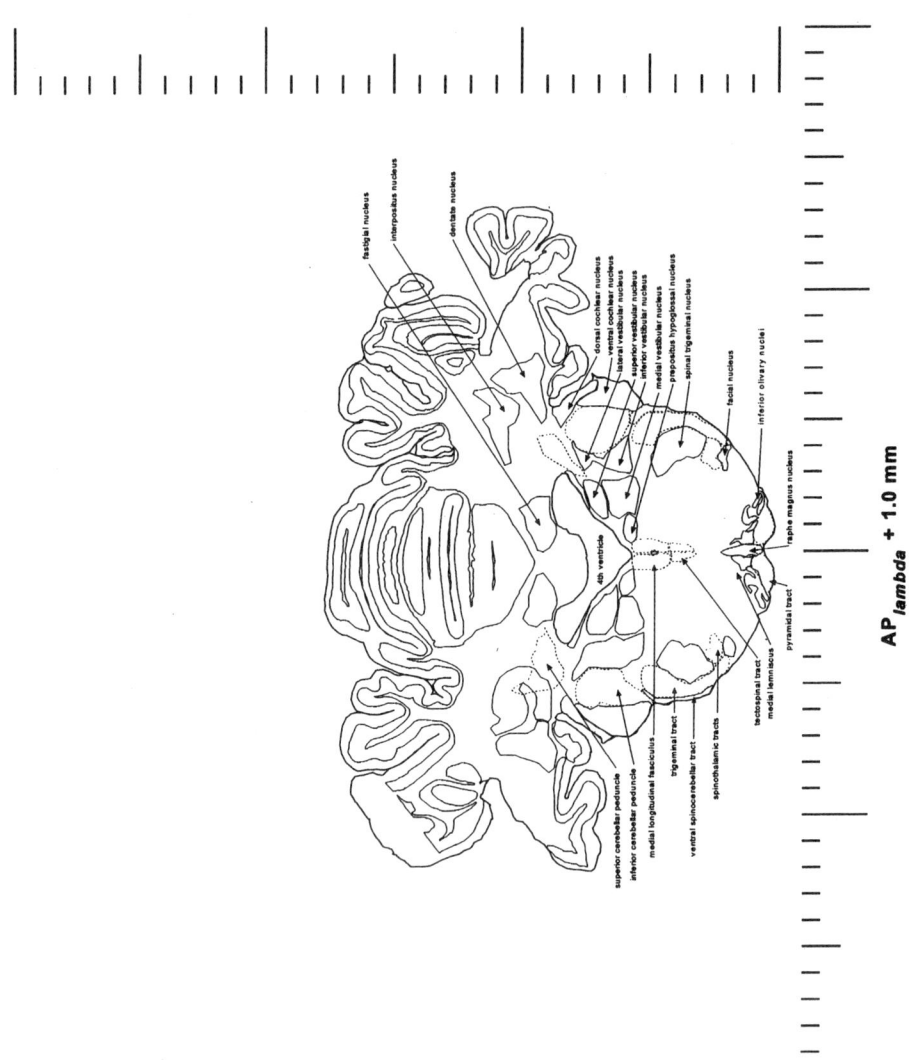

Appendix B 413

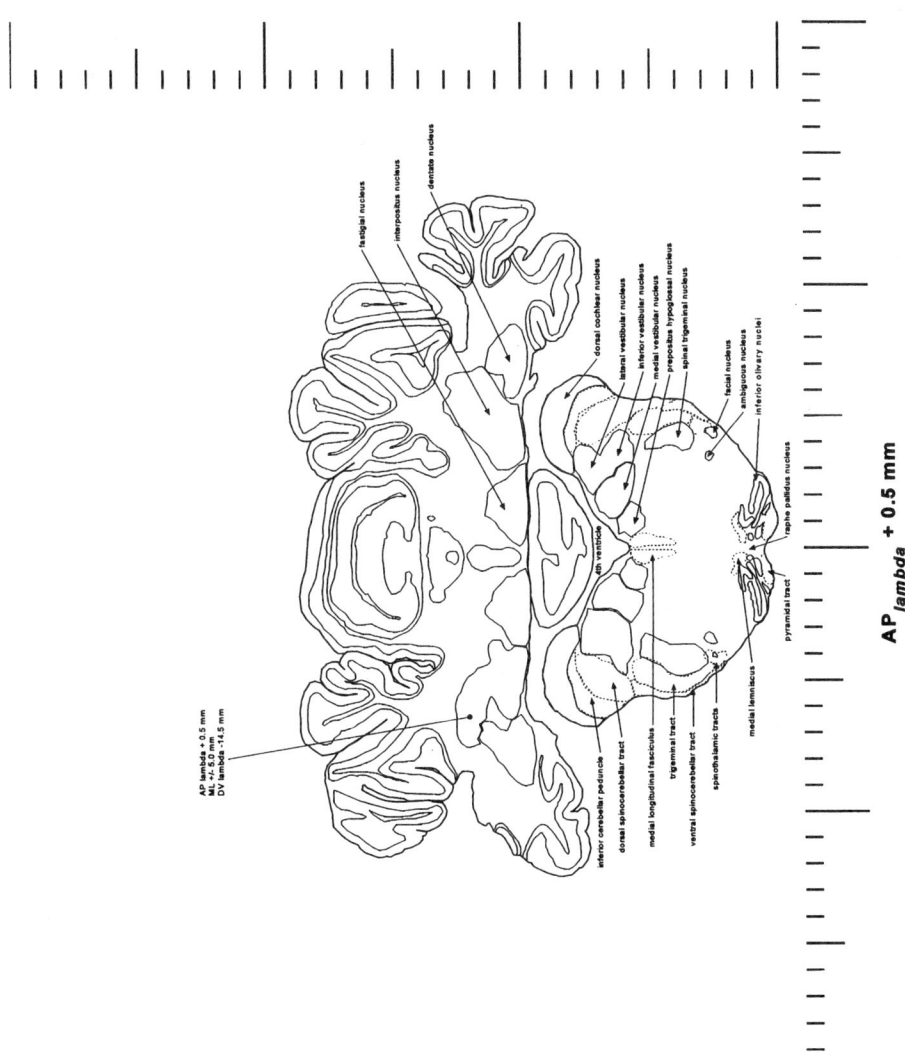

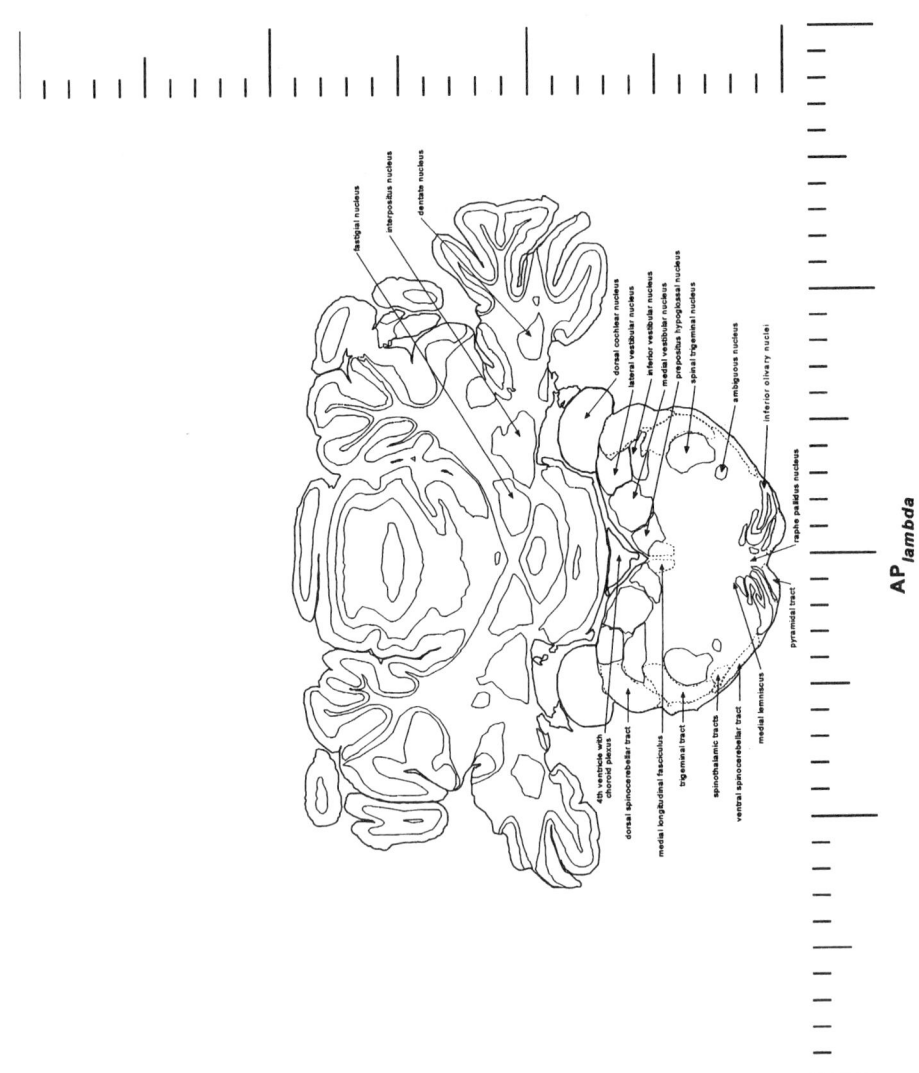

Appendix B

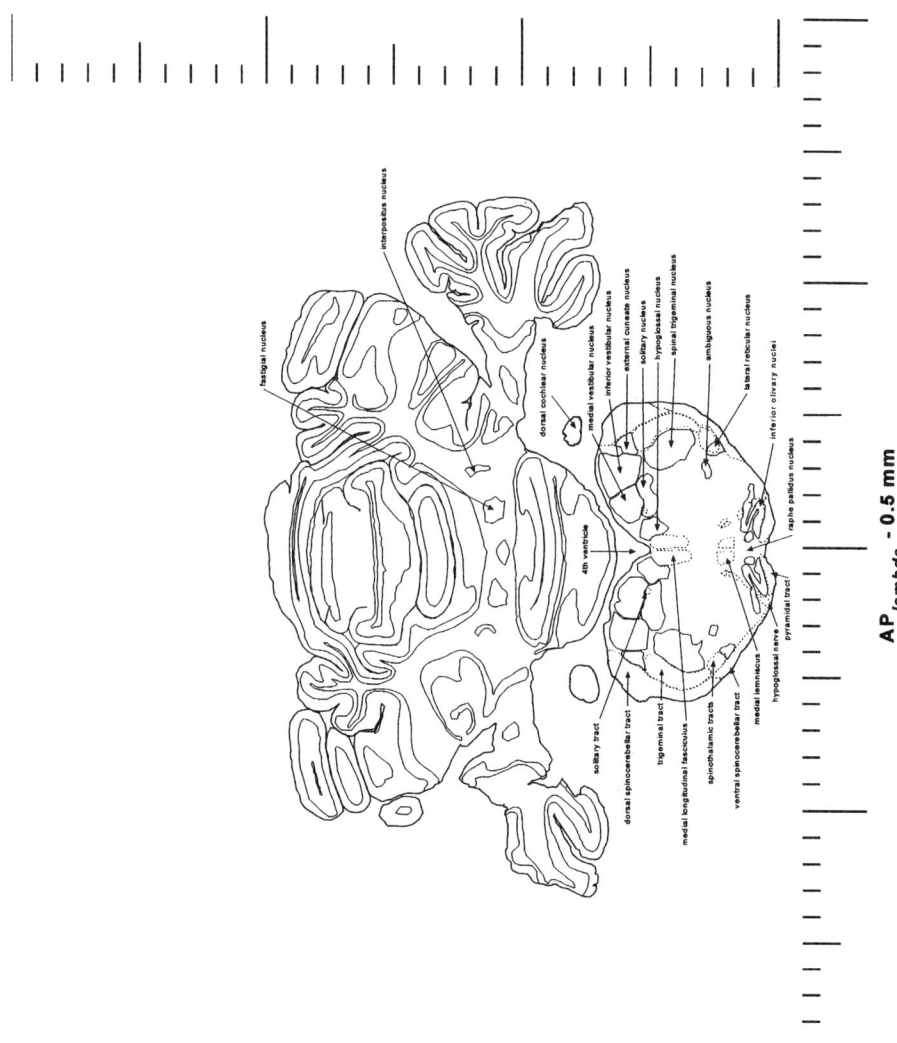

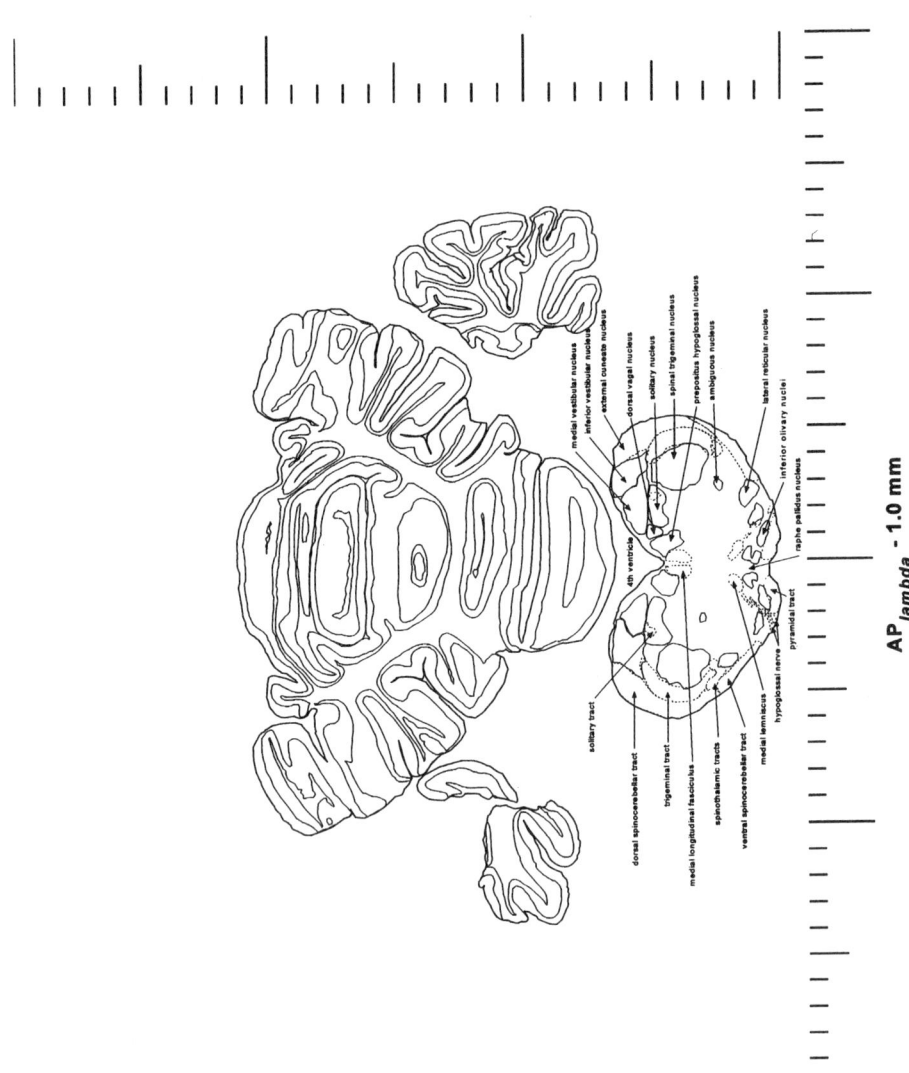

Appendix B

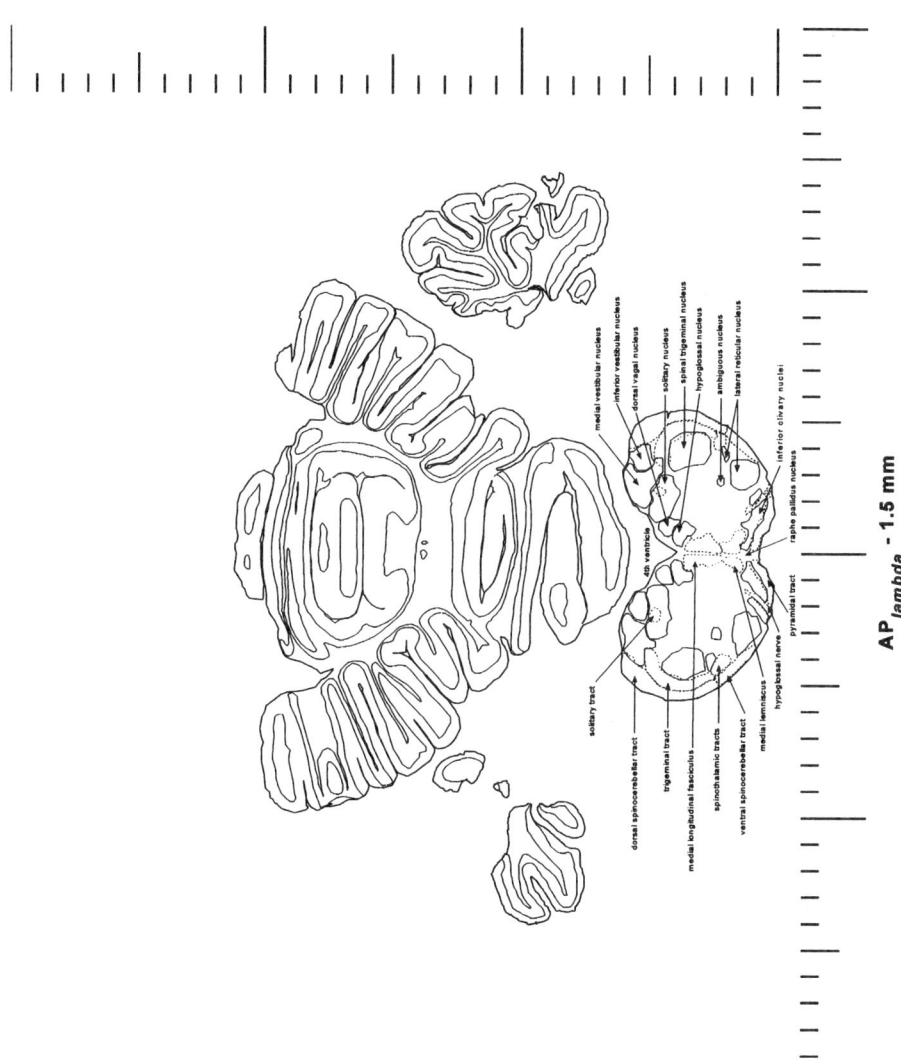

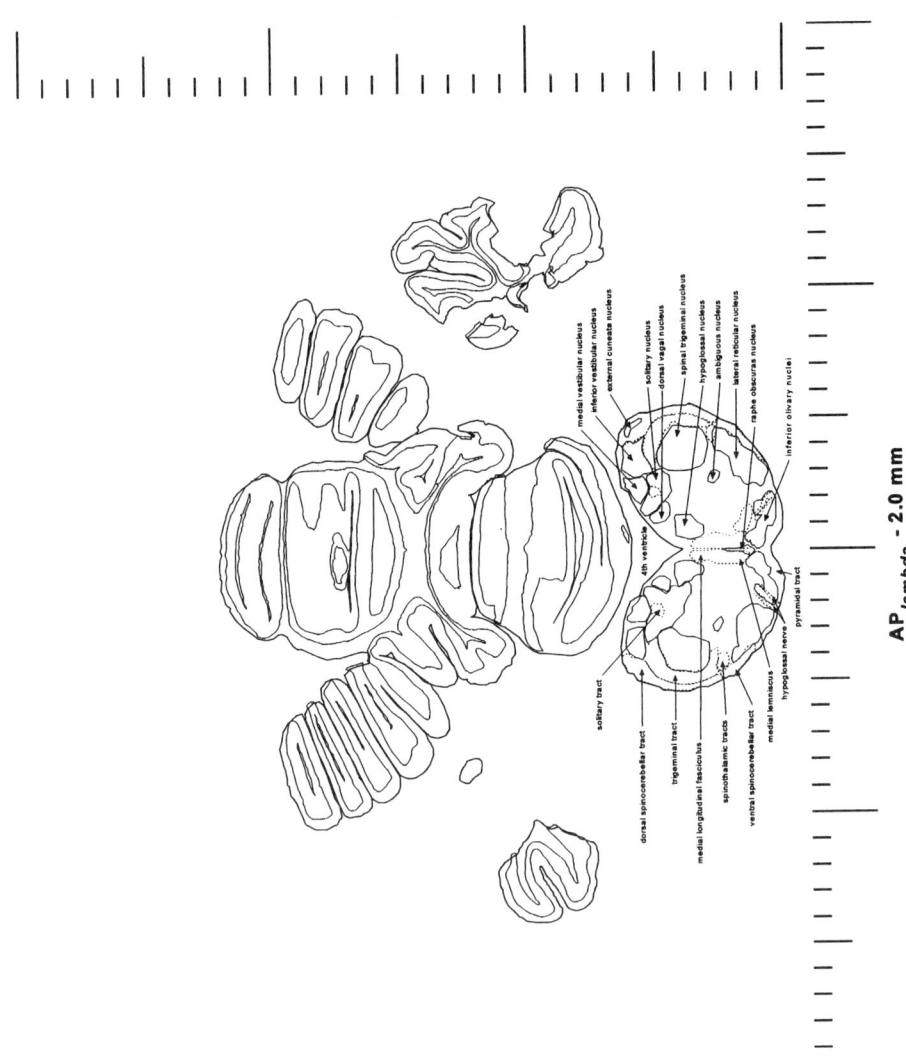

Appendix B

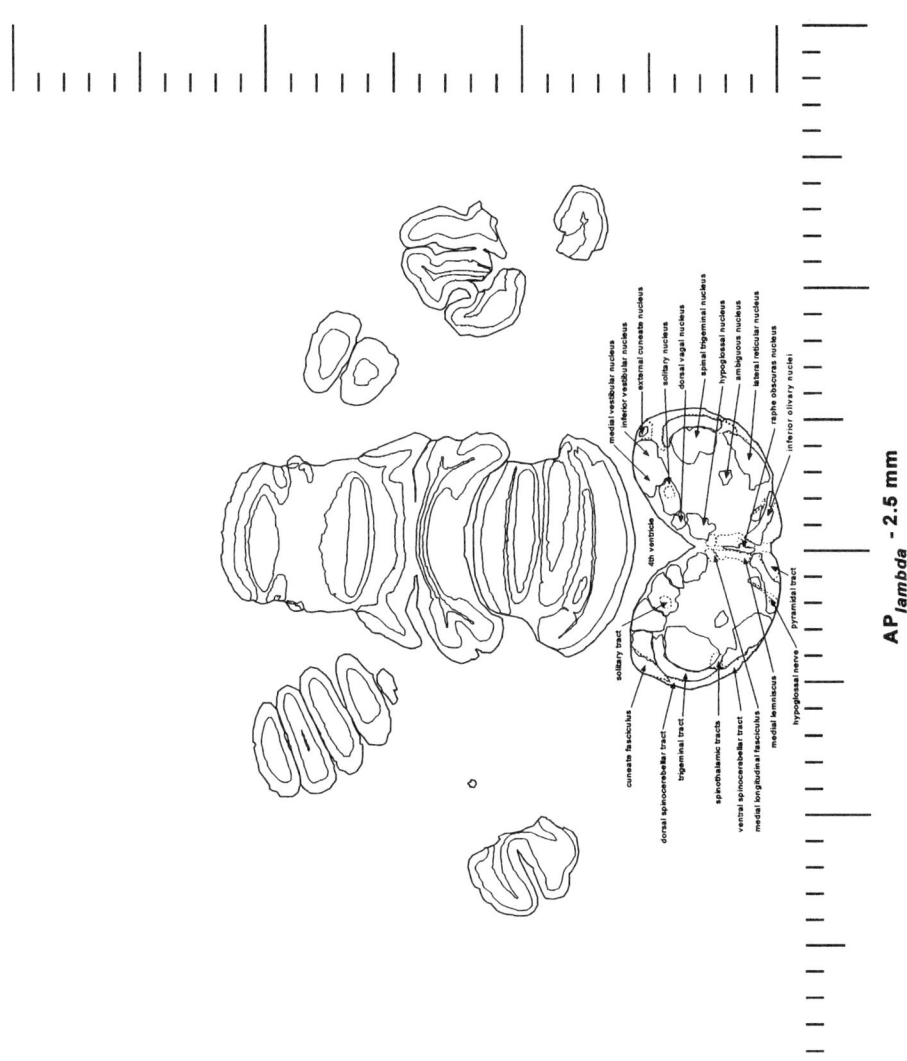

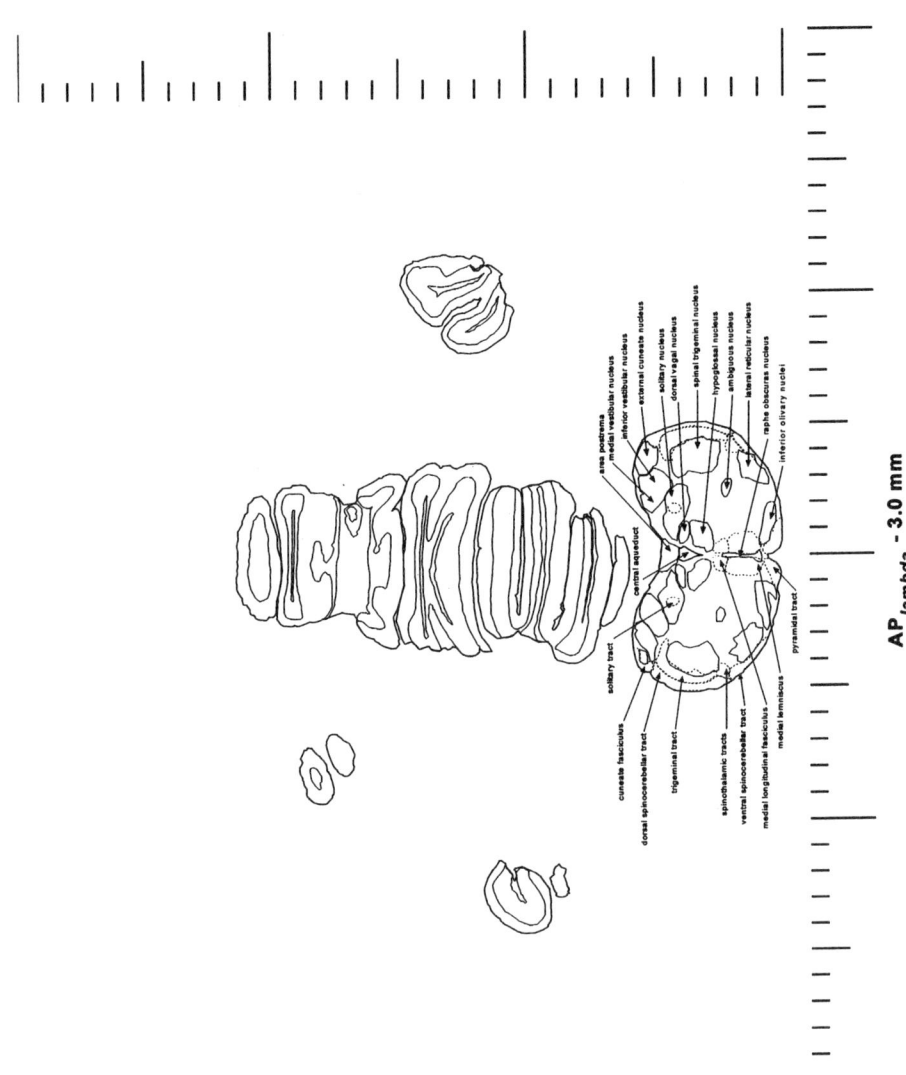

Appendix B

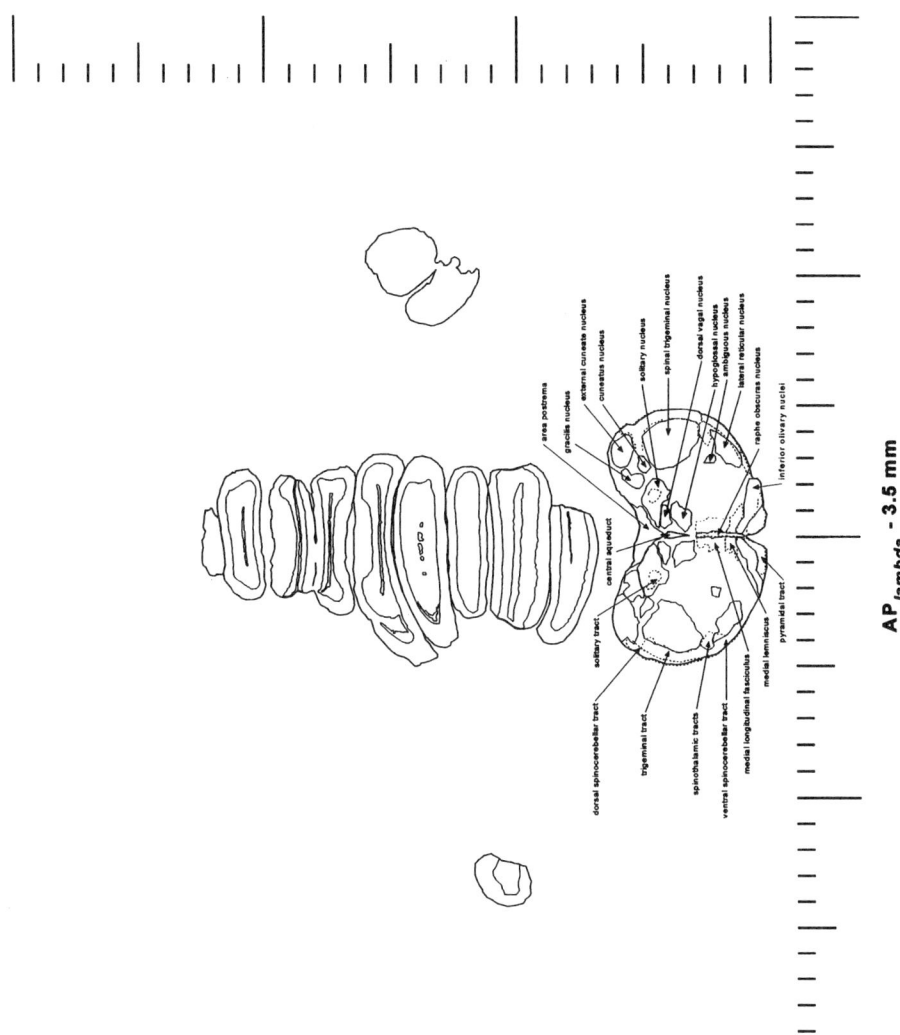

AP_{lambda} - 3.5 mm

REFERENCES

Adams, T. (1963). Hypothalamic temperature in the cat during feedings and sleeping. *Science, 139*, 609-610.

Anagnostaras, S.G., Maren, S. and Fanselow, M.S. (1999). Temporally graded retrograde amnesia of contextual fear after hippocampal damage in rats: Within-subjects examination. *Journal of Neuroscience, 19(3)*, 1106-1114.

Ashley, C.W. (1944). *The Ashley Book of Knots*. New York, Double Day.

Avanzi, V., Castilho, V.M., de Andrade, T.G.C.S. and Brandao, M.L. (1998). Regulation of contextual conditioning by the median raphe nucleus. *Brain Research, 790*, 178-184.

Beggs, A.L., Steinmetz, J.E., Romano, A.G. and Patterson, M.M. (1983). Extinction and retention of a classically conditioned flexor nerve response in acute spinal cats. *Behavioral Neuroscience, 97*, 530-540.

Berger, T. W., Laham, R. I. and Thompson, R. F. (1980). Hippocampal unit-behavior correlations during classical conditioning. *Brain Research, 193*, 229-248.

Berger, T. W., Rinaldi, P. C., Weisz, D. J. and Thompson, R. F. (1983). Single-unit analysis of different hippocampal cell types during classical conditioning of rabbit nictitating membrane response. *Journal of Neurophysiology, 50*, 1197-1219.

Berlin, H.M. (1977). *Design of Active Filters, With Experiments*. Indianapolis, Indiana, Howard W. Sams & Co., Inc.

Berman, A.L. and Jones, E.G. (1982). *The Thalamus and Basal Telencephalon of the Cat: A Cytoarchitectonic Atlas with Stereotaxis Coordinates*. Madison, Wisconsin, University of Wisconsin Press.

Berntson, G.G., Ault, R.T., and Walker, M. (1977). A simple IC pulse-pair stimulator. *Behavior Research Methods & Instrumentation, 9*, 18-21.

Biel, W.C. and Wickens, D.D. (1941). The effects of vitamin B1 deficiency on the conditioning of eyelid responses in the rat. *Journal of Comparative and Physiological Psychology, 32*, 329-340.

Bigon, M. and Regazzoni, G. (1981). *The Morrow Guide to Knots: For Sailing, Fishing, Camping, Climbing*. New York, Quill/William Morrow.

Blanchard, R.J. and Blanchard, D.C. (1969). Crouching as an index of fear. *Journal of Comparative & Physiological Psychology, 67*, 370-375.

Blankenship, M.R., Finn, P.R. and Steinmetz, J.E. (1998). A characterization of approach and avoidance learning in alcohol preferring (P) and non-preferring (NP) rats. *Alcohol: Clini-*

cal and Experimental Research, 22(6), 1227-1233.

Bliss, T.V.P. and Lomo, T. (1973). Long-lasting potentiation of synaptic transmission in the dentate area of anesthetized rabbit following stimulation of the perforant path. *Journal of Physiology (London), 232,* 331-356.

Brakel, S., Babb, T., Mahnke, J. and Verzeano, M. (1971). A compact amplifier for extracellular recording. *Physiology & Behavior, 6,* 731-733.

Broca, P. P. (1861). Remarques sur le siege de la faculté du language articule, suivies d'une observation d'aphemie. *Bulletin de la Société Anatomique, 36,* 330-357.

Brodal, A. (1940). The cerebellum of the rabbit: a topographical atlas of the folia as revealed in transverse sections. *Journal of Comparative Neurology, 72,* 63-81.

Brogden, W.J. and Culler, E. (1936). Device for the motor conditioning of small animals. *Science, 83,* 269-270.

Brooks, V.B. (1983). Study of brain function by local, reversible cooling. *Review of Physiology, Biochemistry and Pharmacology, 95,* 1-109

Bruner, A. (1969). Reinforcement strength during classical conditioning of leg flexion, freezing, and heart rate in cats. *Conditioning Reflex, 4,* 24-31.

Budworth, G. (1997). *The Hamyln Book of Knots.* London, Hamlyn.

Campeau, S. and Davis, M. (1992). Fear potentiation of the acoustic startle reflex using noises of various spectral frequencies as conditioned stimuli. *Animal Learning & Behavior, 20,* 177-186.

Campeau, S. and Davis, M. (1995). Involvement of the central nucleus and basolateral complex of the amygdala in fear conditioning measured with fear-potentiated startle in rats trained concurrently with auditory and visual conditioned stimuli. *Journal of Neuroscience, 15,* 2301-2311.

Capaldi, E.J., Alptekin, S. and Birmingham, K.M. (1996). Instrumental performance and time between reinforcements: Intimate relation to learning or memory retreival? *Animal Learning & Behavior, 24(2),* 211-220.

Cassady, J.M. (1996). Avoidance and classical conditioning of leg flexion in dogs. *Behavioural Brain Research, 77,* 79-99.

Cassady, J.M., Cole, M., Thompson, R.F. and Weinberger, N.M. (1973). Neural correlates of asymptotic avoidance and classical conditioned leg flexion. *Experimental Neurology, 40,* 207-215.

Chapman, P.F., Steinmetz, J.E., Sears, L.L. and Thompson, R.F. (1990). Effects of lidocaine injection in the interpositus nucleus and red nucleus on conditioned behavioral and neuronal responses. *Brain Research, 537,* 149-156.

Chen, G. and Steinmetz, J.E. (1998). A general purpose computer system for behavioral conditioning and neural recording experiments. *Behavior Research Methods, Instruments &*

Computers, 30(3), 384-391.

Chen, G. and Steinmetz, J.E. (2000). Intra-cerebellar infusion of NMDA receptor antagonist AP5 disrupts classical eyeblink conditioning in rabbits. *Brain Research, 887*, 144-156.

Cholewiak, R.W. & Wollowitz, M. (1992). The design of vibrotactile transducers. In I. Summers (Ed.), *Tactile Aids for the Hearing Impaired*. London, Whurr Ltd., pp. 57-82.

Ciancone, M.T. and Rebec, G.V. (1989). A simple device for the reliable production of varnish-insulated, high-impedance tungsten microelectrodes. *Journal of Neuroscience Methods, 27*, 77-79.

Clark, G. and Clark, M.P. (1971). *A Primer in Neurological Staining Procedures*. Springfield, Illinois, Charles C. Thomas.

Clark, G.A., McCormick, D.A., Lavond, D.G., & Thompson, R.F. (1984). Effects of lesions of cerebellar nuclei on conditioned behavioral and hippocampal neuronal responses. *Brain Research, 291*, 125-136.

Clark, R.E. and Lavond, D.G. (1993). Reversible lesions of the red nucleus during acquisition and retention of a classically conditioned behavior in rabbits. *Behavioral Neuroscience, 107*, 264-270.

Clark, R.E. and Squire, L.R. (1998). Classical conditioning and brain systems: The role of awareness. *Science, 280*, 77-81.

Clark, R.E. and Squire, L.R. (2000). Awareness and the conditioned eyeblink response. In D. S. Woodruff-Pak and J.E. Steinmetz (Eds.), *Eyeblink Classical Conditioning, Vol 1: Human Applications*, Boston, Kluwer, pp. 229-251.

Clark, R.E., Zhang, A.A. and Lavond, D.G. (1992). Reversible lesions of the cerebellar interpositus nucleus during acquisition and retention of a classically conditioned behavior. *Behavioral Neuroscience, 106*, 879-888.

Cooley, R.K. and Vanderwolf, C.H. (1979). *The Sheep Brain: A Basic Guide*. London, Ontario, A.J. Kirby Co.

Coolidge, B.J. and Howard, R.M. (1979). *Animal Histology Procedures of the Pathological Technology Section of the National Cancer Institute*. Washington, D.C., NIH Publication.

Craig, J.C., and Sherrick, C.E. (1982). Tactile perception. In W. Schiff and E. Foulke (Eds.), *Tactile Perception: A sourcebook*. Cambridge: Cambridge University Press, pp. 209-233.

Cruikshank, S.J., Edeline, J.-M. and Weinberger, N.M. (1992). Stimulation at a site of auditory-somatosensory convergence in the medial geniculate nucleus is an effective unconditioned stimulus for fear conditioning. *Behavioral Neuroscience, 106*, 471-483.

Davidson, H.L. (1992). *Troubleshooting & Repairing Audio & Video Cassette Players and Recorders*. New York, TAB Books.

Davis, M., Falls, W. A., Campeau, S. and Kim, M. (1993). Fear-potentiated startle: a neural and pharmacological analysis. *Behavioural Brain Research, 58*, 175-198.

Dearmond, S.J., Fusco, M.M. and Dewey, M.M. (1989). *Structure of the Human Brain: A Photographic Atlas*. New York, Oxford University Press.

Deaux, E.B. and Gormezano, I. (1963). Eyeball retraction: classical conditioning and extinction in the albino rabbit. *Science, 141*, 630-631.

Desmedt, A., Garcia, R. and Jaffard, R. (1998). Differential modulation of changes in hippocampal-septal synaptic excitability by the amygdala as a function of either elemental or contextual fear conditioning in mice. *Journal of Neuroscience, 18*, 480-487.

Dicara, L.V., Weaver, L. and Wolf, G. (1974). Comparison of DC and RF for lesioning white and grey matter. *Physiology & Behavior, 12*, 1087-1090.

Disterhoft, J.F., Kwan, H.H. and Lo, W.D. (1977). Nictitating membrane conditioning to tone in the immobilized albino rabbit. *Brain Research, 137*, 127-143.

Domjan, M. (2000). *Principles of Learning and Behavior*, Fourth Edition.. New York: Wadsworth Publishing.

Duarte, L.A. (1996). *The Microcontroller Beginner's Handbook*, Second Edition. Indianapolis, Indiana, Prompt Publications.

Durrant, J.D. and Lovrinic, J.H. (1984). *Bases of Hearing Science*. Baltimore, Maryland, Williams & Wilkins.

Eichenbaum, H., Stewart, C. and Morris, R. G. M. (1990). Hippocampal representation in place learning. *Journal of Neuroscience, 10*, 3531-3542.

Ellison, G.D. (1972). The use of microknives in brain lesion studies and production of isolated brain-stem islands. In R.D. Myers (Ed.), *Methods in Psychobiology* (Vol. 2). New York, Academic Press, pp. 304-312.

Fancher, R.E. (1979). *Pioneers of Psychology*, Third Edition. New York: W.W. Norton & Co.

Fanselow, M.S. (2001). Toward a neurobiology of functional behavioral systems: Contrasting Pavlovian emotional and motor learning. In: J.E. Steinmetz, M.A. Gluck, and P.R. Solomon (Eds.), *Model systems and the neurobiology of associative learning: A festschrift in honor of Richard F. Thompson*. Mahwah, New Jersey: Lawrence Erlbaum Associates, pp. 379-393.

Furedy, J.J. (1992). Pavlov, Ivan. In L. Squire (Ed.), *Encyclopedia of Learning and Memory*. New York, Macmillan, pp. 513-516.

Gabriel, M. and Talk, A.C. (2001). A tale of two paradigms: Lessons learned from parallel studies of discriminative instrumental learning and classical eyeblink conditioning. In: J.E. Steinmetz, M.A. Gluck and P.R. Solomon (Eds.), *Model systems and the neurobiology of associative learning: A festschrift in honor of Richard F. Thompson*. Mahwah, New Jersey: Lawrence Erlbaum Associates, pp. 149-185.

Girgis, M. and Shih-Chang, W. (1981). *A New Stereotaxic Atlas of the Rabbit Brain*. St. Louis, Warren H. Green Inc.

Girgorian, N.A. (1974), Pavlov, Ivan Petrovich. *Dictionary of Scientific Biography, 10,* 431-435, New York, Scribner.

Gormezano, I. (1966). Classical conditioning. In J.B. Sidowski (Ed.), *Experimental Methods and Instrumentation in Psychology.* New York: McGraw-Hill, pp 385-420.

Gormezano, I. (1972). Investigations and defense and reward conditioning in the rabbit. In A. H. Black and W.F. Prokasy (Eds.), *Classical Conditioning II: Current Research and Theory.* New York, Appleton-Century-Croft, pp 151-181.

Gormezano, I., Kehoe, E.J. and Marshall, B.S. (1983). Twenty years of classical conditioning with the rabbit. *Progress in Psychobiology and Physiological Psychology, 10,* 197-275.

Gormezano, I., Schneiderman, N., Deaux, E. and Fuentes, I. (1962). Nictitating membrane: classical conditioning and extinction in the albino rabbit. *Science, 138,* 33-34.

Gould, T.J., Sears, L.L. and Steinmetz, J.E. (1991a). Techniques used for neural recording from awake, behaving or anesthetized animals: Part I. *DKI Carrier, 29,* 1-5.

Gould, T.J., Sears, L.L. and Steinmetz, J.E. (1991b). Techniques used for neural recording from awake, behaving or anesthetized animals: Parts II. *DKI Carrier, 30,* 1-5.

Gould, T.J., Sears, L.L. and Steinmetz, J.E. (1993). Possible CS and US pathways for rabbit classical eyelid conditioning: Electrophysiological evidence for projections from the pontine nuclei and inferior olive to cerebellar cortex and nuclei. *Behavioral and Neural Biology, 60,* 172-185.

Gould, T.J. and Steinmetz, J.E. (1996). Changes in rabbit cerebellar cortical and interpositus nucleus activity during acquisition, extinction and backward classical conditioning. *Neurobiology of Learning and Memory, 65,* 17-34.

Green, J.T., Rogers, R.F., Goodlett, C.R. and Steinmetz, J.E. (2000). Impairment in eyeblink classical conditioning in adult rats exposed to ethanol as neonates. *Alcoholism: Clinical and Experimental Research,, 24,* 438-447.

Gustafsson, B. and Jankowski, E. (1976). Direct and indirect activation of nerve cells by electrical impulses applied extracellularly. *Journal of Physiology (London), 258,* 33-61.

Haines, D. (1991). *Neuroanatomy: An Atlas of Structures, Sections and Systems.* Baltimore, Maryland, Williams & Wilkins.

Hart, B.L. (1969). *Experimental Neuropsychology: A Laboratory Manual.* San Francisco, W.H. Freeman and Company.

Hays, W.L. (1988). *Statistics,* Fourth Edition. New York, Holt, Rinehart & Winston.

Hendelman, W.J. and Morrissey, J-P. (1988). *A Student's Atlas of Neuroanatomy,* 2nd edition. Ottawa, Canada, Univeristy of Ottawa Press.

Hilgard, E.R. (1931). Conditioned eyelid reactions to a light stimulus based upon the reflex wink to a sound. *Psychological Monographs, 41,* 184.

Hilgard, E.R., (1933). Modifications of reflexes and conditioned reaction. *Journal of General Psychology, 9*, 210-215.

Hilgard, E.R. and Marquis, D.G. (1936). Conditioned eyelid responses in monkeys with a comparison of dog: monkey and man. *Psychological Monographs, 47*, 186-198.

Horn, D.T. (1994). *Build Your Own Low-Cost Signal Generator*. New York, TAB Books.

Hughes, D. and Schlosberg, H. (1938). Conditioning in the white rat. IV. The conditioned lid reflex. *Journal of Experimental Psychology, 23*, 641-650.

Hughling-Jackson, J. (1931). In J. Taylor (Ed.), *Selected Writing of John Hughling-Jackson*. London: Hodder.

Ivkovich, D., Eckerman, C.O., Krasnegor, N.A. and Stanton, M.E. (2000). Using eyeblink conditioning to assess neurocognitive development in human infants. In: D.S. Woodruff-Pak & J.E. Steinmetz (Eds), *Eyeblink Classical Conditioning, Vol 1: Human Applications*. Boston, Kluwer, pp. 119-142.

Jacobson, C. (1999). *Knots for the Outdoors*. Old Saybrook, Connecticut, The Globe Pequot Press.

Janssen, R. (1992). Thermal influences on nervous system function. *Neuroscience and Biobehavioral Reviews, 16*, 399-413.

Jung, M.W., Wiener, S.I. and McNaughton, B.L. (1994). Comparison of spatial firing characteristics of units in dorsal and ventral hippocampus of the rat. *Journal of Neuroscience, 14*, 7347-7356.

Katz, D.B. and Steinmetz, J.E. (1995a). Analyzing behavior-related neural activity via correlation of distribution shape: Part I. *DKI Carrier, 42*, 1-4.

Katz, D.B. and Steinmetz, J.E. (1995b). Analyzing behavior-related neural activity via correlation of distribution shape: Part II. *DKI Carrier, 43*, 1-5.

Katz, D.B. and Steinmetz, J.E. (1997). Single-unit evidence for eyeblink conditioning in cerebellar cortex is altered, but not eliminated, by interpositus nucleus lesions. *Learning & Memory, 4(1)*, 88-104.

Kim, J.J., Decola, J.P., Landeria-Fernandez, J. and Fanselow, M.S. (1991). N-methyl-D-asparate receptor antagonist APV blocks acquisition but not expression of fear conditioning. *Behavioral Neuroscience, 105*, 126-133.

Krupa, D.J. and Thompson, R.F. (1995). Inactivation of the superior cerebellar peduncle blocks expression but not acquisition of the rabbit's classically conditioned eye-blink response. *Proceedings of the National Academy of Sciences of the United States of America, 92*, 5097-5101.

Krupa, D.J. & Thompson, R.F. (1997). Reversible inactivation of the cerebellar interpositus nucleus completely prevents acquisition of the classically conditioned eye-blink response. *Learning & Memory, 3*, 545-556.

References

Krupa, D.J., Thompson, J.K. and Thompson, R.F. (1993). Localization of a memory trace in the mammalian brain. *Science, 260,* 989-991.

LaBossiere, E. and Glickstein, M. (1976). *Histological Processing for the Neural Sciences.* Springfield, Illinois, Charles C. Thomas.

Lancaster, D. (1975). *Active-Filter Cookbook.* Indianapolis, Indiana, Howard W. Sams & Co., Inc.

Lancaster, D.E. (1988). *CMOS Cookbook.* Revised by Howard M. Berlin. Carmel, Indiana, Howard W. Sams & Co.

Lattal, K.M. and Abel, T. (2001). Different Requirements for protein synthesis in acquisition and extinction of spatial preferences and context-evoked fear. *Neuroscience, 21,* 5773-5780.

Lauber, S.M. (1970). Brain histology for neuropsychologists. *Psychonomic Monograph Supplements, 3,* 95-102.

Lavond, D.G., Logan, C.G., Sohn, J.H., Garner, W.D.A. and Kanzawa, S.A. (1990). Lesions of the cerebellar interpositus nucleus abolish both nictitating membrane and eyelid EMG conditioned responses. *Brain Research, 514,* 238-248.

Lavond, D.G. and Steinmetz, J.E. (1989). An interface for the IBM-XT/AT and compatibles. *Behavior Research Methods, Instruments, & Computers, 21,* 435-440.

Lee, H.J., Choi, J.-S., Brown, T.H. and Kim, J.J. (2001). Amygdalar NMDA receptors are critical for the expression of multiple conditioned fear responses. *Journal of Neuroscience, 21,* 4116-4124.

Mahoney, W.J. and Ayres, J.J.B. (1976). One-trial simultaneous and backward fear conditioning as reflected in conditioned suppression of licking in rats. *Animal Learning & Behavior, 4,* 357-362.

Mai, J.K. (1997). *Atlas of the Human Brain.* New York, Academic Press.

March, R.R., Hoffman, H.S. and Stitt, C.L. (1979). Eye blink elicitation and measurement in the human infant. *Behavior Research Methods & Instrumentation, 11,* 489-502.

Maren, S. (1998). Overtraining does not mitigate contextual fear conditioning deficits produced by neurotoxic lesions of the basolateral amygdala. *Journal of Neuroscience, 18(8),* 3088-3097.

Markham, M.R. (1993). An interface for controlling external devices via the IPB PC/XT/AT parallel port. *Behavior Research Methods, Instruments, & Computers, 25,* 477-478.

Marston, R. (1988). Working with OTA's: How to use operational transconductance amplifiers in your designs and projects. *Radio Electronics,* May, 63-68.

Marston, R. (1988). Working with OTA's: The LM13600 is a second-generation dual operational transconductance amplifier that can be used as a voltage-controlled amplifier, resistor, filter, or oscillator. *Radio Electronics,* July, 61-65.

Martin, J.H. (1989). *Neuroanatomy: Text and Atlas*. New York, Elsevier.

Mauk, M.D., Steinmetz, J.E. and Thompson, R.F. (1986). Classical conditioning using stimulation of the inferior olive as the unconditioned stimulus. *Proceedings of the National Academy of Sciences, 83*, 5349-5353.

McBride, R.L., and Klemm, W.R. (1968). Stereotaxic Atlas of Rabbit Brain, Based on the Rapid Method of Photography of Frozen, Unstained Sections. *Communications in Behavioral Biology, Part A, 2*, 179-215.

McGregor, I.S. (1996). Using Strawberry Tree Workbench Mac and Workbench PC software for data acquisition and control in the animal learning laboratory. *Behavior Research Methods, Instruments, & Computers, 28*, 38-48.

Meesen, H. & Olszewski, J. (1949). *A Cytoarchitectonic Atlas of the Rhombencephalon of the Rabbit*. New York, Karger.

Merrill, K.A., Steinmetz, J.E., Viken, R.J. and Rose, R.J. (1999). Genetic influences on human conditionability: A twin study of the conditioned eyeblink response. *Behavioral Genetics, 29*, 95-101.

Meyer, D.R., Cho, C. and Wesemann, A.F. (1960). On problems of conditioning disciminated lever-press avoidance responses. *Psychological Review, 67(4)*, 224-228.

Mims, F.M. (1982). *Engineer's Notebook*. Fort Worth, Texas, Tandy Corporation.

Mims, F.M. (1992). *Engineer's Notebook*. Solana Beach, California, Hightext Publications, Inc.

Mintz, M., Lavond, D.G., Zhang, A.A., Yun, Y. and Thompson, R.F. (1994). Unilateral inferior olive NMDA lesion leads to unilateral deficit in acquisition and retention of eyelid classical conditioning. *Behavioral and Neural Biology, 61*, 218-224.

Myers, R.D. (Ed.) (1982). *Methods in Psychobiology, Volume 1: Laboratory Techniques in Neuropsychology and Neurobiology*. New York, New York, Academic Press.

McCormick, D.A., Lavond, D.G. and Thompson, R.F. (1982). Concomitant classical conditioning of the rabbit nictitating membrane and eyelid responses: Correlations and implications. *Physiology & Behavior, 28*, 769-775.

Nicolelis, M.A.L. (Ed.) (1998). *Methods for Neural Ensemble Recordings*. New York: CRC Press.

Nieuwenhuys, R., Voogd, J. and van Huijzen, C. (1988). *The Human Central Nervous System: A Synopsis and Atlas*. New York, Springer-Verlag.

Oakley, B. and Schafer, R. (1978). *Experimental Neurology: A Laboratory Manual*. Ann Arbor, Michigan, The University of Michigan Press.

Oleksy, J.E. (1981). *Practical Solid-state Circuit Design*, Second Edition. Indianapolis, Indiana, Howard W. Sams & Co., Inc., 133-134.

Olszewski, J & Baxter, D. (1954). *Cytoarchitecture of the Human Brain Stem.* New York, Karger.

Olton, D.S., Becker, J.T. and Handelmann, G.E. (1979). Hippocampus, space and memory. *Behavioral and Brain Sciences, 2*, 313-365.

Overmier, J.B. (1966). Differential transfer of control of avoidance responses as a function of UCS duration. *Psychonomic Science, 5*, 25-26.

Palya, W.L. and Walter, D.E. (1993). A powerful, inexpensive experiment controller for IBM PC interface and experiment control language. *Behavior Research Methods, Instruments, & Computers, 25*, 127-136.

Palya, W.L., Walter, D.E. and Chu, J.Y.M. (1995). An inexpensive 1-millisecond experiment control interface for IBM PCs and its user-friendly control language. *Behavior Research Methods, Instruments, & Computers, 27*, 129-130.

Papka, M., Ivry, R.B. and Woodruff-Pak, D.S. (1995). Selective disruption of eyeblink classical conditioning by concurrent tapping. *NeuroReport, 6*, 1493-1497.

Pavlov, I.P. (1927). *Conditioned Reflexes.* Oxford: Oxford University Press.

Paxinos, G. and Watson, C. (1982). *The Rat Brain in Stereotaxic Coordinates.* New York, Academic Press.

Penttonen, M. and Korhonen, T. (1991). Asymmetries in classically conditioned head movements and cingulate cortex slow potentials in cats. *International Journal of Neuroscience, 61*, 121-134.

Penttonen, M., Salmi, M., Hamalainen, P. and Meriluoto, J. (1994). A microcomputer system for controlling classical conditioning experiments. *Behavior Research Methods, Instruments, & Computers, 26*, 447-453.

Polenchar, B.E., Patterson, M.M., Lavond, D.G. and Thompson, R.F. (1985). Cerebellar lesions abolish an avoidance response in rabbit. *Behavioral and Neural Biology, 44*, 221-227.

Port, R., Curtis, K., Inoue, C., Briggs, J. and Seybold, K. (1993). Hippocampal damage does not impair instrumental appetitive conditioning with delayed reinforcement. *Brain Research Bulletin, 30*, 41-44.

Presnell, J.K. and Schreibman, M.P. (1997). *Humason's Animal Tissue Techniques.* Baltimore, Maryland, Johns Hopkins University Press.

Prince, J.H. (Ed.) (1964). *The Rabbit in Eye Research.* Springfield, Illinois, Charles C. Thomas.

Rebec, G.V. (1998). Real-time assessments of dopamine function during behavior: Single-unit recording, iontophoresis, and fast-scan cyclic voltammetry in awake unrestrained rats. *Alcoholism: Clinical & Experimental Research, 22(1)*, 32-40.

Reynolds, R.W. (1965). An irritative hypothesis concerning the hypothalamic regulation of food intake. *Psychological Review, 72*, 105-116.

Rieke, F., Warland, D., de Ruyter van Steveninck, R. and Bialek, W. (1997). *Spikes: Exploring the Neural Code.* Cambridge, MA: MIT Press.

Roemer, R.A., Cegavske, C.F., Thompson, R.F. and Patterson, M.M. (1975). An acquisition and analysis system for on-line experiements on the neurophysiology of learning. *Behavior Research Methods & Instrumentation, 7,* 157-161.

Romano, A.G. and Patterson, M.M. (1987). The rabbit in Pavlovian conditioning. In I Gormezano, W. F. Prokasy and R. F. Thompson (eds.), *Classical Conditioning*, Third Edition. Hillsdale, NJ, Lawrence Erlbaum Associates, pp. 1-36.

Romano, A.G., Steinmetz, J.E. and Patterson, M.M. (1980). A versatile cat-restraint system. *Behavior Research Methods & Instrumentation, 12,* 455-458.

Rosen, D.J., Steinmetz, J.E. and Thompson, R.F. (1989). Classical discrimination conditioning of the rabbit's eyelid response using pontine stimulation as a conditioned stimulus. *Behavioral & Neural Biology, 52,* 51-62.

Scandrett, J. and Gormezano, I. (1980). Microprocessor control and A/D data acquisition in classical conditioning. *Behavior Research Methods & Instrumentation, 12,* 120-125.

Schafe, G.E. and Le Doux, J.E. (2000). Memory consolidation of auditory pavlovian fear conditioning requires protein synthesis and protein kinase A in the amygdala. *Journal of Neuroscience, 20,* 1-5.

Schaltenbrand, G. & Wahren, W. (1977). *Atlas for Stereotaxy of the Human Brain.* Stuttgart, Georg Thieme Publishers.

Schmajuk, N.A. and Christiansen, B.A. (1990). Eyeblink conditioning in rats. *Physiology & Behavior, 48,* 755-758.

Sears, L.L. and Steinmetz, J.E. (1990). Acquisition of classically conditioned-related activity in the hippocampus is affected by lesions of the cerebellar interpositus nucleus. *Behavioral Neuroscience, 104,* 681-692.

Sharp, P.E. and La Regina, M.C. (1998). *The Laboratory Rat.* New York, CRC Press.

Sheehan, D.C. and Hrapchak, B.A. (1980). *Theory and Practice of Histotechnology.* Columbus, Battelle Press.

Simmons, D.M. and Swanson, L.W. (1993). The Nissl stain. *Brain Research Protocols, 93-050-12,* 1-7.

Skinner, J.E. (1971). *Neuroscience: A Laboratory Manual.* Philadelphia, W.B. Saunders Company.

Skinner, J.E. and Lindsley, D.B. (1967). Electrophysiological and behavioral effects of blockade of the nonspecific thalamo-cortical system. *Brain Research, 6,* 95-118.

Skinner, J.E. and Lindsley, D.B. (1968). Reversible cryogenic blockade of neural function in the brain of unrestrained animals. *Science, 161*, 595-597.

Solomon, P.R. and Babcock, B.A. (1979). KIM and the rabbit: The use of the KIM-1 microprocessor to control classical conditioning of the rabbit's nictitating membrane response. *Behavior Research Methods & Instrumentation, 11*, 67-70.

Solomon, P.R., Lewis, J.L., LoTurco, J.J., Steinmetz, J.E. and Thompson, R.F. (1986). The role of the middle cerebellar peduncle in acquisition and retention of the rabbit's nictitating membrane response. *Bulletin of the Psychonomic Society, 24*, 75-78.

Solomon, P.R., Weisz, D.J., Clark, G.A., Hall, J. and Babcock, B.A. (1983). A microprocessor control system and solid state interface for controlling electrophysiological studies on conditioning. *Behavior Research Methods & Instrumentaiton, 15*, 57-65.

Steinmetz, J.E. (1990). Neuronal activity in the cerebellar interpositus nucleus during classical NM conditioning with a pontine stimulation CS. *Psychological Science, 1*, 378-382.

Steinmetz, J.E., Blankenship, M.R., Green, J.T., Smith, G.B. and Finn, P.R. (2000). Evaluation of behavioral disinhibition in P/NP and HAD/LAD rats. *Progress in Neuro-Psychopharmacology & Biological Psychiatry, 24*, 1025-1039

Steinmetz, J.E., Gluck, M.A. and Solomon, P.R. (2001). *Model systems and the Neurobiology of Associative Learning: A festschrift in honor of Richard F. Thompson.* Mahwah, New Jersey: Lawrence Erlbaum Associates.

Steinmetz, J.E., Lavond, D.G. and Thompson, R.F. (1985). Classical conditioning of the rabbit eyelid response with mossy fiber stimulation as the conditioned stimulus. *Bulletin of the Psychonomic Society, 23*, 245-248.

Steinmetz, J.E., Lavond, D.G. and Thompson, R.F. (1989). Classical conditioning in rabbits using pontine nucleus stimulation as a conditioned stimulus and inferior olive stimulation as an unconditioned stimulus. *Synapse, 3*, 225-233.

Steinmetz, J.E., Logue, S.F. and Miller, D.P. (1993). Using signalled bar-pressing tasks to study the neural substrates of appetitive and aversive learning in rats: Behavioral manipulations and cerebellar lesions. *Behavioral Neuroscience, 107*, 941-954.

Steinmetz, J.E. and Patterson, M.M. (1985). Fixation of spinal reflex alterations in rats by sensory nerve stimulation. *Behavioral Neuroscience, 99*, 97-108.

Steinmetz, J.E., Rosen, D.J., Chapman, P.F., Lavond, D.G. and Thompson, R.F. (1986). Classical conditioning of the rabbit eyelid response with a mossy fiber stimulation CS. I. Pontine nuclei and middle cerebellar peduncle stimulation. *Behavioral Neuroscience, 100*, 871-880.

Steinmetz, J.E., Rosen, D.J., Woodruff-Pak, D.S., Lavond, D.G. and Thompson, R.F. (1986). Rapid transfer of training occurs when direct mossy fiber stimulation is used as a conditioned stimulus for classical conditioning. *Neuroscience Research, 3*, 606-616.

Suckow, M.A. and Douglas, F.A. (1997). *The Laboratory Rabbit.* New York, CRC Press.

Swanson, L.W. (1992). *Brain Maps: Structure of the Rat Brain.* New York, Elsevier.

Tasaki, I. (1959). Conduction of the nerve impulse. In J. Field (Ed.), *Handbook of Physiology: Volume I, Neurophysiology*. Washington, D.C., American Physiological Society, pp. 75-121

Thompson, L.T., Moyer, J.R., Akase, E. and Disterhoft, J.F. (1994). A system for quantitative analysis of associative learning. Part 1. Hardware interfaces with cross-species applications. *Journal of Neuroscience Methods, 54*, 109-117.

Thompson, R.F., Berger, T.W., Cegavske, C.F., Patterson, M.M., Roemer, R., Teyler, T.J. and Young, R.A. (1976). The search for the engram. *American Psychologist, 31*, 209-227.

Tracy, J.A., Ghose, S.S., Stetcher, T., McFall, R.M and Steinmetz, J.E. (1999). Classical conditioning in a nonclinical obsessive-compulsive population. *Psychological Science, 10(1)*, 9-13.

Van Way III, C.W. and Buerk, C.A. (1986). *Pocket Manual of Basic Surgical Skills*. St. Louis, Missouri, The C.V. Mosby Company.

Vincent, S.B. (1912). The function of the vibrissae in the behavior of the white rat. *Behavior Monographs, 1*, 1-81.

Wallis, D. I. (Ed.) (1993). *Electrophysiology: A Practical Approach*. New York: IRL Press.

Weiss, O. (1910). Die zeitliche Dauer Des Lidschlages. *Zsch F Psychol U Physiol D Sinnesorg, 45, Abstract II*, 307-312.

Weisz, D.J., Harden, D.G. and Xiang, Z. (1992). Effects of amygdala lesions on reflex faciliation and conditioned response acquisition during nictitating membrane response conditioning in rabbit. *Behavioral Neuroscience, 106*, 262-273.

Wickens, D., Meyers, P. and Sullivan, S. (1961). Classical GSR conditioning, conditioned discrimination, and interstimulus intervals in cats. *Journal of Comparative and Physiological Psychology, 54*, 572-576.

Wickens, D.D., Nield, A.F., Tuber, D.S. and Wickens, C.D. (1973). Stimulus selection as a function of CS1-CS2 interval in compound classical conditioning of cats. *Journal of Comparative and Physiological Psychology, 85*, 295-303.

Windholz, G. (1991). I.P. Pavlov as a youth. *Integrative Physiological and Behavioral Science, 26(1)*, 51-67.

Woodruff-Pak, D.S. and Steinmetz, J.E. (2000a). *Eyeblink Classical Conditioning, Vol 1: Human Applications*, Boston: Kluwer.

Woodruff-Pak, D.S. and Steinmetz, J.E. (2000b). *Eyeblink Classical Conditioning, Vol 2: Animal Models*, Boston: Kluwer.

Woody, C.D., Vassilevsky, N.N. and Engel, J., Jr. (1970). Conditioned eye blink: Unit activity at coronal-pericruciate cortex of the cat. *Journal of Neurophysiology, 33*, 851-864.

Wolf, G. (1971). Elementary histology for neuropsychologists. In R.D. Myers (Ed.), *Meth-*

ods in Psychobiology: Laboratory Techniques in Neuropsychology and Neurobiology. New York: Academic Press, pp. 281-300.

Yeomans, J.S. (1990). *Principles of Brain Stimulation.* New York, Oxford University Press.

Zhang, J., Ni, H. and Harper, R. M. (1986). A miniaturized cryoprobe for functional neuronal blockade in freely moving animals. *Journal of Neuroscience, 16,* 79-87.

Zwaardemaker, H. and Lans, L.J. (1899). Uber ein Stadium relativer Unerregparkeit als Ursache des intermitterenden Charakters des Lidschlagreflexes. *Centbl F Physiology, 13,* 325-329.

INDEX

Ablation lesion, 250–51
Accelerometer, 203, 204
Active avoidance, 213
A/D converter, 126n1, 141, 163, 164, 168
  A/D values and, 88–89, 90
  analog to digital conversion and, 350–52, 365–66
  interface card and, 338–41, 340f
  waveform-matching routines and, 150–51, 366
Address decoder, 45, 70n11
Air puff, 118, 208, 379
  for humans v. non-humans, 63–64
  portable system, 65
  rabbit's reaction to, 72, 73
  US and delivery of, 63–65, 63f
Airpuff (airon/arioff), 17–18, 359
Akase, E., 367
Alpha response, 23, 28, 35, 56
A-M Systems, 389
Amplification, 54, 55f, 84, 85
  for auditory stimulus, 36–39, 38f, 41–42, 42–43, 54, 55f
  biological amplifier and, 138, 151
  Brakel or Verzeano, 369–70
  differential amplifier and, 85, 94, 138–39
  filtering and, 137–39
  for neuronal data recording, 137–39
  operational amplifier and, 147, 160, 295, 384
  parallel amplification and, 140
  quad operational amplifier and, 93, 155–56
  741 operational amplifier and, 42–43, 51–52, 94
  3080 operational amplifier for, 39, 41–42, 69n1
  3130/3160 operational amplifiers and, 89–90
  (dual) transconductance operational amplifier and, 138, 148
Analgesics, 233
Analog
  to digital conversion, 350–52
  digital v., 383
Analog controller, 328–30, 329f
Analog switch, 52, 70n13
Analysis, of behavioral data
  baseline for, 166–67, 168
  CR and UR amplitudes and, 167–68, 167f, 169
  CR and/or UR magnitude measures, CR/UR risetime and, 168–69
  CR occurring in trial as, 166
  good v. bad trials and, 166–67
  neural data and, 170
  of other behaviors, 170
  polygraph paper, 325
  session-wide number crunching for, 169–70
  three point moving average for, 174–75
  within-trial number crunching for, 165–69
Analysis, of neural data
  baseline and, 178–79
  behavior and, 171
  Berthier and Moore's algorithm for, 197–98
  cluster analysis and, 187–94, 189f, 190f, 191f
  cluster analysis computer program for, 194–95
  cluster analysis' perspective and, 195–96
  collection/storage using discriminator in, 171–72
  collection/storage using waveform matching algorithms in, 172–73, 366
  cross-correlation drawbacks and, 185–87, 186t
  cross-correlation methods, brain-behavior relationships and, 180–85, 182f, 184f
  cross-correlation program and, 185
  PSTH and, 175–98, 176f, 177t, 179f, 182f, 184f, 186t, 189f, 190f, 191f
  (further) reduction of unit data for, 173, 174f, 175
  t-score, z-score analysis and, 175–76, 176f, 177t, 178–80, 179f
AND gate, 334
Anesthesia
  after, 233
  choice of, 229–30
  kinds of, 229–30
  local, 230, 274
  ophthalmic, for eyelid, 78–79, 246
  perfusion and overdosing of, 299–300, 301, 302
  as restraint, 79, 111, 113
Animal killing, 300, 302
Animal restraints issues, 104–11, 107f, 109f, 110f, 112f, 113, 114f, 115f
Animal restraints issues. See Restraints, 104–13
Annala, A.J., 118
Anodal lesion, 283
Anterior-posterior (AP)
  corrections for, 241, 241f, 242f, 243
Appropriateness, 189–93
Aseptic surgery, 226, 231
Aspiration lesion, 251–53, 251f
Associative learning, 3, 6, 8, 21, 60
Astable, 273, 330
Atlases, 321–22, 393–421
Attenuator, 37, 341, 376
  modification of rise/fall switch and, 52, 54, 55f
  summing operational amplifier, audio split and, 51–52, 53f
Auditory stimuli
  amplification and, 36–39, 38f, 41–42, 42–43, 54, 55f
  attenuator and, 37, 51–52, 53f, 54, 55f
  buying off-the-shelf for, 55–56
  calibrating auditory signal (to 1 V peak-to-peak) in, 42, 49, 52
  calibrating sound intensity for, 37, 43–44
  cost of, 44
  CS and, 36–56, 38f, 40f, 45f, 47f, 48f, 50f, 51f, 53f, 55f, 57f
  for CS or US, 37
  elements need for tone in, 37, 38f
  frequency of, 37, 46, 47f, 48f

intensity of, 37, 42–43
morse code circuit and, 37–44, 38f
as most frequently used, 36
pure tone/noise and, 36, 46
rise/fall switch for, 39, 40–42, 49, 51f
(other) sound sources for, 49
source of, 37, 39
speaker and, 37, 42–43, 54, 55f, 103
timing of, 37
tuning circuit and, 45–46, 47f, 48f
types of, 36–37
versatile tone circuit and, 45, 45f
white noise circuits and, 36, 46, 48, 50f
Autonomic responses, 93–97, 95f, 97f
Aversive training sessions, 210–11
Avnet Electronics Marketing, 389, 391
Avoidance
active, 213
discriminative, 212
passive, 212–13
procedures, 9

Babb, T., 137
Backward conditioning, 10
Bad trials, 27–28, 31, 33–34, 33f
baseline activity unacceptable for, 31, 32f, 33
exclusion criteria for, 31, 32f, 33
good v., 166–67
short latency and, 33–34, 33f
Badtime, 31
Badun, 31
Bags, 108
Baltimore Model, 109
Baseline
for bad trials, 31, 32f, 33
for behavioral data, 166–67, 168, 328
for neural data, 178–79
Baseline frequency, 193–94
Battery power supplies, 369–71, 370f
dual, system, 140
Behavior measurements
animal restraints issues with, 104–11, 107f, 109f, 110f, 112f, 114f, 115f
computer-compatible input converted from eyeblinks measurements for, 86–90, 91f
drinkometers, lickometers and, 97–99, 100f, 102f
EMG techniques for, 84–86, 87f
external eyelids in, 74–75
eye clips and, 113, 115–18, 117f, 119f
eyeblinks measuring of, 23, 24f, 25, 26f, 27–28, 32f, 72–86, 74f, 77f, 79f, 82f, 83f, 84f, 87f
goal of, 71
of headstages, 118–21, 122f, 123f, 124f
of heart rate and other autonomic responses, 93–97, 95f, 97f
LED/phototransistor eyeblink device for, 81–83, 83f
lever pressing for, 99, 101, 103f, 170
nervous system activity and, 71
nictitating membrane and, 72–75
plexiglass restraint boxes for, 92–93, 105–8, 107f, 109f, 110f
rabbit transport carrier and, 121, 123–25, 125f
reflected infrared light for, 80–81, 82f
reflected visible light for, 83–84, 84f

of respiration responses, 96
of skin conductance responses, 96–97
of (other) somatic responses, 92–93, 94f
torque measuring devices for, 75–80, 77f, 79f
for understanding laws guiding behavior, 71
videotaping as, 101, 103–4, 201
wraps, bags, slings, hammocks and, 108–10, 112f
Behavior measurements. See also Collection, of data
Behavioral paradigms
of fear conditioning, 96, 104, 199–203
instrumental conditional procedures for, 204–11, 206f
instrumental conditioning, runways, mazes, etc. for, 211–19, 215f, 216f, 217f
monitoring general activity and, 219–20
of potentiated startle, 203–4
Behavioral neuroscience, xvii, 35, 71
Bentson, G., xviii
Berger, T.W., 180, 327
Berlin's book, 143, 146–47
Berry, S., 92, 255
Berthier and Moore, 197–98
Best, P., 219
Beta response, 28, 35
Biela, Josef, 234
Bin, 332, 352
Binwidth timing (8253), 352–54
Blocking effects, 12
Blood loss. See Homeostasis
Brain
atlases, 321–22, 393–421
histological methods and injection into, 320–21
inertia during motion for, 131
location/landmarks for surgery on, 236–38, 237f
stereotaxic surgery on, 238–44, 241f, 242f
Brain embedding
chrome-alum solution for, 313
embedding gelatin for, 311–12
microtome or cryostat, brain slicing and, 311, 312
mounting solution for, 313
subbing of glass slides in, 312
sucrose-formalin solution, cryoprotection and, 310–11
Brain staining
common stains for, 313
cresyl violent stain for cell bodies with optional Prussian blue for electrolytic marking lesion for, 314–16
modified Kluver-Barrera stain for, 317–18
modified Weil-Weigert fiber stain for, 316–17
radiolabeling and, 320–21
water-based v. lipid-based stains for, 314
Brain stimulation techniques
applications of, 288–90
basic fundamentals of, 278–79, 279f
bipolar, 284, 285, 287f, 297, 296286
block orientation of, 286
of cathodal stimulation, 283
cerebellum, CS-US pathways and, 289–90
choosing source of, 280, 280f, 281f, 282f, 297
classical eyeblink conditioning and, 277
constant current units for, 282
differential recording and, 291–92

Index

electrode orientation of, 286
electrodes and, 278, 279f, 284–85
equipment acting like antenna in, 293–94
exciting neurons for, 278–84, 279f, 280f281f, 282f
extracellular, 278–79
flexor reflex response and, 289
frequency and, 281, 281f, 283, 287–88
ground loops for, 294, 295f, 296
keeping at distance and, 293
learning, memory and, 289
monopolar, 284–85, 297
neural recordings, removing stimulation, environmental artifacts and, 290–96, 295f, 297
orientation and, 292
overview of, 277–78, 297
parameters of electrical, 283–84, 297
power supply noise of, 296
sample-hold features and, 294–95
saturation and, 292
shielding and, 292–93, 295f
sophistication v. untrained for, 278
special effects of, 285–88, 287f, 288f
stimulation artifacts, filtering and, 291
stimulation isolation units for, 282–83, 282f
stimulation parameters for, 280–81, 281f, 283–84
stimulus isolation for, 294, 295f, 296
train, 283–84
turning off transmitter and, 293
Brain-behavior relationship
cross-correlation methods and, 180–85, 182f, 184f, 187
latency of neural activity and, 187–88
lesion method and, 249–50, 262
Brakel, S., 137
Bregma, 237f, 238–39
Brogden, W., 324
Brooks, V., 263
Brown, T.H., 202

C++ program, 209, 327, 362–66
computer-human interface and, 363
data file format for, 366–67
data structures of, 363–65
EXP_Box and, 334
signal control, data collection and, 364–65
trial control, data collection and, 365–66
Calibration
for auditory signal (to 1 V peak-to-peak), 42, 49, 52
dual battery system for, 140
for neuronal data recording, 139–40, 140f
for sound intensity, 37, 43–44
California Fine Wire, 392
Campeau, S., 203, 204
Capacitor, 39, 69n3, 142–43, 147, 370, 372
Case, G., 108
Cathodal stimulation, 283
Causation, for cluster analysis, 189–93, 190f, 191f
Cegavske, C., 327
Cerebellum, 289–90, 393
Chambers, K., 60, 98, 264, 267
Chapman, P., 263, 271
Chemical lesions

inactivation methods, 274–76
Lidocaine for, 274–75
Muscimol for, 275–76
permanent, 250, 260–62
Tetrodotoxin (TTX) for, 275
Choi, J., 202
Christy, 27
Chrome-alum solution, 313
Circulatory system, 307
Clark, B., 83–84, 84f, 264, 270f, 271
Classical conditioning
as associative learning, 3
bad trials and, 27–28, 31, 32f, 33–34
behavior measurements in, 23, 24f, 25, 26f, 27–28, 32f
C++ program, data collection and, 364–67
computer-based controller systems for, 338–67
CR development and, 21–23, 24f
designing and implementing methods for, 13–34
expansion of, 9
history of controller/technology and, 323–28
instrumental conditioning v., 9
language of, 7–8
measuring learning criteria for, 28–30
methods of, 4
neuronal data recording of, 129–30
operant conditioning v., 113
simple controller for, 328–37, 329f, 331f, 333f, 334t, 336f, 337t
timing relationships in, 15–20
types of, 9–11
variations of procedures of, 8–9, 11–13
Classical discrimination/reversal conditioning, 11
Classical eyeblink conditioning, xx
electrical brain stimulation and, 77–78
human, 4–6, 5f, 21, 63–64
non-human, 6–7, 21, 63–64
Classical ISI discrimination conditioning, 11
Click stimulus, 277
Clicking, elimination of, 39, 40
Cluster analysis, 176, 187–94, 189f, 190f, 191f
baseline frequency and, 193–94
block of trials and, 190–93, 191f
causation problem for, 189–93, 190f, 191f
computer program for, 194–95
definition of, 187–89, 189f
frequency and, 188–89, 191–93
onset and, 188–89
order, appropriateness, consistency of, 189–93
perspective of, 195
single trial and, 189–90, 190f, 193, 196
stimulus event (evoked potentials) and, 189
CMOS, battery and, 369
CMOS dual timer, 98, 99, 126n6, 328–29
CMOS interface, 378, 378f
Collection, of data. *See also* Analysis, of behavioral data; Analysis, of neural data
and representation of behavioral data, 164–65
from subject to computer, 161–64, 162f
Columbus Instruments, 220
Common-mode rejection, 139
Common-mode voltage, 139
Commutators, 110–11, 114f, 159
Comparator, 141, 152, 153, 155–56
quad, 98–99, 126n5

Computer(s)
  analog to digital converter (ADC0808) and, 350–52
  Apple, 326–27, 335, 358
  assembly language routines of, 354, 358
  binwidth timing (8253) and, 352–54
  C++ program for, 209, 327, 362–67
  cluster analysis program for, 194–95
  collection and representation of behavioral data and, 161–64, 162f
  computer-based controllers and, 338–67
  control by, 41
  control of delivery of stimuli by, 14
  cross-correlation program and, 185
  data collection of, 17
  data from subject to, 161–64, 162f
  decimal command for, 345, 346
  digital input/output (8255) and, 344–50
  FIRST program for, 324, 326, 338, 358
  Forth program for, 195, 326–27, 338, 344, 348, 350–51, 352, 356, 362
  hex command for, 345, 346
  history of use of, xvii–xviii, 323–28
  human interface and, 363
  IBM/Clone interface for controllers and, 338–41, 339t, 340f, 342t–344t, 343–67
  IMB-OC, 337
  importance of, 323
  interface card, A/D converter and, 338–41, 340f
  interface cards, memory map of addresses and, 342t–344t, 344
  intertrial interval and, 361, 365
  KIM-1, 324, 326, 335, 358
  off-the-shelf, xvii
  PDP-12, 324, 325–26, 337, 358, 383
  pre-programming of, 335
  (other) programs for, 367
  RAM and, 258, 352
  simple controllers and, 328–37
  speed, memory capacity and, 129–30
  trial of classical conditioning and, 358–61
  unit counting and, 354–58
  unit data, PSTH, 41
  wave form matching system and, 172–73, 366
Computer-compatible input
  A/D values and, 88–89, 90
  data acquisition caution with, 90
  eyeblinks measurements converted for, 86–90, 91f
  NM (nictitating membrane) box for, 88, 91f
Conditional discrimination (occasion setting), 12–13
Conditional inhibition conditioning, 12
Conditioned reflex, 3
Conditioned response (CR), 3, 8
  amplitudes, 167–68, 167f, 169
  CR/UR magnitude measures, CR/UR risetime and, 168–69
  from CS-US pairings for, 72
  occurrence in trial, 166
  onset and peak latencies for, 167f, 168, 170
  predictable and systemic process of, 21–23, 24f
  training breaks over weekend for, 21–22
Conditioned stimulus (CS), 3, 8, 328
  auditory stimuli and, 36–56, 38f, 40f, 45f, 47f, 48f, 50f, 51f, 53f, 55f, 57f

CR as response of, 72
fear conditioning, US and, 199–200
intracranial stimuli and, 61, 63
neutral value (to reflex) by, 35–36
tactile stimuli and, 59–60
taste stimuli and, 60–61, 62f
visual stimuli and, 56–59, 59f, 60f
Conditioned suppression, 9
Conditioned taste aversion (CTA), 60, 62f
Consistency, 189–93
Constant current units, 282
Controller(s)
  analog, 328–30, 329f
  bin, 332–35, 333f, 334t
  computer-based, 338–67
  decoded bin, 330, 331f, 332, 333
  history of, 323–28
  IBM/Clone edge, 338–39, 339t
  IBM/Clone interface for, 338–41, 339t, 340f, 342t–344t, 343–67
  ROM-based, 41, 69n7, 325, 335, 336f, 337t, 358
  6502 micro, 41, 69n6, 335
  software, trial of classical conditioning and, 358–61
  summary of early, 336–37
Cooling technique, 262
  controlling, 271–73, 273f, 274f
  cooling probe for, 264–73, 265f, 268f, 269f, 270f, 272f
  full version of probe for, 269–71, 269f, 272f
  heater coil of probe for, 270–71, 270f, 272f
  inner shaft of probe for, 267–69
  low-tech Korhonen solution for, 272–73
  overview for, 263
  soldering and, 266–67
  without soldering and, 267, 268f
  Y-joint for, 264–67, 265f
CR. See Conditioned response (CR)
Cresyl violet stain, 320
  for cell bodies, 314–16
Criterion. See Learning, 28
Critlevel, 25
Cross-correlation methods
  best correlation judgment for, 181–82
  brain-behavior relationships and, 180–85, 182f, 184f
  calculation for, 181
  drawbacks of, 185–87, 186t
  Pearson product-moment correlation of, 180
  simple, 182, 182f, 184f
  time-lagged, 181
  z-scores and, 181
Cross-correlation program, 185
Cruikshank, S., 202
Cryoprotection, 310
Cryostat, 311, 312
CS. See Conditioned stimulus (CS)
CS-Alone extinction, 11
Cs-meas, 27

DAT recorders, 387
Daunton, N., xviii
Davis, M., 203, 204
Decoded bin controller, 330, 331f, 332, 333
Defense conditioning, 8

Index

Delay conditioning
   learning/training progression of, 22–23, 24*f*
   long, 9, 18–20, 20*f*, 30
   short, 9, 13–14, 14*f*, 15–18, 16*f*, 30
   timing relationships of, 15–20, 16*f*, 20
Digestive secretions
   cue-elicited, 3
   increase of, 3
   reflexive, 3
Digital
   analog to conversion in, 350–52
   analog v., 383
   input/output (8255), 344–50
   noise and, 375
Diodes, 49, 70n12
Discriminative avoidance conditioning, 212
Discriminators
   collection/storage of neural data using, 171–72
   gain, polarity and comparator functions for, 152–53, 154*f*, 155–57, 156*f*
   multiplexing display for, 155–57, 156*f*
   for neuronal data recording, 139, 151–53, 152*f*, 154*f*, 155–57, 156*f*
   simple, 151–53, 152*f*
   window, 153
Disterhoft, J., 81, 83*f*, 108–9, 367
Distortion, minimum, 39, 40, 89
Domjan, M., 205
Dorsal-ventral (DV), 238
   corrections for, 241, 241*f*, 242*f*, 243
Douglas, F.A., 248
Dragana, 27
Dremel drill, 218
Drinkometers, 97–99, 100*f*, 102*f*
   suppression of drinking and, 98
DV. *See* Dorsal-ventral (DV)

Edeline, J., 202
Edmund Scientific, 389, 390
EEG (electroencephalogram), 383–84
   freely moving rats and, 85
   neuronal data recording and, 130
Electrical shock, 118
   electrical stimulation and, 64*f*, 67–68, 67*f*
   shock leads and, 66
   shocker design for, 64*f*, 66–67, 68
   shorting out electrodes between stimulation for, 64*f*, 68
   US and, 64*f*, 66–68, 67*f*
Electrode(s), 151, 163
   acetylene torch shaping of, 135
   bending/shaping of, 133–34
   bipolar stimulating, 284, 285, 286, 287*f*, 296, 297
   for brain stimulation techniques, 278, 279*f*, 284–85
   failure during lesioning, 258
   grinding/shaping of, 134–35
   implanted v. moveable, 137
   implanting of, 239–41, 241*f*, 242*f*, 243–44
   inhibitory v. excitatory patterns in, 132
   insect pins and, 285
   lesion, 259
   micro, 235
   monopolar configuration of, 284–85, 297
   multiple-unit procedures for, 131–33, 139, 148
   for neuronal data recording, 131–37
   orientation in brain stimulation, 286
   single v. multiple-unit procedures for, 131–32
   single-unit procedures for, 133, 139
   soldering and, 133
   stainless steel and, 133, 134*f*, 285
   (other) techniques of making, 136–37
   zapper removal of insulation from tip of, 135–36, 135*f*
Electrode impedance tester, 136
Electrolytic lesions
   design improvements for, 258
   lesion electrodes and, 259
   lesion maker and, 255–57, 256*f*, 257*f*, 370
   location of lesion electrodes and, 259
   radio frequency lesion techniques v., 260
   theory behind, 250–55, 254*f*
Electronics
   CMOS interface, 378, 378*f*
   digital v. analog, 383
   interface circuits, 376–77
   optoisolator, 41, 70n9, 294, 296, 380–83, 382*f*
   power supplies, 296, 369–76
   (interface for) recording data to tape by, 383–88, 384*f*, 385*f*, 386*f*
   TTL interface, 41, 69n8, 94, 152–53, 155, 171, 379–80, 379*f*, 386–87
   voltage divider, 377–78, 377*f*
Ellison, G., 251
Embedding gelatin, 311–12
EMG techniques, 84–86, 87*f*, 117, 162–63, 204
   integrator and, 86, 87*f*
   nictitating membrane-like DC signal and, 86, 87*f*
Engel, J., Jr., 131
Environmental artifacts, 290–91
Epoxyline Corp., 391–92
Equipment. *See* Electronics
Euthanasia, 300, 302
Evoked response potentials (ERP)
   neuronal data recording and, 130
Exclusion criteria
   of bad trials, 31, 32*f*, 33
Extinction training, 118
Eye clips, 116
   no, 117–18
   no restraint v., 113, 115
   ophthalmic, 116–17, 119*f*
   standard, 116, 117*f*
Eyeblink Classical Conditioning, Volume I: Human Applications, 5–6
Eyeblink measurements, 72–86, 74*f*, 77*f*, 79*f*, 82*f*, 83*f*, 84*f*, 87*f*
   computer-compatible input converted from, 86–90, 91*f*
   EMG techniques for, 84–86, 87*f*
   external eyelids in, 74–75
   eye clips and, 113, 115–18, 117*f*, 119*f*
   LED/phototransistor eyeblink device for, 81–83, 83*f*
   nictitating membrane and, 72–75
   reflected infrared light for, 80–81, 82*f*
   reflected visible light for, 83–84, 84*f*
   schematic overview of, 162–63, 162*f*
   torque measuring devices for, 75–80, 77*f*, 79*f*
Eyeblinks

high spontaneous human, 6
low spontaneous rabbit, 28
Eyelid closure
A/D values and, 89

Falls, W., 203, 204
Fear conditioning, 96, 104, 204
equipment for, 200–201
response measured in, 200–201
Fear, freeze in, 105, 200–201, 202
FET transistors, 139
Filter(s), 174
on, 174
active, 143–47, 144*f*, 145*f*, 146*f*
amplitude and, 149
best delay, 145
cutoff frequency and, 141, 142–43, 147, 149, 150
data possibly changed through, 149–50
feedback, damping resistors and, 145
4th order, high pass, 143, 145*f*, 147
4th order, low pass, 143, 145*f*, 147
frequency shifts from, 149–50
high frequency, 141
high pass, 138, 141, 142*f*
higher order, 147
low frequency, 141
low pass, 138, 141, 142*f*
low pass v. high pass, 146–47
for neuronal data recording, 141–50, 142*f*, 144*f*, 145*f*, 146*f*, 148*f*
notch, 138
pass band between, 147
passive, 141–43, 142*f*
perspectives/misunderstandings of, 149–50
2nd order, high pass, 143–44, 144*f*, 147
2nd order, low pass, 143–44, 144*f*, 147
6th order, high pass, 143, 146*f*, 147
6th order, low pass, 143, 146*f*, 147
transconductance, 147–49, 148*f*
tunable, 138
Fine Science Tools, 389
Flex-Field Animal Activity System, 220
Flexor reflex response, 289
Flip flop, 98–99, 126n7, 333, 334–35
Food reinforcement, 207, 209–11, 214
Food shock, 210
Force transducers, 93
Formalin
clean up for, 308, 310
induced movements, 301
perfusion's infusion of, 307–8
perfusion's recipe for, 308, 310
records of, 310
soaking, 308
sucrose in, 308, 310–11
Forth (program), 195, 250–351, 326–27, 338, 344, 348, 352, 356, 362
Freely moving animals
EEG and, 85
heart rate recording of, 95–96, 97*f*
implanted recording electrodes in, 159
neuronal data recording of, 158–60, 158*f*, 159*f*
no restraints and, 111–12, 114*f*, 115*f*
Frequency, 376, 378
auditory stimuli and, 37, 46, 47*f*, 48*f*

baseline, 193–94
brain stimulation, 281, 281*f*, 283, 287–88
cluster and, 188–89, 191–93
cutoff, 141, 142–43, 147, 148, 149, 150
filters and shifts in, 149–50
modulation, 384
tone, 341
Fry's Electronics, 389, 390
Furedy, J.J., 3

Gabriel, M., 110, 212
Gain, 152–53, 154*f*, 155
Gantt, W.H., 324
Gate, 37, 49
Gate. *See also* AND gate; NOR gate
Gaussian (curve), 48, 196, 197
General Valve, 389, 391
Generator
precision function, 139
precision waveform, 45, 70n10, 384
Get-buffer, 352, 358
Gloves, 303
Goesling, W., xviii
Gormezano, I., 6, 25, 73, 80, 104, 105, 113, 115, 324, 325, 326, 338
Gould, T.J., 427
Grant, D., 324
Grass Instruments, 202, 280, 282
Grass lesion maker, 58, 255–56
Ground loops
brain stimulation techniques and, 294, 295*f*, 296
power supplies and, 370, 372

Haer, Frederick, 389
Hamalainem, P., 367
Hammocks, 109–10
Harper, R.M., 263, 264, 269, 270, 272, 273
Hayes, U., 61, 98
Headstages, 118–21, 122*f*
definition of, 118
harness of, 120
neural recording with, 119
plug, 119, 121, 123*f*, 124*f*, 160
Heart rate
fear response and, 200–202
freely moving rats and recording of, 95–96, 97*f*
recording of, 93–96, 95*f*, 97*f*, 200–201
Hildreth and Marr, 197
Hilgard, E., 5, 5*f*, 324, 325
Hippocampus, 160
Histological methods
brain atlases and, 320–21, 393–421
brain embedding, sectioning techniques and, 310–13
brain lesion assessing with, 319–20
brain staining and, 313–18
injection into brain and assessment by, 320
perfusion, fixing brain for sectioning and, 299–310
processed tissue inspected with, 318
radiolabeling and, 320–21
Homeostasis, 251, 253
Hothersall, D., xviii
Hull, C., 5*f*, 324
Human classical eyeblink conditioning, 4–6, 5*f*

Index

air puffs for, 63–64
associative learning and, 21
cognitive effects on, 6
high spontaneous blinking and, 6
human and nonhuman closeness in, 5
human learning and memory and, 5
reduction then resurgence of, 6

IBM/Clone
  analog to digital conversion and, 350–52
  binwidth timing (8253) and, 352–54
  C++ program for, 362–67
  digital input/output (8255), 344–50
  edge connector contacts for, 338–39, 339t
  interface card, 338–41, 340f, 342t–344t, 344
  memory map, interface card and, 342t–344t, 344
  (other) programs for, 367
  software, trial of classical conditioning and, 358–61
  trial of classical conditioning and, 358–61
  unit counting and, 354–58
Ibotenic acid, 260–61
Impedance, 378
Infrared light, reflected, 80–81, 82f
Inhibitory Post-Synaptic Potentials (IPSP), 275
Instrumental conditioning. *See also* Operant conditioning
  active/passive avoidance and, 212–13
  aversive training sessions in, 210–11
  classical conditioning v., 9
  of differential reinforcement of low rates (DRL), 214
  discriminative avoidance conditioning and, 212
  eyeblink conditioning and, 207–8
  food reinforcement and, 207, 209–11, 214
  food shock and, 210
  mazes for, 215–19, 216f, 217f
  neurobiology, learning, memory and, 208–9
  for nonreflexive behaviors, 205
  (simple) runway task of, 214–15, 215f
  Skinner box for, 206–7, 206f
  videotaping of, 201, 213, 217, 218, 219, 220
Integrator, 86
Intensity, 37, 42–43, 281, 281f
Interface card
  A/D converter and, 338–41, 340f
  digital input/output (8255) and, 344–50
  memory map of addresses for, 342t–344t, 344
  software programs and, 342t–344t, 344
Interface circuits, 376–77
Interpositus nucleus, 179–80
Interstimulus interval (ISI), 8, 16–17, 328, 329, 330
Intertrial interval (ITI), 8, 21, 89, 329
  computers and, 361, 365
Intracranial stimuli
  CS and, 61, 63
Iron hematoxylin, 314
Isolation
  optoisolator, input/output and, 381–82, 381f, 382f
  stimulus, 282–83, 294, 295f, 296
ITC, 391
Ivkovich, D., 27

Jameco Electronics, 389, 390
Janseen, R., 263
JNT Manufacturing, 389, 391

Kainic acid, 260–61
Katz, D.B., 161
Kent, M., xviii
Kim, J.J., 202
Kim, M., 203, 204
KIM-1 computer, 324, 326, 335, 358
King, D., 358
Kluver-Barrera stain, 314
  modified, 317–18
Knot(s)
  angler's, 245f, 247
  double overhead, 245f, 246–47
  eyelid, 245f, 247
  granny, 245f, 246
  overhead, 245f, 246
  surgeon, 245f, 247
Kolmogorov-Smirnov test, 197
Kopf, David, 234, 389
Korhonen, T., 78, 116, 119, 134, 264, 267, 272–73
Krupa, D., 276

La Regina, M.., 248
Laboratory, creation of, xix–xx
Laboratory Rabbit, The (Suckow and Douglas), 248
Laboratory Rat, The (Sharp and La Regina), 248
Lafayette, 389
Laham, R.I., 180, 327
Lambda, 237f, 238–39
Lancaster's book, 143, 146–47
Language, of classical conditioning, 7–8
Latency, 25, 27, 188, 192
  bad trials and short, 33–34, 33f
  of neural activity, 187–88
  onset, 25, 27, 170, 197
  peak, 27, 167f, 168, 170
  short, 33–34, 33f
Latent inhibition conditioning, 11–12
Lavond, D.G., xviii–xix, xxi, 80, 85, 106, 110, 113, 118, 138, 149, 174, 178, 188, 208, 241, 289, 323, 326–27, 358, 376
Learning
  active and passive avoidance, 213
  associative, 3, 6, 8, 21, 60
  CR evidence for, 23
  delay conditioning and, 22–23, 24f
  measuring criteria of, 28–30
  memory and, 208–2209
  spatial, 215, 217
Learning criterion
  conditioning procedures influence on, 30
  definition of, 28
  features of, 29
  performance criterion required for, 29–30
  training for, 30
LED, 58, 76, 103, 126n2, 153, 155, 213, 294, 381
  phototransistor eyeblink device, 81–83, 83f
  videotaping and, 103, 104
Lee, H.J., 202

Lesion (techniques), 262. *See also* Cooling technique
  ablation, 250–51
  anodal, 283
  aspiration, 251–53, 251*f*
  brain-behavior relationship and, 249–50, 262
  chemical inactivation methods for, 274–76
  constant current, 253–55, 254*f,* 256*f,* 257, 257*f*
  controlling cooling for, 271–73, 273*f,* 274*f*
  cooling probe for, 264–73, 265*f,* 268*f,* 269*f,* 270*f,* 272*f*
  cooling technique overview for, 263
  electrolytic, 250–60, 254*f,* 256*f,* 257*f*
  fiber *en passant,* 250, 260, 276
  Grass lesion maker and, 58, 255–56
  histological methods of assessing, 319–20
  integrated on/off switch against electrode failure in, 258
  lesion effect/problem with brain-behavior relationship and, 249–50
  lesion maker and, 255–57, 256f, 257*f,* 258, 302–3, 309, 370
  perfusion, 302–3, 309
  permanent chemical, 250, 260–62
  Prussian blue for electrolytic marking lesion and, 314–16
  scoop technique and, 250
  temporary inactivation/reversible, 262–73, 265*f,* 268*f,* 269*f,* 270*f,* 272*f,* 273f, 274*f*
Lever pressing, 99, 101, 103*f,* 170
  reward, 206, 209–10, 214
Libkumen, T., xviii
Lickometers, 97–99, 100*f,* 102*f*
Lidocaine, 274–75
Life Science Laboratory Animal Series, 248
Light
  reflected infrared, 80–81, 82*f*
  reflected visible, 83–84, 84*f*
Linear power supplies, 371*f,* 373*f*
  AC to DC for, 371, 374
  floating, 372
  problems with, 374–75
  variable voltage regulators with, 374, 374*f,* 375*f*
Logan, C., 27
Logan, S.F., 208
Long-delayed conditioning, 9
  timing relationship of, 18–20, 20*f*
Longer timer, 153

Mahnke, J., 137
Mamounas, L., 264
Maren, S., 202
Markham, M., 367
Marr and Hildreth, 197
Maze
  interamaze cues for, 215
  Morris water, 217–19, 218*f*
  radial arm, 216–17, 217*f*
  spatial learning and, 215, 217, 218
  T-, 206, 215–16, 216*f*
McCormick, D., 105, 393
McGregor, I., 367
McNaughton, B., 219
Memory map, 342*t*–344*t,* 344
Meriluoto, J., 367

Merrill, K.A., 185
Micro controller, 6502, 41, 69n6
Micromanipulator, 235
Microtome, 311, 312
Miller, D., 208
Moore and Berthier, 197–98
Morse code circuit, 37, 38*f,* 44
  components of, 37–44, 38*f*
Moyer, J.., 367
Multiplexer, 155–57, 156*f*
Muscimol, 275–76, 320

NAND gate, 99, 127n9
Narishige, 389
Neuronal data recording. *See also* Analysis, of neural data
  amplification, filtering and, 137–39
  bioelectrical signals from neurons for, 130
  brain stimulation, removing stimulation, environmental artifacts and, 290–96, 295*f,* 297
  calibration for, 139–40, 140*f*
  of classical conditioning, 129–30
  data signal processed for, 141
  differential recording and, 138–39, 291–92
  discriminators for, 151–53, 152*f,* 154*f,* 155–57, 156*f*
  EEG and, 130
  electrodes for, 131–37, 134*f,* 135*f*
  equipment acting like antenna in, 293–94
  evoked response potentials (ERP) and, 130
  extracellular unit, 130
  filters for, 141–50, 144*f,* 145*f,* 146*f,* 148*f*
  free moving animals and, 158–60, 158*f,* 159*f*
  ground loops for, 294, 295*f,* 296
  inhibitory v. excitatory patterns in, 132
  isolating action potentials for, 150–53, 152*f,* 154*f,* 155–57, 156*f*
  keeping at distance and, 293
  lesion techniques and, 250
  multiple-unit procedures for, 131–33, 139, 148
  orientation and, 292
  power supply noise of, 296
  sample-hold features and, 294–95
  saturation and, 292
  shielding and, 292–93, 295*f*
  short sampling trail for, 129
  single-ended, 138
  single-unit procedures for, 131–32, 133, 139, 196
  stimulation artifacts, filtering and, 291
  stimulus isolation for, 294, 295*f,* 296
  surgical preparation considerations for, 157–58
  theory of, 130–31
  waveform matching routines for, 150–51, 366
Neurons, 290
  brain stimulation fundamentals and, 278–79, 279*f*
  choosing source of brain stimulation for, 280, 280*f,* 281*f,* 282*f*
  depolarization of, 279
  parameters of electrical brain stimulation for, 283–84
  special effects of brain stimulation on, 285–88, 287*f,* 288*f*
New York Hardware Trading Co., 389, 392

Index

Ni, H., 263, 264, 269, 270, 272, 273
Nicolelis, M., 160
Nictitating membrane, 72–74, 115–16, 324
   box (NM), 88, 91*f,* 143
   like DC signal and, 86, 87*f*
   potentiometers and, 77–80, 79*f*
NM (Nictitating membrane) box, 88, 91*f,* 143
Noise
   high frequency, 81
   of power supplies, 296, 369, 372, 375
Non-human classical eyeblink conditioning, 6–7
   air puffs for, 63–64
   animal preparation improvements for, 6
   rabbit and rats associative learning and, 21
   rabbit as ideal subject for, 6–7
   rabbit brain size and, 7
   rabbit NM/eyelid procedure for, 6
   rat, 6–7
   switching from human to, 6
   variety of animals for, 6–7
NOR gate, 98–99, 127n8, 332, 334

Ohm's Law, 253
Onset, 25, 188, 192, 198
Onset latency, 25, 27, 170, 197
Operant box (chamber), 206–7, 206*f,* 209–10
Operant conditioning
   classical conditioning v., 113
   neuronal data recording of, 129–30
Operant conditioning. *See also* Instrumental conditioning
Optoisolator, 41, 70n9, 294, 296
   input/output, isolation and, 380–81
   safety and, 380–81
   versatility for range of output with, 381*f,* 382, 382*f*
Order, 189–90, 192
Oscilloscope, 30, 140, 153, 157, 296, 372
Out of, 28

Passive avoidance, 212–13
Patterson, M., xviii, 80, 92, 106, 116, 208, 325, 326, 327
Pavlov, Ivan, 1, 2*p,* 3–4, 200, 201
   background of, 1
   classical (Pavlovian) conditioning of, 3, 324
   digestive system studies of, 1, 3
   dog salvation experiments and, 3–4
   labeling of stimuli by, 8
   Nobel Prize for, 3
   unconditioned stimulus, unconditioned response and, 3–4
PCs-units, 31
PDP-12 computer, 324, 325–26, 337, 358, 383
Peak, H., 5*f,* 324
Peak latency, 167*f,* 168, 170
Pentonnen, M., 367
Perfusion. *See also* Brain embedding
   circulatory system and, 307
   disposal with, 310
   euthanasia and, 299–300, 302
   formalin infused in, 307–8
   formalin recipe for, 308, 310
   formalin-induced movements and, 301
   gloves and, 303
   hood and, 307
   initial check for, 302
   lesion maker and, 302–3, 309
   needle, 305
   overdosing of animal for, 299–300, 301, 302
   overview process of, 301
   postfix of, 310–11
   protocol, 302–10
   rack, 303
   retractors for, 304
   rig, 305–6
   ronguers for, 309–10
   saline for, 300, 301, 306–7
   scalpel for, 303, 309
   scissors for, 303–4, 306, 309
   sponge forceps for, 304–5
   sucrose-formalin recipe for, 308, 310–11
   training for, 300–301
   vice grips for, 309
Peripheral Programmable Interface (PPI) chip, 8255, 94, 126n4
Photobeams detectors, 219
Phototransistor eyeblink device, LED, 81–83, 83*f,* 126n3
   disadvantages of, 82–83
   non-invasive value of, 82
Plastic cooler, 124–25, 125*f*
Plexiglass restraint boxes, 92–93, 105–8, 107*f,* 109*f,* 110*f*
Polarity, 152–53, 154*f,* 155
Polencar, B., 208
Polling code, 365
Polygraph paper, 325
Potentiated startle, 203–4
Potentiometers, 39, 69n4, 90, 92, 330, 378
   attaching (loop) to eyelid with, 78–79, 244, 246
   force transducers used for, 93
   linkage to nictitating membrane with, 77–80, 79*f*
   minitorque, 74*f,* 75–76
   polarized light, 76–77, 77*f*
   setting of, 75–76, 77
   spring purpose for, 80
Power amplification, 54, 55*f*
Power supplies
   battery, 369–71, 370*f*
   floating, 372
   ground loops and, 370, 372
   linear, 371–75, 371*f,* 373*f,* 374*f,* 375*f*
   noise of, 296, 369, 372, 375
   safety with, 372
   switching, 375–76
   tape recording for, 386–87, 386*f*
Precision function generator, 139
Precision waveform generator, 45, 70n10, 384
Prim, M., 263
Prince, J.H., 115
Principles of Brain Stimulation (Yeomans), 283
Principles of Learning & Behavior (Domjan), 205
Prussian blue
   for electrolytic marking lesion, 314–16
   solution, 259
PSTH (peristimulus time histograms), 173
   Berthier and Moore's algorithm for, 197–98
   cluster analysis and, 187–94, 189*f,* 190*f,* 191*f*

cluster analysis computer program for, 194–95
cluster analysis' perspective and, 195–96
cross-correlation drawbacks and, 185–87, 186t
cross-correlation methods, brain-behavior
 relationships and, 180–85, 182f, 184f
cross-correlation program and, 185
t-score, z-score analysis and, 175–76, 176f,
 177t, 178–80, 179f
Psychic secretions, 3
 as conditioned response (CR), 3
Pure tone, 36, 46

Quad comparator, 98–99, 126n5

Rabbit(s), 248
 air puff reaction of, 72, 73
 analgesics for, 233
 anesthetics for, 229–30
 brain atlases of, 393–421
 as ideal subject for eyeblink conditioning, 6–7
 location/landmark on brain of, 236–38, 237f
 low spontaneous eyeblinks of, 28
 nictitating membrane and, 72–75, 77–80, 79f,
 115–16
 not jump from heights, 123–24
 restraint accepted by, 105, 158–59
 transport carrier, 121, 123–25, 125f
 wrap around, 108–9
Radio frequency lesion techniques, 260
Radio Shack, 389, 390
Radiolabeling, 320–21
RAM, 352, 358
Ranvier (of axons), 286
Rat(s), 6–7, 159, 248
 anesthetics for, 230
 associative learning and, 21
 EEG and freely moving, 85
 freely moving, 95–96, 111, 114f
 learning, memory and, 208–9
 skull/brain, 237–39, 237f
Rebec, G., 220
Recording, data to tape. See Tape recording, data
 to
Reflected infrared light, 80–81, 82f
Reflected visible light, 83–84, 84f
Resistance, 378
Resistors, 39, 69n2
 feedback, damping, 145
Respiration responses, 96
Response
 alpha, 23, 28, 35, 56
 beta, 28, 35
 conditioned, 3, 8, 21–23
 unconditioned, 3, 8, 15
Restraints
 anesthesia as, 111, 113
 animal issues of, 104–5
 no, used, 110–11, 114f, 115f
 plexiglass box, 92–93, 105–8, 107f, 109f, 110f
 rabbits acceptance of, 105, 158–59
 wraps, bags, slings, hammocks and, 108–10,
 112f
Reward
 food, 214
 lever pressing, 206, 209–10, 214

Reward conditioning, 8–9
Rinaldi, P., 327
Rise/fall switch, 39, 40–42, 49, 51f
Risetime, 168–69
Rodewald, H.K., xviii
Roemer, R., 325, 327
ROM-based controller, 41, 69n7, 325, 335.336f,
 337t, 358
Ronguers, 309–10
Rorick, L., 96, 97f, 201
Rose, R,J., 185
R-score, 181, 185
Runway, simple runway task of, 214–15, 215f

Safety, 372, 380–81
Saline, for perfusion, 300, 301, 306–7
Salmi, M., 367
San Diego Instruments, 220
Scalpel, 303, 309
Scandrett, J., 326, 338
Schmajuk, Nestor, 106
Scissors, 306, 309
 large, 303–4
Sears, L.L., 427
Second order conditioning, 12
Sensory preconditioning, 12
Sharp, P., 248
Shielding, 292–93, 295f
Shipley, W.C., 5f, 324
shkoff (shock offset), 17, 359
shkon (shock onset), 17–18, 359–60, 361
Short-delay conditioning, 9, 13–14, 14f, 23
 learning little in, 30
 timing relationship of, 15–18, 16f
Shuttle boxes, 213
Sign-tracking, 9
Simmons, D., 314
Simultaneous conditioning, 10
Skeletal muscle responses, 5
Skin conductance responses, 96–97
Skinner, B.F., 99, 205, 206, 221
Skinner box. See Operant box
Slings, 109–10
Small Parts, 389, 390
Smooth, routine, 174
Snarsky, Anton, 3
Soldering, 133, 266–67
 without, 267, 268f
Solenoids, 272–73
Solomon, P., 81, 324, 326
Somatic responses
 behavior measurements and other, 92–93, 94f
South Bay Plastics, 392
Speaker, 37, 42–43, 54, 55f
Sponge forceps, 304–5
Stabilimeter, 203, 219
Stanton, M., 101, 103
Steinmetz, J.E., xviii–xix, 80, 85, 99, 103–4, 108,
 110, 111, 117, 118, 138, 151, 158, 159, 161,
 172, 185, 198, 201, 204, 208, 255, 288–89,
 323, 326–27, 358
Stereotaxic surgery
 after, 244

Index

Biela rabbit stereotaxic headholder for, 234–35, 234f
brain atlases and, 321
corrections for AP and DV positions in, 241, 241f, 242f, 243
drilling for, 238
example of, 238–44, 241f, 242f
implanting electrode with, 239–41, 241f, 242f, 243–44
Kopf stereotaxic headholder for, 235, 235f, 239
location/landmark on brain found for, 236–38, 237f
location/landmarks brain for, 236–38, 237f
positioning for, 238–39
Vernier scale for, 235–36, 236f
Sterilization
  clean v., 228–29
  cold, 228
  difficulties of, 229
  prepackaged items of, 228
  steam, 227–28
Stimulation isolation unit (SIU), 282–83, 282f
Stimulators, 278
  biphasic, 280, 280f
  Glass Instruments and WPI, 280, 282
  monophasic, 280, 280f
  trigger output for, 281
Stimulus
  conditioned, 3, 8
  delivery of, 35–70
  stimulus (S-S), 9
  timing of, 9
  unconditioned, 3, 8, 15
Stimulus isolation, 282–83, 294, 295f, 296
Suckow, M. A., 248
Sucrose, in formalin, 308, 310–11
Suppliers, 389–92
Surgery
  analgesics after, 233
  anesthetics for, 229–30
  animal's biological knowledge before, 233
  aseptic surgery and, 226, 231
  assistant at, 231–32
  basic procedures of, 232
  consecutive, 232
  disinfectants for, 228–29, 232
  general preparation for, 157–58, 226–27, 227t
  post-surgical cleanup after, 232–33
  post-surgical monitoring after, 233
  preparing animal for, 229–31, 230–31
  Standard Operating Procedure (SOP) for, 223–24
  stereotaxic surgery techniques for, 234–44, 234f, 235f, 236f, 237f, 241f, 242f
  sterilization and, 227–29
  surgeon's preparation for, 231
  surgical instruments for, 227–29
  surgical records of, 224–26, 224f
  surgical suite for, 225f, 226
  suture knot-tying techniques for, 244, 245f, 246–47
  training of, 224–25
Swanson, L., 314
Synchronization, 384–85, 384f, 385f

Tactile stimuli, CS and, 59–60, 59f, 60f, 99

Tape recording, data to, 383–88, 384f, 385f, 386f
  accurate, 386
  off-line data analysis and, 386
  power supply and, 386–87, 386f
  storing of data for, 387–88
  synchronization and, 384–85, 384f, 385f
  tape's quality for, 387
Taste stimuli, CS and, 60–61, 62f
Temporal conditioning, 10
Temporary inactivation/reversible lesion techniques, 262
  controlling cooling for, 271–73, 273f, 274f
  cooling probe for, 264–73, 265f, 268f, 269f, 270f, 272f
  cooling technique overview for, 263
Tetrodotoxin (TTX), 275
Teyler, T., 327
Thermocouple, 264–65
Thompson, J.K., 276
Thompson lesion maker, 255–57, 256f, 257f
Thompson, L.T., 367
Thompson, R., xviii, 80, 104, 108, 110, 137, 143, 175, 180, 208, 255, 263, 264, 276, 323, 324–27, 337, 358, 369–70, 383, 393
Tiemann, George, 389
Timer
  dual, 98–99, 126n6, 155, 328–29
  longer, 153
Timing relationship, 15–20, 35
  intertrial interval and, 361–62
  of long delay conditioning, 18–20, 20f
  of short delay conditioning, 15–18, 16f
  of trace conditioning, 18, 19f
Tone trial, 18, 21
Toneon/toneoff, 18, 21, 358–59, 361
Torque measuring devices
  linkage of, 76–80, 79f
  minitorque potentiometers and, 74f, 75–76
  polarized light potentiometers and, 76–77, 77f
Trace conditioning, 9–10, 30
  timing relationship of, 18, 19f
Trace, eyeblink more with higher, 14
Training
  aversive, 210–11
  breaks over weekend for, 21–22
  delay conditioning, progression of, 22–23, 24f
  extinction, 118
  for learning criterion, 30
  shorter, 105
  12 blocks of, 178, 179f
Transistor, 40, 69n5
  FET, 139
Trial(s). *See also* Intertrial interval
  average v. accumulation of, 183
  bad, 27–28, 31, 33–34, 33f
  blocks of, 169, 170, 183
  cluster analysis and block of, 190–92, 191f
  cluster analysis and single, 189–90, 190f
  computer of classical conditioning, 358–61
  computers, data collection and control of, 365–66
  continuous presentation, 328
  CR occurrence in, 166
  duration, 328
  good, 169–70

good v. bad, 166–67
short sampling, 129
single, 182, 183, 189–90, 190f, 193, 196
tone, 18, 21
T-score, PSTH and analysis of, 175–76, 176f, 177t, 178, 196
TTL interface, 41, 69n8, 94, 152–53, 155, 171, 379–80, 379f, 386–87
Tuber, D., 110
Tuning circuit, 45–46, 47f, 48f
Tuning curve, 45

Ucs-meas, 28
Unconditioned response (UR)
 amplitudes, 167–68, 167f
 definition of, 3, 8, 15, 72
 onset and peak latencies for, 167f, 168
 UR/CR magnitude measures, CR/UR risetime and, 168–69
Unconditioned stimulus (US), 328
 air puff delivery and, 63–65, 63f
 definition, 3, 8, 15
 electrical shock, stimulation and, 64f, 66–68, 67f
 fear conditioning, CS and, 199–200
 reflexive response of, 35–36
 UR as response of, 72
Unit counting, 354–58
Unpaired conditioning, 10
UR. *See* Unconditioned response (UR)
US. *See* Unconditioned stimulus (US)

Vacuum, 251–52
 without, 273
Variac, 66, 70n14
Vassilevsky, N.N.., 131
VCR recorders, 386, 387
Vector graphics, 153
Vernier scale, 235–36, 236f, 293
Versatile tone circuit, 45, 45f
Verzeano, M., 137
Vetter device, 384
Vice grips, 309
Videotaping
 behavior measurements and, 101, 103–4
 of instrumental conditioning, 201, 213, 217, 218, 219, 220
Viken, R.J., 185

Vincentizing data, 30
Visual stimuli
 alternative light sources for, 58–59
 CS and, 56–59, 59f, 60f
 equipment less for, 56–58
 as second most used, 56
Voltage
 brain stimulation and, 281–82
 common-mode, 139
 follower, 148
Voltage divider, 377–78, 377f
Voltage regulators
 variable, 374, 374f, 375f
VWR, 392

Walker, M., xviii
Wallis, D, 160
Waltke, J., 105
Wang, Y., 264
Ward's Biology, 389, 391
Waveform matching routines
 collection/storage of neural data using, 172–73, 366
 for neuronal data recording, 150–51
Weil-Weigart stain, 314
 modified, 316–17
Weinberger, N., 202
Weitz, D., 80
White noise circuits, 36, 46, 48, 50f
Wickens, D., 110
Wiesz, D., 247, 327
Window discrimination/comparator, 95
WinForth, 195
Wolsohn, S., 3
Woodruff-Pak, D., 18, 65
Woody, C.D., 131
Wraps, 108–9, 112f

Yeomans, J., 283, 285–86
Young, R., 327

Zhang, A., 264
Zhang, J., 263, 264, 269, 270, 272, 273
Z-score
 cross-correlation and, 181
 PSTH and analysis of, 175–76, 176f, 177t, 178–80, 179f